The Audio Expert

The Audio Expert is a comprehensive reference book covering all aspects of audio, with both practical and theoretical explanations. It is written for people who want to understand audio at the deepest, most technical level, but without needing an engineering degree. *The Audio Expert* explains how audio really works in much more depth than usual, using common sense plain-English explanations and mechanical analogies, with minimal math. It uses an easy to read conversational tone, and includes more than 400 figures and photos to augment the printed text.

However, this book goes beyond merely explaining how audio works. It brings together the concepts of audio, aural perception, musical instrument physics, acoustics, and basic electronics, showing how they're intimately related. It also describes in great detail many practices and techniques used by recording and mixing engineers, including video production and computers. This book is meant for intermediate to advanced recording engineers and audiophiles who want to become experts. There's plenty for beginners too.

One unique feature is explaining how audio devices such as equalizers, compressors, and A/D converters work internally, and how they're spec'd and tested, rather than merely describing how to use them. There's plenty of myth-busting and consumerism too. The book doesn't tell readers what brand power amplifier to buy, but it explains in great detail what defines a good amplifier so people can choose a first-rate model wisely without over-paying.

Most explanations throughout the book are platform-agnostic, applying equally to Windows and Mac computers, and to most software and hardware. Many audio and video examples are included to enhance the written text.

The new edition offers many updates and improvements throughout. New sections on coding an equalizer, comparing microphone preamps, testing results of loudspeaker isolation devices, new online video content on music theory, plus incorporated chapters on MIDI basics, computers, video production, plus new myth-busters, and much more!

Ethan Winer has, at various times, earned a living as a professional musician, computer programmer, circuit designer, recording engineer, composer/arranger, technical writer, acoustician, and college instructor. Ethan has more than 150 feature articles published in various computer and audio magazines. He has produced dozens of educational and music videos, and composed three pieces for full orchestra, all of which have been performed. He now co-owns RealTraps, a manufacturer of acoustic treatment.

The Audio Expert

Everything You Need to Know About Audio

Second Edition

Ethan Winer

Routledge
Taylor & Francis Group

LONDON AND NEW YORK

First edition published by Focal Press in 2012.

Second edition published 2018
by Routledge
2 Park Square, Milton Park, Abingdon, Oxon, OX14 4RN
52 Vanderbilt Avenue, New York, NY 10017

Routledge is an imprint of the Taylor & Francis Group, an informa business

Library of Congress Cataloging-in-Publication Data
A catalog record for this book has been requested

ISBN 13: 978-0-415-78883-0 (hbk)
ISBN 13: 978-0-415-78884-7 (pbk)

Typeset in Times New Roman
by Apex CoVantage, LLC

Visit the companion website: www.TheAudioExpertBook.com/
www.routledge.com/cw/winer

For my Wife, Ilene,
who for reasons I'll never understand has
stuck with me for more than 40 years.

Contents

Acknowledgments

It's important to thank the people who were most influential over the years in teaching me "how technology works." First was Marty Yolles, who, when I was about 12, spent many Saturday afternoons helping me overhaul an old lawnmower engine I would later use for a home-made go-cart. Then later in my teens and early 20s, Cliff Mills, Allen Podell, Marvin Fleischman, and others graciously gave their valuable time to answer my endless beginner questions about audio and radio electronics. Later, Leo Taylor and I designed and built several analog synthesizers and other audio devices, with Leo doing most of the designing and me doing all of the learning. Leo worked as an engineer at Hewlett-Packard, so I got to play with every type of audio test gear imaginable as we created and refined our designs. Finally, my thanks especially to Bill Eppler, who for more than 30 years has been my friend, mentor, and advisor on all technical matters.

I mention these influences because, aside from lawnmower engines and electronics, these people also taught me what I call the "ham radio ethic" of sharing information freely and with endless patience. When I grew up in the 1950s and 1960s, ham radio was popular, and many towns had a local ham radio club where you could learn about electronics for free in a friendly and informal atmosphere. This ethic of sharing has been a driving force throughout my entire life. It was why I wrote my first book in 1991—a manual for programmers to explain the unpublished internal workings of Microsoft's DOS BASIC compilers—even though at the time I owned a software company that sold products based on the same information! My company was unique because we included all of the assembly language source code so people would not only buy a packaged solution but could also learn how the programs they bought worked.

It always bothers me when people are secretive or protective of their knowledge, as if sharing what they know means they'll no longer be as valued or even needed. This is why today I continue—through my posts in audio forums and articles on my personal website—to help people build acoustic treatment, even though my current business manufactures and sells acoustic treatment products.

Many people contributed and helped me over the ten months it took to write this book. I'd like to thank Catharine Steers of Focal Press for recognizing the value of this book and helping to make it a reality. Carlin Reagan, also of Focal Press, guided me throughout the process. Expert audio engineer and journalist Mike Rivers did the technical review, offering many incredibly helpful comments and suggestions. My friend, cellist, and electronics engineer, Andy Woodruff, provided formulas for the Decibels.xls spreadsheet, and also performed the cello demonstration video. My friend, black belt electronics engineer Bill Eppler, contributed his considerable wisdom and helped with fact checking, as did microphone expert Bruce Bartlett, video expert Mark Weiss, acoustics expert Wes Lachot, and loudspeaker maven Floyd Toole.

I must also thank Collin Wade for the saxophone demo; David Gale for his violin expertise; Steve Isaacson and Terry Flynn, who demonstrated the piano; expert arranger Arnie Gross for years of musical tutoring; luthier Bob Spear for his string instrument expertise; and composer Ed Dzubak, who contributed several pieces for the plug-in effects demos. I offer my special thanks to John Roberts, a professional electronics engineer and friend since we worked together in the 1970s, for his many valuable suggestions.

For this second edition, I offer my thanks to Lara Zoble and Kristina Ryan at Taylor & Francis for their enthusiasm and assistance, and to technical reviewer Michael Lawrence for his many excellent suggestions. Project manager Tina Cottone was especially helpful, and the work of copyeditor Rodney Williams and proofreader Sioned Jones is also greatly appreciated. Finally, I also need to thank my long-time friend, master recording engineer Peter Hodgson for his wisdom, advice, and support.

About the Author

Ethan Winer is a reformed rock 'n' roll guitar and bass player who sold his successful software business in 1992 at the age of 43 to take up the cello. Ethan has, at various times, earned a living as a recording engineer, studio musician, computer programmer, circuit designer, composer/arranger, technical writer, and college instructor. In addition to a best-selling book about computer programming, more than 200 of his feature articles have been published in audio and computer magazines, including *Mix, PC Magazine, Electronic Musician, EQ Magazine, Audio Media, Sound on Sound, Computer Language, Microsoft Systems Journal, IBM Exchange, Strings, Keyboard, Programmers Journal, Skeptic, The Strad, Pro Sound News*, and *Recording*. Ethan is also famous (some might say infamous) for his no-nonsense posts about audio science in online audio forums.

Besides his interest in audio and computers, Ethan produced two popular Master Class videos featuring renowned cellist Bernard Greenhouse and five CDs for Music Minus One, including a recording of his own cello concerto. His *Cello Rondo* music video has received nearly two million views on YouTube and other websites. Besides writing, playing, and recording pop tunes, Ethan has composed three pieces for full orchestra, all of which have been performed publicly. He has also played in several amateur orchestras, often as principal cellist.

Ethan lives in New Milford, Connecticut, with his wife Ilene and cats Emma, Lily, and Bella. In 2002 he started the company RealTraps to manufacture bass traps and other acoustic treatment, which he continues to this day. When he's not watching reruns of *The Simpsons* or writing music, Ethan enjoys cooking and playing his collection of vintage pinball machines.

Introduction

Hundreds of books about audio have been published over the years, so you might well ask why we need another book on the topic. I'll start with what this book is *not*. This book will not explain the features of some currently popular audio software or describe how to get the most out of a specific model of hard disk recorder. It will not tell you which home theater receiver or loudspeakers to buy or how to download songs to your MP3 player. This book assumes you already know the difference between a woofer and a tweeter, and line versus loudspeaker signal levels. But it doesn't go as deep as semiconductor physics or writing a sound card device driver. Those are highly specialized topics that are of little use to most recording engineers and audio enthusiasts. However, this book is definitely not for beginners!

The intended audience is intermediate- to advanced-level recording engineers—both practicing and aspiring—as well as audiophiles, home theater owners, and people who sell and install audio equipment. This book will teach you advanced audio concepts in a way that can be applied to *all* past, current, and future technology. In short, this book explains how audio really "works." It not only tells you what, but *why*. It delves into some of the deepest aspects of audio theory using plain English and mechanical analogies, with minimal math. It explains signal flow, digital audio theory, room acoustics, product testing methods, recording and mixing techniques, musical instruments, electronic components and circuits, and much more. Therefore, this book is for everyone who wants to truly understand audio but prefers practical rather than theoretical explanations. Using short chapter sections that are easy to digest, every subject is described in depth using the clearest language possible, without jargon. All that's required of you is a genuine interest and the desire to learn.

Equally important are dispelling the many myths that are so prevalent in audio and explaining what really matters and what doesn't about audio fidelity. Even professional recording engineers, who should know better, sometimes fall prey to illogical beliefs that defy what science knows about audio. Most aspects of audio have been understood fully for 50 years or more, with only a little added in recent years. Yet people still argue about the value of cables made from silver or oxygen-free copper or believe that ultra-high digital sample rates are necessary even though nobody can hear or be influenced by ultrasonic frequencies.

In this Internet age, anyone can run a blog site or post in web forums and claim to be an "expert." Audio magazines print endless interviews with well-intentioned but clueless pop stars who lecture on aspects of audio and recording they don't understand. The amount of public misinformation about audio science is truly staggering. This book, therefore, includes healthy doses of skepticism and consumerism, which, to me, are intimately related. There's a lot of magic and pseudo-science associated with audio products, and often price has surprisingly little to do with quality.

Hopefully you'll find this book much more valuable than an "audio cookbook" or buyer's guide because it gives you the knowledge to separate fact from fiction and teaches you how to discern real science from marketing hype. Once you truly understand how audio works, you'll be able to recognize the latest fads and sales pitches for what they are. So while I won't tell you what model power amplifier to buy, I explain *in great detail* what defines a good amplifier so you can choose a first-rate model wisely without overpaying.

Finally, this book includes audio and video examples for many of the explanations offered in the text. If you've never used professional recording software, you'll get to see compressors and equalizers and other common audio processing devices in action and hear what they do. When the text describes mechanical and electrical resonance, you can play the demo video to better appreciate why resonance is such an important concept in audio. Although I'll use software I'm familiar with for my examples, the basic concepts and principles apply to *all* audio software and hardware. Several examples of pop tunes and other music are mentioned throughout this book, and most can be found easily by searching YouTube if you're not familiar with a piece.

As with every creative project, we always find something to improve or add after it's been put to bed. So be sure to look for addendum material on my personal website linked below. I'm also active on Facebook, where I post new ideas and experiments related to audio, and I've posted many educational videos to my YouTube channel:

> http://ethanwiner.com/book.htm
> www.facebook.com/ethan.winer.1
> www.youtube.com/user/EthanWiner/videos

Bonus Web Content

This book provides a large amount of additional material online. There are many audio and video demos available that enhance the explanations in the printed text, and spreadsheets and other software are provided to perform common audio-related calculations. All of this additional content is on the book's website: www.theaudioexpertbook.com/. In addition, the book's website contains links to the external YouTube videos, related articles and technical papers, and the software mentioned in the text, so you don't have to type them all yourself.

PART 1

Audio Defined

If a tree falls in the forest and no one is there to hear it, does it make a sound?

I hope it's obvious that the answer to the preceding question is yes, because sounds exist in the air whether or not a person, or a microphone, is present to hear them. At its most basic, audio is sound waves—patterns of compression and expansion that travel through a medium such as air—at frequencies humans can hear. Therefore, audio can be as simple as the sound of someone singing or clapping her hands outdoors or as complex as a symphony orchestra performing in a reverberant concert hall. Audio also encompasses the reproduction of sound as it passes through electrical wires and circuits. For example, you might place a microphone in front of an orchestra, connect the microphone to a preamplifier, which then goes to a tape recorder, which in turn connects to a power amplifier, which is finally sent to one or more loudspeakers. At every stage as the music passes through the air to the microphone, and on through the chain of devices, including the connecting wires in between, the entire path is considered "audio."

CHAPTER 1

Audio Basics

When you can measure what you are speaking about, and express it in numbers, you know something about it; but when you cannot measure it, when you cannot express it in numbers, your knowledge is of a meager and unsatisfactory kind; it may be the beginning of knowledge, but you have scarcely in your thoughts advanced to the state of science.
—Lord Kelvin (Sir William Thomson), nineteenth-century physicist

Volume and Decibels

When talking about sound that exists in the air and is heard by our ears (or picked up by a microphone), volume level is referred to as sound pressure level, or SPL. Our ears respond to changing air pressure, which in turn deflects our eardrums, sending the perception of sound to our brains. The standard unit of measurement for SPL is the *decibel*, abbreviated dB. The "B" refers to Alexander Graham Bell (1847–1922), and the unit of measure is actually the Bel. But one Bel is too large for most audio applications, so one-tenth of a Bel, or one decibel, became the common unit we use today.

By definition, decibels express a *ratio* between two volume levels, but in practice SPL can also represent an *absolute* volume level. In that case there's an implied reference to a level of 0 dB SPL—the softest sound the average human ear can hear, also known as the *threshold of hearing*. So when the volume of a rock concert is said to be 100 dB SPL when measured 20 feet in front of the stage, that means the sound is 100 dB louder than the softest sound most people can hear. Since SPL is relative to an absolute volume level, SPL meters must be calibrated at the factory to a standard acoustic volume.

For completeness, 0 dB SPL is equal to a pressure level of 20 micropascals (millionths of 1 Pascal, abbreviated µPa). Like pounds per square inch (PSI), the Pascal is a general unit of pressure—not only air pressure—and it is named in honor of the French mathematician Blaise Pascal (1623–1662).

Note that decibels use a *logarithmic* scale, which is a form of numeric "compression." Adding dB values actually represents a *multiplication* of sound pressure levels, or voltages when it relates to electrical signals. Each time you add some number of decibels, the underlying change in air pressure, or volts for audio circuits, increases by a multiplying factor:

+6 dB = 2 times the air pressure or volts
+20 dB = 10 times the air pressure or volts

+40 dB = 100 times the air pressure or volts
+60 dB = 1,000 times the air pressure or volts
+80 dB = 10,000 times the air pressure or volts

Likewise, subtracting decibels results in division:

−6 dB = 1/2 the air pressure or volts
−20 dB = 1/10 the air pressure or volts
−40 dB = 1/100 the air pressure or volts
−60 dB = 1/1,000 the air pressure or volts
−80 dB = 1/10,000 the air pressure or volts

So when the level of an acoustic source or voltage increases by a factor of 10, that increase is said to be 20 dB louder. But increasing the original level by 100 times adds only another 20 dB, and raising the volume by a factor of 1,000 adds only 20 dB more. Using decibels instead of ratios makes it easier to describe and notate the full range of volume levels we can hear. The span between the softest sound audible and the onset of extreme physical pain is about 140 dB. If that difference were expressed using normal (not logarithmic) numbers, the span would be written as 10,000,000,000,000 to 1, which is very unwieldy! Logarithmic values are also used because that's just how our ears hear. An increase of 3 dB represents a doubling of power,[1] but it sounds only a little louder. To sound twice as loud, the volume needs to increase by about 8 to 10 dB, depending on various factors, including the initial volume and the frequencies present in the source.

Note that distortion and noise specs for audio gear can be expressed using either decibels or percents. For example, if an amplifier adds 1 percent distortion, that amount of distortion could be stated as being 40 dB below the original signal. Likewise, noise can be stated as a percent or dB difference relative to some output level. Chapter 2 explains how audio equipment is measured in more detail.

You may have read that the smallest volume change people can hear is 1 dB. Or you may have heard it as 3 dB. In truth, the smallest level change that can be noticed depends on several factors, including the frequencies present in the source. We can hear smaller volume differences at midrange frequencies than at very low or very high frequencies. The room you listen in also has a large effect. When a room is treated with absorbers to avoid strong reflections from nearby surfaces, it's easier to hear small volume changes because echoes don't drown out the loudspeaker's direct sound. In a room outfitted with proper acoustic treatment, most people can easily hear level differences smaller than 0.5 dB at midrange frequencies.

It's also worth mentioning the *inverse square law*. As sound radiates from a source, it becomes softer with distance. This decrease is due partly to absorption by the air, which affects high frequencies more than low frequencies, as shown in Table 1.1. But the more important reason is simply because sound radiates outward in an arc, as shown in Figure 1.1.

**Table 1.1: Frequencies over Distances at 20°C
(68°F) with a Relative Humidity of 70%**

Frequency	Attenuation Over Distance
125 Hz	0.3 dB/Km
250 Hz	1.1 dB/Km
500 Hz	2.8 dB/Km
1,000 Hz	5.0 dB/Km
2,000 Hz	9.0 dB/Km
8,000 Hz	76.6 dB/Km

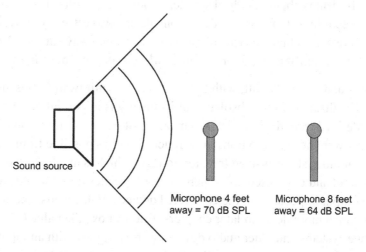

Sound source

Microphone 4 feet
away = 70 dB SPL

Microphone 8 feet
away = 64 dB SPL

Figure 1.1: Sound radiates outward from a source in an arc, so its volume is reduced by 6 dB with each doubling of distance. As you can see in Table 1.1, very high frequencies are reduced over distances due to absorption by the air. This attenuation is in addition to losses caused by the inverse square law, which applies equally to all frequencies.

Each time the distance from a sound source is doubled, the same amount of energy is spread over an area twice as wide. Therefore, the level reduces by a corresponding amount, which in this case is 6 dB.

Standard Signal Levels

As with acoustic volume levels, the level of an audio signal in a wire or electrical circuit is also expressed in decibels, either relative to another signal or relative to one of several common standard reference levels. An amplifier that doubles its input voltage is said to have a gain of 6 dB. This is the same amount of increase whether the input is 0.001 volts or 5 volts; at any given moment, the output voltage is twice as large as the input voltage. But, as with SPL, the dB is also used to express absolute levels for electronic signals using a chosen reference. The most common metrics for expressing volume levels in audio devices are dBu, dBm, dBV, and dBFS.

Most professional audio gear currently manufactured specifies input and output levels as dBu, with 0 dBu corresponding to 0.775 volts. This seemingly unusual voltage is not arbitrary because it relates to the earlier dBm standard described next. The "u" in dBu stands for "unloaded" (or, in some earlier usage, "unterminated") because the levels are not dependent on a specific load impedance, or termination. Other dBu values describe levels either lower or higher than the 0.775 volt 0 dBu reference. For example, 20 dBu is 10 times larger at 7.75 volts, and −10 dBu is about one third smaller at only 0.245 volts. The volume unit (VU) meters in a recording studio or broadcast console are also offset by convention. When calibrating VU meters, a technician typically uses a 1 KHz sine wave at +4 dBu (1.23 volts). So when that level appears at an input or output connector, the console's meter reads 0 VU.

The "m" in dBm stands for milliwatt, with 0 dBm equal to 1 milliwatt (thousandth of a watt) of power. The dBm is mostly obsolete, but its history is important because the current dBu standard is derived from dBm, which is a measure of power rather than voltage. Years ago audio systems were designed to transfer as much power as possible from one device to another. Modern audio systems instead transfer voltage, which is much more efficient and uses a different input and output scheme. Therefore, dBu relates only to volts, regardless of the input and output impedance of the connected devices. Volts, watts, and impedance are explained in much more depth in later chapters. But for now, the value 0.775 is used because telephone systems and older audio devices were designed with an input and output impedance of 600 ohms. In that case 1 milliwatt of power is dissipated when 0.775 volts is applied to a 600 ohm load. So dBm and dBu values are often the same, but dBm applies only when the input and output devices have a 600 ohm impedance.

Another standard is dBV, where the "V" stands for volts. References to dBV are less common with professional devices, but they're sometimes used for consumer electronics. With this standard, 0 dBV equals 1 volt, so by extension 20 dBV equals 10 volts and −6 dBV is half a volt. Since a value of 0 dBV equals 1.0 volt versus 0.775 volts for 0 dBu, the 0.225-volt disparity yields a constant difference of 2.21 dB. And since dBV is referenced to a larger value, it will always be a smaller number than dBu for the same signal level.

The unit dBFS is specific to digital audio, where FS means Full Scale. This is the maximum level a digital device can accommodate or, more accurately, the largest equivalent digital *number* a sound card or A/D/A converter can accept or output. Once an audio signal has been digitized, no reference voltage level is needed or implied. Whatever input and output voltage your particular sound card or outboard converter is calibrated for, 0 dBFS equals the maximum internal digital level possible before the onset of gross distortion.

One more item related to basic signal levels is the difference between a *peak level* and an *average level*, which is also called *RMS level*. The voltage of any waveform, or any varying or static audio signal, can be expressed as either a peak or average value. Figure 1.2 shows three basic waveforms: sine, square, and pulse. I added shading inside each wave to convey the mathematical concept of "area under the curve," which represents the amount of *energy* each wave shape contains. When recording audio we usually care most about the maximum or peak level, because that's what determines where the audio will become distorted. But the perceived *loudness* of music or other sound is related to its average level.

The square wave in this example always contains either plus 1 volt or minus 1 volt, where the pulse wave is active only about 20 percent of the time. So the square wave sounds louder than the pulse wave even though both have the same peak value of 1 volt. The average level for the pulse wave is about 8 dB lower too. In this case "average" is in part a theoretical concept because it requires averaging every possible voltage that exists over time, which in theory is an infinite number of values. But the concept should be clear enough even if you consider only a few of the values within the shaded areas. The average level of the pulse wave depends entirely on its peak voltage and "on and off" times, but the sine wave changes constantly over time, so that requires averaging all the voltages. And since energy and loudness are independent of polarity, positive and negative voltages are both treated as positive values.

Finally, RMS is an abbreviation for the math term *root-mean-square*. This is a type of average that's calculated differently than a normal average and gives a slightly different result. The difference between RMS and average is more important to electrical engineers than to mixing engineers and audiophiles, but for the sake of completeness it's worth explaining briefly: When some level of RMS voltage is applied to a resistive load, the *amount of heat dissipated* is the same as if the same level of DC voltage were applied.

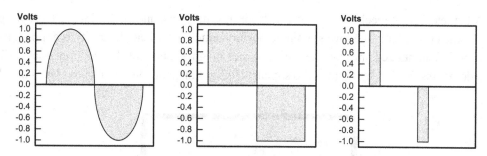

Figure 1.2: If the maximum voltage of a sine wave (left) peaks at 1 volt, the average of the voltage for all points in between is slightly lower. Only a square wave (center) has the same peak and average values, and a pulse wave (right) can have an even greater disparity if the pulse width is very narrow.

A regular average is obtained by adding up all the values of interest, then dividing the result by the quantity of values. So the sum of the four values 1, 3, 4, and 12 is 1 + 3 + 4 + 12 = 20. Then divide 20 by the quantity of values, which is 4, and that yields the average, which is 20 / 4 = 5. An RMS average is calculated by squaring every value, then averaging the squared values as usual, then taking the square root of the average. So the RMS of those same four values is about 6.5 instead of 5, calculated as follows:

First square all the values:

$$1 * 1 = 1$$
$$3 * 3 = 9$$
$$4 * 4 = 16$$
$$12 * 12 = 144$$

Then calculate the average:

$$1 + 9 + 16 + 144 = 170$$
$$170 / 4 = 42.5$$

Finally, calculate the square root of 42.5 which is 6.519.

Signal Levels and Metering

Level meters are an important part of recording and mixing because every recording medium has a limited range of volume levels it can accommodate. For example, when recording to analog tape, if the audio is recorded too softly, you'll hear a "hiss" in the background when you play back the recording. And if the music is recorded too loudly, an audible distortion can result. The earliest type of audio meter used for recording (and broadcast) is called the VU meter, where VU stands for *volume units*. This is shown in Figure 1.3.

Early VU meters were mechanical, made with springs, magnets, and coils of wire. The spring holds the meter pointer at the lowest volume position, and then when electricity is applied, the coil becomes magnetized, moving the pointer. Since magnets and coils have a finite mass, VU meters do not respond instantly to audio signals. Highly transient sounds such as claves or other

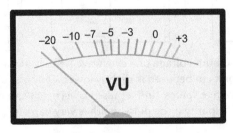

Figure 1.3: Standard VU meters display a range of levels from −20 to +3 dB.

percussion instruments can come and go before the meter has a chance to register its full level. So when recording percussive instruments and instruments that have a lot of high-frequency content, you need to record at levels lower than the meter indicates to avoid distortion.

Adding "driver" electronics to a VU meter offers many potential advantages. One common feature holds the input voltage for half a second or so, giving the meter time to respond to the full level of a transient signal. Another useful option is to expand the range displayed beyond the typical 23 dB. This is often coupled with a logarithmic scale, so, for example, a meter might display a total span of 40 or even 50 dB, shown as 10 dB steps equally spaced across the meter face. This is different from the VU meter in Figure 1.3, where the nonlinear log spacing is incorporated into the dB scale printed on the meter's face.

Modern digital meters use either an LED "ladder" array as shown in Figure 1.4 or an equivalent as displayed on a computer screen by audio recording software. Besides showing a wider range of volumes and holding peaks long enough to display their true level, many digital meters can also be switched to show either peak or average volumes. This is an important concept in audio because our ears respond to a sound's average loudness, where computer sound cards and analog tape distort when the peak level reaches what's known as the *clipping point*. Mechanical VU meters inherently average the voltages they receive by nature of their construction. Just as it takes some amount of time (50 to 500 milliseconds) for a meter needle to deflect fully and stabilize, it also takes time for the needle to return to zero after the sound stops. The needle simply can't keep up with the rapid changes that occur in music and speech, so it tends to hover around the average volume level. Therefore, when measuring audio whose level changes constantly—which includes most music—VU meters are ideal because they indicate how loud the music actually sounds. But mechanical VU meters won't tell you whether audio exceeds the maximum allowable peak level unless extra circuitry is added.

Modern digital meters often show both peak and average levels at the same time. All of the lights in the row light up from left to right to show the average level, while single lights farther to the right blink one at a time to indicate the peak level, which is always higher. Some digital meters can even be told to hold peaks indefinitely. So you can step away and later, after the recording finishes, you'll know if the audio clipped when you weren't watching. The difference between a signal's peak and average levels is called its *crest factor*. By the way, the crest factor relation between peak and average levels applies equally to acoustic sounds in the air.

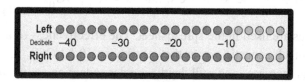

Figure 1.4: Digital meters feature instant response times, a wide range of display levels, and often peak-hold capability.

The concept of peak versus average levels also applies to the output ratings of power amplifiers. Depending on their design, some amplifiers can output twice as much power for brief periods than they can provide continuously. Many years ago, it was common practice for amplifier makers to list only peak power output in their advertisements, and some of the claims bordered on fraud. For example, an amplifier that could output only 30 watts continuously might claim a peak power output of hundreds of watts, even if it could provide that elevated power for only one millisecond. Thankfully, the US Federal Trade Commission passed a law (FTC Rule 46 CFR 432) in 1974 making this practice illegal. I can think of several current audio practices the FTC should consider banning!

Calculating Decibels

The included Excel spreadsheet Decibels.xls calculates decibel values from voltages or ratios or percents, as well as computing decibel changes when combining two identical signals having opposite polarities. This is useful because it lets you determine the extent of peaks and nulls caused by acoustic reflections having a known strength. It also works the other way around, letting you use room testing software to derive absorption coefficients of acoustic materials based on the measured strength of reflections at various frequencies. That type of testing is explained in Chapter 20.

The spreadsheet clearly shows what you enter for each section and what information is returned, so I won't elaborate here. All input and output values are in Column B, with the input fields you enter shown in bold type. Simply replace the sample values in those bold fields. The first section accepts two voltages and tells you the dB difference between them. The second section is similar but accepts a voltage or SPL difference as a ratio and returns the difference in decibels. The third section does the opposite: It accepts a decibel difference and returns the equivalent voltage or SPL difference as a ratio. The fourth section computes percent distortion from a decibel relationship, and vice versa.

Although this book aims to avoid math as much as possible, for completeness the following formulas show how to calculate decibels, where the asterisk (*) signifies multiplication:

$$\text{dB difference between two voltages} = 20 * LOG_{(10)} \frac{\text{voltage1}}{\text{voltage2}}$$

$$\text{dB difference between two wattages} = 10 * LOG_{(10)} \frac{\text{wattage1}}{\text{wattage2}}$$

Both of these formulas are used in the spreadsheet. The effectiveness of acoustic materials is measured by how much acoustic *power* they absorb, so $10 * LOG_{(10)}$ is used for those calculations. But peak increases and null depths are dependent on sound *pressure* differences, which are calculated like voltages, so those cells instead use $20 * LOG_{(10)}$. I encourage you to look at the formulas in the various result cells to see how they work.

Frequencies

The unit of frequency measurement is the hertz, abbreviated Hz, in honor of German physicist Heinrich Hertz (1857–1894). However, before 1960, frequencies were stated as cycles per second (CPS), kilocycles (KC = 1,000 Hz), megacycles (MC = 1,000,000 Hz), or gigacycles (GC = 1,000,000,000 Hz). As with volume, frequencies are also heard and often expressed logarithmically. Raising the pitch of a note by one octave represents a doubling of frequency, and going down one octave divides the frequency in half. Figure 1.5 shows all of the A notes on a standard 88-key piano. You can see that each unit of one octave doubles or halves the frequency. This logarithmic frequency relation also corresponds to how our ears naturally hear. For an A note at 440 Hz, the pitch might have to be shifted by 3 to 5 Hz before it's perceived as out of tune. But the A note at 1,760 Hz two octaves higher would have to be off by 12 to 20 Hz before you'd notice that it's out of tune. Musical notes can also be divided into *cents*, where 1 cent equals a pitch change equal to 1 percent of the difference between adjacent half-steps. Using a percent variance for note frequencies also expresses a ratio, since the number of Hz contained in 1 cent depends on the note's fundamental frequency.

Again, for completeness, I'll mention that for equal tempered instruments such as the piano, the distance between any two musical half-steps is equal to the 12th root of 2, or 1.0595. We use the 12th root because there are 12 half-steps in one musical octave. This divides the range of one octave logarithmically rather than into equal Hz steps. Therefore, for the A note at 440 Hz, you can calculate the frequency of the Bb a half-step higher as follows:

$$440 * 2^{1/12} = 466.16 \text{ Hz}$$

or

$$440 * 1.0595 = 466.16 \text{ Hz}$$

To find the frequency that lies musically halfway between two other frequencies, multiply one times the other, then take the square root:

$$Center = \sqrt{High * Low}$$

Table 1.2 lists all of the note frequencies you're likely to encounter in real music. You'll probably never encounter a fundamental frequency in the highest octave, but musical instruments and other sound sources create *overtones*—also called *harmonics* or

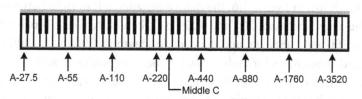

Figure 1.5: A span of one musical octave corresponds to a doubling, or halving, of frequency.

Table 1.2: Standard Frequencies for Musical Notes

A	27.5	55.0	110.0	220.0	440.0	880.0	1760.0	3520.0	7040.0
A#	29.1	58.3	116.5	233.1	466.2	932.3	1864.7	3729.3	7458.6
B	30.9	61.7	123.5	246.9	493.9	987.8	1975.5	3951.1	7902.1
C	32.7	65.4	130.8	261.6	523.3	1046.5	2093.0	4186.0	8372.0
C#	34.6	69.3	138.6	277.2	554.4	1108.7	2217.5	4434.9	8869.8
D	36.7	73.4	146.8	293.7	587.3	1174.7	2349.3	4698.6	9397.3
D#	38.9	77.8	155.6	311.1	622.3	1244.5	2489.0	4978.0	9956.1
E	41.2	82.4	164.8	329.6	659.3	1318.5	2637.0	5274.0	10548.1
F	43.7	87.3	174.6	349.2	698.5	1396.9	2793.8	5587.7	11175.3
F#	46.2	92.5	185.0	370.0	740.0	1480.0	2960.0	5919.9	11839.8
G	49.0	98.0	196.0	392.0	784.0	1568.0	3136.0	6271.9	12543.9
G#	51.9	103.8	207.7	415.3	830.6	1661.2	3322.4	6644.9	13289.8

Note: The C at 261.6 Hz is middle C.

partials—that can extend that high and beyond. Indeed, cymbals and violins can generate overtones extending to frequencies much higher than the 20 KHz limit of human hearing.

Graphing Audio

As we have seen, both volume levels and frequencies with audio are usually expressed logarithmically. The frequency response graph in Figure 1.6 is typical, though any decibel and frequency ranges could be used. This type of graph is called *semi-log* because only the horizontal frequency axis is logarithmic, while the vertical decibels scale is linear. Although the dB scale is linear, it's really not because decibels are inherently logarithmic. If the vertical values were shown as volts instead of decibels, then the horizontal lines would be spaced logarithmically one above the other, rather than spaced equally as shown. And then the graph would be called *log-log* instead of semi-log.

Standard Octave and Third-Octave Bands

As we have seen, audio frequencies are usually expressed logarithmically. When considering a range of frequencies, the size of each range, or band, varies rather than contains a constant number of Hz. As shown in Figure 1.6, the octave distance left to right between 20 and 40 Hz is the same as the octave between 200 and 400 Hz. Likewise for the 10-to-1 range (called a *decade*) between 100 and 1,000 Hz, and the decade spanning 1,000 to 10,000 Hz. Table 1.3 shows the standard frequencies for audio based on octave (boldface) and third-octave bandwidths. These bands are used for measuring the frequency response of microphones and other gear, for specifying the absorption of acoustic products, as well as for the available

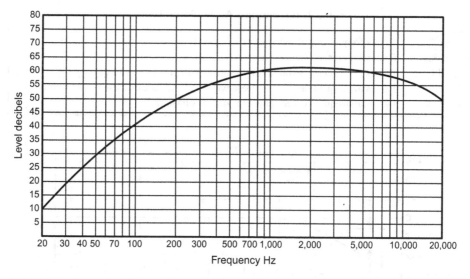

Figure 1.6: This is the typical layout for a semi-log audio graph, where the horizontal frequency axis is logarithmic and the vertical dB volume axis is linear (equal step distances).

Table 1.3: Standard Octave and Third-Octave Audio Bands

16	20	25	**31.5**	40	50	**63**	80		100	**125**	160	200	**250**	315		400	**500**
630	800	**1K**	1.25K	1.6K	**2K**	2.5K	3.15K	**4K**	5K	6.3K	**8K**	10K	12.5K	**16K**	20K		

frequencies in most graphic equalizers. The stated frequency is at the center of a band that encompasses a range of frequencies, and so it is called the *center frequency*.

Filters

An audio filter is a device that selectively passes, or suppresses, a range of frequencies. A common filter type familiar to all audio enthusiasts is the equalizer, though in practice most equalizers are more complex than the basic filters from which they're created. Table 1.4 shows the five basic filter types, which are named for how they pass or suppress frequencies.

Besides the stated cutoff frequency, high-pass and low-pass filters also have a fall-off property called *slope*, which is specified in *dB per octave*. The cutoff for these filter types is defined as the frequency at which the response has fallen by 3 dB, also known as the *half-power point*. The slope is the rate in dB at which the level continues to fall at higher or lower frequencies. The high-pass filter in Figure 1.7 has a cutoff frequency of 125 Hz, which is where the response is 3 dB below unity gain. At 62 Hz, one octave below 125 Hz, the response is therefore at −9 dB. An octave below that, it's −15 dB at 31 Hz. The low-pass filter

Table 1.4: Common Audio Filter Types

High-Pass	Passes frequencies above the stated cutoff
Low-Pass	Passes frequencies below the stated cutoff
Band-Pass	Passes frequencies within a range surrounding the center frequency
Band-Stop	Passes all frequencies except those within a range around the center frequency
All-Pass	Passes all frequencies equally, but applies phase shift

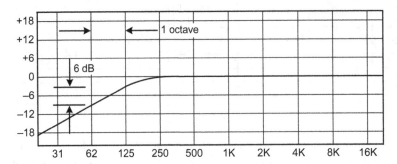

Figure 1.7: This high-pass filter has a −3 dB cutoff frequency of 125 Hz, with a slope of 6 dB per octave.

in Figure 1.8 has a cutoff frequency of 1 KHz, which again is where the response is 3 dB below unity gain. At 2 KHz, one octave above 1 KHz, the response is therefore at −9 dB. An octave higher, it's −15 dB at 4 KHz.

Although the official names for these filters describe the frequencies they pass, I prefer to call them by the frequencies they actually affect. For example, when a high-pass filter is set to a low frequency to remove rumble picked up by a microphone, I think of that as a low-*cut* filter because that's how it's being used. Likewise for a low-pass filter with a cutoff frequency in the treble range. Yes, it passes frequencies below the treble range cutoff, but in practice it's really reducing high frequencies in relation to the rest of the audio range. So I prefer to call it a high-cut filter. Again, this is just my preference, and either wording is technically correct.

Band-pass and band-stop filters have a center frequency rather than a cutoff frequency. Like high-pass and low-pass filters, band-pass filters also have a slope stated in dB per octave. Both of these filter types are shown in Figures 1.9 and 1.10. Determining the slope for a band-stop filter is tricky because it's always very steep at the center frequency where the output level is reduced to near zero. Note that band-stop filters are sometimes called *notch* filters due to the shape of their response when graphed.

Most filters used for audio equalizers are variants of these basic filter types, and they typically limit the maximum amount of boost or cut. For example, when cutting a treble frequency range to remove harshness from a cymbal, you'll usually apply some amount of dB cut at

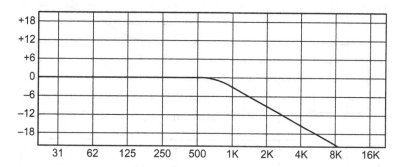

Figure 1.8: This low-pass filter has a cutoff frequency of 1 KHz, with a slope of 6 dB per octave.

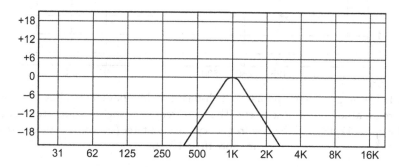

Figure 1.9: This band-pass filter has a center frequency of 1 KHz, with a slope of 18 dB per octave.

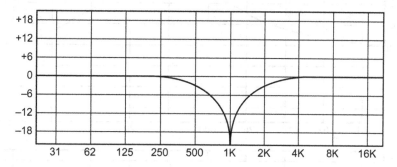

Figure 1.10: This band-stop filter has a cutoff frequency of 1 KHz, with a slope of 6 dB per octave. In practice, the response at the center frequency approaches zero output. Therefore, the slope becomes much steeper than 6 dB per octave at the center frequency.

the chosen frequency rather than reduce it to zero, as happens with a band-stop filter. When boosting or cutting a range above or below a cutoff frequency, we call them *shelving* filters because the shape of the curve resembles a shelf, as shown in Figure 1.11. When an equalizer boosts or cuts a range by some amount around a center frequency, it's called a *peaking* filter or a *bell* filter because its shape resembles a bell, like the one shown in Figure 1.12.

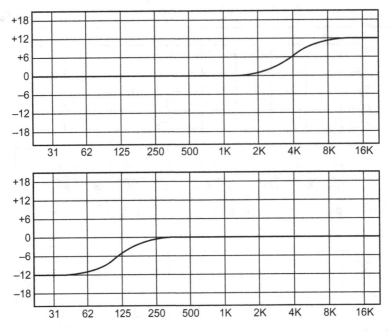

Figure 1.11: Shelving filters are similar to high-pass and low-pass filters, but they level out at some maximum amount of boost or cut. The high-frequency shelving filter (top) boosts high frequencies by up to 12 dB, and the low-frequency shelving filter (bottom) cuts low frequencies by no more than 12 dB.

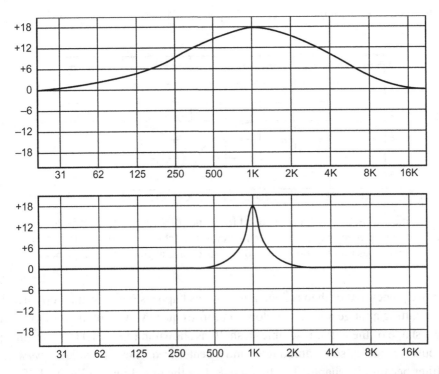

Figure 1.12: Peaking-style EQ has three basic properties: center frequency, amount of boost or cut in dB, and Q, or bandwidth. Both of these screens show an equalizer adding 18 dB boost at 1 KHz, but the top EQ has a Q of 0.5, while the bottom Q is 6.0.

At the circuit level, filters are made from passive components—capacitors, inductors, and resistors—and are inherently cut-only. So to obtain boost requires active electronics. In practice, you may find equalizers that claim to be passive, but they usually include active circuitry to raise the entire signal level either before or after the passive filter, or both.

Filter slopes are inherently multiples of 6 dB per octave, which is the same as 20 dB per decade. As mentioned earlier, a decade is a range of 10 to 1, also referred to as an *order of magnitude*. A filter made from a single capacitor and resistor falls off at a rate of 6 dB per octave, as does a filter made from one inductor and one resistor. To get a slope of 12 or 18 dB per octave, or even larger, requires multiple filter sections, where each section contributes one *pole* to the response. Therefore, a three-pole low-pass filter has a slope of 18 dB per octave.

Another important filter parameter is its *Q*, which stands for *quality*. How the Q of a filter is interpreted depends on the filter type. Q usually applies to band-pass filters, but it can also be used for peaking filters whether they're set to boost or to cut. The equalizer response graphs shown in Figure 1.12 both have 18 dB of boost at 1 KHz, but the boost shown in the top graph has a fairly low Q of 0.5, while the bottom graph shows a higher Q of 6.0. EQ changes made using a low Q are more audible simply because a larger span of frequencies is affected. Of course, for equalization to be audible at all, the source must contain frequencies within the range being boosted or cut. Cutting 10 KHz and above on an electric bass track will not have much audible affect, because most basses have little or no content at those high frequencies.

High-pass, low-pass, and shelving filters can also have a Q property, which affects the response and slope around the cutoff frequency, as shown in Figure 1.13. As you can see, as the Q is increased, a peak forms around the cutoff frequency. However, the slope eventually settles to 6 dB per octave (or a multiple of 6 dB). Applying a high Q to a low-pass filter is the basis for analog synthesizer filters, as made famous by early Moog models. For example, the low-pass filter in a MiniMoog has a slope of 24 dB per octave; the sharp slope coupled with a resonant peak at the cutoff frequency creates its characteristic sound. These days, digital filters are often used to create the same type of sounds using the same slope and Q.

Again, to be complete, Figure 1.14 shows the mathematical relation between frequencies, bandwidth, and Q. The cutoff frequency of low-pass and high-pass filters is defined as the frequency at which the response has fallen by 3 dB, and the same applies to band-pass, band-stop, and peaking EQ filters.

Phase Shift and Time Delay

Table 1.4 shown earlier lists the *all-pass* filter, which might seem nonsensical at first. After all, what good is an audio filter that doesn't boost or cut any frequencies? In truth, a filter that applies phase shift without changing the frequency balance has several uses in audio.

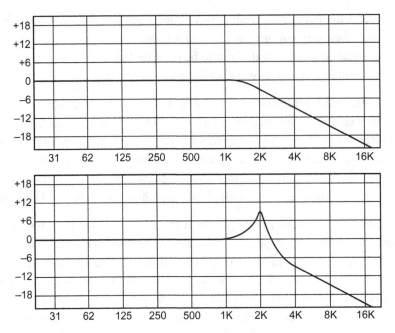

Figure 1.13: High-pass and low-pass filters can also have a Q parameter. Both of these low-pass filters have a cutoff frequency of 2 KHz and an eventual slope of 6 dB per octave, but the top filter has a Q of 1.4, while the bottom filter's Q is 6.0.

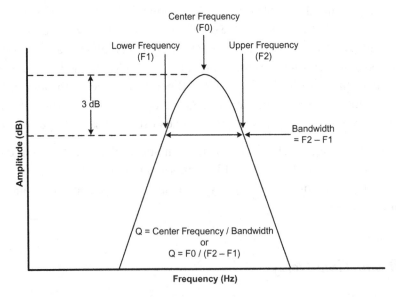

Figure 1.14: Bandwidth is the reciprocal (opposite) of Q, with higher Q values having a narrower bandwidth.

For example, all-pass filters are at the heart of phase shifter effects. They can also be used to create artificial stereo from a mono sound source. Flanger effects are similar, using a simple time delay rather than phase shift. But let's first consider what phase shift really is, since it's at the heart of *every* filter and equalizer.

Like a circle, one complete cycle of a sine wave is divided into 360 degrees. The upper sine wave in Figure 1.15 shows the wave starting at a level of zero at an arbitrary point in time called Time Zero. Since one cycle contains 360 degrees, after 90 degrees—one-quarter of the way through the cycle—the wave has reached its peak positive amplitude. After 180 degrees the level is back to zero, and at 270 degrees the wave reaches its maximum negative level. At 360 degrees it's back to zero again. Note that the span between the maximum positive level at 90 degrees and the maximum negative level at 270 degrees is a *difference* of 180 degrees. The significance of this will become obvious soon.

Now let's consider the lower sine wave, which started at the same time as the upper wave but was sent through an all-pass filter that delays this particular frequency by 90 degrees. Viewing both waves together you can see the time delay added by the all-pass filter. The phase shift from an all-pass filter is similar to a simple time delay, but not exactly the same. Time delay shifts all frequencies by the same amount of time, where phase shift delays some frequencies longer than others. In fact, an all-pass filter's center frequency is defined as the frequency at which the phase shift is 90 degrees.

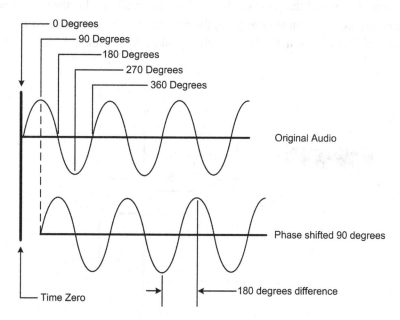

Figure 1.15: Phase shift is similar to time delay in that certain frequencies exit an all-pass filter later than they arrived at its input.

To put this theory into practice, let's see what happens when music—which typically contains many frequencies at once—is sent through an all-pass filter or time delay, and the delayed audio is mixed with the original. When you combine audio with a delayed version of itself, the frequency response is altered. As one cycle of the wave is rising, the delayed version is falling, or perhaps it hasn't yet risen as high. So when the two are combined, they partially cancel *at some frequencies only*. This is the basis for all analog equalizers. They shift the phase for a range of frequencies and then combine the phase-shifted audio with the original.

An all-pass filter can also be used to create a pseudo-stereo effect. This is done by combining both the original and phase-shifted audio through two separate paths, with the polarity of one path reversed. A block diagram is shown in Figure 1.16, and the output showing the response of one channel is in Figure 1.18. One path is combined with the original, as already explained, and the other path is combined with a reversed polarity at the same time. This creates equal but opposite comb filter responses such that whatever frequencies are peaks at the left output become nulls at the right output, and vice versa. The different frequency responses at the left and right channel outputs are what create the pseudo-stereo effect.

A different method for using an all-pass filter to create fake stereo is to simply apply different amounts of phase shift to the left and right channels without mixing the original and shifted versions together. In this case the frequency response is not altered, but the sound takes on an exaggerated sense of width and dimension. This technique can also make sounds seem to come from a point beyond the physical location of the speakers. Further, if the amount of phase shift is modulated to change over time, the audible result is similar to a Leslie rotating speaker. In fact, this is pretty much what a Leslie speaker does, creating phase shift and

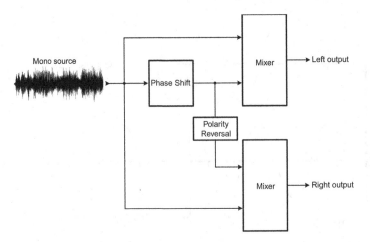

Figure 1.16: A stereo synthesizer is similar to a phaser effect unit, except it combines the phase-shifted output twice, with the polarity of one path reversed.

constantly varying time delays via the motion of its speaker driver. With a real Leslie speaker, the Doppler effect causes the pitch to rise and fall as the rotating horn driver moves toward and away from you. But this also happens when phase shift is varied over time.

Early digital equalizers mimicked the behavior of analog equalizers, though with a totally different circuit design. Instead of using capacitors or inductors to shift phase, they use taps on a digital delay line. A digital delay line is a series of memory locations that the digitized audio samples pass through. The first number that arrives is stored in Address 0. Then, at the next clock cycle (44,100 times per second for a 44.1 KHz sample rate), the number now in Address 0 is shifted over to Address 1, and the next incoming sample is stored at Address 0. As more numbers enter the input, they are all shifted through each memory location in turn, until they eventually arrive at the output. This is the basis for a digital delay. You can alter the delay time by changing the total number of addresses the numbers pass through or the clock rate (shift speed), or both. Indeed, a block of memory addresses used this way is called a *shift register* because of the way numbers are shifted through them in turn.

To create an equalizer from a digital delay line, you tap into one of the intermediate memory addresses, then send that at some volume level back to the input. It's just like the feedback control on an old EchoPlex tape-based delay, except without all the flutter, noise, and distortion. You can also reverse the polarity of the tapped signal, so a positive signal becomes negative and vice versa, before sending it back to the input to get either cut or boost. By controlling which addresses along the delay route you tap into and how much of the tapped signal is fed back into the input and with which polarity, an equalizer is created. With an analog EQ the phase shift is created with capacitors and inductors. In a digital EQ the delays are created with a tapped shift register. Actual computer code for a simple digital EQ filter will be shown in Chapter 25, but the key point is that all equalizers rely on phase shift unless they use special trickery.

Finally, phase shift can also be used to alter the ratio of peak to average volume levels without affecting the sound or tone quality. The top waveform in Figure 1.17 shows one sentence from a narration Wave file I recorded for a tutorial video. The bottom waveform shows the same file after applying phase shift using an all-pass filter.

The Orban company sells audio processors to the broadcast market, and its Optimod products contain a phase rotator feature that's tailored to reduce the peak level of typical male voices without lowering the overall volume. Just as the maximum level you can pass through a tape recorder or preamp is limited by the waveform peaks, broadcast transmitters also clip based on the peak level. By reducing the peak heights with phase shift, broadcasters can increase the overall volume of an announcer without using a limiter, which might negatively affect the sound quality. Or they can use both phase shift and limiting to get even more volume without distortion.

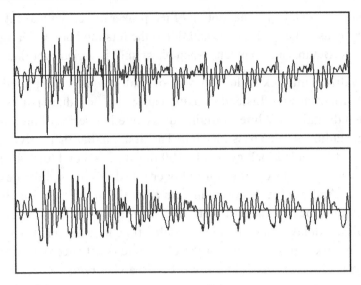

Figure 1.17: Phase shift can alter the peak level of a source (top) without changing its average level.

Comb Filtering

A *comb filter* is a unique type of filter characterized by a series of peaks and deep nulls that repeat at equally spaced (not logarithmic) frequency intervals. As with other filter types, a comb filter is created by combining an audio signal with a delayed version of itself. If time delay is used rather than phase shift, the result is an infinite number of peaks and nulls, as shown in Figure 1.18. A comb filter response also occurs naturally when sound reflects off a wall or other surface and later combines in the air with the original sound. Indeed, comb filtering will make an appearance many times throughout this book.

The *flanging effect* is the classic implementation of a comb filter and is easily recognized by its characteristic hollow sound. The earliest flanging effects were created manually using two tape recorders playing the same music at once but with one playback delayed a few milliseconds compared to the other. Analog tape recorders lack the precision needed to control playback speed and timing accurately to within a few milliseconds. So recording engineers would lay their hand on the tape reel's flange (metal or plastic side plate) to slightly slow the speed of whichever playback was ahead in time. When the outputs of both recorders were then mixed together at equal volume, the brief time delay created a comb filter response, giving the hollow sound we all love and know as the flanging effect.

Comb filtering peaks and nulls occur whenever an audio source is combined with a delayed version of itself, as shown in Figure 1.19. For any given delay time, some frequency will be shifted exactly 180 degrees. So when the original wave at that frequency is positive,

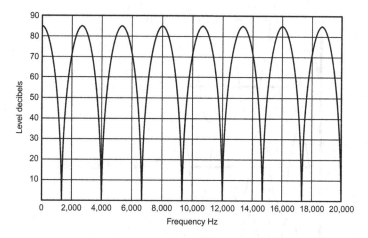

Figure 1.18: Comb filtering is characterized by a repeating pattern of equally spaced peaks and deep nulls. Because the frequencies are at even Hz multiples, comb filtering is usually graphed using a linear rather than logarithmic frequency axis.

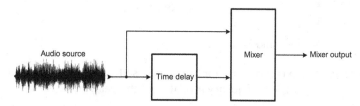

Figure 1.19: The flanging effect is created by sending audio through a time delay, then combining the delayed output with the original audio.

the delayed version is negative, and vice versa. If both the original and delayed signals are exactly the same volume, the nulls will be extremely deep, though the peaks are boosted by only 6 dB. When used as an audio effect, the comb filter frequency is usually swept slowly up and down to add some animation to, for example, an otherwise static-sounding rhythm guitar part. Faster speeds can also be used to create a warbling or vibrato effect.

Figure 1.20 shows a single frequency tone delayed so its phase is shifted first by 90 degrees, then by a longer delay equal to 180 degrees. If the original tone is combined with the version shifted by 180 degrees, the result is complete silence. Other frequencies present in the audio will not be canceled unless they are multiples of the same frequency. That is, a delay time that shifts 100 Hz by 180 degrees will also shift 300 Hz by one full cycle plus 180 degrees. The result therefore is a series of deep nulls at 100 Hz, 300 Hz, 500 Hz, and so forth. Please understand that the severely skewed frequency response is what creates the hollow "swooshy" sound associated with flanger and phaser effect units. You are *not* hearing the phase shift itself.

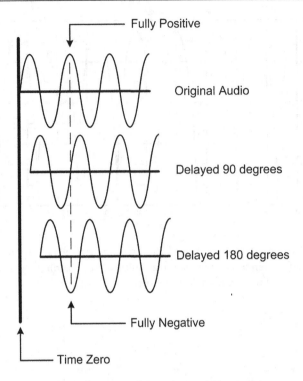

Figure 1.20: Delaying audio is similar to applying phase shift, and the amount of phase shift at any given frequency is related to the delay time.

For comb filter type effects, the delay is typically just a few milliseconds. Most of us will simply turn the delay knob until we get a sound we like, though it's simple to determine the first (lowest frequency) peak from the delay time:

$$\text{Lowest peak frequency Hz} = 1/\text{Time in seconds}$$

or

$$\text{Lowest peak frequency KHz} = 1/\text{Time in milliseconds}$$

The lowest null frequency is always half the lowest peak frequency. Both then repeat at multiples of the first peak's frequency. So for a delay of 1 millisecond, the first peak is at 1 KHz, with subsequent peaks at 2 KHz, 3 KHz, 4 KHz, and so forth. The lowest null is at 500 Hz, and subsequent nulls are at 1,500 Hz, 2,500 Hz, 3,500 Hz, and so forth. You can see this behavior in Figure 1.18.

Getting the strongest effect requires mixing the original and delayed sounds together at precisely equal volumes. Then the peak frequencies are boosted by 6 dB, and the nulls become infinitely deep. Nearby frequencies that are shifted less, or more, than 180 degrees also cancel, but not as much. Likewise, when the two signal levels are not precisely equal, the peaks and nulls are less severe. This is the same as using a lower Strength or Mix setting on phaser or flanger effect units. The Decibels.xls calculator described earlier can tell you the maximum extent of peaks and nulls when original and delayed sounds are mixed together at various levels.

There's a subtle technical difference between flanging and phasing effects. A flanger effect creates an infinite series of peaks and nulls starting at some lower frequency. But a phaser creates a limited number of peaks and nulls, depending on how many stages of phase shift are used. Many phaser guitar pedal effects use six stages, though some use more. Each stage adds up to 90 degrees of phase shift, so they're always used in pairs. Each pair of phase shift stages (0 to 180 degrees) yields one peak and one null. Phaser hardware effects are easier to design and build than flanger effects, because they use only a few simple components. Compare this to a flanger that requires A/D and D/A converters to implement a time delay digitally. Of course, with computer plug-ins the audio they process is already digitized, avoiding this extra complication.

This same hollow sound occurs acoustically in the air when reflections off a wall or floor arrive delayed at your ears or a microphone. Sound travels at a speed of about 1.1 feet per millisecond, but rounding that down to a simpler 1 foot = 1 millisecond is often close enough. So for every foot of distance between two sound sources, or between a sound source and a reflecting room surface, there's a delay of about 1 millisecond. Since 1 millisecond is the time it takes a 1 KHz tone to complete one cycle, it's not difficult to relate distances and delay times to frequencies without needing a calculator.

Figure 1.21 shows that for any frequency where the distance between a listener (or a microphone) and a reflective wall is equal to one-quarter wavelength, a null occurs. The delay-induced phase

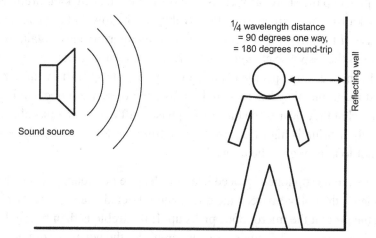

Figure 1.21: Reflections off a nearby wall or other surface create the same type of comb filter response as a flanger effect unit.

shift occurs at predictable distances related to the frequency's wavelength, because at that point in space your ear hears a mix of both the direct and reflected sounds. The depth of the notch depends on the strength of the reflection at that frequency. Therefore, hard, reflective surfaces create a stronger comb filter effect.

Understand that a one-quarter wavelength distance means the total round trip is one-half wavelength, so the reflection arrives after 180 degrees of phase shift, not 90. Nulls also occur at related higher frequencies where the distance is equal to three-quarters wavelengths, one and one-quarter wavelengths, and so forth. This is why the frequency response has a series of peaks and nulls instead of only one. Note that comb filtering also occurs at lower frequencies where the distances are larger, causing peaks and nulls there, too. The Frequency-Distance Calculator for Windows described in Chapter 19 will tell you the relationship between frequencies and distances in one-quarter wavelength increments.

To make this easier to visualize and hear, the video "comb_filtering" shows pink noise playing through a loudspeaker that's pointed at a reflecting window about two feet away. I held a DPA 4090 measuring microphone and then moved it slowly toward and away from the window. The audio you hear in this video is what the microphone captured, and the sweeping comb filter frequencies are very easy to hear. It's not always easy to hear this comb filter effect when standing near a boundary because we have two ears. So the peak and null frequencies in one ear are often different in the other ear, which dilutes the effect. Just for fun, I ran the recorded signal through the Room EQ Wizard measuring software, to show a live screen capture of its Real Time Analyzer at the same time in the video.

Besides reflections, comb filtering can also occur when mixing tracks that were recorded with multiple microphones due to arrival time differences. For example, a microphone near the snare drum picks up the snare, as well as sound from the nearby kick drum. So when the snare and kick drum mics are mixed, it's possible for the low end to be reduced—or boosted—because of the arrival time difference between microphones. Again, while phase shift is the cause of the response change, it's the response change that you hear and not the phase shift itself. Indeed, some people claim they can hear phase shift in equalizers because when they boost the treble, they hear a sound reminiscent of phaser effects units. So they wrongly assume what they hear is the damaging phase shift everyone talks about. In truth, what they're really hearing is high-frequency comb filtering that was already present in the recording, but not loud enough to be noticed.

For example, when a microphone is placed near a reflective boundary such as the wooden lid of a grand piano, the delay between the direct and reflected sounds creates a comb filter acoustically in the air that the microphone picks up. If the treble is then boosted with EQ, the comb filtering already present becomes more apparent. So the equalizer did not add the comb filtered sound, but merely brought it out. The "problems" caused by phase shift have been repeated so many times by magazine writers and audio salespeople that it's now commonly accepted, even though there's not a shred of truth to it.

Comb filtering also intrudes in our lives by causing reception dropouts and other disturbances at radio frequencies. If you listen to AM radio in the evening, you'll sometimes notice a hollow sound much like a flanger effect. In fact, it *is* a flanger effect. The comb filtering occurs when your AM radio receives both the direct signal from the transmitting antenna and a delayed version that's been reflected off the ionosphere.

Likewise, FM radio suffers from a comb filtering effect called *picket fencing*, where the signal fades in and out rapidly as the receiving antenna travels through a series of nulls. As with audio nulls, radio nulls are also caused by reflections as the waves bounce off nearby large objects such as a truck next to your car on the highway. The signal fades in and out if either you or the large object is moving, and this is often noticeable as you slow down to approach a stoplight in your car. The slower you travel, the longer the timing between dropouts.

Reception dropouts also occur with wireless microphones used by musicians. This is sometimes called *multi-path fading* because, just as with acoustics, comb filtering results when a signal arrives through two paths and one is delayed. The solution is a *diversity* system having multiple receivers and multiple antennas spaced some distance apart. In this case it's the transmitter (performer) that moves, which causes the peak and null locations to change. When one receiving antenna is in a null, the other is likely not in a null. Logic within the receiver switches quickly from one antenna to another to ensure reception free of the dropouts that would otherwise occur as the performer moves around.

The same type of comb filtering also occurs in microwave ovens, and this is why these ovens have a rotating base. (Or the rotor is attached to the internal microwave antenna hidden from view.) But even with rotation, hot and cold spots still occur. Wherever comb filter peaks form, the food is hot, and at null locations, the food remains cold. As you can see, comb filtering is prevalent in nature and has a huge impact on many aspects of our daily lives; it's not just an audio effect!

Fourier and the Fast Fourier Transform

Joseph Fourier (1768–1830) showed that *all* sounds can be represented by one or more sine waves having various frequencies, amplitudes, durations, and phase relations. Conversely, any sound can be broken down into its component parts and identified completely using a Fourier analysis; one common method is the Fast Fourier Transform (FFT). Fourier's finding is significant because it proves that audio and music contain no unusual or magical properties. Any sound you hear, and any change you might make to a sound using an audio effect such as an equalizer, can be known and understood using this basic analysis of frequency versus volume level.

FFT is a valuable tool because it lets you assess the frequency content for any sound, such as how much noise and distortion are added at every frequency by an amplifier or sound card. The FFT analysis in Figure 1.22 shows the spectrum of a pure 1 KHz sine wave after being played back and recorded through a modestly priced sound card (M-Audio Delta 66) at 16 bits. A sine wave is said to be "pure" because it contains only a single frequency, with no noise or distortion components. In this case the sine wave was generated digitally inside an audio editor program (Sound Forge), and its purity is limited only by the precision of the math used to create it. Creating a sine wave this way avoids the added noise and distortion that are typical with hardware signal generators used to test audio gear. Therefore, any components you see other than the 1 KHz tone were added by the sound card or other device being tested.

In Figure 1.22, you can see that the noise floor is very low, with small "blips" at the odd-number harmonic distortion frequencies of 3, 5, 7, and 9 KHz. Note the slight rise at 0 Hz on the far left of the graph, which indicates the sound card added a small amount of DC offset to the recording. Although the noise floor of 16 bit digital audio used for this test is at −96 dB,

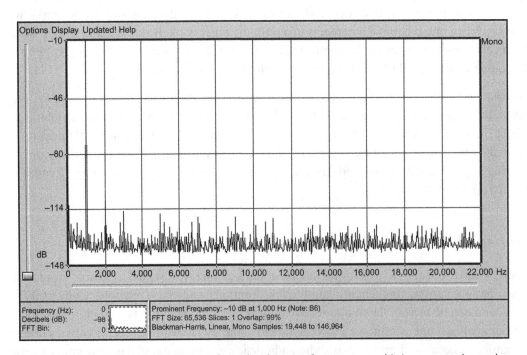

Figure 1.22: An FFT analysis shows volume level versus frequency, and it is commonly used to assess both the noise and distortion added by audio equipment.

the noise in the FFT screen appears lower—well below the −114 dB marker line. This is because the −96 dB noise level of 16 bits is really the sum of the noise at all frequencies.

A complete explanation of FFT spectrum analysis quickly gets into very deep math, so I've addressed only the high points. The main settings you'll deal with when using an FFT display are the upper and lower dB levels, the start and end frequencies (often 20 Hz to 20 KHz), and FFT resolution. The resolution is established by the FFT Size setting, with larger sizes giving more accurate results. I recommend using the largest FFT size your audio editor software offers.

Sine Waves, Square Waves, and Pink Noise—Oh My!

Fourier proved that all sounds comprise individual sine waves, and obviously the same applies to simple repeating waveforms such as sine and square waves. Figure 1.23 shows the five basic waveform types: sine, triangle, sawtooth, square, and pulse. A sine wave contains a single frequency, so it's a good choice for measuring harmonic distortion in audio gear. You send a single frequency through the amplifier or other device being tested, and any additional frequencies at the output must have been added by the device. Triangle waves contain only odd-numbered harmonics. So if the fundamental pitch is 100 Hz, the wave also contains 300 Hz, 500 Hz, 700 Hz, and so forth. Each higher harmonic is also softer than the previous one. The FFT display in Figure 1.24 shows the spectrum for a 100 Hz triangle wave having a peak level of −1 dB, or 1 dB below full scale (dBFS). Note that the 100 Hz fundamental frequency

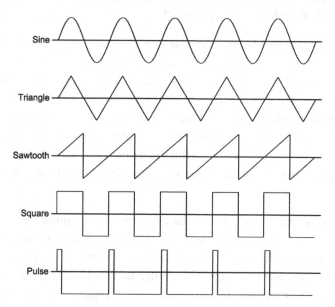

Figure 1.23: The five basic waveforms are sine, triangle, sawtooth, square, and pulse.

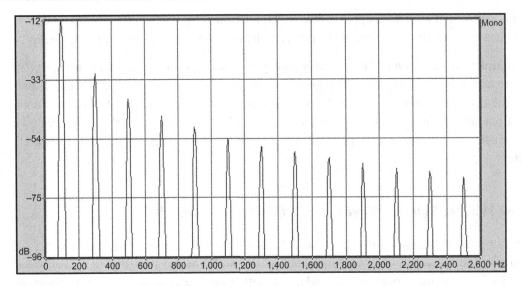

Figure 1.24: Triangle waves contain only odd-numbered harmonics, with each higher harmonic at a level lower than the one before.

has a level of −12, well below the file's peak level of −1 dBFS. This is because the total energy in the file is the sum of the fundamental frequency plus all the harmonics. The same applies to other waveforms, and indeed to any audio data being analyzed with an FFT. So when analyzing a music Wave file whose peak level is close to full scale, it's common for no one frequency to be higher than −10 or even −20.

Sawtooth waves contain both odd and even harmonics. Again, the level of each progressively higher harmonic is softer than the one before, as shown in Figure 1.25. Triangle waves and square waves have only odd-numbered harmonics because they are symmetrical; the waveform goes up the same way it comes down. But sawtooth waves contain both odd and even harmonics because they are not symmetrical. As you can see in Figure 1.26, the level of each harmonic in a square wave is higher than for a triangle wave because the rising and falling slopes are steeper. The faster a waveform rises or falls—called its *rise time*—the more high-frequency components it contains. This principle applies to all sounds, not just static waveforms.

Pulse waves are a subset of square waves. The difference is that pulse waves also possess a property called *pulse width* or *duty cycle*. For example, a pulse wave that's positive for 1/10th of the time, then zero or negative for the rest of the time, is said to have a duty cycle of 10 percent. So a square wave is really just a pulse wave with a 50 percent duty cycle, meaning the voltage is positive half the time and zero or negative half the time. Duty cycle is directly related to crest factor mentioned earlier describing the difference between average

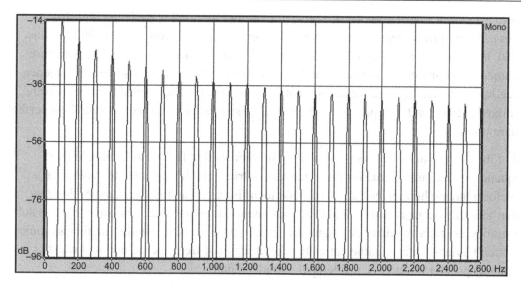

Figure 1.25: Sawtooth waves contain both odd and even harmonics, with the level of each falling off at higher frequencies.

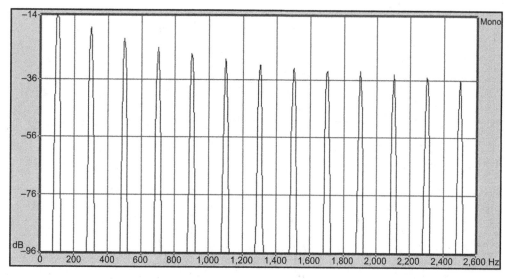

Figure 1.26: Square waves contain only odd-numbered harmonics because they're symmetrical, and, again, the level of each harmonic becomes progressively softer at higher frequencies.

and peak levels. As the duty cycle of a pulse wave is reduced from 50 percent, the peak level remains the same, but the average level—representing the total amount of energy—becomes lower and lower. Since non-square pulse waves are asymmetrical, they contain both odd and even harmonics.

Just to prove that Fourier is correct, Figure 1.27 shows a square wave being built from a series of harmonically related sine waves. The top sine wave is 100 Hz, then 300 Hz was added, then 500 Hz, and finally 700 Hz. As each higher harmonic is added, the waveform becomes closer and closer to square, and the rising and falling edges also become steeper, reflecting the added high-frequency content. If an infinite number of odd harmonics were all mixed together at the appropriate levels and phase relations, the result would be a perfect square wave.

All of the waveforms other than sine contain harmonics that fall off in level at higher frequencies. The same is true for most musical instruments. It's possible to synthesize a waveform having harmonics that fall off and then rise again at high frequencies, but that doesn't usually occur in nature. It's also worth mentioning that there's no such thing as sub-harmonics, contrary to what you might have read. Some musical sounds contain harmonics that aren't structured in a mathematically related series of frequencies, but the lowest frequency present is always considered the fundamental.

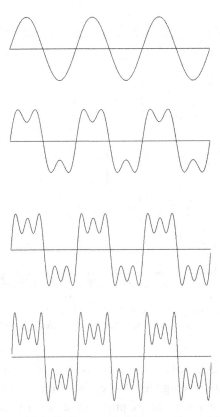

Figure 1.27: A square wave can be built from an infinite number of frequency-related sine waves.

Just as basic waveforms and most musical instruments create waveforms having a fundamental frequency plus a series of numerically related harmonics, the same happens when audio circuits create distortion. Some amplifier circuits tend to create more odd-numbered harmonics than even-numbered, and others create both types. But the numerical relation between the distortion frequencies added by electronic gear is essentially the same as for basic waveforms and most musical instruments. The FFT graph in Figure 1.22 shows the odd harmonics added by a typical sound card. Of course, the level of each harmonic is much lower than that of a square wave, but the basic principle still applies.

One exception to the numerically sequential harmonic series is the harmonic content of bells, chimes, and similar percussion instruments such as steel drums. The FFT in Figure 1.28 shows the spectrum of a tubular bell tuned to a D note. Even though this bell is tuned to a D note and sounds like a D note, only two of the five major peaks (at 1,175 Hz and 2,349 Hz) are related to either the fundamental or harmonics of a D note. The harmonic content of cymbals is even more complex and dense, containing many frequencies all at once with a sound not unlike white noise (hiss). However, you can coax a more normal series of harmonics from a cymbal by striking it near the bell portion at its center. Chapter 26 examines harmonic and inharmonic sound sources in more detail.

One important difference between the harmonics added by electronic devices and the harmonics present in simple waveforms and musical instruments is that audio circuits also add non-harmonic components known as *intermodulation distortion* (IMD). These are *sum*

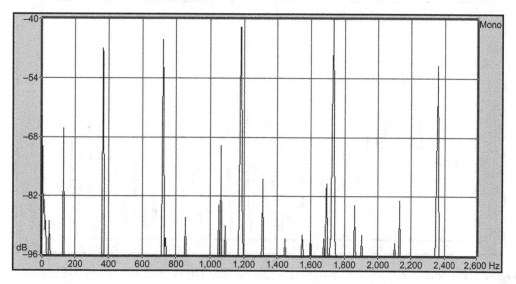

Figure 1.28: The harmonic series for a tubular bell is not a linear sequence as occurs with basic waveforms or non-percussive musical instruments such as violins and clarinets.

and difference frequencies created when two or more frequencies are present in the source. For example, if audio contains an A note at 440 Hz as well as the B note above at 494 Hz, audio circuits will add a series of harmonics related to 440 Hz, plus another series related to 494 Hz, plus an additional series related to the sum of 440 + 494 = 934 Hz, plus another series related to the difference of 494 − 440 = 54 Hz. Distortion is impossible to avoid in any audio circuit, but most designers aim for distortion that's too soft to hear. Figure 1.29 shows the spectrum of a Wave file containing 440 Hz and 494 Hz mixed together, after adding distortion. The many closely spaced peaks are multiples of the 54 Hz difference between the two primary frequencies, as well as the difference between the overtones of the primary frequencies.

The last wave type I'll describe is noise, which contains all frequencies playing simultaneously. Noise sounds like hiss or falling rain and comes in several flavors. The two noise types most relevant to audio are white noise and pink noise, both of which are common test signals. White noise has the same amount of energy at every frequency, so when displayed on an FFT, it appears as a straight horizontal line. Pink noise is similar, but it falls off at higher frequencies at a rate of 3 dB per octave. Pink noise has two important advantages for audio testing: Because it contains less energy at treble frequencies, it is less irritating to hear at loud levels when testing loudspeakers, and it is also less likely to damage your tweeters. The other advantage is it contains equal energy per octave rather than per fixed number of Hz, which corresponds to how we hear, as explained at the beginning of this chapter. Therefore,

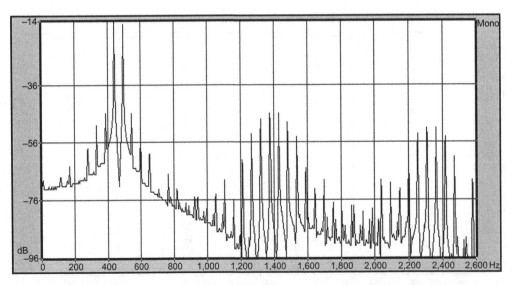

Figure 1.29: Adding distortion to a signal containing only 440 Hz and 494 Hz creates harmonics related to both frequencies, plus intermodulation components related to the sum and difference of those frequencies.

the octave between 1 KHz and 2 KHz contains the same total amount of energy as the octave between 100 Hz and 200 Hz. Compared to white noise that has the same amount of energy within every 100 Hz or 1 KHz span.

Earlier I mentioned that audio filters come in multiples of 6 dB per octave, because that's the minimum slope obtainable from a single capacitor or inductor. So obtaining the 3 dB per octave slope needed to build a pink noise generator is actually quite complex. The article "Spectrum Analyzer and Equalizer Designs" listed on the Magazine Articles page of my website ethanwiner.com shows a circuit for a pink noise generator, and you'll see that it requires four resistors and five capacitors just to get half the roll-off of a single resistor and capacitor!

Resonance

Resonance is an important concept in audio because it often improves the perceived quality of musical instruments, but it harms reproduction when it occurs in electronic circuits, loudspeakers, and listening rooms. Mechanical resonance results when an object having a finite weight (mass) is coupled to a spring having some amount of tension. One simple example of resonance is a ball hanging from a spring (or rubber band), as shown in Figure 1.30. Pendulums and tuning forks are other common mass-spring devices that resonate.

In all cases, when the mass is set in motion, it vibrates at a frequency determined by the weight of the mass and the stiffness of the spring. With a pendulum, the resonant frequency

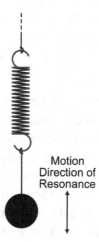

Motion
Direction of
Resonance

Figure 1.30: A ball hanging from a spring will resonate vertically at a frequency determined by the weight of the ball and the stiffness of the spring.

depends on the length of the pivot arm and the constant force of gravity. So pushing the ball in Figure 1.30 horizontally will most likely vibrate at a different frequency. A tuning fork's natural resonant frequency depends on the mass of its tines and the springiness of the material it's made from. The resonant frequency of a singer's vocal cords depends on the mass and length of the tissues, as well as their tension, which is controlled by the singer.

A close cousin of resonance is called *damping*, which is generally caused by friction that converts some of the motion energy to heat. The more friction and damping that are applied, the sooner the motion slows to a stop. The shock absorbers in your car use a viscous fluid to damp the car's vibration as it rests on the supporting springs. Otherwise, every time you hit a bump in the road, your car would bounce up and down many times before eventually stabilizing.

As relates to audio, resonance occurs in mechanical devices such as loudspeaker drivers and microphone diaphragms. In a speaker driver, the mass is the cone with its attached wire coil, and the spring is the foam or rubber *surrounds* that join the cone's inner and outer edges to the metal frame. The same applies to microphone diaphragms and phonograph cartridges. Most musical instruments also resonate, such as the wood front and back plates on a cello or acoustic guitar and the head of a snare or timpani drum. Air within the cavity of a musical instrument also resonates, such as inside a violin or clarinet. With a violin, the resonant frequencies are constant, but with a clarinet or flute, the pipe's internal resonance depends on which of the various finger keys are open or closed.

Another way to view damping is by its opposite property, Q. This is the same Q that applies to filters and equalizers as described earlier. For example, a plank of wood has a low Q due to friction within its fibers. If you strike a typical wooden 2 by 4 used for home construction with a hammer, you'll hear its main resonant frequency, but the sound will be more like a thud that dies out fairly rapidly. However, denser types of wood, such as rosewood or ebony used to make xylophones and claves, have less internal friction and thus have a higher Q. So the tone from a xylophone or marimba is more defined and rings out for a longer time, making it musically useful.

Rooms also resonate naturally, and they too can be damped using absorbers made from rigid fiberglass or acoustic foam. As sound travels through the fissures in the fiberglass or foam, friction inside the material converts the acoustic energy into heat. A rectangular-shaped room has three resonant frequencies, with one each for its length, width, and height. If you clap your hands in a small room that's totally empty, you'll generally hear a "boing" type of sound commonly known as *flutter echo*. This is due to sound bouncing repeatedly between two or more opposing surfaces. With nothing on the walls or floor to absorb the reflections, the sound waves continue to bounce back and forth for several seconds. But rooms also resonate at very low frequencies, and hand claps don't contain enough low-frequency energy to excite the resonances and make them audible. However, these resonances are definitely present, and Chapter 19 explains room resonance in much more detail, including how to control it.

One key point about resonance with audio is understanding when you want it and when you don't. Musical instruments often benefit from resonance, and many instruments such as violins and cellos *require* resonance to create their characteristic sound. The primary differences between a cheap violin and a Stradivarius are the number, strength, Q, and frequency of their natural body resonances. But resonance is always damaging in playback equipment and listening rooms because it adds a response peak at each resonant frequency, and those frequencies also continue to sound even after the music source has stopped. If I record myself playing a quarter note on my Fender bass, then press my hand on the string to stop the sound, that note should not continue for a longer time when played back later through loudspeakers.

Earlier I mentioned the natural resonance of loudspeaker drivers. This resonance tends to be stronger and more damaging with larger speakers used as woofers, because they have much more mass than the small, lightweight cones of typical tweeter drivers. Some speaker drivers use viscous fluid for cooling and to help damp vibration and ringing, but a useful amount of damping is also obtained from the power amplifier that drives the speaker. Loudspeaker electrical damping is similar to mechanical damping, but it uses electromagnetic properties instead of viscous fluids or shock absorber type devices. This is explained fully in Chapter 23.

To illustrate the resonant frequencies in speaker drivers, I created the short video "loudspeaker_resonance" that compares three speaker models: Yamaha NS-10M, Mackie HR624, and JBL 4430. I placed a Zoom H2 portable recorder near each speaker's woofer to record the sound, then tapped each woofer's cone with my finger. You can hear that each speaker has a different resonant frequency, and the video also shows an FFT of each recording on-screen. Note that the JBL's self-resonance is very low at 50 Hz, so you won't hear it if you listen on small speakers.

Audio Terminology

Earlier I mentioned that I prefer the term "low-cut" rather than "high-pass" when changing the frequency response in the bass range. Both are technically correct, but some common audio terms make less sense. For example, "warm," "cold," "sterile," "digital," "forward," "silky," and so forth are not useful because they don't mean the same thing to everyone. On the other hand, "3 dB down at 200 Hz" is precise and leaves no room for misinterpretation. Of course, "warm" and "cold" or "sterile" could describe the relative amount of high-frequency content. But saying "subdued or exaggerated highs" is still better than "sterile" in my opinion. However, while many of the terms I see are nonsensical, I gladly make an exception if a word sounds like the effect—this is called an *onomatopoeia*—because "buzzy" distortion sounds like the word "buzz," and hiss type noise sounds like the word "hiss."

Sometimes people refer to a piece of gear as being "musical" sounding or "resolving," but what does that really mean? What sounds musical to you may not sound musical to me. Some people like the added bass you get from a hi-fi receiver's Loudness switch. To me that usually makes music sound tubby, unless the music is already too thin sounding. The same goes for a slight treble boost to add sheen or a slight treble cut to reduce harshness. Whether these response changes sound pleasing or not is highly dependent on the music being played, the specific frequencies being boosted or cut, and personal preference.

I don't think we need yet more adjectives to describe audio fidelity when we already have perfectly good ones. Some audiophile words are even sillier, such as "fast bass," which is an oxymoron. The common audiophile terms "PRaT" (Pace, Rhythm, and Timing) take this absurdity to new heights, because these words already have a specific *musical* meaning unrelated to whatever audiophiles believe they are conveying. Some of the worst examples of nonsensical audio terms I've seen arose from a discussion in a hi-fi audio forum. A fellow claimed that digital audio misses capturing certain aspects of music compared to analog tape and LP records. So I asked him to state some specific properties of sound that digital audio is unable to record. Among his list were tonal texture, transparency in the midrange, bloom and openness, substance, and the organic signature of instruments. I explained that those are not legitimate audio properties, but he remained convinced of his beliefs anyway. Perhaps my next book will be titled *Scientists Are from Mars, Audiophiles Are from Venus*.

Another terminology pet peeve of mine relates to the words "hum," "buzz," and "whine." To me, hum is a low-frequency tone whose frequency is 60 Hz or 120 Hz, or a mix of both. In Europe, AC power is 50 Hz, but the same principle applies. Buzz is also related to the AC power frequency, but it has substantial high-frequency content. Whine to me is any frequency *other* than those related to AC power—for example, the sound of a car engine revving at a high RPM.

I'm also amazed when people confuse the words "front" and "rear" when talking about the orientation of their listening rooms. To me, the front wall of a room is the wall you face while listening. It's common for people to call that the rear wall because it's behind their speakers. But it's the *front* wall!

Finally, many people confuse phase with polarity. I see this often in audio magazines and even on the front panel labels of audio gear. As explained earlier, phase shift is related to frequency and time. Any phase shift that's applied to audio will delay different frequencies by different amounts. Polarity is much simpler, and it is either positive or negative at all frequencies. One example of polarity reversal is swapping the wires that connect to a loudspeaker. With the wires connected one way, the speaker cone pushes outward when a positive voltage is applied to the speaker's plus terminal. When reversed, a positive voltage instead causes the cone to pull inward. When you reverse the polarity of audio, all frequencies are inverted. So a "phase" switch on a mixing console or outboard mic preamp doesn't really affect the phase, but rather simply reverses the polarity. To be perfectly clear, phase shift always involves time delay, while polarity has no delay component. So saying "reversing the polarity is the same as 180 degrees of phase shift" is also incorrect.

The Null Test

The last topic I'll address in this chapter is the *null test*, which is an important concept for audio testing. Earlier I mentioned the forum poster who believed that digital recording somehow misses capturing certain aspects of audio. Other claims might seem difficult to prove or disprove, such as whether two competent speaker wires can sound different, or the common claim that the sound of wires or solid state electronics change over time, known as *break-in*. Audio comparisons are often done using *blind tests*. With a blind test, one person switches playback between two audio sources, while another person tries to identify which source is playing by listening alone, without watching.

Blind tests are extremely useful, but they're not always conclusive. For example, if you blind test two CD decks playing the same CD, there may be slightly different amounts of distortion. But the distortion of both CD players could be too soft to hear even if it can be measured. So no matter how often you repeat the test, the result is the same as a coin toss, even if one CD player really does have less distortion. Another potential fault of blind testing is it applies only to the person being tested. Just because *you* can't hear a difference in sound quality doesn't mean that nobody can. A proper blind test will test many people many times each, but that still can't prove conclusively that nobody can hear a difference. Further, some people believe that blind tests are fundamentally flawed because they put stress on the person being tested, preventing him or her from noticing real differences they could have heard if only they were more relaxed. In my opinion, even if a difference is real but so small that you can't hear it when switching back and forth a few times, how important is that difference, really?

I'll have more to say about blind testing in Chapter 3 and null tests in Chapter 24. But the key point for now is that a null test is absolute and 100 percent conclusive. The premise of a null test is to *subtract* two audio signals to see what remains. If nothing remains, then the signals are by definition identical. If someone claims playing Wave files from one hard drive sounds different than playing them from another hard drive, a null test will tell you for certain whether or not that's true.

Subtracting is done by reversing the polarity of one source, then mixing it with the other at precisely the same volume level. If the result is total silence when viewed on a wide-range VU meter that displays down to total silence (also called *digital black*), then you can be confident that both sources are identical. Further, if a residual difference signal does remain, the residual level shows the extent of the difference. You can also assess the nature of a residual difference either by ear or with an FFT analysis. For example, if one source has a slight low-frequency roll-off, the residual after nulling will contain only low frequencies. And if one source adds a strong third harmonic distortion component to a sine wave test tone, then the difference signal will contain only that added content. So when that forum fellow claimed that digital recording somehow misses certain aspects of audio, a null test can easily disprove that claim. Of course, whether or not this proof will be accepted is another matter!

Summary

This chapter explains the basic units of measurement for audio, which apply equally for both acoustic sounds in the air and for signal voltages that pass through audio equipment. Although the decibel always describes a ratio difference between two signal levels, it's common for decibels to state an absolute volume level using an implied reference. Decibels are useful for audio because they can express a very large range of volume levels using relatively small numbers. You also saw that both decibels and frequencies are usually assessed using logarithmic relationships.

Phase shift and time delay are important properties of audio, whether created intentionally in electronic circuits or when caused by reflections in a room. Indeed, phase shift is the basis for all filters and equalizers, including comb filters. Although phase shift is often blamed in the popular press for various audio ills, phase shift in usual amounts is never audible. What's really heard is the resultant change in frequency response when the original and delayed sounds are combined. Further, using the findings of Fourier, we know that all sound comprises one or more sine waves, and likewise all sound can be analyzed and understood fully using a Fourier analysis.

This chapter also explains the harmonic series, which is common to all simple waveforms, and describes the harmonic and intermodulation distortion spectrum added by audio equipment. Simple waveforms that are perfectly symmetrical contain only odd-numbered harmonics, while asymmetrical waves contain both odd and even harmonics. Further, where harmonic distortion adds frequencies related to the source, IM distortion is more audibly damaging because it contains sum and difference frequencies. Resonance is another important property of audio because it's useful and needed for many musical instruments, but it's not usually wanted in audio gear and listening rooms.

I also mentioned several terminology pet peeves, and I listed better wording for some commonly misused subjective audio terms. Finally, you saw that the null test is absolute because it shows all differences between two audio signals, including distortion or other artifacts you may not have thought to look for when measuring audio gear.

Note

1 You may know that doubling the *power* of an audio signal gives an increase of 3 dB rather than 6 dB shown above for voltages. When the voltage is doubled, twice as much current is also drawn by the connected device such as a loudspeaker. Since both the voltage and the current are then twice as large, the amount of power consumed actually quadruples. Hence, doubling the voltage gives a 6 dB increase in power.

CHAPTER 2

Audio Fidelity, Measurements, and Myths

Science is not a democracy that can be voted on with the popular opinion.
—Earl R. Geddes, audio researcher

In this chapter I explain how to assess the fidelity of audio devices and address what can and cannot be measured. Obviously, there's no metric for personal preference, such as intentional coloration from equalization choices or the amount of artificial reverb added to recordings as an effect. Nor can we measure the quality of a musical composition or performance. While it's easy to tell—by ear or with a frequency meter—if a singer is out of tune, we can't simply proclaim such a performance to be bad. Musicians sometimes slide into notes from a higher or lower pitch, and some musical styles intentionally take liberties with intonation for artistic effect. So while you may not be able to "measure" Beethoven's Symphony #5 to learn why many people enjoy hearing it performed, you can absolutely measure and assess the fidelity of audio equipment used to play a recording of that symphony. The science of audio and the art of music are not in opposition, nor are they mutually exclusive.

High Fidelity Defined

By definition, "high fidelity" means the faithfulness of a copy to its source. However, some types of audio degradation can sound pleasing—hence the popularity of analog tape recorders, gear containing tubes and transformers, and vinyl records. As with assessing the quality of music or a performance, a preference for intentional audio degradation cannot be quantified in absolute terms, so I won't even try. All I can do is explain and demonstrate the coloration added by various types of audio gear and let you decide if you like the effect or not. Indeed, the same coloration that's pleasing to many people for one type of music may be deemed unacceptable for others. For example, the production goal for most classical (and jazz or big band) music is to capture and reproduce the original performance as cleanly and accurately as possible. But many types of rock and pop music benefit from intentional distortion ranging from subtle to extreme.

The Allnic Audio's bottom end was deep, but its definition and rhythmic snap were a bit looser than the others. However, the bass sustain, where the instrumental textures reside, was very, very good. The Parasound seemed to have a 'crispy' lift in the top octaves.

The Ypsilon's sound was even more transparent, silky, and airy, with a decay that seemed to intoxicatingly hang in the air before effervescing and fading out.

—Michael Fremer, comparing phonograph preamplifiers in the
March 2011 issue of Stereophile *magazine*

Perusing the popular hi-fi press, you might conclude that the above review excerpt presents a reasonable way to assess and describe the quality of audio equipment. It is not. Such flowery prose might be fun to read, but it's totally meaningless because none of those adjectives can be defined in a way that means the same thing to everyone. What is rhythmic snap? What is a "crispy" lift? And how does sound hang in the air and effervesce? In truth, only four parameters are needed to define *everything* that affects the fidelity of audio equipment: noise, frequency response, distortion, and time-based errors. Note that these are really parameter *categories* that each contain several subsets. Let's look at these categories in turn.

The Four Parameters

Noise is the background hiss you hear when you raise the volume on a hi-fi receiver or microphone preamp. You can usually hear it clearly during quiet passages when playing cassette tapes. A close relative is *dynamic range*, which defines the span in decibels (dB) between the residual background hiss and the loudest level available short of gross distortion. CDs and DVDs have a very large dynamic range, so if you hear noise while playing a CD, it's from the original master analog tape, it was added as a by-product during production, or it was present in the room and picked up by the microphones when the recording was made.

Subsets of noise are AC power-related hum and buzz, vinyl record clicks and pops, between-station radio noises, electronic crackling, tape modulation noise, left-right channel bleed-through (cross-talk), doors and windows that rattle and buzz when playing music loudly, and the triboelectric cable effect. Tape modulation noise is specific to analog tape recorders, so you're unlikely to hear it outside of a recording studio. Modulation noise comes and goes with the music, so it is usually drowned out by the music itself. But you can sometimes hear it on recordings that are not bright sounding, such as a bass solo, as each note is accompanied by a "pfft" sound that disappears between the notes. The triboelectric effect is sometimes called "handling noise" because it happens when handling stiff or poor-quality cables. The sound is similar to the rumble you get when handling a microphone. This defect is rare today, thanks to the higher-quality insulation materials used by wire manufacturers.

Frequency response describes how uniformly an audio device responds to various frequencies. Errors are heard as too much or too little bass, midrange, or treble. For most people, the audible range extends from about 20 Hz at the low end to slightly less than 20 KHz at the high end. Some youngsters can hear higher than 20 KHz, though many senior citizens cannot hear much past 10 KHz. Some audiophiles believe it's important for audio equipment to pass frequencies far beyond 20 KHz, but in truth there's no need to reproduce ultrasonic content because nobody will hear it or be affected by it. Subsets of frequency response are physical microphonics (mechanical resonance), electronic ringing and oscillation, and acoustic resonance. Resonance and ringing will be covered in more detail later in this and other chapters.

Distortion is a layman's word for the more technical term *nonlinearity*, and it adds new frequency components that were not present in the original source. In an audio device, nonlinearity occurs when a circuit amplifies some voltages more or less than others, as shown in Figure 2.1. This nonlinearity can result in a flattening of waveform peaks, as at the left, or a level shift near the point where signal voltages pass from plus to minus through zero, as at the right. Wave peak compression occurs when electrical circuits and loudspeaker drivers are pushed to levels near their maximum limits.

Some circuits compress the tops and bottoms equally, which yields mainly odd-numbered harmonics—3rd, 5th, 7th, and so forth—while other circuit types flatten the top more than the bottom, or vice versa. Distortion that's not symmetrical creates both odd and even harmonics—2nd, 3rd, 4th, 5th, 6th, and so on. *Crossover distortion* (shown at right in Figure 2.1) is also common, and it's specific to certain power amplifier designs. Note that some people consider any change to an audio signal as a type of distortion, including frequency response errors and phase shift. My own preference is to reserve the term "distortion" for when nonlinearity creates new frequencies not present in the original.

When music passes through a device that adds distortion, new frequencies are created that may or may not be pleasing to hear. The design goal for most audio equipment is that all distortion be so low in level that it can't be heard. However, some recording engineers and audiophiles like the sound of certain types of distortion, such as that added by vinyl records,

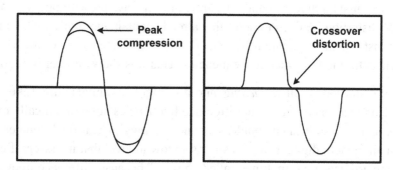

Figure 2.1: Two types of nonlinearity: peak compression at the top and/or bottom of a wave (left), and crossover distortion that affects electrical signals as they pass through zero volts (right).

transformers, or tube-based electronics, and there's nothing wrong with that. My own preference is for gear to be audibly transparent, and I'll explain my reasons shortly.

The two basic types of distortion are *harmonic* and *intermodulation*, and both are almost always present together. Harmonic distortion adds new frequencies that are musically related to the source. Ignoring its own inherent overtones, if an electric bass plays an A note whose fundamental frequency is 110 Hz, harmonic distortion will add new frequencies at 220 Hz, 330 Hz, 440 Hz, and subsequent multiples of 110 Hz. Some audio devices add more even harmonics than odd, or vice versa, but the basic concept is the same. In layman's terms, harmonic distortion adds a thick or buzzy quality to music, depending on which specific frequencies are added. The notes created by most musical instruments include harmonics, so a device whose distortion adds more harmonics merely changes the instrument's character by some amount. Electric guitar players use harmonic distortion—often *lots* of it—to turn a guitar's inherent plink-plink sound into a singing tone that has a lot of power and sustains.

Intermodulation (IM) distortion requires two or more frequencies to be present, and it's far more damaging audibly than harmonic distortion because it creates new sum and difference frequencies that aren't always related musically to the original frequencies. For example, if you play a two-note A major chord containing an A at 440 Hz and a C# at 277 Hz through a device that adds IM distortion, new frequencies are created at the sum and difference frequencies:

$$\text{Sum:} \quad 440 \text{ Hz} + 277 \text{ Hz} = 717 \text{ Hz}$$
$$\text{Difference:} \quad 440 \text{ Hz} - 277 \text{ Hz} = 163 \text{ Hz}$$

A frequency of 717 Hz is about halfway between an F and F# note, and 163 Hz is slightly below an E note. Neither of these are related musically to A or C#, nor are they even standard note pitches. Therefore, even in relatively small amounts, IM distortion adds a dissonant quality that can be unpleasant to hear. Again, both harmonic and intermodulation distortion are caused by the same nonlinearity and thus are almost always present together. What's more, when IM distortion is added to notes that already contain harmonics, which is typical for all musical instruments, sum and difference frequencies related to all of the harmonics are created, as well as for the fundamental frequencies. This was shown earlier in Figure 1.29.

Another type of distortion is called *aliasing*, and it's unique to digital audio. Like IM distortion, aliasing creates new sum and difference frequencies not harmonically related to the original frequencies, so it can be unpleasant and irritating to hear if it's loud enough. Fortunately, in all modern digital gear, aliasing is so low in level that it's rarely if ever audible. Aliasing artifacts are sometimes called "birdies" because difference frequencies that fall in the 5–10 KHz range change pitch in step with the music, which sounds a little like birds chirping. An audio file letting you hear what aliasing sounds like is in Chapter 3.

Transient intermodulation distortion (TIM) is a specific type of distortion that appears only in the presence of transients—sounds that increase quickly in volume such as snare drums, wood blocks, claves, or other percussive instruments. This type of distortion may not show up in a standard distortion test using static sine waves, but it's revealed easily on an oscilloscope connected to the device's output when using an impulse-type test signal such as a pulse wave. TIM will also show up as a residual in a null test when passing transient material. Negative feedback is applied in amplifiers to reduce distortion by sending a portion of the output back to the input with the polarity reversed. TIM occurs when stray circuit capacitance delays the feedback, preventing it from getting back to the input quickly enough to counter a very rapid change in input level. In that case the output can distort briefly. However, modern amplifier designs include a low-pass filter at the input to limit transients to the audible range, which effectively solves this problem.

Time-based errors are those that affect pitch and tempo. When playing an LP record whose hole is not perfectly centered, you'll hear the pitch rise and fall with each revolution. This is called *wow*. The pitch instability of analog tape recorders is called *flutter*. Unlike the slow, once per revolution pitch change of wow, flutter is much faster and adds a warbling effect. Digital recorders and sound cards have a type of timing error called *jitter*, but the pitch deviations are so rapid they instead manifest as added noise. With all modern digital audio gear, jitter is so soft compared to the music that it's almost always inaudible. The last type of time-based error is *phase shift*, but this too is inaudible, even in relatively large amounts, unless the amount of phase shift is different in the left and right channels. In that case the result can be an unnaturally wide sound whose location is difficult to identify.

Room acoustics could be considered an additional audio parameter, but it really isn't. When strong enough, acoustic reflections from nearby boundaries create the comb filtered frequency response described in Chapter 1. This happens when reflected sound waves combine in the air with the original sound and with other reflections, enhancing some frequencies while canceling others. Room reflections also create audible echoes, reverb, and resonance. In an acoustics context, resonance is often called modal ringing at bass frequencies, or flutter echo at midrange and treble frequencies. But all of these are time-based phenomena that occur outside the equipment, so they don't warrant their own category.

Another aspect of equipment quality is *channel imbalance*, where the left and right channels are amplified by different amounts. I consider this to be a "manufacturing defect" caused by an internal trimmer resistor that's set incorrectly, or one or more fixed resistors that are out of tolerance. But this isn't really an audio parameter either, because the audio *quality* is not affected, only its volume level.

The preceding four parameter categories encompass *everything* that affects the fidelity of audio equipment. If a device's noise and distortion are too soft to hear, with a response that's sufficiently uniform over the full range of audible frequencies, and all time-based errors are too small to notice, then that device is considered *audibly transparent* to music and other sound passing through it. In this context, a device that is transparent means you will not hear a change in quality after audio has passed through it, even if small differences could be measured. For this reason, when describing audible coloration, it makes sense to use only words that represent what is actually affected. It makes no sense to say a power amplifier possesses "a pleasant bloom" or has a "forward" sound when "2 dB boost at 5 KHz" is much more accurate and leaves no room for misinterpretation.

Chapter 1 explained the concept of resonance, which encompasses both frequency and time-based effects. Resonance is not so much a parameter as it is a property, but it's worth repeating here. Resonance mostly affects mechanical transducers—loudspeakers and microphones—which, being mechanical devices, must physically vibrate. Resonance adds a boost at some frequency and *also* continues a sound's duration over time after the source has stopped. Resonance in electrical circuits generally affects only one frequency, but resonances in rooms occur at multiple frequencies related to the spacing between opposing surfaces. These topics will be examined in more depth in the sections that cover transducers and room acoustics.

When assessing frequency response and distortion, the finest loudspeakers in the world are far worse than even a budget electronic device. However, clarity and stereo imaging are greatly affected by room acoustics. Any room you put speakers in will exaggerate their response errors, and reflections that are not absorbed will reduce clarity. Without question, the room you listen in has much more effect on sound quality than any electronic device. However, the main point is that measuring these four basic parameters is the correct way to assess the quality of amplifiers, preamps, sound cards, loudspeakers, microphones, and every other type of audio equipment. Of course, to make an informed decision, you need *all* of the relevant specs, which leads us to the following.

Lies, Damn Lies, and Audio Gear Specs

> Jonathan: "You lied first."
> Jack: "No, you lied to me first."
> Jonathan: "Yes, I lied to you first, but you had no knowledge I was lying. So as far as you knew, you lied to me first."
> —**Bounty hunter Jack Walsh (Robert De Niro) arguing with white-collar criminal Jonathan Mardukas (Charles Grodin) in the movie** Midnight Run

When it comes to audio fidelity, the four standard parameter categories can assess any type of audio gear. Although published product specs *could* tell us everything needed to evaluate a device's transparency, many specs are incomplete, misleading, and sometimes even fraudulent. This doesn't mean that specs can't tell us everything needed to determine transparency—we just need all of the data. However, getting complete specs from audio manufacturers is another matter. Often you'll see the frequency response given but without a plus/minus dB range. Or a power amp spec will state harmonic distortion at 1 KHz, but not at higher or lower frequencies where the distortion might be much worse. Or an amplifier's maximum output power is given, but its distortion was spec'd at a much lower level such as 1 watt.

Lately I've seen a dumbing down of published gear reviews, even by contributors in pro audio magazines, who, in my opinion, have a responsibility to their readers to aim higher than they often do. For example, it's common for a review to mention a loudspeaker's woofer size but not state its low-frequency response, which is, of course, what really matters. Audio magazine reviews often include impressive-looking graphs that imply science but are lacking when you know what the graphs actually mean. Much irrelevant data is presented, while important specs are omitted. For example, the phase response of a loudspeaker might be shown but not its distortion or off-axis frequency response, which are far more important. I recall a hi-fi magazine review of a very expensive tube preamplifier so poorly designed that it verged on self-oscillation (a high-pitched squealing sound). The reviewer even acknowledged the defect, which was clearly visible in the accompanying frequency response graph. Yet he summarized by saying, "Impressive, and very highly recommended." The misguided loyalty of some audio magazines is a huge problem in my opinion.

Even when important data is included, it's sometimes graphed at low resolution to hide the true performance. For example, a common technique when displaying frequency response graphs is to apply *smoothing*, also called *averaging*. Smoothing reduces the frequency resolution of a graph, and it's justified in some situations. But for loudspeakers you really do want to know the full extent of the peaks and nulls. Another trick is to format a graph using large, vertical divisions. So a frequency response line may look reasonably straight, implying a uniform response, yet a closer examination shows that each vertical division represents a substantial dB deviation.

The graphs in Figures 2.2 through 2.4 were all derived from the same data but are presented with different display settings. For this test I measured the response of a single Mackie HR624 loudspeaker in a fairly large room with my precision DPA 4090 microphone about a foot away pointed directly at the tweeter. Which version looks more like what speaker makers publish?

Test Equipment

Empirical evidence trumps theory every time.

Noise measurements are fairly simple to perform using a sensitive voltmeter, though the voltmeter must have a flat frequency response over the entire audible range. Many budget models are not accurate above 5 or 10 KHz. To measure its inherent noise, an amplifier or

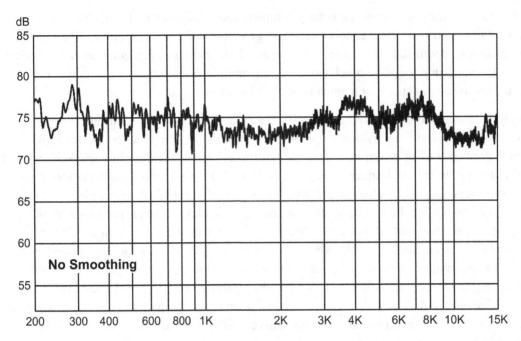

Figure 2.2: This graph shows the loudspeaker response as measured, with no smoothing.

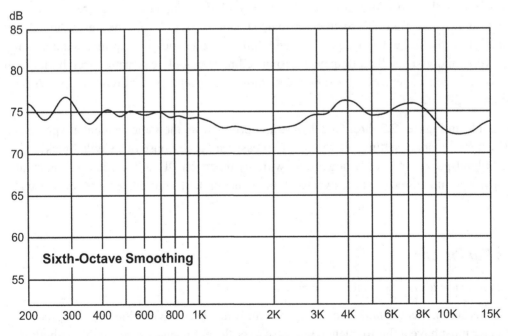

Figure 2.3: This graph shows the exact same data but with sixth-octave smoothing applied.

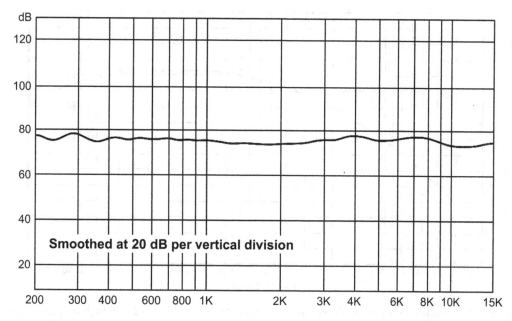

Figure 2.4: This graph shows the same smoothed data as in Figure 2.3, but at 20 dB per vertical division instead of 5 dB, making the speaker's response appear even flatter.

other device is powered on but with no input signal present; then the residual voltage is measured at its output. Usually a resistor or short circuit is connected to the device's input to more closely resemble a typical audio source. Otherwise, additional hiss or hum might get into the input and be amplified, unfairly biasing the result. Some power amplifiers include a volume control, so you also need to know where that was set when the noise was measured. For example, if the volume control is typically halfway up when the amplifier is used but was turned way down during the noise test, that could make the amplifier seem quieter than it really is.

Although it's simple to measure the amount of noise added by an audio device, what's measured doesn't necessarily correlate to its audibility. Our ears are less sensitive to very low and very high frequencies when compared to the midrange, and we're especially sensitive to frequencies in the treble range around 2 to 3 KHz. To compensate for this, many audio measurements employ a concept known as *weighting*. This intentionally reduces the contribution of frequencies where our ears are less sensitive. The most common curve is A-weighting, as shown in Figure 2.5. Note that A-weighting corresponds to the frequency balance we hear at low to moderate volume, because at very loud levels our hearing is closer to flat. Chapter 3 explains this in more detail.

In the old days before computers were common and affordable, harmonic distortion was measured with a dedicated analyzer. A distortion analyzer sends a high-quality sine wave,

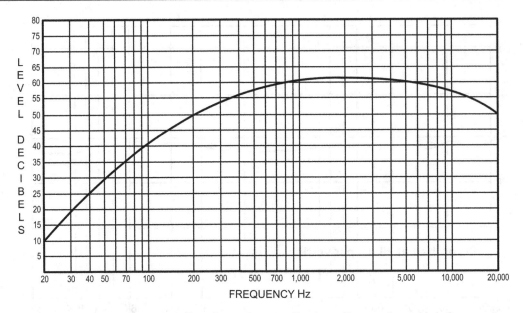

Figure 2.5: A-weighting intentionally reduces the contribution of low and very high frequencies, so noise measurements will correspond more closely to their audibility. This curve shows the response of an A-weighting filter.

containing only the single desired frequency with minimal harmonics and noise, through the device being tested. Then a notch filter is inserted between the device's output and a voltmeter. Notch filters are designed to remove a very narrow band of frequencies, so what's left is the distortion and noise generated by the device being tested. Figure 2.6 shows the basic method, and an old-school Hewlett-Packard distortion analyzer is shown in Figure 2.7.

Intermodulation distortion is measured using two test tones instead of only one, and there are two standard methods. One method sends 60 Hz and 7 KHz tones through the device being tested, with the 60 Hz sine wave being four times louder than the 7 KHz sine wave. The analyzer then measures the level of the 7,060 Hz and 6,940 Hz sum and difference frequencies that were added by the device. Another method uses 19 KHz and 20 KHz at equal volume levels, measuring the amplitude of the 1 KHz difference tone that's generated.

Modern audio analyzers like the Audio Precision APx525 shown in Figure 2.8 are very sophisticated and can measure more than just frequency response, noise, and distortion. They are also immune to human hearing foibles such as masking,[1] and they can measure noise, distortion, and other artifacts reliably down to extremely low levels, far softer than anyone could possibly hear.

Professional audio analyzers are very expensive, but it's possible to do many useful tests using only a Windows or Mac computer with a decent-quality sound card and suitable software.

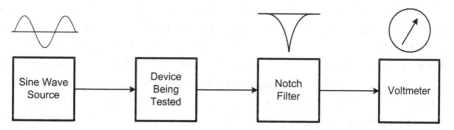

Figure 2.6: To measure a device's harmonic distortion, a pure sine wave is sent through the device at a typical volume level. Then a notch filter removes that frequency. Anything that remains is the distortion and noise of the device being tested.

Figure 2.7: The Hewlett-Packard Model 334A Distortion Analyzer. *Photo courtesy of Joe Bucher.*

Figure 2.8: The Audio Precision Model APx525 Audio Analyzer is a sophisticated device that can measure every aspect of audio fidelity. *Photo courtesy of Audio Precision.*

I use the FFT feature in the Sound Forge audio editing program to analyze frequency response, noise, and distortion. For example, when I wanted to measure the distortion of an inexpensive sound card, I created a pure 1 KHz sine wave test signal in Sound Forge. I sent the tone out of the computer through a high-quality sound card having known low distortion, then back into the budget sound card, which recorded the 1 KHz tone. The result is shown in Figure 2.9. Other test methods you can do yourself with a computer and sound card are described in Chapter 24. As you can see in Figure 2.9, a small amount of high-frequency distortion and noise above 2 KHz was added by the sound card's input stage. But the added artifacts are all more than 100 dB softer than the sine wave and so are very unlikely to be audible.

Low distortion at 1 KHz is easy to achieve, but 30 Hz is a different story, especially with gear containing transformers. Harmonic distortion above 10 KHz matters less because the added harmonics are higher than the 20 KHz limit of most people's hearing. However, if the distortion is high enough, audible IM difference frequencies below 20 KHz can result. Sadly, many vendors publish only total harmonic distortion (THD) measured at 1 KHz, often at a level well below maximum output. This ignores that distortion in power amplifiers

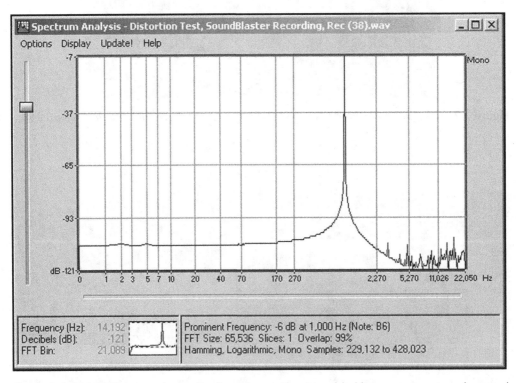

Figure 2.9: This FFT screen shows the distortion and noise added by a consumer-grade sound card when recording a 1 KHz sine wave.

and gear containing transformers usually increases with rising output level and at lower frequencies.

The convention these days is to lump harmonic distortion, noise, and hum together into a single THD + Noise spec and express it as either a percentage or some number of dB below the device's maximum output level. For example, if an amplifier adds 1 percent distortion, that amount can be stated as 40 dB below the original signal. A-weighting is usually applied because it improves the measurement, and this is not unfair. There's nothing wrong with combining noise and distortion into a single figure either when their sum is safely below the threshold of audibility. But when distortion artifacts *are* loud enough to be audible, it can be useful to know their specific makeup. For example, artifacts at very low frequencies are less objectionable than those at higher frequencies, and harmonics added at frequencies around 2 to 3 KHz are especially noticeable compared to harmonics at other frequencies. Again, this is why A-weighting is usually applied to noise and distortion measurements and why using weighting is not unreasonable.

Audio Transparency

As we have seen, the main reason to measure audio gear is to learn if a device's quality is good enough to sound *transparent*. All transparent devices by definition sound the same because they don't change the sound enough to be noticed even when listening carefully. But devices that add an audible amount of distortion can sound different, even when the total measured amount is the same. A-weighting helps relate what's measured to what we hear, but some types of distortion are inherently more objectionable (or pleasing) than others. For example, harmonic distortion is "musical," whereas IM distortion is not. But what if you *prefer* the sound of audio gear that is intentionally colored?

In the 1960s, when I became interested in recording, ads for most gear in audio magazines touted their flat response and low distortion. Back then, before the advent of multi-layer printed circuit boards, high-performance op-amps, and other electronic components, quality equipment was mostly handmade and very expensive. In those days design engineers did their best to minimize the distortion from analog tape, vacuum tubes, and transformers. Indeed, many recordings made in the 1960s and 1970s still sound excellent even by today's standards. But most audio gear is now mass-produced in Asia using modern manufacturing methods, and very high quality is available at prices even hobbyists can easily afford.

Many aspiring recording engineers today appreciate some of the great recordings from the mid-twentieth century. But when they are unable to make their own amateur efforts sound as good, they wrongly assume they need the same gear that was used back then. Of course, the real reason so many old recordings sound wonderful is because they were made by very good recording engineers in great (often very large) studios having excellent acoustics. That some of those old recordings still sound so clear today is *in spite of* the poorer-quality recording gear available back then, not because of it!

Somewhere along the way, production techniques for popular music began incorporating intentional distortion and often extreme EQ as creative tools. Whereas in the past, gear vendors bragged about the flat response and low distortion of their products, in later years we started to see ads for gear claiming to possess a unique character, or *color*. Some audio hardware and software plug-ins claim to possess a color similar to specific models of vintage gear used on famous old recordings. Understand that "color" is simply a skewed frequency response and/or added distortion; these are easy to achieve with either software or hardware, and in my opinion need not demand a premium price. For example, distortion similar to that of vacuum tubes can be created using a few resistors and a diode, or a simple software algorithm.

The key point is that adding color in the form of distortion and EQ is proper and valuable when recording and mixing. During the creative process, anything goes, and if it sounds good, then it *is* good. But in a playback system the goal must be for transparency—whether a recording studio's monitors or the loudspeakers in a consumer's living room. In a studio setting the recording and mixing engineers need accurate monitoring to know how the recording really sounds, including any coloration they added intentionally. With a consumer playback system you want to hear exactly what the producers and mix engineers heard; you'll hear their artistic intent *only* if your own system adds no further coloration of its own.

Common Audio Myths

> *I thought cables didn't matter, so I tried running my system without them. Huge difference!*
> —**Posted in a hi-fi audio forum**

Now that we understand what to measure and how, and know that a null test can prove our measurements valid, let's use that knowledge to bust some common audio myths. The earliest audio myth I can recall is the benefit of fancy wire for connecting loudspeakers, and it's still going strong. Some vendors claim their wire sounds better than normal wire, and, of course, it's more expensive than normal wire. In truth, the most important property of speaker wire is resistance, which is directly related to its thickness. The wire's resistance must be small to pass the high-current signals a power amplifier delivers, and this is exactly analogous to a large pipe letting more water flow through it than a small pipe. For short distances—say, up to 5 or 10 feet—16-gauge wire of any type is adequate, though thicker wire is required for longer lengths. When heavier gauges are needed—either for longer runs or when connecting high-power amplifiers and speakers—Romex wire typically used for AC power wiring is a fine choice for loudspeakers.

The three other wire parameters are inductance, capacitance, and skin effect, and these will be explained in more detail in the section of this book that covers electronics. But these parameters are not important with usual cable lengths at audio frequencies, especially when connecting speakers to a power amplifier. Low-capacitance wire can be important in some cases, such as between a phonograph cartridge or high-impedance microphone and a preamp. But high-quality, low-capacitance wire is available for pennies per foot. Unscientific and even impossible claims for wire products are common because wire is a low-tech device that's simple to manufacture, and the profit margin for manufacturers and retailers is very high. I could devote this entire section to debunking wire claims, but instead I'll just summarize that any audio (or video) cable costing more than a few dollars per foot is a poor value.

Bi-wiring is a more recent myth, and it's a pretend relative to *bi-amping*, which is legitimate. No single-speaker driver can reproduce the entire range of audible frequencies, so speaker makers use two or three drivers—woofers and tweeters—to handle each range. Bi-amping splits the audio into low/high or low/mid/high frequency ranges, and each range goes to a separate amplifier that in turn powers each speaker driver. This avoids passive crossovers that lose some of their power as heat and can add distortion. Bi-wiring uses two separate speaker wires, but they're both connected to the *same* power amplifier that then feeds a passive crossover! So the only person who benefits from bi-wiring is the dealer, because he gets to sell you twice as many wires.

A related myth is cable elevators—small devices that prevent your speaker wires from touching the floor. Like so many audiophile "tweak" products, the claim that cable elevators improve sound quality by avoiding damaging static electric and mechanical vibration is unfounded. If vibration really affected electricity as it passed through wires, airplanes—with their miles of wire subject to shock and extreme vibration—would fall from the sky daily. Indeed, it would be trivial for vendors to prove that audio passing through wire is affected by vibration, thus establishing a real need for their products. To my knowledge, no vibration-control product vendor has ever done that.

Even less likely to improve sound than replacement speaker wire is after-market AC power cords and most other "power conditioner" products. The sales claims seem reasonable: Noise and static can get into your gear through the power line and degrade sound quality. In severe cases it's possible for power-related clicks and buzzes to get into your system, but those are easily noticed. The suggestion that power products subtly increase "clarity and presence," or stereo width and fullness, is plain fraud. Indeed, every competent circuit designer knows how to filter out power line noise, and such protection is routinely added to all professional and consumer audio products. Buying a six-foot replacement power cord ignores the other hundred-odd feet of regular AC wiring between the wall outlet and power pole—likewise for replacement AC outlets, and even more so for replacement AC outlet *cover plates* that claim to improve audio quality. Again, this would be easy for vendors to prove with hard science, but they never do. Power conditioner vendors sometimes show an oscilloscope display of the power line noise before and after adding their product. But they never show the change at the output of the connected equipment, which, of course, is what really matters.

The last wire myth I'll mention is the notion that boutique USB and HDMI cables avoid or reduce the degradation of audio (and video) compared to standard wires. The signals these wires pass are digital, not analog, so the usual wire properties that can lose high frequencies don't apply except in extreme cases. For the most part, digital data either arrives at the other end intact or it doesn't. And many digital protocols employ some type of error checking to verify the integrity of the received data. So generally, if you hear any sound at all through a digital cable, you can be confident that nothing was lost or changed along the way.

Among devoted audiophiles, one of the most hotly debated topics is the notion that reproducing ultrasonic frequencies is necessary for high fidelity reproduction. But no human can hear much past 20 KHz, and few microphones respond to frequencies beyond that. Even fewer loudspeakers can reproduce those high frequencies. If recording and capturing ultrasonic frequencies were free, there'd be little reason to object. But in this digital age, storing frequencies higher than necessary wastes memory, media space, and bandwidth. The DVD format accommodates frequencies up to 96 KHz, but then lossy[2] data compression, which *is* audibly degrading, is needed to make it fit! Record companies and equipment manufacturers were thrilled when we replaced all our old LPs and cassettes with CDs back in the 1980s and 1990s. Now, with newer "high-resolution" audio formats, they're trying hard to get us to buy all the same titles again, and new devices to play them, with the false promise of fidelity that exceeds CDs.

Another myth is the benefit of mechanical isolation. The claims have a remote basis in science but are exaggerated to suggest relevance where none is justified. If you ever owned a turntable, you know how sensitive it is to mechanical vibration. Unless you walk lightly, the record might skip, and if you turn up the volume too high, you may hear a low-frequency feedback howl. A turntable is a mechanical device that relies on physical contact between the needle and the record's surface. But CDs and DVDs work on an entirely different principle that's mostly immune to mechanical vibration. As a CD or DVD spins, the data is read into a memory buffer, and from there it's sent to your receiver or headphones. The next few seconds of music is already present in the player's buffer, so if the transport is jostled enough to make the CD mis-track, the player continues to send its data stream from the buffer until the drive finds its place again. For this reason, large buffers were common on CD players sold to joggers before MP3 players took over.

Mechanical isolation is not useful for most other electronic gear either. However, mechanical isolation with loudspeakers is sometimes valid because they're mechanical devices that vibrate as they work. When a speaker rests on a tabletop, the table may vibrate in sympathy and resonate if the loudspeaker's cabinet is not sufficiently massive or rigid. Loudspeaker isolation will be described in detail in Chapter 18. Electronic devices that contain vacuum tubes may also be sensitive to vibration because tubes can become microphonic. If you tap a tube with a pencil while the amplifier is turned on, you might hear a noise similar to tapping a microphone. But microphonic tubes are excited mainly by sound waves in the air that strike the tube. Placing a tube amplifier on a cushion reduces only vibrations that arrive from the floor.

Vinyl records and vacuum tube equipment are very popular with devoted audiophiles who believe these old-school technologies more faithfully reproduce subtle nuance. There's no question that LPs and tubes sound different from CDs and solid state gear. But are they really better? The answer is, not in any way you could possibly assess fidelity. Common to both formats is much higher distortion. LPs in particular have more inherent noise and a poorer high-frequency response, especially when playing the inner grooves. I'm convinced that some people prefer tubes and vinyl because the distortion they add sounds pleasing to them. In the audio press this is often called *euphonic distortion*. Adding small amounts of distortion can make a recording sound more cohesive, for lack of a better word. Distortion can seem to increase clarity, too, because of the added high-frequency content. Recording engineers sometimes add distortion intentionally to imitate the sound of tubes and analog tape, and I've done this myself many times. Simply copying a song to a cassette tape and back adds a slight thickening that can be pleasing if the instrumentation is sparse. But clearly this is an effect, no matter how pleasing, and not higher fidelity.

Other common audio myths involve very small devices that claim to improve room acoustics. You can pay a hundred dollars each for small pieces of exotic wood the size and shape of hockey pucks. Other common but too-small acoustic products are metal bowls that look like sake cups and thin plastic dots the size and thickness of a silver dollar. Sellers of these devices suggest you put them in various places around your room to improve its acoustics. But with acoustics, what matters is covering a sufficient percentage of the room's surface. Real acoustic treatment must be large to work well, and that's not always conducive to a domestic setting. Some people want very much to believe that something small and unobtrusive can solve their bad acoustics, without upsetting the decor. Sadly, such products simply do not work. Worse, an acoustic device that purports to be a "resonator" can only add unwanted artifacts, assuming it really is large enough to have an audible effect. There's a type of bass trap called a Helmholtz resonator, but that works as an *absorber* rather than adding the sound of resonance into the room.

Another myth is that the sound of vinyl records and CDs can be improved by applying a demagnetizer. There's no reason to believe that the vinyl used for LP records could be affected by magnetism. Even if plastic could be magnetized, there's no reason to believe that would affect the way a diamond needle traces the record's grooves. A change in sound quality after demagnetizing a CD is even less likely because CDs are made from plastic and aluminum, and they store digital data! Again, for the most part, digital audio either works or it doesn't. Although digital audio might possibly be degraded when error checking is not employed, degradation is never due to a CD becoming "magnetized."

As an audio professional I know that $1,000 can buy a very high-quality power amplifier. So it makes no sense to pay, say, $17,000 for an amplifier that is no better and may in fact be worse. However, some myths are more like urban legends: No products are sold, but they're still a waste of time. For example, one early hi-fi myth claims you can improve the sound of a CD by painting its edge with a green felt marker. (Yes, it must be green.) A related myth is that cables and electronic devices must be "broken in" for some period of time before they achieve their final highest fidelity. Speaker and headphone drivers can change slightly over time due to material relaxation. But aside from a manufacturing defect, the idea that wire or solid state circuits change audibly over time makes no sense and has never been proven. This myth becomes a scam when a vendor says that for best results you must break in the product for 90 days. Why 90 days? Because most credit card companies protect your right to a refund for only 60 days.

The Stacking Myth

The last audio myth I'll debunk is called stacking. The premise is that audio gear such as a microphone preamp or sound card might measure well and sound clean with a single source, but when many separate tracks are recorded through that same preamp or sound card and later mixed together, the degradation "stacks" and becomes more objectionable. In this sense, stacking means the devices are used in parallel, versus sending one source through multiple devices in series with the output of one sent to the input of the next. Stacking theory also presumes that when many tracks are recorded through a device having a non-flat frequency response, such as a microphone's presence boost, the effect of that skewed response accumulates in the final mix more than for each separate track. However, this type of accumulated coloration is easy to disprove, as shown in Figure 2.10.

As an extreme example, let's say the preamp used for every track of a recording has a 4 dB boost at 1 KHz. The result is the same as using a flat preamp and adding an equalizer with 4 dB boost on the output of the mixer. Of course, no competent preamp has a frequency response that skewed. Even modest gear is usually flat within 1 or 2 dB from 20 Hz to 20 KHz. But even if a preamp did have such a severe response error—whether pleasing or not—it could be countered exactly using an opposite equalizer setting. So no matter how many tracks are mixed, only 4 dB of EQ cut would be needed to counter the response of the preamp.

Now let's consider distortion and noise—the other two audio parameters that affect the sound of a preamp or converter. Artifacts and other coloration from gear used in parallel do not add

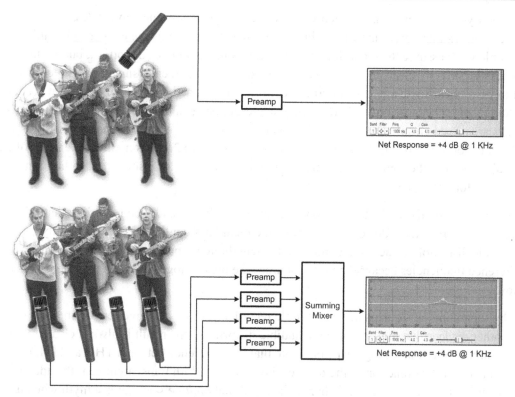

Figure 2.10: If a microphone or preamp has a skewed frequency response, shown here as a 4 dB boost at 1 KHz, the net response is the same no matter how many microphones or preamps are used. And whatever frequency response error the microphone or preamp may have, it can be countered with equalization.

the same as when the devices are connected in series. When connected in series, it is *far* more damaging because noise and coloration accumulate. Related, some people believe that two pieces of gear might sound and measure exactly the same, but it's easier or faster to get a good sounding mix if all the tracks had been recorded through one device versus the other. In truth, recording multiple tracks repeatedly through the same device and then mixing those tracks together later actually *reduces* distortion compared to mixing the tracks first and going through the device only once. Even then, any difference between stacking or not is audible only if the device's distortion is loud enough to hear in the first place.

As we learned earlier, where harmonic distortion adds new harmonically related frequencies, IM distortion creates sum and difference frequencies and thus is more dissonant and audibly damaging. Further, whenever harmonic distortion is added by a device, IM distortion is also added. Both are caused by the same nonlinearity and so are inseparable except in special contrived circuits.

Let's say you have three tracks, each with a different frequency sine wave. (Yes, music is more complex than three sine waves, but this more easily explains the concept.) For this example we'll assume the recording medium adds some amount of distortion, but the mixing process is perfectly clean and is not part of the equation. When each sine wave is recorded on its own track, some amount of harmonic distortion is added. But no IM distortion is added by the recorder because only one frequency is present on each track. So when the recorder's tracks are mixed cleanly, the result is three sine waves, each with its own harmonically related distortion frequencies added. This is shown in Figure 2.11, using the three notes of an A major chord as the source frequencies. For simplicity, only the first two added harmonics are listed for each tone.

Compare that to mixing the three sine waves together cleanly and then recording that mix onto a single track that adds distortion. Now the recorder's nonlinearity adds not only harmonic distortion to each of the three fundamental pitches but *also* adds IM sum and difference frequencies because the three sources are present together when recorded. This is shown in Figure 2.12.

So by separating out sources across multiple recorder tracks—or converters or preamps or any other devices that might contribute audible distortion—the result is always cleaner than when mixing the sources together first. Note that the difference between THD and IMD amounts is purely a function of the device's nonlinearity. With transparent gear the added IM products are not audible anyway—hence the proof that *audible* stacking is a myth when using high-quality gear. And even when gear is not transparent, stacking can only *reduce* distortion, which is the opposite of what's claimed.

This brings us to *coherence*. Noise and distortion on separate tracks do not add coherently. If you record the same mono guitar part on two analog tape tracks at once, when played

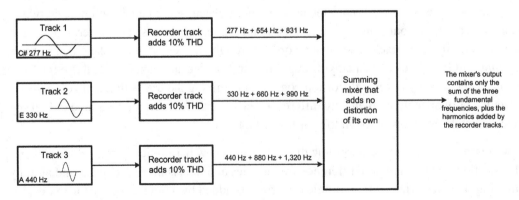

Figure 2.11: Recording multiple single-frequency sources onto separate recorder tracks adds new distortion products created within the recorder, but only at frequencies harmonically related to each source.

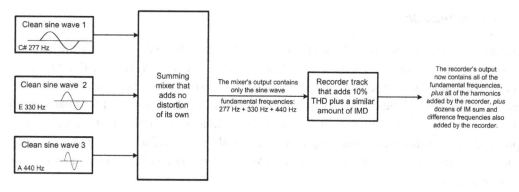

Figure 2.12: Recording audio sources onto a single recorder track after they're combined adds harmonics related to each source *and* adds sum and difference frequencies related to all of the sources.

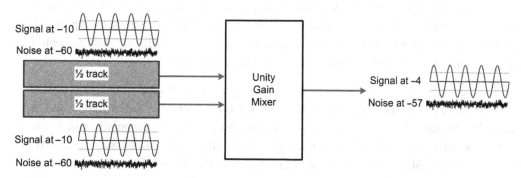

Figure 2.13: Coherent signals add by 6 dB, but noise is random and increases only 3 dB.

back, the signals combine to give 6 dB more output. But the tape noise is different on each track and so rises only 3 dB. This is the same as using a tape track that's twice as wide, or the difference between eight tracks on half-inch tape versus eight tracks on one-inch tape. Figure 2.13 shows this in context, where recording the same source to two tracks at once yields a 3 dB improvement in the signal to noise ratio.

The same thing happens with distortion. The distortion added by a preamp or converter on a bass track has different content than the distortion added to a vocal track. So when you combine them cleanly in a mixer, the relative distortion for each track remains the same. Thus, there is no "stacking" accumulation for distortion either. If you record a DI bass track through a preamp having 1 percent distortion on one track and then record a grand piano through the same preamp to another track, the mixed result will have the same 1 percent distortion from each instrument.

Myth-Information

One key to identifying many audio myths is the high prices charged. Another is the lack of any supporting data, or offering only "user testimonials" as evidence. It's one thing for a vendor to claim improved sound, but quite another to prove it. If one brand of speaker wire really is better than all the others, it can be easily proven using the standard four parameters. When a vendor offers flowery wording instead of test data or says only "Just listen," that's a pretty good sign that the claims are probably not truthful. I imagine some vendors actually believe their own claims! But that's irrelevant. What really matters is that *you* know how to separate truth from fiction.

Many of the myths I've described do have a factual basic in science, but the effects are so infinitesimal that they can't possibly be audible. I often see "subjectivists" proclaim that science has not yet found a way to identify and measure things they are certain they can hear, such as a change in sound after a solid state power amp has warmed up for half an hour. I've also heard people state that audio gear can measure good but sound bad, or vice versa. But if a device measures good yet sounds bad—and sounding bad is confirmed by a proper blind test—then clearly the wrong things were measured. This is very different from the belief that what is heard as sounding bad (or good) can't be measured at all.

In truth it's quite the other way around. We can easily measure digital jitter that's 120 dB below the music, which is a typical amount and is about 1,000 times softer than could be audible. It's the same for distortion, frequency response, and noise, especially when you factor in the ear's susceptibility to the masking effect. Many audiophiles truly believe they hear a change in quality when science and logic suggest that no audible difference should exist. But this is easy to disprove: If there were more to audio than the four basic parameters, it would have been revealed by now as a residual in a null test. Hewlett-Packard distortion analyzers going back to the mid-twentieth century use nulling to remove the test signal and reveal any artifacts that remain. The beauty of nulling is that it reveals *all* differences between two signals, including distortion or other artifacts you might not have even thought to look for.

The Big Picture

Keeping what truly matters in perspective, it makes little sense to obsess over microscopic amounts of distortion in a preamp or computer sound card, when most loudspeakers have at least ten times more distortion. Figure 2.14 shows the first five individual components measured from a loudspeaker playing a 50 Hz tone at a moderately loud volume. When you add them up, the total THD is 6.14 percent, and this doesn't include the IM sum and difference products that would also be present had there been two or more source frequencies, as is typical for music.

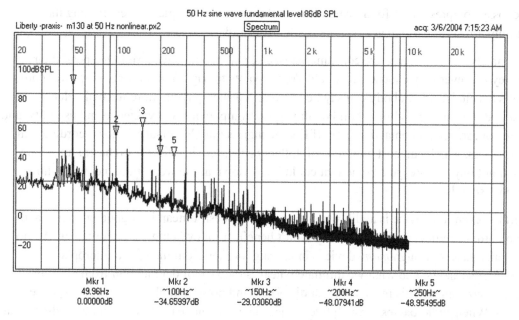

Figure 2.14: This graph shows the fundamental plus first four distortion frequencies measured from a loudspeaker playing a single 50 Hz tone. *Courtesy of audioheuristics.org.*

Midrange and treble speaker drivers often have less distortion than woofers, mostly because woofer cones have to move much farther to create similar volume levels. But even high-quality tweeters playing at modest volumes typically have more distortion than modern electronic devices.

Summary

This chapter explains the four parameter categories that define everything affecting audio fidelity, as well as important specs that vendors sometimes hide, and ways vendors skew data in published graphs to appear more favorable. We also busted a number of common audio myths and learned the correct terminology to define fidelity.

The quality of audio gear can be measured to a much higher resolution than human ears can hear, and those measurements are more accurate and reliable than hearing. Although transparency can be defined and determined conclusively through measuring, color is more difficult to quantify because it involves preference, which cannot be defined. Likewise, degradation caused by lossy MP3-type audio compression is difficult to measure because it doesn't lend itself to traditional fidelity tests. Even with bit-rates high enough to not audibly harm the music, a null test will always reveal residual artifacts. In that case, blind tests—requiring many trials with many test subjects—are the only way to assess how objectionable the lossy compression is for a given bit-rate. But there's no magic, and everything that audibly affects *electronic* gear can be easily measured.

Ultimately, many of these are consumerist issues, and people have a right to spend their money however they choose. If Donald Trump wants to pay $6,000 for an AC power cord, that's his choice and nobody can say he's wrong. Further, paying more for real value is justified. Features, reliability, build quality, good components, convenience and usability, and even appearance all demand a price. If I'm an engineer at Universal Studios recording major film scores, which can cost hundreds of dollars *per minute* just for the orchestra musicians, I will not buy the cheapest brand that could break down at the worst time, no matter how clean it sounds.

Further, even if a device is audibly transparent, that doesn't mean it's "good enough," and so recording engineers and consumers won't benefit from even higher performance. Audio typically passes through *many* devices in its long journey from the studio microphones to your loudspeakers, and what we ultimately hear is the *sum* of degradation from all of the devices combined. This means not just distortion and noise, but also frequency response errors. When audio passes through five devices in a row that each have a modest 1 dB loss at 20 Hz, the net response is a 5 dB reduction at 20 Hz.

The goal of this chapter is to explain what affects audio fidelity, to what degree of audibility, and why. But one important question remains: Why do people sometimes believe they hear a change in audio quality—for example, after replacing one competent wire with another— even when measurements prove there is no audible difference? This long-standing mystery will be explored fully in Chapter 3.

Notes

1 The masking effect refers to the ear's inability to hear a soft sound in the presence of a louder sound. For example, you won't hear your wristwatch ticking at a loud rock concert, even if you hold it right next to your ear. Masking is strongest when both the loud and soft sounds contain similar frequencies, and this is described more fully in Chapter 3.

2 Lossy compression is applied to audio data to reduce its size for storage or transmission. Lossy methods allow for a substantial size reduction—MP3 files are typically compressed about 10 to 1—but the original content is not restored exactly upon playback. When using a sufficiently high bit-rate, the small loss in quality is usually acceptable and may not even be audible. Contrast this to less effective but lossless compression that reduces the size of computer files, where the data must not change. Lossy compression is also used with JPG images and video files to reduce their size, though the specific methods differ from those that reduce the size of audio data.

CHAPTER 3

Hearing, Perception, and Artifact Audibility

I wasn't present in 1951 when the Pultec equalizer was designed, but I suspect the engineers were aiming for a circuit that affects the audio as little as possible beyond the response changes being asked of it. I'm quite sure they were not aiming for a 'vintage' sound. The desire for 'warmth' and a 'tube sound' came many years later, as a new generation of engineers tried to understand why some old-school recordings sound so good. Failing to understand the importance of good mic technique in a good-sounding room coupled with good engineering, they assumed (wrongly IMO) that it must be the gear that was used. Personally, I want everything in my recording chain to be absolutely clean. If I decide I want the sound of tubes, I'll add that as an effect later.

—Ethan, posting in an audio forum

I agree with this in every respect.

—George Massenburg, famous recording engineer, and designer of the first parametric equalizer, response to Ethan's comment

Chapter 2 listed a number of common audio myths, such as special speaker wire that claims to sound better than common wire having very similar electrical properties. It's also a myth that vibration isolation devices placed under solid state electronics or wires improve the sound by avoiding resonance because those components are mostly immune to vibration. The same goes for too-small acoustic treatments that supposedly improve clarity and stereo imaging but can't possibly work simply because they don't cover enough surface area to affect a room's frequency response or decay time. Indeed, if you visit audiophile web forums, you'll see posters who claim all sorts of improvements to their audio systems after applying various "tweak" products or procedures. Some of these tweaks are like folk medicine, such as taping a small piece of quartz crystal to the top of a speaker cabinet, though others are sold as commercial products. Besides fancy wire and isolation devices, "power" products claiming to cleanse the AC mains feeding your audio devices are also popular. Another tweak product is "mod" (modification) services, where sellers replace existing resistors, capacitors, and integrated circuits with supposedly higher-quality components. Others cryogenically treat (deep-freeze) wires, fuses, circuit boards, and even entire amplifiers and power supplies for a fee.

It's easy to prove through measuring and null tests whether sound passing through a CD player or other device changed after applying a tweak or mod, or after being "broken in" for some period of time, as is often claimed. Yet even when a difference cannot be measured, or defies the laws of physics, some people still insist they can hear an improvement. Beliefs, expectation bias, and the placebo effect are very strong. When people argue about these things on the Internet, they're called "religious" arguments, because opinions seem based more on faith than facts and logic. I've even heard people argue against blind tests, claiming they stress the listener and "break the mood," thus invalidating the results. Blind testing is an important tool used by all branches of science, and it's equally necessary for assessing audio equipment. As explained in Chapter 1, even if blind listening did hide subtle differences that might be audible in special situations, is a difference so small that you can't hear it when switching between two sources really that important in the grand scheme of things?

This chapter uses logic and audio examples to explain what types of quality changes, and added artifacts, can be heard and at what volume levels. It also addresses the fallibility of human hearing and perception, which are closely related. Before we get to the frailty of hearing, let's first examine the thresholds for audibility. If a musical overtone or background sound is so soft that you can just barely hear it, you might think you hear it when it's not present, and vice versa. So it's important to learn at what levels we can hear various sounds in the presence of other sounds to help distinguish what's real from what's imagined.

Some people are more sensitive to soft sounds and subtle musical details than others. One reason is due to high-frequency hearing loss that occurs with age. Most teenagers can hear frequencies up to 18 KHz or higher, but once we reach middle age, it's common not to hear much higher than 14 KHz, or even 10 KHz. So if an audio circuit has a small but real loss at the very highest audio frequencies, some people will notice the loss, while others won't. Likewise, distortion or artifacts containing only very high frequencies may be audible to some listeners but not others. Further, for frequencies we can hear, learning to identify subtle detail can be improved through ear training. This varies from person to person—not only due to physical attributes such as age, but also with hearing acuity that improves with practice.

Even though I had been a musician for more than 30 years before I started playing the cello, after practicing for a few months, I realized that my sense of fine pitch discrimination had improved noticeably. I could tell when music was out of tune by a very small amount, whether hearing myself play or someone else. We can also learn to identify artifacts, such as the swishy, swirly sound of lossy MP3 compression and digital noise reduction. It helps to hear an extreme case first, such as orchestral music at a low bit-rate like 32 kilobits per second. Then, once you know what to listen for, you're able to pick out that artifact at much lower levels.

I created two examples of low bit-rate encoding to show the effect: "cymbal.wav" is a mono file containing a cymbal strike, and "cymbal_compressed.mp3" is the same file after applying lossy MP3 compression at a very low bit-rate. The files "music.wav" and "music_compressed. mp3" are similar, but they play music instead of just one cymbal. Encoding music at low bit-rates also discards the highest frequencies, so you'll notice that the compressed MP3 files are not as bright sounding as the originals. But you'll still hear the hollow effect clearly, as various midrange frequencies are removed aggressively by the encoding process. Note that the "compression" process for encoding MP3 files is totally different from compression used to even out volume changes. Only the names are the same.

To better understand what lossy compression removes from the music, I created an MP3 file of a pop tune encoded at 192 kbps, which is fairly high quality. Then I nulled that against the original rendered Wave file to obtain the difference. The original music is in the file "mp3. mp3" and the nulled difference is "mp3_null.mp3." I didn't raise the level of the residual in the difference file, so it's about 22 dB softer than the source audio and therefore lets you hear exactly what was removed by the compression.

It's impossible for me to tell someone else what they can and cannot hear, so I won't even try. At the time of this writing I'm 63 years old, and I can hear well up to about 14 KHz. I have two different audio systems—one based around modest but professional grade gear in my large home studio and one a 5.1 surround system in my living room home theater. I consider both systems to be very high quality, and both rooms are very well treated acoustically, but I don't own the most expensive gear in the world either. So when discussing audibility issues in audio forums, it's common for someone to claim, "Your system is not revealing enough, old man, so of course you can't hear the difference." Therefore, the best way to show what does and does not affect audio fidelity is with examples you'll listen to on your own system. You can play the sound clips whenever you want, as often as you want, and never feel pressured to "perform" in front of someone else. Of course, for the A/B-type comparison clips, you must be honest with yourself because you know which version is playing.

All of the audibility examples for this chapter were created entirely in software to avoid passing the audio signals through electronics that could potentially mask (or add) subtle sounds. Obviously, the music used for these examples was recorded and sent through microphones and preamps and other electronic devices. But for assessing the audibility of *changes and additions* to the sound, all the processing was done using high-resolution software that's cleaner and more accurate than any audio hardware.

Fletcher-Munson and the Masking Effect

The *masking effect* influences the audibility of artifacts. Masking is an important principle because it affects how well we can hear one sound in the presence of another sound. If you're standing next to a jackhammer, you won't hear someone talking softly ten feet away. Masking is strongest when the loud and soft sounds have similar frequency ranges. So when playing an old Led Zeppelin cassette, you might hear the tape hiss during a bass solo but not when the cymbals are prominent. Likewise, you'll easily hear low-frequency AC power line hum when only a tambourine is playing, but maybe not during a bass or timpani solo.

Low-frequency hum in an audio system is the same volume whether the music is playing or not. So when you stop the CD, you can more easily hear the hum because the music no longer masks the sound. Some artifacts like tape modulation noise and digital jitter occur only while the music plays. So unless they're fairly loud, they won't be audible at all. Note that masking affects our ears only. Spectrum analyzers and other test gear can easily identify any frequency in the presence of any other frequency, even when one is 100 dB below the other. In fact, this is the basis for lossy MP3-type compression, where musical data that's deemed inaudible due to masking is removed, reducing the file size.

When I first became interested in learning at what level distortion and other unwanted sounds are audible, I devised some experiments that evolved into the example clips that accompany this book. For one test I created a 100 Hz sine wave in Sound Forge, then mixed in a 3 KHz tone at various levels below the 100 Hz tone. I picked those two frequencies because they're very far apart and so minimize masking. Research performed by Fletcher-Munson shows that our hearing is most sensitive around 2 to 3 KHz. So using these two frequencies biases the test in favor of being able to hear soft artifacts. Further, I inserted the 3 KHz tone as a series of pulses that turn on and off once per second, making it even easier to spot.

Figure 3.1 shows the *Equal Loudness* curves of hearing sensitivity versus frequency as determined by Harvey Fletcher and W. A. Munson in the 1930s. More recent research by D. W. Robinson and R. S. Dadson in 1956 produced similar results. You can see that the response of our ears becomes closer to flat at louder volumes, which is why music sounds fuller and brighter when played more loudly. But even at high volume levels, our hearing favors treble frequencies. Note that this graph shows how loud sounds must be at various frequencies to be perceived as the same volume. So frequencies where our ears are more sensitive show as lower SPLs. In other words, when compared to a 4 KHz tone at 43 dB SPL, a 30 Hz tone must be at 83 dB SPL to sound the same volume. This is why many consumer receivers include a *loudness* switch, which automatically boosts the bass at low volumes. Some receivers also boost the treble to a lesser extent.

For this test I played the combined tones quite loudly, listening through both loudspeakers and earphones. When the 3 KHz tone was 40 dB softer than the 100 Hz tone, it was easy to hear it pulse on and off. At −60 dB it was *very* soft, but I could still hear it. At −80 dB I was unable to hear the tone at all. Even if someone can just barely pick out artifacts at such low levels, it's difficult to argue that distortion this soft is audibly damaging or destroys the listening experience. But again, the accompanying example files let you determine your own audibility thresholds through any system and loudspeakers you choose. The four files for this example are named "100hz_and_3khz_at_-40.wav" through "100hz_and_3khz_ at_-80.wav."

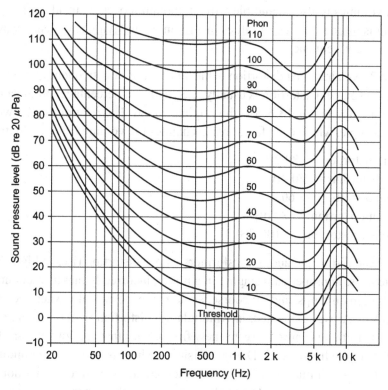

Figure 3.1: The Fletcher-Munson *Equal Loudness Curves* show the volume that different frequencies must be to sound equally loud. *Graph reproduced by permission of Eddy Brixen.*

Note that when running tests that play high-frequency sine waves through loudspeakers, it helps to move your head slightly while you listen. This avoids missing a high-frequency tone that's present and audible but in an acoustic null. Even when a room has acoustic treatment, deep nulls are often present at high frequencies every few inches, especially when playing the same mono source through two loudspeakers at once. You can hear this easily by playing a 3 KHz tone by itself, then moving your head a few inches in any direction. That file is named "3khz_alone.wav."

Distortion and Noise

Everyone reading this book understands what distortion is and why minimizing distortion is an important goal for high fidelity, even if it's sometimes useful as an effect. But there are several types of distortion and many causes of distortion. Further, audio gear can create audible artifacts that are not distortion. For example, there's hiss, hum and buzz, ringing in crossover filters, and digital aliasing. The rattle of a buzzing window or rack cabinet at high playback volumes is also an audible artifact, though it occurs outside of the equipment. Indeed, nothing is more revealing of rattles in a room than playing a loud sine wave sweep starting at a very low frequency.

Chapters 1 and 2 explained that distortion is the addition of new frequency components not present in the source. Distortion is usually specified as a percentage, but it can also be expressed as some number of dB below the original sound. The relationship between dB and percent is very simple, where each factor of 10 changes the volume level by 20 dB:

    10 percent distortion = −20 dB
    1 percent distortion = −40 dB
    0.1 percent distortion = −60 dB
    0.01 percent distortion = −80 dB
    0.001 percent distortion = −100 dB

Many things create distortion: audio transformers whose distortion rises at high signal levels and low frequencies, improper gain staging in a mixer, incorrectly biased vacuum tubes, slew rate limiting in op-amps, and, of course, analog tape—especially when recording at levels above 0 VU. As was explained, intermodulation distortion (IMD) is more damaging than harmonic distortion because it creates new frequencies that are not necessarily related musically to the source frequencies. Digital aliasing—which is rarely if ever audible with modern converters—is similar to IMD because it too creates new frequencies not musically related to the source. Therefore, IMD and aliasing can usually be heard at lower levels than harmonic distortion simply because the artifact frequencies do not blend as well with the source frequencies.

Many types of distortion and noise can be added by audio gear, and all four of the audio parameters listed in Chapter 2 are important. But what matters most is their *magnitude*, because that alone determines how audible they are. If the sum of all distortion is 80 dB or more below the music, it's unlikely to be heard while the music plays. Further, some types of distortion are masked more by the music. For example, let's say the third harmonic of a low note played on a Fender bass is 10 dB softer than the fundamental. An amplifier that adds 0.1 percent third harmonic distortion will increase the note's own harmonic content by a very small amount. Since 0.1 percent is the same as −60 dB, the 0.03 dB increase after adding distortion at −60 dB to a natural harmonic at −10 dB is so small it can't possibly be heard. Not to mention that most loudspeakers add 10 to 100 times more distortion than any competent amplifier.

Some people insist that because of masking, the amount of harmonic distortion is irrelevant, and all that matters is the nature of the distortion. If an amplifier adds 0.1 percent distortion at the 3rd harmonic, the distortion is not only much softer than the original sound, but its frequency is also 1.5 octaves away. Some types of distortion are more trebly, adding a "buzzy" sound quality whose components are even farther away from the source frequencies. So with trebly distortion added to an A bass note at 110 Hz, the harmonics many octaves away will be more audible than harmonics nearer to the fundamental. But again, once artifacts are −80 dB or softer, their spectrum doesn't matter simply because they're too soft to hear.

Jitter

> *It's amazing to me that nobody ever complained about analog recording like they do about digital recording. I'm doing a project right now completely in Pro Tools 24 bit / 48 KHz. The musicians were great, everything sounds great, so that's all I care about. The TDM buss doesn't sound thin to me. What is a thin TDM buss supposed to sound like? I've done a dozen albums completely in Pro Tools, including three Grammy-winning Bela Fleck albums.*
> —**Roger Nichols, famous recording engineer and an early proponent of digital recording**

One artifact that's often cited as detrimental to audio clarity is *jitter*, a timing error specific to digital audio. The sample rate for CD-quality audio is 44.1 KHz, which means that 44,100 times per second a new data sample is either recorded or played back by the sound card. If the timing between samples varies, the result is added noise or artifacts similar to noise or IM distortion. The more the timing between samples deviates, the louder these artifacts will be. Modern sound cards use a highly stable quartz crystal or ceramic resonator to control the flow of input and output data. So even if the timing interval between samples is not at precisely 44.1 KHz, it is *very* close and also varies very little from one sample to the next.

Note that there are two types of frequency deviation. One is simply a change in frequency, where the sample rate might be 44,102 Hz instead of precisely 44,100 Hz. There's also clock drift, which is a slow deviation in clock frequency over long periods of time. This is different from jitter, which is a difference in timing from one sample to the next. So the time between one pair of samples might be 1/44,100th of a second, but the next sample is sent 1/44,100.003 seconds later. Therefore, jitter artifacts are caused by a *change* in timing between adjacent samples.

With modern digital devices, jitter is typically 100 dB or more below the music, even for inexpensive consumer-grade gear. In my experience that's too soft to be audible. Indeed, this is softer than the noise floor of a CD. Even though jitter is a timing issue, it manifests either as noise or as FM side bands added to the music, similar to IM distortion. Depending on the cause and nature of the jitter, the side bands may be harmonically related or unrelated to the music. The spectral content can also vary. All of these could affect how audible the jitter will be when music is playing because of the masking effect. In fairness, some people believe there's more to jitter than just added artifacts, such as a narrowing of the stereo image, though I've never seen compelling proof.

Similarly, an artifact called *truncation distortion* occurs when reducing 24-bit audio files to 16 bits if dither is not applied, and some people believe this too can affect things like fullness and stereo imaging. However, fullness is a change in frequency response that's easily verified. And good imaging, in my opinion, is related more to room acoustics and avoiding early reflections than low-level distortion or microscopic timing errors. (Dither will be explained shortly.)

One obstacle to devising a meaningful test of the audibility of jitter and some other artifacts is creating them artificially in controlled amounts. Real jitter occurs at extremely high frequencies. For example, one nanosecond of jitter equates to a frequency of 1 GHz. Yes, GHz—that is not a typo. The same goes for distortion, which adds new frequencies not present in the original material. It's easy to generate a controlled amount of distortion with one sine wave but impossible with real music that contains many different frequencies at constantly changing volume levels.

Audibility Testing

Since many people don't have the tools needed to prepare a proper test, I created a series of CD-quality Wave files to demonstrate the audibility of artifacts at different levels below the music. Rather than try to artificially generate jitter and the many different types of distortion, I created a nasty-sounding treble-heavy noise and added that at various levels equally to both the left and right channels. The spectrum of the noise is shown in Figure 3.2. Since this noise has a lot of treble content at frequencies where our ears are most sensitive, this biases the test in favor of those who believe very soft artifacts such as jitter are audible. This noise should be at least as noticeable as distortion or jitter that occurs naturally, if not more audible. So if you play the example file containing noise at −70 dB and can't hear the noise, it's unlikely that naturally occurring jitter at the same volume or softer will be audible to you.

To make the noise even more obvious—again favoring those who believe very soft artifacts matter—the noise pulses on and off rather than remains steady throughout the music. In all of the example files, the noise pulse is about 3/4 second long and restarts every 2 seconds. The first pulse starts 2 seconds into each file and lasts for 3/4 second. The next pulse starts 4 seconds in, and so forth.

- The "noise.wav" file is the noise burst by itself, so you can hear it in isolation and know what to listen for when the music is playing. The level is at −20 dB rather than 0 because it sounds *really* irritating. I don't want you to lunge for the volume control when you play it at a normal volume level!

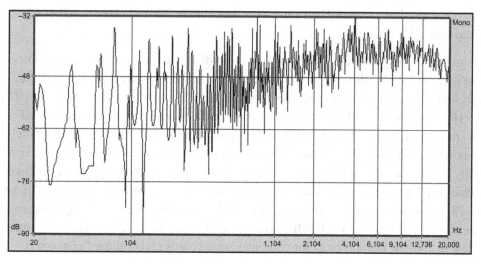

Figure 3.2: This noise signal has a lot of energy at frequencies where our hearing is most sensitive.

- The "concerto-40.wav" file is a gentle passage from my cello concerto with the noise mixed in at −40 dB. Since this passage is very soft, mostly around −25 and peaking at −15 dB, the noise is only 15 to 25 dB below the music. Everyone will easily hear where the noise starts and stops.
- The files "concerto-50.wav," "concerto-60.wav," and "concerto-70.wav" are similar, with the noise mixed in at −50, −60, and −70 dB, respectively. In the −70 dB version the noise is 45 to 55 dB below the music. Note that a slight noise occurs naturally in this piece at around 8 seconds in. I believe it's the sound of a musician turning a page of music during the recording. The noise is in the original recording, and at this low volume it just happens to sound like my intentional noise.
- The file "men_at_work_1–40.wav" is a section from one of my pop tunes, *Men At Work*, with the noise mixed in at −40 dB. I had planned to create other versions with the noise at ever-softer levels as above, but it's barely audible (if at all) even at this relatively high level, so I didn't bother.
- The "men_at_work_2–40.wav" file is a different section from the same pop tune that's more gentle sounding, which potentially makes the noise at −40 dB a little easier to notice.

It's worth mentioning one situation where jitter can be so severe that it really does border on being audible. In February 2009, the British magazine *Hi-Fi News* published an article showing that audio from HDMI connections often have much higher jitter levels than when output via S/PDIF. HDMI is a protocol used by digital televisions, consumer receivers, and Blu-ray disk players. It's popular with content providers because it supports copy protection, and it's convenient for consumers because it carries both audio and video over a single wire.

Modern receivers that handle HDMI can switch both the audio and video together when you choose to watch TV or a DVD.

The magazine's report measured the jitter from four different receiver models, and in every case the jitter was worse through the HDMI connection. The best receiver tested was a Pioneer, with 37 picoseconds (ps) of jitter from its S/PDIF output and 50 ps through HDMI. But the HDMI audio from all the other receivers was far worse than from their S/PDIF outputs, ranging from 3,770 ps to 7,660 ps. Compare that to 183 to 560 ps for the same receivers when using S/PDIF, which is certainly too soft to hear. To relate jitter timing with audibility, 10,000 picoseconds (10 nanoseconds) is about equal to the −96 dB noise floor of 16-bit audio at 1 KHz. But the noise produced at 10 KHz from that same amount of jitter is about −78 dB, which is potentially audible in certain situations. Again, this higher-than-usual amount of jitter applies only to HDMI audio as used by consumer equipment at the time of the article in 2009. It is not typical for computer sound cards and outboard converters.

Some people believe that *correlated* artifacts such as added harmonics or IM products are more audible than *uncorrelated* artifacts like random noise. Jitter can be either correlated or not, depending on its cause. But if the jitter noise is more than 100 dB below the music, which is always the case except for HDMI audio, it's unlikely to be audible regardless of its spectrum or correlation to the music.

It's clear to me that the burden of proof is on those who believe jitter is audible. This subject has been discussed endlessly in audio forums, and someone will inevitably insist audibility tests such as those presented here are not conclusive. So I always ask them to make their own example files using any musical source and any distortion or other soft artifact they believe best makes their case and then post it for everyone to hear. As far as I'm aware nobody has ever risen to the challenge.

Dither and Truncation Distortion

Conventional audio wisdom says that dither is required to eliminate *truncation distortion* whenever you reduce the bit-depth of an audio file, and it's most often used when reducing a 24-bit mix file to 16 bits for putting onto a CD. Dither is a very soft noise whose level is at the lowest bit—around −90 dB when reducing digital audio to 16 bits. Most people would have a hard time hearing noise that's 60 dB below the music, since the music masks the noise. Yet if you ask a dozen audio recording engineers if dither is necessary when going from 24 to 16 bits, every one of them will say yes. Even the manual for Sound Forge claims dither is important:

> *If you want to burn a 24-bit audio file to an audio CD, dithering will produce a cleaner signal than a simple bit-depth conversion.*

Some engineers even argue over which type of dither is best, claiming that this algorithm is more airy or full sounding than that one, and so forth. But just because everyone believes this, does that make it true? To be clear, using dither is never a bad thing, and it can reduce distortion on soft material recorded at very low levels. So I never argue against using dither! But I've never heard dither make any difference when applied to typical pop music recorded at sensible levels. Not using dither is never the reason an amateur's mixes sound bad.

To put this to the test, I created a set of eight files containing both truncated and dithered versions of the same sections of my pop tune *Lullaby*. These are the exact steps I followed to create the files named "lullaby_a.wav" through "lullaby_h.wav": I started by rendering *Lullaby* from SONAR at 24 bits, then extracted four short sections. I alternately dithered and truncated each section down to 16 bits and renamed the files to hide their identity. So "lullaby_a" and "lullaby_b" are the same part of the tune, with one dithered and the other truncated. The same was done for the file pairs "c/d," "e/f," and "g/h." Your mission is to identify which file in each pair is dithered and which is truncated. The dithering was done in Sound Forge using *high-pass triangular* dither with *high-pass contour* noise shaping. You're welcome to send me your guesses using the email link on my home page ethanwiner. com. When I have enough reader submissions, I'll post the results on my website or on my Facebook page.

Hearing Below the Noise Floor

It's well known that we can hear music and speech in the presence of noise, even if the noise is louder than the source. I've seen estimates that we can hear music or speech when it's 10 dB below the noise floor, which in my experience seems about right. Of course, the spectral content of the noise and program content affects how much the noise masks the program, and disparate frequencies will mask less. So I was surprised when an audio design engineer who works for a famous company claimed in an audio forum that he can hear artifacts 40 dB below the noise floor of analog tape while music plays. To test this for myself—and for you—I created this series of CD-quality files you can play through your own system:

- The file "tones_and_noise.wav" contains pink noise and a pair of test tones mixed at equal levels. The test tones portion contains both 100 Hz and 3 KHz at the same time to be more obvious to hear when they turn on and off.
- The "tones-10.wav" file is the same pink noise and test tones, but with the tones 10 dB below the noise. It's still easy to hear where the tones start and stop in this file.
- The "tones-20.wav" and "tones-30.wav" files are similar, but with the tones 20 and 30 dB below the noise, respectively. I can just barely hear the tones in the −20 file, and it seems unlikely anyone could hear them in the −30 version.

- For a more realistic test using typical program material, I also created files using speech and pop music mixed at low levels under pink noise. The file "speech_and_noise.wav" contains these two signals at equal levels, and it's easy to understand what is said.
- The "speech-10.wav" and "speech-20.wav" files are similar, but with the speech 10 and 20 dB below the noise, respectively. When the speech is 10 dB below the noise, you can hear that someone is talking, but it's difficult to pick out what's being said. When the speech is 20 dB lower than the noise, it's all but inaudible.
- The last two files mix pop music softly under the noise. The files "music_10db_below_ noise.wav" and "music_20db_below_noise.wav" are self-descriptive. It's not difficult to hear that music is playing in the −10 dB version, but can you hear it when it's 20 dB below the noise floor?

I'm confident that these test files bust the myth that anyone can hear music or speech that's 40 dB below a typical noise floor. However, there are many types of noise. The audibility of program content softer than noise depends directly on the frequencies present in the noise versus the frequencies present in the program. Noise containing mostly high frequencies will not mask low-frequency sounds, and vice versa.

Frequency Response Changes

Chapter 1 explained that applying a broad low-Q boost or cut with an equalizer is more audible than a narrow boost or cut simply because a low Q affects more total frequencies. Some people believe that very narrow peaks and nulls are not audibly damaging, especially nulls. While a narrow bandwidth affects a smaller range of frequencies, and thus less overall energy than a wide bandwidth, EQ changes using very narrow bandwidths can still be audible. What matters is if the frequencies being boosted or cut align with frequencies present in the program. The response graphs in Figure 3.3 show an equalizer set to cut 165 Hz by 10 dB with a Q of 2 and 24.

I believe the notion that narrow EQ cuts are not damaging arises from a 1981 paper[1] by Roland Bücklein in the *Journal of the Audio Engineering Society*, describing his tests of boost and cut audibility at various bandwidths. Some of the tests used speech and white noise, while others used music. White noise contains all frequencies in equal amounts, so a wide bandwidth boost adds more energy than a narrow boost, and it is more audible. The same is true for broad cuts that reduce more content than narrow cuts.

But for the *music* tests, the frequencies boosted and cut in Mr. Bücklein's experiments did not align with the frequencies in the music being played. Instead, he used the standard third-octave frequencies listed in Table 1.3 from Chapter 1, and *none* of those align exactly with any of the standard music note frequencies in Table 1.2. Music consists mostly of single tones and harmonics that are also single tones, so the correlation between the frequencies

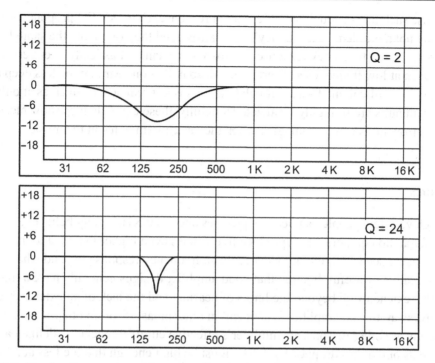

Figure 3.3: Cutting frequencies with an equalizer having a low Q (top) affects more total sound energy than the same amount of cut with a high Q (bottom).

changed with EQ and the frequencies present in the music is very important. An extremely narrow bandwidth may miss a particular frequency of interest, but a boost or cut of all the frequencies in a third-octave band is bound to change the music audibly. If a mix engineer needs to reduce the level of a particular frequency—for example, a single bass note that's 10 dB louder than other notes—he'd use a parametric equalizer with a high Q to zero in on that one frequency to avoid affecting other nearby notes.

To illustrate the potential audibility of very narrow boosts and cuts, I created a series of three Wave files. The first clip, "men_at_work.wav," is an excerpt of the tune as I mixed it, but it is reduced in volume to allow adding EQ boost without distorting. The second file, "men_at_work_boost.wav," is the same clip with 10 dB of very narrow EQ boost (Q = 24) applied at 165 Hz. The third file, "men_at_work_cut.wav," is the original clip but with a 10 dB very narrow cut (Q = 24) at 165 Hz. I chose 165 Hz because that's an E note, which is the key of the tune. So in these examples the narrow boost and cut are very obvious because they correspond to notes the bass plays.

This is directly related to the folly of expecting EQ to improve room acoustics at low frequencies. One problem with using EQ for acoustics is it's not possible to counter deep nulls. Nulls of 20 to 30 dB, or even deeper, are common, and you'll blow up your speakers trying to raise such nulls enough to achieve a flat response. I've seen EQ proponents claim that nulls are not a problem because they're so narrow, and they often cite the same Bücklein article! However, the frequency response in a room can change drastically over very small distances, even at low frequencies. Therefore, a deep null at one ear may be less deep at the other ear, so the total volume heard through both ears is not reduced as much as at only one ear. But not all nulls are so highly localized. Hopefully, these example files show clearly that even very narrow nulls can be damaging when they align with notes in the music.

Ultrasonics

Even though very few people can hear frequencies above 20 KHz, many believe it's important for audio equipment to reproduce frequencies even higher than that to maintain clarity. I've never seen compelling evidence that a frequency response beyond what humans can hear is audible or useful. It's true that good amplifier designs generally have a frequency response that extends well beyond the limits of hearing, and the lack of an extended response can be a giveaway that an amplifier is deficient in other areas. If for no other reason, though there certainly are other reasons, an amplifier's effective cutoff frequency—defined as the point where its output has dropped by 3 dB—must be high enough that the loss at 20 KHz is well under 1 dB. So it's common for the −3 dB point of good-quality amplifiers to be 50 KHz or even higher.

With microphones and speakers, their cutoff frequency can be accompanied by a resonant peak, which can add ringing as well as a level boost at that frequency. Therefore, designing a transducer to respond beyond 20 KHz is useful because it pushes any inherent resonance past audibility. This is one important feature of condenser microphones that use a tiny (less than ½-inch) diaphragm designed for acoustic measuring. By pushing the microphone's self-resonance to 25 KHz or even higher, its response can be very flat with no ringing in the audible range below 20 KHz.

It's easy to determine, for once and for all, if a response beyond 20 KHz is noticeable to you. All you need is a sweepable low-pass filter. You start with the filter set to well beyond 20 KHz, play the source material of your choice, and sweep the filter downward until you can hear a change. Then read the frequency noted on the filter's dial. I've used a set of keys jingling in front of a high-quality, small-diaphragm condenser mic, but a percussion instrument with extended high-frequency content such as a tambourine works well, too. Most people don't have access to suitable audio test gear, but you can do this with common audio editing software. Record a source having content beyond 20 KHz using a sample rate of 88.2 or 96 KHz, then sweep a filter plug-in as described above. I suggest you verify that ultrasonic frequencies are present using an FFT or Real Time Analyzer plug-in to be sure your test is valid.

So you won't have to do it, I recorded the file "tambourine.wav" at a sample rate of 96 KHz through a precision DPA microphone. As you can see in Figure 3.4, this file contains energy beyond 35 KHz, so it's a perfect source for such tests. It's only 7 seconds long, so set it to loop continuously in your audio editor program as you experiment with a plug-in EQ filter.

Years ago there was a widely publicized anecdote describing one channel in a Neve recording console that was audibly different than other channels, and the problem was traced to an oscillation at 54 KHz. I'm sure that channel sounded different, but it wasn't because Rupert Neve or Beatles engineer Geoff Emerick was hearing 54 KHz. When an audio circuit oscillates, it creates hiss and "spitty" sounds and IM distortion in the audible range. So obviously that's what Geoff heard, not the actual 54 KHz oscillation frequency. Further, no professional studio monitor speakers I'm aware of can reproduce 54 KHz anyway.

There was also a study by Tsutomu Oohashi et al. done in 2000 that's often cited by audiophiles as proof that we can hear or otherwise perceive ultrasonic content. The problem with this study is they used one loudspeaker to play many high-frequency components at once, so IM distortion in the tweeters created difference frequencies within the *audible* range. When the Oohashi experiment was repeated by Shogo Kiryu and Kaoru Ashihara using six separate speakers,[2] none of the test subjects was able to distinguish the ultrasonic content. This is from their summary:

> *When the stimulus was divided into six bands of frequencies and presented through six loudspeakers in order to reduce intermodulation distortions, no subject could detect any ultrasounds. It was concluded that addition of ultrasounds might affect sound impression by means of some nonlinear interaction that might occur in the loudspeakers.*

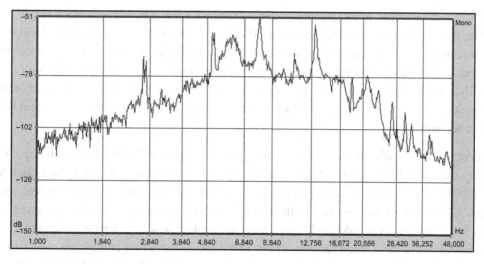

Figure 3.4: This 96 KHz recording of a tambourine has content beyond 35 KHz, so it's a great test signal for assessing your own high-frequency hearing.

I've also seen claims proving the audibility of ultrasonic content where a 15 KHz sine wave is played, then switched to a square wave. Proponents believe that the quality change heard proves the audibility of ultrasonic frequencies. But this doesn't take into account that loudspeakers and power amplifiers can be nonlinear at those high frequencies, thereby affecting the audible spectrum. Further, most hardware generators used to create test tones output a fixed peak level. When the peak (not average) levels are the same, a square wave has 2 dB more energy at the fundamental frequency than a sine wave. So, of course, the waves could sound different.

Finally, it's worth mentioning that few microphones, and even fewer loudspeakers, can handle frequencies much higher than 20 KHz. Aside from tiny-diaphragm condenser microphones meant for acoustic testing, the response of most microphones and speakers is down several dB by 20 KHz if not lower.

Ringing

Chapter 1 explained the concepts of resonance and ringing, which occur in both mechanical and electronic devices. Whenever a peaking boost is added with an equalizer, some amount of ringing is also added. This sustains the boosted frequency after the original sound has stopped. The higher the Q of the boost, the longer the ringing sustains for. However, the source must contain energy at the frequency being boosted in order for ringing to be added. So depending on its loudness, ringing is another potentially audible artifact. Note that ringing is different from distortion and noise that add new frequency components. Rather, ringing merely sustains existing frequencies.

To illustrate the concept of EQ adding ringing, I created a Wave file containing a single impulse, then applied 18 dB of EQ boost at 300 Hz with two different Q settings. A single impulse contains a wide range of frequencies whose amplitudes depend on the duration and rise times of the impulse. Figure 3.5 shows an impulse about 3 milliseconds long that I drew manually into Sound Forge using the pencil tool. I set the time scale at the top of each screen to show seconds, so 0.010 on the timeline means that marker is at 10 milliseconds, and 0.500 is half a second.

I then copied the impulse twice, half a second apart, and applied EQ boost to each copy. If you play the audio file "impulse_ringing.wav," you'll hear the sound switch from a click to a partial tone, then to a more sustained tone. Figure 3.6 shows the entire file in context, and Figures 3.7 and 3.8 show close-ups of the impulses after ringing was added, to see more clearly how they were extended. Although these examples show how ringing is added to a single impulse, the same thing happens when EQ is applied to audio containing music or speech. Any frequency present in the source that aligns with the frequency being boosted is affected the same way: amplified and also sustained.

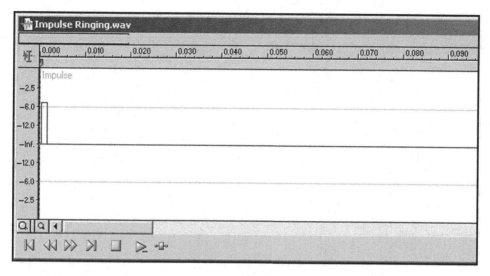

Figure 3.5: This short impulse contains all the frequencies between DC and the 22 KHz limit of a 44.1 KHz sample rate.

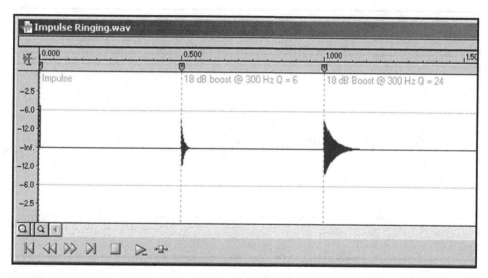

Figure 3.6: This Wave file contains three brief impulses in a row. The second version has 18 dB of EQ boost applied at 300 Hz with a Q of 6, and the third applied the same boost with a Q of 24, making it sustain longer.

Earlier I explained that ringing doesn't add new frequencies, and it can only sustain frequencies that already exist. In truth, ringing can add new frequencies, which could be seen as a type of distortion or the addition of new content. This can occur in a room that has a strong resonance at a frequency near, but not exactly at, a musical note frequency. If a

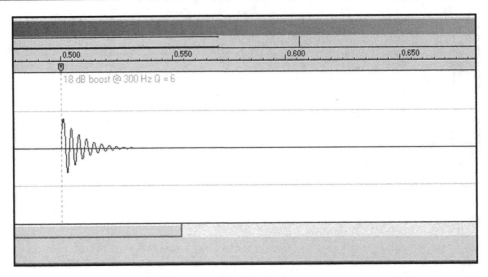

Figure 3.7: Zooming in to see the wave cycles more clearly shows that a Q of 6 brings out the 300 Hz component of the impulse and sustains it for about 40 milliseconds after the impulse ends.

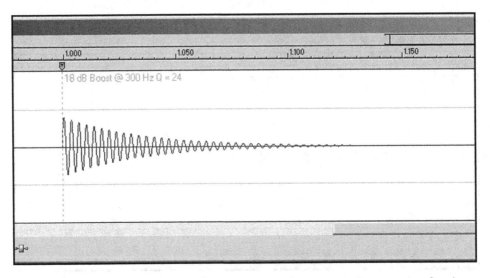

Figure 3.8: When the Q is increased to 24, the impulse now continues to ring for about 130 milliseconds.

room has a strong resonance at 107 Hz and you play an A bass note at 110 Hz, that note will sound flat in the room. How flat the note sounds depends on the strength and Q of the room's resonance.

The same thing happens with EQ boost when the boosted frequency is near to a frequency present in the source. Figure 3.9 shows an FFT of the example file "claves_original.wav."

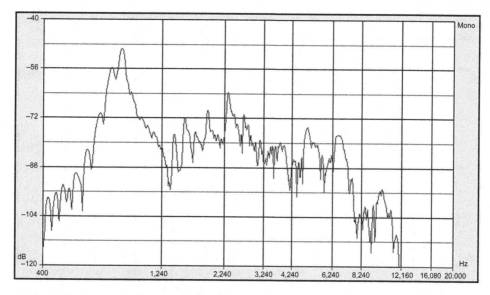

Figure 3.9: This is the spectrum of a pair of claves whose fundamental pitch is 853 Hz.

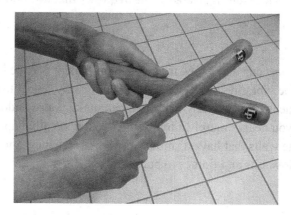

Figure 3.10: Claves are a popular percussion instrument used in Salsa and other types of Latin music. When struck together, they produce a distinctive pitched sound that's heard easily, even over loud music.

The most prominent peak near the left of the screen corresponds to the primary pitch of 853 Hz for these particular claves. For those not familiar with Latin percussion instruments, a pair of claves is shown in Figure 3.10.

After adding 18 dB of high-Q boost at 900 Hz, one musical half-step higher, the original 853 Hz is still present, but a new *much louder* component is added at 900 Hz. This file is named "claves_boost.wav," with its spectrum shown in Figure 3.11. Playing the two audio files side by side, you can hear the prominent higher pitch in the EQ'd version. This boost also makes the tone of the claves sound more pure than the original, with less of a wood click sound.

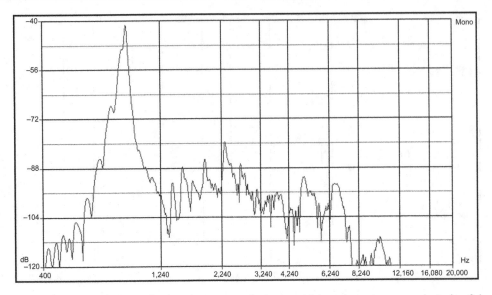

Figure 3.11: After adding a high Q boost at a frequency higher than the natural pitch of the claves, a new tone is created at the frequency that was boosted.

The reason a high-Q boost can add a new frequency component is called *sympathetic resonance*, whereby one vibrating object causes another nearby object to also vibrate. This is why playing low-frequency tones loudly in a room can make your windows rattle, though the trebly buzzing sound you hear is merely a by-product. What's really happening is the low-frequency tones vibrate walls that have a similar resonant frequency, and *that* vibration then causes any loose window sills and frames to buzz.

Aliasing

Aliasing is a type of artifact specific to digital recording, and it's similar to IM distortion because new sum and difference frequencies are created. In this case, one of the source frequencies is the sample rate of the sound card. I haven't heard this problem in many years, but it could happen with old converters having poor input filters or that were made before the advent of oversampling. Converter filters and oversampling are explained in Chapter 8, but it's worth hearing what aliasing sounds like now.

The file "quartet.wav" is a short section of a string quartet, and "quartet_aliasing.wav" is the same file after adding artificial aliasing using a plug-in. Aliasing artifacts are sometimes called "birdies" because difference frequencies that fall in the range around 5–10 KHz come and go in step with the music, which sounds a bit like birds chirping. For these examples

I made the aliasing louder than would ever occur even with poor converters, just to make it easier to identify.

Phase Shift

Phase shift is another frequent target of blame, accused by the audio press repeatedly of "smearing" clarity and damaging stereo imaging. To disprove this myth, I created the file "phase_shift.wav," which contains the same two bars from one of my pop tunes twice in a row. The first version is as I mixed it, and the second is after applying four stages (360 degrees) of phase shift at 1,225 Hz. This is more phase shift than occurs normally in audio gear, and it's in the midrange where our ears are very sensitive. Phase shift in audio gear usually happens only at the frequency extremes—below 20 Hz or above 20 KHz—due to coupling capacitors at the low end or a natural roll-off at the high end caused by various factors.

In my experience, the only time phase shift is audible in normal amounts is either when it's changing or when its amount or frequency is different in the left and right channels. With digital signal processing (DSP), it's possible to generate thousands of degrees of phase shift, and eventually that will be audible because some frequencies are delayed half a second or more relative to other frequencies. But this never happens with normal analog gear, and certainly not with wire, as I've seen claimed. As explained previously, sometimes when boosting high frequencies with an equalizer, you'll hear the hollow sound of a phase shifter effect. But what you're hearing is comb filtering already present in the source that's now brought out by the EQ. You're not hearing the phase shift itself.

I learned that phase shift by itself was relatively benign in the 1970s, when I built a phase shifter outboard effect unit. As explained in Chapter 1, phaser effects work by shifting the phase of a signal, then combining the original source with the shifted version, thus yielding a series of peaks and nulls in the frequency response. When testing this unit, I listened to the phase-shifted output only. While the Shift knob was turned, it was easy to hear a change in the apparent "depth" of the track being processed. But as soon as I stopped turning the knob, the sound settled in and the static phase shift was inaudible.

Absolute Polarity

Another common belief is that *absolute polarity* is audible. While nobody would argue that it's okay to reverse the polarity of one channel of a stereo pair, I've never been able to determine that reversing the polarity of a mono source—or both channels if stereo—is audible. Admittedly, it would *seem* that absolute polarity could be audible—for example, with a kick drum. But in practice, changing the absolute polarity has never been audible to me.

You can test this for yourself easily enough: If your Digital Audio Workstation (DAW) software or console offers a polarity-reverse switch, listen to a steadily repeating kick or snare drum hit, then flip the switch. It's not a valid test to have a drummer play in the studio hitting the drum repeatedly while you listen in the control room, because every drum hit is slightly different. The only truly scientific way to compare absolute polarity is to audition a looped drum sample to guarantee that every hit is identical.

Years ago my friend and audio journalist Mike Rivers sent me a Wave file that shows absolute polarity can sometimes be audible. The "polarity_sawtooth.wav" file is 4 seconds long and contains a 20 Hz sawtooth waveform that reverses polarity halfway through. Although you can indeed hear a slight change in low-end fullness after the transition point, I'm convinced that the cause is nonlinearity in the playback speakers. When I do the test using a 50 Hz sawtooth waveform played through my large JBL 4430 loudspeakers, there's no change in timbre. (Timbre is tone quality, pronounced **tam**-bur.)

To test this with more typical sources, I created two additional test files. The "polarity_kick. wav" file contains the same kick drum pattern twice in a row, with the second pattern reversed. I suggest you play this example a number of times in a row to see if you reliably hear a difference. The "polarity_voice.wav" file is me speaking because some people believe that absolute polarity is more audible on voices. I don't hear any difference at all. However, I have very good loudspeakers in a room with proper acoustic treatment. As just explained, if your loudspeakers (or earphones) don't handle low frequencies symmetrically, that can create a difference. That is, if the loudspeaker diaphragm pushes out differently than it pulls in, that will account for the sound changing with different polarities.

Besides nonlinearity in loudspeakers, our ears are also nonlinear, as explained in the next section. This came up in an audio forum when a well-known mastering engineer bought very expensive new speakers and said they let him "clearly and obviously hear absolute polarity differences." He credited this to the speaker's claim to be "phase coherent," even though phase and polarity are very different as explained in Chapter 1. When I pointed out that what he heard was more likely due to speaker nonlinearity, another well-known audio expert, a plug-in developer, posted a Wave file whose polarity reversal was indeed audible. His file also used a sawtooth wave, but at 220 Hz the frequency was much higher than Mike's, and he applied three bands of EQ to emphasize the wave's harmonics. To test this for myself I created a similar file "polarity_saw2.wav" which uses the EQ shown following. The volume was adjusted to prevent clipping from all that boost, and the resulting waveform is shown in Figure 3.12:

    18 dB boost at 220 Hz with a Q of 16
    18 dB boost at 440 Hz with a Q of 16
    18 dB boost at 880 Hz with a Q of 16

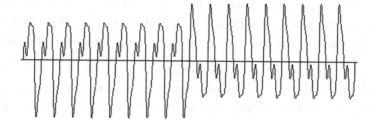

Figure 3.12: This EQ'd sawtooth wave changes polarity halfway through, and you can clearly see more signal voltage in one direction than the other. This asymmetry in turn changes the amount of speaker driver excursion in each direction. If loudspeaker distortion flattens the sharp peaks a little differently in each direction, the harmonic content will also change.

If you play this clip at a moderate volume through either loudspeakers or headphones, you'll probably hear a slight change in tonality and possibly pitch too as the polarity is reversed. But then play this clip at a very soft volume, and you'll notice the sound *doesn't* change at the transition!

IM distortion in our ears is a well-known phenomenon, and the levels of distortion are a lot higher than people might expect! A sawtooth wave (or any non-sine wave) contains multiple frequencies that create IM distortion in the ear. For this test file, adding EQ makes the harmonics as prominent as the fundamental, and that creates the asymmetry needed to hear the polarity switch. So absolute polarity can be audible in contrived situations like this. Further, asymmetrical waveforms also occur naturally, such as with the trumpet and other brass instruments. So it's possible absolute polarity can change their sound too. But clearly the effect is caused by nonlinearity, not an innate ability to perceive polarity.

Ears Are Not Linear!

Even though few people can hear frequencies above 20 KHz, it might be possible to be affected by those higher frequencies. This is not because of bone conduction or other means of perception, as is sometimes claimed. It's because our ears are nonlinear, especially at loud volumes. As explained in the previous chapters, whenever an audio pathway is nonlinear, sum and difference frequencies are created.

One of my favorite examples derives from when I played percussion in a local symphony. A glockenspiel is similar to a vibraphone but smaller, and the metal bars are higher pitched and have no resonating tubes underneath. A glockenspiel can also play very loud! I noticed while playing a chord of two very high notes that I heard a low frequency difference tone. That is, I'd hear a low note even though both of the notes played were very high pitched.

But the distortion was created entirely within my ears. It's not that my ears are defective; anyone can hear this. But you have to be very close to the glockenspiel in order for the volume to be loud enough to distort your ears. I've never noticed this from out in the audience.

You can even hear IM products from a single note because the various overtones of one note can create audible sum and difference frequencies. Unlike square waves, the overtones from bell-type instruments are not necessarily harmonically related. It's possible for a very high note to contain overtones closer together than the fundamental frequency. If one harmonic is 15 KHz and another is 18 KHz, the result is the perception of a 3 KHz tone. So it's possible to be influenced by ultrasonic content, but only because the IM tones are generated inside your ear. Further, I *don't want* to hear that when I'm listening to a recording of my favorite music. If a recording filters out ultrasonic content that would have created inharmonic distortion inside my ears at loud volumes, I consider that a feature!

To prove the point I created the audio file "ear_imd.wav." This file contains 1,000 Hz and 1,200 Hz mixed together at equal volume to show how our ears create a new phantom tone at the 200 Hz difference frequency. First 1,000 Hz is played, then 1,200 Hz, then 200 Hz at a low level so you'll know what to listen for. Then finally the two higher frequencies are played together. If you listen at a moderate to loud volume you'll clearly hear 200 Hz in the fourth (last) segment even though only the two higher frequencies are actually present. In fact, I suspect that IM distortion in our ears is behind the long-standing belief that "the brain" creates a phantom fundamental pitch if only the harmonics of a musical tone are present. Using 500 Hz as an example, the second and third harmonics are 1,000 Hz and 1,500 Hz, and the IM difference frequency is the missing 500 Hz fundamental!

Blind Testing

> *High-end audio lost its credibility during the 1980s, when it flatly refused to submit to the kind of basic honesty controls (double-blind testing, for example) that had legitimized every other serious scientific endeavor since Pascal. [This refusal] is a source of endless derisive amusement among rational people and of perpetual embarrassment for me, because I am associated by so many people with the mess my disciples made of spreading my gospel.*
> —*J. Gordon Holt, audio journalist and founder of* **Stereophile** *magazine*

Informal self-tests are a great way to learn what you can and cannot hear when the differences are large. I don't think anyone would miss hearing a difference between AM and FM radio, or between a cassette recording and one made using a modern digital recorder. But when differences are very small, it's easy to think you hear a difference even when none exists. Blind testing is the gold standard for all branches of science, especially when evaluating perception that is subjective. Blind tests are used to assess the effectiveness of pain medications, as well as for food and wine tasting comparisons. If a test subject knows what he's eating or drinking, that knowledge can and does affect his perception.

The two types of blind tests are single-blind and double-blind. With a single-blind test, the person being tested does not know which device he's hearing or which brand of soda she's drinking, but the person administering the test knows. So it's possible for a tester to cue the test subject without meaning to. Double-blind tests solve this because nobody knows which product was which until after the tests are done and the results are tabulated.

Even though measuring audio gear is usually conclusive, blind tests are still useful for several reasons. One reason is the listener is part of the test, and not all listeners hear the same. Since the specific makeup of distortion and other artifacts affects how audible they are, it can be difficult to pin down exact numbers relating dB level to audibility. A blind test can tell exactly how loud a particular sound must be before *you'll* hear it on *your* system. Further, blind testing is the only practical way to assess the damage caused by lossy MP3-type compression. That process doesn't lend itself to traditional static fidelity measurement because the frequency response changes from moment to moment with the music. Blind testing can also be used to satisfy those who believe that "science" hasn't yet found a way to measure what they're certain they can hear. At least, you'd think participating in a blind test *should* satisfy them!

As mentioned earlier, some people argue against blind tests because they believe these tests put the listener on the spot, making it more difficult to hear differences that really do exist. This can be avoided with an *ABX* test. ABX testing was co-developed by Arny Krueger in the 1980s, and it lets people test themselves as often as they want, over a period as long as they want, in the comfort of their own listening environment. The original ABX tester was a hardware device that played one of two audio sources at random each time you pressed the button. The person being tested must identify whether the "X" currently playing is either source A or source B. After running the same test, say, 10 times, you'd know with some certainty whether you really can reliably identify a difference. These days greatly improved ABX testers are available as software, and Arny's own freeware test program is shown in Figure 3.13.

Another useful self-test that's very reliable is simply closing your eyes while switching between two sources with software. When I want to test myself blind, I set up two parallel tracks in SONAR and assign the Mute switches for those tracks to the same Mute Group while the Mute switches are in opposite states. That is, one track plays while the other is muted, and vice versa. Each time the button is clicked, the tracks switch. This lets me change smoothly from one track to the other without interruption or clicks. I put the mouse cursor over either track's Mute button, close my eyes, then click a bunch of times at random without paying attention to how many times I clicked. This way, I don't know which version will

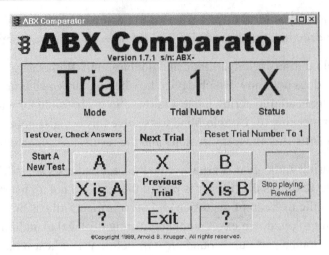

Figure 3.13: The freeware ABX Comparator lets you test yourself to determine whether you can reliably identify a difference between two audio sources.

play first. Then I press the space bar to start playback, still with my eyes closed, and listen carefully to see if I can really tell which source is which as I switch back and forth. When I open my eyes, I can see which track is currently playing.

Whether you're using a single-blind, double-blind, or ABX test, it's important to understand a few basic requirements. First, the volume of both sources much be matched exactly, to within 0.1 dB if possible, or at least to 0.25 dB. When all else is equal, people generally pick the louder (or brighter) version as sounding better, unless, of course, it was already too loud or bright. Indeed, people sometimes report a difference even in an "A/A" test, where both sources are the same. And just because something sounds "better," it's not necessarily higher fidelity. Boosting the treble and bass often makes music sound better, but that's certainly not more faithful to the original source material.

It's also important to test using the same musical performance. A common mistake I see is comparing microphones or preamps by recording someone playing a guitar part with one device, then switching to the other device and performing again. The same subtle details we listen for when comparing gear also change from one performance to another—for example, a bell-like attack of a guitar note or a certain sheen on a brushed cymbal. Nobody can play or sing exactly the same way twice or remain perfectly stationary. So that's not a valid way to test microphones, preamps, or anything else. Even if you could sing or play the same, a change in microphone position of even half an inch is enough to make a real difference in the frequency response the microphone captures.

One solution is a technique known as *re-amping*. Rather than recording live performances that will surely vary, you instead record a single performance, then play that recording

Figure 3.14: When comparing microphones and preamps, using a loudspeaker to play the same source repeatedly is more valid than expecting a musician to sing or play exactly the same several times in a row.

through a loudspeaker. Figure 3.14 shows a re-amp test setup in my home studio using a JBL 4430 monitor and an Audio-Technica AT4033 microphone. This is a great way to audition preamps and microphone wires, though for comparing microphones it's critical that each microphone be in exactly the same place as all the others. Re-amping is described in more detail in Chapter 7.

Another solution is to record several times in a row to obtain an average. Many people try to compare microphones by recording separate performances, but that's not valid because nobody can sing or play exactly the same way twice. Splitting one microphone to two or more preamps is also a problem because the very act of splitting might affect how the mic interacts with the preamps. But if you record multiple performances you can listen later to each several times and get an average. Of course, you need to listen to the files blind, with another person playing the files from each group several times at random to get a significant sample of choices. If you can distinguish Preamp #1 from Preamp #2 at least a dozen times or more, you can be pretty sure the difference between preamps really is audible. Of course, the record levels must be matched exactly. Below are specific steps to do a proper subjective comparison of two preamps.

Calibrate the Preamp Gain Controls

- Put a microphone a foot in front of a loudspeaker in your recording room, leaving it in the exact same place while calibrating both preamp levels.
- Play a 1 KHz tone through the speaker at a moderate volume level.

- Plug the microphone into Preamp #1 and set the gain to a reasonable level such as −6 dB in your DAW.
- Plug the microphone into Preamp #2, then connect the output of Preamp #2 to your DAW.
- Play the 1 KHz tone through the speaker again, then set Preamp #2 to the same −6 dB record level.

Now you can swap the input and output wires to go back and forth between preamps as you record someone singing several times in a row on different tracks. I suggest recording the same short vocal segment six times through Preamp #1, then again six times through Preamp #2. Or record an acoustic guitar, or saxophone, or any other source you think will reveal a difference between preamps. It's okay to move the microphone into an appropriate recording position after setting the levels, because the preamp gain settings will remain the same. But tell the performer to stay the same distance from the microphone, and move as little as possible from one take to the next.

Prepare and Play the Tracks

- Now you have 12 tracks in your DAW, clearly labeled of course! Slide the recorded clips around if needed so they all start at the same time.
- Have a friend play the clips, solo'ing one track at a time at random while you're unable to see which track is playing. Tell your friend it's okay to play the same track more than once, and even twice in a row, so you're choosing which preamp you think is playing a total of 20–30 times.
- After each playback, tell your friend which preamp you think you're hearing. Have him or her note your choice and also note which preamp/track was actually playing.
- After you've heard all 12 tracks a few times each, in random order, look at your friend's notes to see how many times you were right.

In 2008, I did a comparative test of 10 small diaphragm measuring microphones and needed to ensure that each microphone was in the exact same spot in front of the speaker playing sweep tones. A full explanation of that test and the results are in Chapter 22, and you'll need the same precision placement if you intend to compare microphones.

It can be difficult to prove or disprove issues like those presented here because human auditory perception is so fragile and our memory is so short. With A/B testing—where you switch between two versions to audition the difference—it's mandatory that the switch be performed very quickly. If it takes you 15 minutes to hook up a replacement amplifier or switch signal wires, it will be very difficult to tell if there really was a difference, compared to switching between them instantly.

Finally, it's important to understand that, logically, a blind test cannot prove that added artifacts or changes to the frequency response are *not* audible. All we can hope to prove is that the specific people being tested were or were not able to discern a difference with sufficient statistical validity. This is why blind tests typically use large groups of people tested many times each. But blind tests are still very useful, regardless of whether or not a difference can be measured. If a large number of trained listeners are unable to hear a difference in a proper test between, say, a converter using its own clock versus an outboard clock, there's a pretty good chance you won't be able to hear a difference either.

Psychoacoustic Effects

Psychoacoustics is the field that studies human perception of sound, which is different from the physical properties of sound. For example, playing music very loudly makes it sound sharp in pitch because the eardrum and its supporting muscles tighten, raising their resonant frequencies. This can be a real problem for singers and other musicians when using earphones at loud volumes in a recording studio. There are many examples of audio perception illusions on the Internet, not unlike optical illusions, so I won't repeat them here. But it's worth mentioning a few key points about hearing perception.

One important principle is the *Haas Effect*, which is closely related to the *Precedence Effect*. When two versions of the same sound arrive at your ears within about 20 milliseconds of each other, the result is a slight thickening of the sound rather than the perception of a separate echo. A comb filtered frequency response also occurs, as explained earlier. Further, if the two sounds arrive from different directions, the sound will seem to come from the direction of whichever source arrives first. This location illusion occurs even if the latter sound is as much as 10 dB louder than the first sound. You won't perceive the delayed sound as an echo unless it arrives at least 25 milliseconds later.[3] For this reason, PA systems in auditoriums and other large venues often use delay lines to delay sound from the speakers farther from the stage. As long as people in the rear hear sound first from the speakers close to the stage, that's where the sound will seem to come from. Even though the speakers farther back are closer and louder, they won't draw attention to themselves.

In the 1970s, a product called the Aphex Aural Exciter was introduced for use by recording studios. This device claimed to improve clarity and detail in a way that can't be accomplished by boosting the treble with an equalizer. At the time, Aphex refused to sell the device, instead *renting* it to studios for use on albums. The company charged a per-minute fee based on song length. The Aural Exciter was used on many important albums of the time, most notably Fleetwood Mack's *Rumours*. The device's operation was shrouded in mystery, and at the time Aphex claimed phase shift was at the heart of its magic. Sound familiar? I recall a review in *Recording Engineer/Producer Magazine* at the time that included a graph showing phase shift versus frequency, which, of course, was a ruse. Eventually it was revealed that what the device really does is add a small amount of trebly distortion, above 5 KHz only.

In principle, this is a very clever idea! For program material that has no extended high frequencies at all, using distortion to synthesize new treble content could add sparkle that can't be achieved any other way. Even when high frequencies are present, applying substantial EQ boost will increase the background hiss, too, which is a real problem when using analog tape. I consider adding trebly distortion a psychoacoustic process because it seems to make music sound clearer and more detailed, even though in truth clarity is *reduced* by the added distortion. To illustrate the effect, I created a patch in SONAR that mimics the Aural Exciter, shown in Figure 3.15.

I put the same tune on two tracks, then inserted a high-pass filter set for 5 KHz at 12 dB per octave on the second track. The Sonitus EQ I use offers only 6 dB per octave, so I used two bands to get a steeper slope. The EQ was followed by a distortion plug-in. You can hear short excerpts in the files "light_pop.wav" and "light_pop_distorted.wav." Track 2 is mixed in at a very low level—around 20 dB softer—to add only a subtle amount of the effect. But it really does make the high-hat snap and stand out more clearly.

Some years after the Aphex unit was released, BBE came out with a competing product called the Sonic Maximizer. This unit also claims to increase clarity via phase shift, though when I tried one, it sounded more like a limiter that affects only treble frequencies. What bothered me at the time was BBE's claim that clarity is increased by applying phase shift. According to their early literature, the Sonic Maximizer counters the "damaging phase shift" inherent in all loudspeaker crossovers. But speakers use wildly different crossover

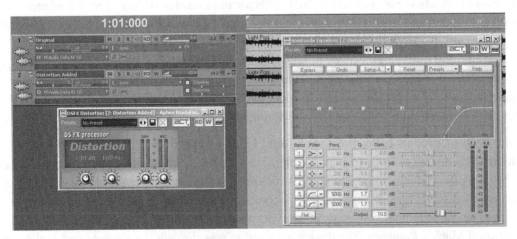

Figure 3.15: To emulate an Aphex Aural Exciter, the same stereo mix is placed on two tracks, with high-frequency distortion added to the second track at a very low level.

frequencies, and some speakers are two-way designs with a woofer and tweeter, while others have midrange drivers with three or even four bands. It's impossible for phase shift of one amount and frequency to do what's claimed for all speakers. So even if the BBE device does add phase shift, that's not the effect you actually hear.

I'm convinced that the seeming increase in clarity after adding small amounts of distortion is the real reason some people prefer the sound of LP records and analog tape. But while small amounts of distortion might be subjectively pleasing when added to some types of music, it's certainly not higher fidelity. I find it surprising and even amusing that recording engineers who are highly vocal about preferring analog tape and tubes are often highly critical of digital converters, claiming that none are transparent enough.

I've also noticed many times that music sounds better when accompanied by visuals. In the 1960s, Joshua White used colored oil and other liquids over plastic film to project psychedelic patterns onto large screens at live rock concerts. MTV launched their cable channel in the 1980s, and music videos became an instant hit. Today, both classical and pop music concerts on DVD and Blu-ray are hugely popular. I've theorized that music sounds better with visuals because our perception is split between the two senses, and the "criticism" part of each sense is reduced slightly. Merely closing your eyes while listening to music can change the way it sounds.

Finally, no discussion of hearing and perception would be complete without a mention of mixing music after drinking alcohol or taking drugs. Alcohol suppresses hearing sensitivity, forcing you to raise the playback level to achieve the sensation of a satisfying volume. In the long term this can damage your hearing because your ears are physically harmed by the louder volume, even if your senses perceive a softer level. The loss of hearing sensitivity after drinking ranges from 2 to 7 dB,[4] depending on frequency and sex (male versus female). In my experience, alcohol also dulls the senses, making it more difficult to discern fine detail. Aspirin also affects high-frequency perception and is best avoided when mixing.

Marijuana tends to have the opposite effect, often making your hearing *more* sensitive such that soft instruments and details become clearer. But this isn't useful either because important elements in a mix can end up too soft if you hear them too clearly. Further, different strains of marijuana vary quite a lot, from making you sleepy to fidgety—versus alcohol, which has the same basic effect whether it's beer or wine, or clear or brown liquor. So while you can probably learn to mix under the influence of drugs or alcohol, there are risks involved. If nothing else, it will surely take you longer to complete a mix!

The interaction between our hearing "hardware" and brain "software" that interprets sounds is complex and not fully understood. We recognize familiar voices by their fundamental pitches, inflection, timing, vocal tract formants,[5] and the frequency ratio of the fundamental pitch to those formants. This ear/brain interaction also lets us easily tell whether a song on the radio is an original or a remake, often after only the first few notes. But it's important to distinguish between perception, which varies over time and from one person to the next, versus fidelity, which sends the exact same acoustic waves from your speakers each time you press Play.

Placebo Effect and Expectation Bias

Everyone understands and accepts that the placebo effect is real, but for some reason audiophiles think it never happens to them.

Audiophile magazines and web forums are filled with anecdotes about perceived improvements after applying various tweaks. People often think they heard an improvement after replacing a wire or precisely leveling their CD player. But it's more likely that they simply became more familiar with the music after repeated playing and noticed more details. What many listeners overlook is that human hearing is fragile and short-term. If you play a piece of music, then spend five minutes replacing your speaker wires and listen again, it's very difficult to recall the earlier playback's tonality. Is that violin section really brighter now, or does it just seem that way? Every time you play a recording you might hear details you missed previously.

According to former DTS chief scientist James Johnston,[6] hearing memory is valid for about 250 milliseconds. James (he prefers JJ) also explains that we cannot focus on everything in a piece of music all at once; on one playing we might notice the snare drum but ignore the rhythm guitar, and so forth. This makes it very difficult to know if subtle differences are real or imagined. If you play the same section of music five times in a row, the sound reaching your ears will not change unless you move your head, but the way you hear and *perceive* each playing can and will vary.

Psychological factors like expectation and fatigue are equally important. If I brag to a friend how great my home theater sounds and that person comes for a visit, it always sounds *worse* to me while we're both listening. I recall a forum post where a fellow recounted recording several guitar tracks through various amp simulators—hardware or software that emulates the sound of an amplifier and loudspeaker. He said that many of his guitar player friends hate amp sims with a passion, so when he played them his tracks, he told them they were real guitar amps and asked them to pick which they liked best. After they stated their preferences, he told them all the tracks were recorded through amp sims. They were furious. It seems to me they should have thanked him for the education about bias.

I've found that mood is *everything*, both when assessing the quality of audio gear and when enjoying music. When you feel good, music sounds better. When you feel lousy, music *and* audio quality both sound poor. Further, our ears easily acclimate to bad sound such as resonances or a poor tonal balance, especially if you keep raising the volume to make a mix you're working on sound clearer and more powerful. But the same mix will probably sound bad tomorrow.

> *Argument from authority is a common logical fallacy. Just because someone is an expert in one field does not make them an expert in other fields, no matter how famous they may be. And even in their own field, experts are sometimes wrong.*

It's not uncommon for professional mixing engineers who are very good at their craft to believe they also understand the science, even when they don't. Knowing how to turn the knobs to make music sound pleasing is very different from understanding how audio works at a technical level. The same applies to musicians. I love guitarist Eric Johnson, and nobody is more expert than he when it comes to playing the guitar. But he is clearly mistaken with this belief, from a 2011 magazine interview:

> *I once read a story about a guitar player who didn't like his instrument. Rather than changing pickups or configuration, he decided to just will it into sounding different as he practiced. And after months and months of playing, that guitar did in fact sound totally different. I believe that story.*

I'm sure the guitar did sound different after the guy practiced for many months. But any change in sound was obviously from his improved playing, not a physical change to the guitar due to wishful thinking. This is not an uncommon belief. I know a luthier who truly believes she can make her violins sound better just by willing them to do so.

There's also the issue of having spent a lot of money on something, so you expect it to be better for that reason alone. If I just paid $4,500 for a high-end CD player and someone told me I could have gotten the exact same sound from a $50 CD Walkman®, I'd be in denial, too. It's also important to consider the source of any claim, though someone's financial interest in a product doesn't mean the claims are necessarily untrue. But sometimes the sound reaching your ears really does change, if not for the reasons we think. Which brings us to the following.

When Subjectivists Are (Almost) Correct

Frequency response in a room is highly position dependent. If you move your head even one inch, you can measure a real change in the response at mid and high frequencies. Some of this is due to loudspeaker beaming, where different frequencies radiate unequally in different directions. But in rooms without acoustic treatment, the main culprit is comb filtering caused by reflections.

Chapter 2 busted a number of audio myths, yet audiophiles and even recording engineers sometimes report hearing things that defy what is known about the science of audio.

For example, some people claim to hear a difference between electrically similar speaker cables. When pressed, they often say they believe they can hear things that science has not yet learned how to measure. But modern audio test equipment can measure everything known to affect sound quality over a range exceeding 100 dB, and it's difficult, if not impossible, to hear artifacts only 80 dB below the music while it is playing. In my experience, the top 20 to 30 dB matter the most, even if softer sounds can be heard.

Certainly some of these reports can be attributed to the placebo effect and expectation bias. If you know that a $4 cable has been replaced with a cable costing $1,000, it's not unreasonable to expect the sound to improve with the more expensive model. This applies to any expensive audio components. After all, how could a $15,000 power amplifier *not* sound better than one that costs only $150? Yet tests show repeatedly that most modern gear has a frequency response that's acceptably flat—within a fraction of a dB—over the audible range, with noise, distortion, and all other artifacts below the known threshold of audibility. (This excludes products based on lossy compression such as MP3 players and satellite radio receivers.) So what else could account for these perceived differences?

Through my research in room acoustics, I believe acoustic comb filtering is the most plausible explanation for many of the differences people claim to hear with cables, power conditioners, isolation devices, low-jitter external clocks, ultra-high sample rates, replacement power cords and fuses, and so forth. As explained in Chapter 1, comb filtering occurs when direct sound from a loudspeaker combines in the air with reflections off the walls, floor, ceiling, or other nearby objects. Comb filtering can also occur without reflections, when sound from one speaker arrives at your ear slightly sooner than the same sound from the other speaker.

Figure 3.16 shows a simplification of reflections in a small listening room, as viewed from above. The direct sound from the loudspeakers reaches your ears first, followed quickly by *first reflections* off the nearby side walls. First reflections are sometimes called *early reflections* because in most rooms they arrive within the 20-millisecond Haas time gap. Soon after, *secondary reflections* arrive at your ears from the opposite speakers, and these can be either early or late, depending on how long after the direct sound they arrive. Other early first reflections also arrive after bouncing off the floor and ceiling, but those are omitted in this drawing for clarity. Finally, *late reflections* arrive after bouncing off the rear wall behind you. In truth, reflections from the rear wall can be early or late, depending on how far the wall is behind you. If it's closer than about 10 feet, the reflections will be early. Again, this illustration is simplified, and it omits reflections such as those from the rear wall that travel to the front wall and back again to your ears.

The thin panel on the right-side wall is where an absorber should go to reduce the strength of reflections off that wall. Sound at most frequencies doesn't travel in a straight line like a laser beam as shown here. So while we often refer to placing absorbers at *reflection points*, in

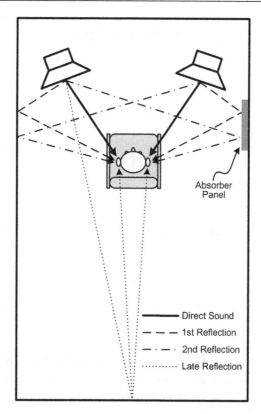

Absorber
Panel

—— Direct Sound

— — — 1st Reflection

— · — · 2nd Reflection

·········· Late Reflection

Figure 3.16: In a typical small room, direct sound from the speakers arrives first, followed by the first reflections, then the second, and finally any late reflections.

practice you'll treat a larger area. How much area requires covering with absorption depends on the distance between your ears and the loudspeakers, and between the speakers and the walls, with greater distances needing larger coverage.

While testing acoustic treatment, I had occasion to measure the frequency response in an untreated room at very high resolution. This room is typical of the size you'll find in many homes—16 by 11½ by 8 feet high. The loudspeakers used were Mackie HR824 active monitors, with a Carver Sunfire subwoofer. The frequency response measurements were made with the R + D software and an Earthworks precision microphone. Besides measuring the response at the listening position, I also measured at a location 4 inches away. This is less distance than the space between an adult's ears. At the time I was testing bass traps, so I considered only the low-frequency response, which showed a surprising change for such a small span.

Conventional wisdom holds that the bass response in a room cannot change much over small distances because the wavelengths are very long. For example, a 40 Hz sound wave is more than 28 feet long, so you might think a distance of 14 feet is needed to create a null

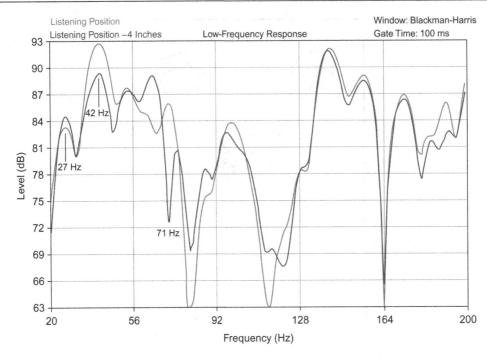

Figure 3.17: This graph shows the low-frequency response in a room 16 by 11½ by 8 feet at two locations 4 inches apart. Even over such a small physical span, the response changes substantially at many frequencies.

180 degrees away. Yet you can see in Figure 3.17 that the large peak at 42 Hz varies by 3 dB for these two nearby locations, and there's still a 1 dB difference even as low as 27 Hz. The reason the frequency response changes so much even at such low frequencies is because many reflections, each having different arrival times and phase offsets, combine at different volume levels at each point in the room. In small rooms, the reflections are strong because all the reflecting boundaries are nearby, further increasing the contribution from each reflection. Also, nulls tend to occupy a relatively narrow physical space, which is why the nulls on either side of the 92 Hz marker have very different depths. Indeed, the null at 71 Hz in one location becomes a peak at the other.

I also examined the same data over the entire audible range, and that graph is shown in Figure 3.18. The two responses in this graph are so totally different that you'd never guess they're from the same room with the same loudspeakers. The cause of these large response differences is comb filtering. Peaks and deep nulls occur at predictable quarter-wavelength distances, and at higher frequencies it takes very little distance to go from a peak to a null. At 7 KHz, one-quarter wavelength is less than half an inch! At high frequencies, reflections from a nearby coffee table or high leather seat back can be significant.

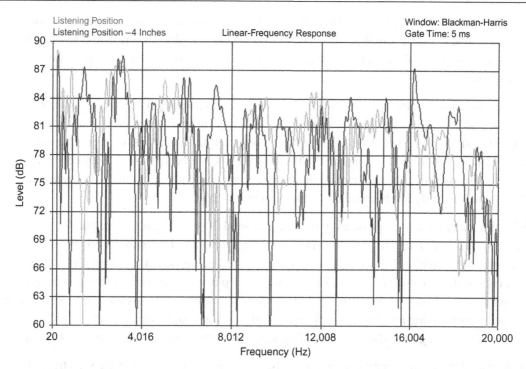

Figure 3.18: This graph shows the full-range response from the same measurements in Figure 3.17. At high frequencies the response difference 4 inches apart is even more substantial than below 200 Hz.

Because of comb filtering, moving even a tiny distance changes the response by a very large amount at mid and high frequencies. This is especially true in small rooms having no acoustic treatment at nearby reflection points. The response at any given cubic inch location in a room is the sum of the direct sound from the speakers, plus many competing reflections arriving from many different directions. So unless you strap yourself into a chair and clamp your head in a vise, there's no way the sound will *not* change while you listen.

Even when absorbers are placed strategically at reflection points, loudspeaker beaming and lobing[7] contribute to response changes with location. One speaker design goal is a flat response not only directly in front of the speaker but also off-axis. But it's impossible to achieve the exact same response at every angle, so that's another factor.

We don't usually notice these changes when moving around because each ear receives a different response. So what we perceive is more of an average. A null in one ear is likely not present or as deep in the other ear. Since all rooms have this property, we're accustomed to hearing these changes and don't always notice them. However, the change in response over distance is very real, and it's definitely audible if you listen carefully. If you cover one ear, it's even easier to notice, because frequencies missing in one ear are not present in the other ear. You can also hear comb filtering more clearly if you record sound from a loudspeaker near a reflecting boundary with only one microphone, then play that back in mono through both speakers.

I'm convinced that comb filtering is at the root of people reporting a change in the sound of cables and electronics, even when no change is likely (at least when they're not using headphones). If someone listens to her system using one pair of cables, then gets up and switches cables and sits down again, the frequency response heard is sure to be very different because it's impossible to sit down again in exactly the same place. So the sound really did change, but probably not because the cables sound different.

With audio and music, frequencies in the range around 2 to 3 KHz are rather harsh sounding. Other frequencies are more full sounding (50 to 200 Hz), and yet others have a pleasant "open" quality (above 5 KHz). So if you listen in a location that emphasizes harsh frequencies, then change a cable and listen again in a place where comb filtering suppresses that harshness, it's not unreasonable to believe the new cable is responsible for the difference. Likewise, exchanging a CD player or power amplifier might seem to affect the music's fullness, even though the change in low-frequency response was due entirely to positioning.

Large variations in frequency response can also occur even in rooms that are well treated acoustically, though bass traps reduce the variation at low frequencies. Figure 3.19 shows the response I measured at seven locations one inch apart, left-to-right. These measurements were made in my large, well-treated home recording studio, not the same room as the graphs in Figures 3.17 and 3.18.

Since this graph displays seven curves at once, I used third octave averaging to show each response more clearly. Without averaging there is too much detail, making it difficult to see the "forest for the trees," so to speak. Besides a fair amount of bass trapping, the listening position in this room is free of strong early reflections, so variations in the response at higher frequencies are due mainly to other factors.

Since the listening position has little reflected energy, one likely cause of the differences is loudspeaker beaming, which alters the response at different angles in front of the speaker drivers. Another is comb filtering due to the different arrival times from the left and right speakers. The test signal was played through both speakers at the same time, and the measuring microphone was moved in one-inch increments from the center toward the right. As the microphone was moved to the right, sound from the right speaker arrived earlier and earlier compared to the left speaker, and the phase differences also caused comb filtering. So even in a well-treated room, the response at higher frequencies can vary a large amount over small distances. This further confirms that comb filtering is the main culprit, even when a room has sufficient acoustic treatment.

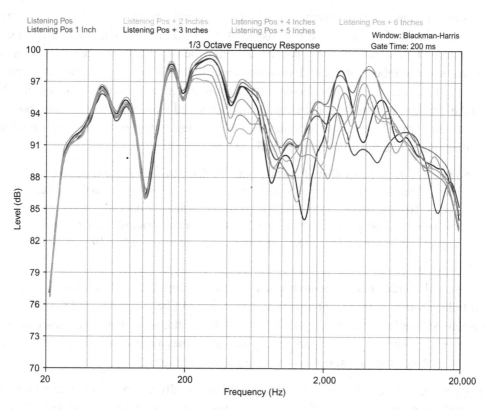

Figure 3.19: This graph shows the full-range response at seven locations one inch apart in a large room well treated with broadband absorption and bass traps. With bass traps, the response at low frequencies varies much less, but the mid and high frequencies still change drastically over very small distances.

It's clear that bass traps make the low-frequency response more consistent across small distances. You can't see that here specifically, because the low frequencies are also averaged in third octaves, which reduces the display resolution. But the seven responses below 200 Hz when viewed at high resolution are similar to the third-octave averaged graphs, so I didn't make a separate graph just for the bass range.

There are other reasons people might think the sound changed even when no change is likely. One is the short-term nature of auditory memory, as mentioned. Crawling around on the floor to change speaker wires will also likely raise your blood pressure, which can affect perception. We also hear differently early in the day versus later when we're tired, and even just listening for a while can change our perception. Does that solid state amplifier really sound different after warming up for half an hour, or is it our perception that changed? Add a couple of cocktails to the mix, and then you really can't tell for sure what you're hearing!

Some people like the sound of certain artifacts such as small amounts of distortion. Preferring the sound of a device that's not audibly transparent is not the same as imagining a change when one doesn't exist, but it's not unrelated. This may be why some people prefer the sound of vinyl records and tube equipment. Everyone agrees that vinyl and tubes sound different from CDs and solid state gear. It's also easy to measure differences in the artifacts each adds. So then the question is, why do some people consider added distortion to be an improvement in sound quality? In a 2006 article for *Sound On Sound* magazine, I wrote about the positive effects of adding subtle distortion intentionally to a recording:

> *Years ago I did a mix in my DAW [Digital Audio Workstation] and made a cassette copy for a friend. I noticed the cassette sounded better—more "cohesive" for lack of a better word—and I liked the effect. A few times I even copied a cassette back into the computer, used a noise reducer program to remove the hiss, then put the result onto a CD. I knew it was an effect, not higher fidelity or the superiority of analog tape, but I had to admit I liked the effect.*

Over the years I've read many arguments in audio newsgroups and web forums. Science-minded folks insist everything that affects audio fidelity can be measured, and things like replacement AC wall outlets cannot possibly affect the sound, no matter what subjectivist "tweakers" claim. The subjectivists argue they're certain they hear a difference and insist that the gear-head objectivists are simply measuring the wrong things.

It now appears that both sides have a point. Some things really are too insignificant to be audible, and sometimes the wrong things are measured, such as total harmonic distortion (THD) without regard to IMD. The room you listen in has far more influence on what you hear than any device in the signal path, including speakers in most cases. It seems reasonable that the one thing neither camp usually considers—acoustic comb filtering—is an important factor.

I've visited the homes of several audiophile reviewers—the same people who write the flowery prose about loudspeaker imaging, and power amplifier presence, and turntable preamps whose sound is "forward" at high frequencies. Sadly, few reviewers have any acoustic treatment at all in their rooms. One professional reviewer I visited had several mid/high-frequency absorbers in his listening room, but none were in the right place! When I moved the panels and showed him how much the stereo imaging improved, he was thankful but not the least bit embarrassed. Think about that.

As you can see, there are many reasons sound quality can seem to change even when it didn't, or when the change is real but due only to room acoustics. It's also possible for real differences to be masked by the listening environment or by a deficiency elsewhere in the playback path other than the component being tested. So after all of these explanations, it's worth mentioning that sometimes replacing one competent cable with another really does change the sound. This can happen when connectors are unplugged, then reseated.

RCA connectors in particular tend to corrode over time, so removing a cable and plugging it in again can improve the sound because the contacts were scraped clean. But this happens even when using the *same* cable. Spraying both the male and female connectors occasionally with contact cleaner is also a good idea.

Summary

When I was a teenager in the 1960s, I loved the *Twilight Zone* and *Outer Limits* TV shows. In later years *The X-Files* was very popular. The appeal of the unknown and mysterious is as old as time. Many people *want* to believe there's more "out there" than meets the eye. Some of us like to believe there's life on other planets. And we're very entertained by stories about unusual ideas and events. So it only follows that people want to believe there's some mysterious force that prevents a double-blind test from revealing subtle differences that can be appreciated only over time. Likewise, the belief that "science" has not yet found a way to measure what they are sure they can hear is appealing to some people. To my mind, this is the same as believing in the supernatural. There's also some arrogance in that position: "I don't need no stupid 'science' because I know what I hear."

As we have seen, what affects the audibility of distortion and other artifacts is their magnitude. If the sum of all artifacts is too soft to hear, then their specific spectrum is irrelevant. The masking effect coupled with the Fletcher-Munson curves often reduces the artifact audibility further compared to isolated sounds at disparate frequencies. But sometimes the sound really did change, due to comb filtering caused by acoustic reflections in the room.

Like the Emperor's New Clothes, some people let themselves be conned into believing that a higher truth exists, even if they cannot personally hear it. It's common to see someone in an audio forum ask if this or that will make a difference in sound quality. When others reply, "Why don't you try it for yourself?," it's common for the original poster to reply that he fears his hearing isn't good enough to tell the difference, and he doesn't want to be embarrassed when others hear the failings in his recording. There's no disputing that hearing can improve with practice, and you can learn to recognize different types of artifacts. But that's not the same as imagining something that doesn't exist at all. And, logically speaking, just because a large number of people believe something does not alone make it true.

Once we understand what really affects audio fidelity, it seems pointless to fret over tiny amounts of distortion in an A/D converter or microphone preamp when most loudspeakers have at least ten times more distortion. And compared to listening rooms that really do have an obvious effect on fidelity, any difference between two competent audio devices is far less important. In the end, the purpose of audio is to present the music. A slamming mix of a great tune played by excellent musicians will sound terrific even on AM radio or a low bit-rate

MP3 file. What makes a mix sound great has much more to do with musicianship and overall frequency balance than a lack of low-level artifacts. Room buzzes and rattles are often even worse. If you slowly sweep a loud sine wave starting around 30 or 40 Hz, you will likely hear rattles at various frequencies as windows and other furnishings resonate sympathetically. Those rattles are present and audible when excited by certain bass notes, but they're often masked by the midrange and treble content in the music.

Ultimately, people on both sides of the "subjectivist versus objectivist" debate want the same thing: to learn the truth. As I see it, the main difference between these two camps is what evidence they consider acceptable. Hopefully, the audio examples presented in this chapter helped you determine for yourself at what level artifacts such as truncation distortion and jitter are a problem. And just because something is very soft doesn't mean it's not audible or detrimental. My intent is simply to put everything that affects audio quality in the proper perspective.

Notes

1 "The Audibility of Frequency Response Irregularities" by R. Bücklein, published in the *Journal of the Audio Engineering Society*, 29(3), 1981 March. www.aes.org/e-lib/browse.cfm?elib=10291

2 "Inaudible High-Frequency Sounds Affect Brain Activity: Hypersonic Effect" by Tsutomu Oohashi et al., published in the *Journal of Neurophysiology*, 83: 3548–3558, 2000. www.aes.org/e-lib/ browse.cfm?elib=10005. "Detection threshold for tones above 22 kHz" by Ashihara Kaoru and Kiryu Shogo, published in the Audio Engineering Society's 110th Convention May 12–15, 2001.

3 How much later an echo must arrive to be heard as a separate sound depends on the nature of the sound. A brief transient such as a snare drum rim shot can be heard as an echo if it arrives only a few milliseconds after the direct sound. But with sounds that change slowly over time, such as a slow, sustained violin section, a delay might not be noticed until it arrives 50 or even 80 milliseconds later.

4 "The acute effects of alcohol on auditory thresholds" by Tahwinder Upile et al., published as Open Access at the web address posted for note vii. www.ncbi.nlm.nih.gov/pmc/articles/PMC2031886/

5 Formants are part of the complex acoustic filtering that occurs in our vocal tracts, and it's described in Chapter 10.

6 "Audio Myths Workshop" by Ethan Winer, www.youtube.com/watch?v=BYTlN6wjcvQ www.aes.org/events/119/press/pr_papers.cfm

7 Beaming and lobing refer to the dispersion pattern of loudspeakers, where various frequencies are sent in varying strengths at different angles. This will be explained further in Chapter 18.

CHAPTER 4

Gozintas and Gozoutas

Years ago, in the 1960s through 1980s, I designed audio circuits both professionally and as a hobby. One day I went to the local surplus electronics store to buy connectors for some project or other. This place was huge and sold all sorts of used electronics devices, wire, and component parts. The two guys who owned the shop—John and Irv—were colorful characters, but they loved what they did and offered great deals. They sold some World War II-era military surplus, but they also had more recent goodies such as used oscilloscopes and other test gear. I once got a great deal on a dozen high-quality audio transformers. When I told John the type of connectors I needed, he said, "Oh, you want some gozintas." I must have looked puzzled for a few seconds, then we both laughed when John saw that I understood. Hence the title of this chapter.

Audio Signals

Audio wiring involves three different issues: signal levels, source and destination impedance, and connector types. We'll consider audio signals first, starting with devices that output very small voltages, then progress to higher levels. In the following descriptions, the term *passive* refers to microphones and musical instrument pickups that have no built-in preamplifier or other electronics.

In most cases, voltage from a dynamic microphone or phonograph cartridge is created via magnetic induction: A coil of wire (or strip of metal) is placed in close proximity to one or more magnets. When either the coil or magnet moves, a small voltage is generated. The more the coil or magnet moves while remaining in close proximity, the larger the voltage that's created. The playback head in an analog tape recorder is also an electromagnetic device; in that case the head contains the coil, and the tape itself serves as the moving magnet. A varying voltage proportional to the tape's magnetism and travel speed is generated in the head's coil as the tape passes by. Piezo guitar and violin pickups work on a different principle: Flexing or squeezing a thin crystal or ceramic plate generates a voltage proportional to the torque that's applied.

Passive magnetic devices such as dynamic and ribbon microphones output extremely small voltages, typically just a few millivolts (thousandths of a volt). The general term for these very small voltages is *microphone level*. Other passive devices that output very small signals include phono cartridges and the playback heads in analog tape recorders. The output level from a typical magnetic guitar or bass pickup is also small, though not as small as the signal from most low-impedance dynamic microphones.

Passive ribbon microphones output even smaller levels, as you can see in Table 4.1. Of course, the exact output voltage at any moment depends on the loudness of sounds reaching the microphone or how hard you strum the strings on a guitar. The number of turns of wire in the coil also affects its output voltage and, simultaneously, its output impedance. For example, the coil in a low-impedance microphone has fewer turns of wire than a guitar pickup, which is high impedance. Some inexpensive dynamic microphones are high impedance, using a built-in audio step-up transformer to increase the output voltage and impedance. Note that low impedance and high impedance are often abbreviated as low-Z and high-Z, respectively.

Because the voltages these devices generate are so small, a preamplifier is needed to raise the signals enough to drive a line-level input on a tape recorder or power amplifier. When dealing with very low signal levels, it's best to place the preamp near to the source. Using a short shielded wire reduces the chance of picking up AC power mains hum or interference from radio stations and nearby cell phones. Depending on the output impedance of the voltage source, using short wires can also minimize signal loss at high frequencies due to cable capacitance that accumulates over longer distances.

Line-level signals are typically around 1 volt, though again the exact level depends on the volume, which can change from moment to moment. There are also two line-level standards: −10 and +4. Chapter 1 explained that signal levels are often expressed as decibels relative to a standard reference. In this case, −10 is actually −10 dBV, where −10 dB is relative to 1 volt. Professional audio gear handles +4 signals, or +4 dBu, which means the nominal level is 4 dB above 1 milliwatt when driving 600 ohms. Some professional gear includes a switch on the rear of the unit to accommodate both standards.

Table 4.1: Output Levels for Passive Transducers

Type	Output Level
Ribbon microphone	0.1 millivolts
Moving coil phono cartridge	0.15 millivolts
Analog tape playback head	2 millivolts
Moving magnet phono cartridge	5 millivolts
150-ohm dynamic microphone	10 millivolts
Fender Precision bass pickup	150 millivolts
Humbucker guitar pickup	200 millivolts
Piezo guitar pickup	0.5 volts

To offer adequate headroom for brief louder bursts of sound, audio gear capable of +4 levels must be able to output up to +18 dBu and even higher. Most such gear can also drive 600-ohm loads, which requires a fair amount of output current. These days, most pro devices don't actually need to drive 600-ohm loads, which originated in the early days of telephone systems. But vintage 600-ohm equipment is still in use, so pro gear usually has that capability. Consumer audio uses a lower level, with no need to drive a low-impedance input, hence the −10 label. Driving a +4 dBu signal into 600 ohms requires a more substantial power supply than sending −10 dBV to a high-impedance load. So the lower −10 level is used with home audio equipment, mainly to reduce manufacturing costs.

Speaker-level signals are much larger than line levels. Indeed, a hefty power amplifier can put out enough voltage to give you a painful shock! For example, a power amplifier sending 500 watts into an 8-ohm speaker outputs more than 60 volts. Because loudspeakers also draw a relatively large amount of current, the wiring and connectors are much more substantial than for microphone and line-level signals. And that brings us to audio wiring.

Audio Wiring

Microphone and line-level signals generally use the same types of wire: one or two conductors with an outer shield. The cord used for an electric guitar carries one signal, so this type of wire uses a single conductor surrounded by a braided or wrapped shield. Wire that isn't handled and flexed frequently often has a simpler but less sturdy foil shield. Stereo and balanced signals require two conductors plus a shield. Several types of signal wires are shown in Figure 4.1.

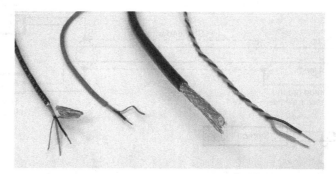

Figure 4.1: Most audio cable has one or two insulated signal wires surrounded by a metal shield to avoid picking up hum and radio interference. From left to right: two-conductor with a foil shield and bare drain wire, single conductor with a wrapped shield, standard coaxial cable ("coax") with a braided shield, and unshielded twisted pair.

Low-level audio cables generally use relatively thin 20- or 22-gauge copper wire because they pass only small amounts of current, and most use stranded conductors rather than a single solid copper core. Wire made from many twisted thinner strands is more expensive to manufacture, but the cable can be flexed many times without breaking, and it also handles better because it's less stiff. Copper conductors are often tin-plated to avoid tarnishing, which would make them difficult to solder after a few years. So even if a bare wire appears silver colored, it's still copper underneath.

Wire that has two active conductors is used for two very different situations: stereo *unbalanced* and mono *balanced*. As you likely know, electricity requires two conductors, with one acting as a return path. A stereo wire used to connect a portable MP3 player to a home receiver has two conductors, plus a third return wire that serves both channels, as shown in Figure 4.2. The return can be either a plain wire or a surrounding shield. Most such wires use a ⅛-inch (or ¼-inch) male phone plug that has a tip, ring, and sleeve, abbreviated TRS. The other end could have either two ¼-inch phone plugs or two RCA connectors, as shown in Figure 4.2. By convention, the tip carries the left channel, the right channel is the ring, and the sleeve carries the common ground connection for both channels.

Two-conductor shielded wire is also used to carry a single balanced channel, as shown in Figure 4.3. Here, the two center conductors carry the signal, and neither signal voltage is referenced to the grounded shield. In this case, the shield serves only to reduce hum and radio interference getting to the active wires within. When used for balanced microphone and line-level signals, XLR connectors are often used, though ¼-inch TRS phone plugs are also common. In that case, the tip is considered the positive connection, and the ring is negative.

Sometimes it's necessary to connect an unbalanced output to a balanced input, or vice versa, for example when connecting balanced gear to a mixer's insert points. Figure 4.4 shows

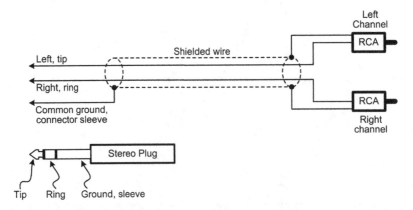

Figure 4.2: With two-conductor shielded wire used for consumer stereos, each conductor carries the active, or "hot" signal for one channel, and the shield carries the return signal for both the left and right channels.

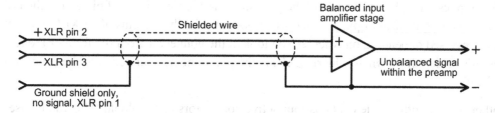

Figure 4.3: When two-conductor shielded wire is used for balanced microphone and line-level signals, the signal voltage is carried by the two active conductors only.

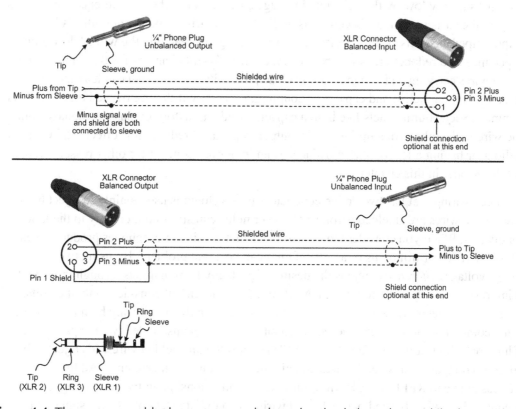

Figure 4.4: These custom cables let you connect balanced and unbalanced gear while also avoiding ground loops.

how to wire custom balanced-to-unbalanced cables that also avoid ground loops. Note that it's best to initially connect the shield at both ends. But if this results in hum or buzz from a ground loop, you can disconnect the shield at the *receiving* end as shown. Also note that some balanced connections use a ¼-inch TRS phone plug instead of an XLR connector. The same wiring applies, though you'll connect the plus signal to the tip instead of pin 2, and the minus signal to the ring instead of pin 3, as shown at bottom. Finally, the pins of an XLR connector may be reversed depending on whether you look at the connector from the front or the rear where it's soldered. So when soldering, check the pin numbers stamped onto the connector carefully!

Another type of audio cable contains four active conductors plus a shield. This can be used to transmit balanced stereo signals down one cable or can be arranged in a "star quad" configuration that offers slightly greater hum rejection when used for one balanced channel. Microphone and line-level signal wires are always shielded, but earphone and speaker wires don't need a shield because the signals are larger, and the output impedance of power amplifiers is very low. With large signal voltages, any hum or other interference that arrives through the air is very small in comparison. Further, the driving power amplifier's low output impedance acts as a short circuit to airborne signals reaching the wire. A low output impedance also reduces cross-talk between the left and right channels for the same reason. When wires carrying audio signals are in close proximity, one channel can leak into the other via both inductive and capacitive coupling, especially if they're twisted together, as is common. Such coupling acts like both a capacitor and a transformer, passing signals from one wire to the other. An amplifier's low output impedance reduces this effect. The value of audio gear having a low output impedance is an important concept for other reasons, too, as you'll see later in this chapter.

Balanced wiring that has two active conductors plus a shield is used mainly to reject hum. Even when wires are shielded, strong AC power fields can still find their way to the inner conductors. When you consider the tiny voltages that passive microphones produce, even a few microvolts (millionths of a volt) of hum can be a real problem. By using two wires whose voltage *difference* contains the desired signal, any hum that gets through the shield is impressed equally onto both wires. A *differential* input circuit considers only the voltage difference between the two signal wires, so it is mostly unaffected by any hum or interference that's common to both wires. The same principle applies to unshielded twisted pair wires. If hum or radio frequency interference (RFI) reaches the unshielded wires, it's rejected by a differential input because both wires contain the same hum or interference. Adding a shield reduces hum and RFI further, but for noncritical applications, plain twisted wiring is often adequate. Likewise, twisted pair balanced wiring also radiates less of its own signal out into the air where it could be picked up by other wires, and shielding reduces such radiation further.

Humbucking guitar pickups are similar in concept to using balanced wiring to reject hum arriving through the air. A guitar pickup is especially prone to hum pickup because its coil acts as a highly efficient antenna at the 60 Hz mains frequency. A humbucking pickup is built from two separate coils, each with its own magnets. The two coils are wired in series, but with the polarity of one coil reversed. When hum in the air reaches both coils, the two signals cancel each other out. To avoid canceling the desired signal from the guitar strings, the magnets in each coil are oriented in opposite directions. That is, where one coil has the north pole at the top closer to the strings, the other coil has the north pole at the bottom. This way the desired signal comes out twice as strong as with only one coil, while the hum is canceled. This is why humbucking pickups are usually louder than single coil pickups. Very clever!

Another important reason to use balanced wiring is to avoid ground loops between two pieces of equipment. When a single conductor shielded cable carries audio, the grounded shield serves as one of the two conductors. In a bedroom studio, all of the gear is likely plugged into a single AC outlet or power strip, so each device has the same ground connection. But in larger installations it's possible for the ground potential (voltage) to vary slightly at different outlets. In theory, a ground should always be zero volts. But voltage losses due to resistance of the power wires in the walls prevents a ground connection from being exactly zero volts. So ground at one outlet may be 2 millivolts but 7 millivolts at another, creating a hum signal of 5 millivolts. Disconnecting the grounded shield at the receiving end of the wire avoids this problem, while still passing the desired signal present in the two hot wires. In this case, 5 millivolts of hum is only 46 dB below a nominal line level of 1 volt.

As mentioned, even a few microvolts of hum can be a real problem at microphone signal levels. Using a balanced input with balanced wiring avoids hum because the desired signal is the difference between the two active conductors, unrelated to the ground voltage. So balanced wiring avoids hum pickup through the air like an antenna and also hum caused by a difference in the ground voltages at the sending and receiving equipment. The spec for how well an input rejects hum common to both signal wires is called its *common mode rejection ratio*, or CMRR, expressed as some number of dB below the difference voltage that contains the desired signal.

Many balanced outputs provide the same signal on both the plus and minus connections, but with opposite polarity. Older devices having output transformers work that way, and active balanced outputs (no transformer) often do this as well. If the plus output on pin 2 of an XLR output is +1 volt and the minus output on pin 3 is −1 volt, the total output between pins 2 and 3 is 2 volts. This provides a 6 dB increase compared to either pin alone, which in turn increases the signal to noise ratio by 3 dB. As explained in Chapter 2, noise is random and doubles by 3 dB, where coherent signals double by 6 dB. So the signal increases 3 dB more than the noise, which of course is useful. However, not all balanced outputs provide an active signal on both wires. Sometimes to reduce manufacturing cost, one output is unpowered and simply remains at 0 volts. However, the signal is still balanced, and the circuit will still reject hum caused by ground loops at the receiving end, as long as both wires have the same impedance when referred to ground. This simpler arrangement is often called *impedance balanced* as opposed to being *truly balanced* with both outputs active and opposite.

Another factor with audio wiring is its capacitance, which becomes progressively like a short circuit between the conductors at higher frequencies. A signal source such as a passive guitar pickup has a high output impedance, which limits the amount of current it can provide. It's equivalent to placing a large value resistor in series with the pickup's output. So for guitar and bass pickups, the capacitance of the connecting wire is very important, unless the guitar contains active electronics to better drive its output signal to the amplifier.

Later sections will cover capacitance in more detail, but briefly for now, a capacitor is similar to a battery. When you put a battery into a charging station, the power supply provides current to charge the battery. The more current that's available, the faster the battery will charge. It's not practical to charge a battery very quickly, because the battery would draw too much current from the power supply and overheat. So a resistor is wired in series to limit the amount of current that can flow. The larger the resistor, the less current that can pass through, and the longer the battery takes to charge. The same thing happens when a high output impedance is coupled with wire having a large amount of capacitance. The available output current from the pickup is limited by its high impedance, so the wire's inherent capacitance can't be charged quickly enough to follow rapid voltage changes, and in turn the high-frequency response suffers.

This is another reason shielded wire is not usually used for loudspeakers. Adding a shield can only increase the capacitance between each conductor and the grounded shield. The amplifier then has to work harder at high frequencies to charge the cable's capacitance at each wave cycle. Low-capacitance wire is important for long lengths with digital audio, too, because the frequencies are at least twice as high as the audio signals represented. That is, achieving a response out to 20 KHz with digital audio requires passing frequencies equal to the sample rate of 44.1 KHz or higher. A digital signal traveling through a wire is actually an analog square wave, so cable capacitance can affect the signal by rounding the steep edges of the waves, blurring the transition between ones and zeros.

Another type of digital audio cable uses fiber-optics technology, which sends the signals as light waves, so it is immune to capacitance and other electrical effects. The S/PDIF format was developed jointly by Sony and Philips—the S/P part of the name—and DIF stands for digital interface. Most people pronounce it as "**spid**-iff." S/PDIF uses either 75-ohm coaxial cable with a BNC or RCA connector or fiber-optic cable with TOSLINK (Toshiba Link) connectors. Fiber-optic cables are sometimes called *light pipes* because light is transmitted down the cable rather than electrical voltages. One huge advantage of using fiber-optics versus copper wire is there's no possibility of ground loops or picking up airborne hum or radio interference. Another advantage is that a single cable can transmit many audio channels at once. However, the downside is that optical cables are limited to a length of about 20 feet unless a repeater is used. S/PDIF is popular for both professional and home audio systems, and many consumer devices include connectors for both coax and fiber-optic cables.

Speaker wires are rarely shielded, but the conductors must be thick to handle the high currents required. For short runs (less than 10 feet) that carry up to a hundred watts or so, 16-gauge wire is usually adequate. I generally use "zip cord" lamp wire for short, low-power applications. But for longer runs, the conductors must be thicker, depending on the length and amount of current the wire will carry. Romex used for AC power wiring is a good choice for high-powered speaker applications up to 50 feet or even longer. Romex is commonly available in 14- to 10-gauge and even thicker, where lower-gauge numbers represent thicker conductors.

I've purposely omitted AC power wiring and connectors because they vary around the world. Plus, AC wires and connectors are mostly self-evident and not complicated. Likewise, there's little point in including FireWire, USB, and CAT5 wiring and connectors as used for audio because a computer or other digital device handles the signals at both ends, and users do not determine the specific voltages or frequencies. For the most part, a digital connection either works or it doesn't.

Audio Connectors

There are many types of audio connectors, but I'll cover only the common types. For example, some devices use proprietary connectors that carry many channels at once over 25 or even more separate pins and sockets. For most connector styles, the "male" and "female" designations are self-evident. Male connectors are also called plugs, though it's not clear to me why female connectors are called jacks.

The ¼-inch *phone plug* and corresponding jack is used for electric guitars, but it's sometimes used with loudspeakers at lower power levels. This type of connector comes in one- and two-conductor versions having only a tip, or a tip and a ring. As with two-conductor wire, the two-conductor phone plug and jack shown in Figure 4.5 are used for both stereo unbalanced and mono balanced signals. They're also used for unbalanced insert points in mixing consoles, to allow routing a mixer channel to outboard gear and back. In that case, the tip is generally used for the channel's output, and the ring is the return path back into the mixer. But you should check the owner's manual for your mixer to be sure.

Phone plugs are also available in 3.5 mm (⅛ inch) and 2.5 mm sizes for miniature applications. The 3.5 mm stereo plug in Figure 4.6 was wired as a stereo adapter, with separate ¼-inch left and right channel phone plugs at the other end to connect an MP3 player to a professional mixer. The smaller 2.5 mm type (not shown) is commonly used with cell phones and is also available as mono or stereo. In fact, some 2.5 mm plugs and jacks have three hot conductors, with two for the left and right earphones, plus a third for a cell phone headset's microphone.

Another common audio connector is the RCA plug (left) and jack (right), shown in Figure 4.7. These connectors are also meant to be soldered to wire ends. RCA connectors are mono only and are used mostly with consumer audio equipment because they're not reliable enough for professional use compared to better connector types. In fact, when the RCA connector was invented in the 1940s, it was meant for use only *inside* televisions sets, so a technician could remove modules for servicing without having to unsolder anything. RCA connectors were never intended for general purpose use! But they caught on and were adapted by the consumer audio industry anyway because they're so inexpensive. RCA connectors are also called *phono* connectors because they're commonly used with phonograph turntables, as opposed to ¼-inch phone connectors that are similar to the ¼-inch plugs and jacks used in early telephone systems.

For completeness, panel-mounted ¼-inch and RCA jacks are shown in Figure 4.8. Note the switch contact on the ¼-inch jack at the right. This is the thin, flat blade with a small dimple that touches the curved tip contact. When a plug is inserted, the connection between the switch blade and tip contact opens. Some RCA connectors have a similar switch that's normally connected to the active tip contact but disconnects when a plug is inserted.

Figure 4.5: The ¼-inch stereo phone plug (left) and jack (right) shown here are meant to be soldered to the end of a shielded wire. The plastic-coated paper sleeve at the left prevents the plug's soldered connections from touching the grounded metal outer sleeve.

Figure 4.6: I wired up this ⅛-inch stereo plug as an audio adapter to connect an MP3 player to my home studio system.

Figure 4.7: The RCA plugs and jacks here are mono only, though some jacks include a switch that opens when a plug is inserted fully.

Figure 4.8: Panel-mounted jacks like these are commonly used for audio equipment. From left to right: stereo ¼-inch jack with switch contacts for each channel, dual RCA jacks for left and right inputs or outputs, and ¼-inch mono jack with a switch.

Whoever first thought to add a switch to a connector was a genius, because it opens up many possibilities. One common use is to disconnect a loudspeaker automatically when an earphone is plugged in, as shown in the schematic diagram in Figure 4.9. This shows only a mono source and speaker, but the same principle can be used with stereo jacks like the one at the left in Figure 4.8. That's why that phone jack has five solder points: one for the common ground, one each for the active left and right channel conductors, plus one each for the left and right switch contacts.

Another useful switch arrangement sends a mono signal to both stereo channels when plugging into only the left channel input jack. This is common with audio mixers that have separate inputs for the left and right channels, as shown in Figure 4.10. Note that switches coupled to jacks are not limited to simple one-point contacts as shown here. One or more switches can be physically attached to a connector to engage several unrelated circuits at once when a plug is inserted.

Another clever arrangement uses a ¼-inch stereo phone jack to automatically turn on battery-powered electronics inside an electric guitar or bass only when the signal wire is plugged in. Rather than require a separate power switch, this method instead uses the ring contact

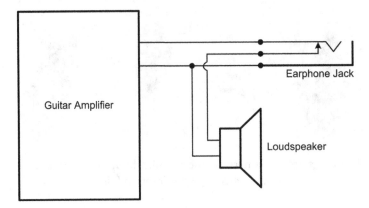

Figure 4.9: When nothing is plugged into the earphone jack, the amplifier's output passes through the jack's switch to the loudspeaker. But when an earphone is plugged in, the switch opens, disconnecting the loudspeaker, and only the earphone receives the signal.

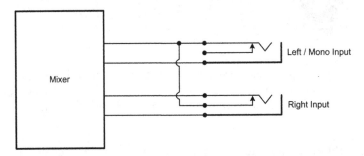

Figure 4.10: When a plug is inserted into only the left channel input, the signal passes through the right channel jack's switch into the mixer, sending the mono signal to both the left and right inputs. But when a plug is inserted into the right channel, the left channel's signal is interrupted, and the right channel then goes to the mixer's right input.

as a ground return for the battery, as shown in Figure 4.11. A standard ¼-inch mono phone plug has a solid metal barrel, so the grounded barrel touches the jack's ring contact when it's plugged in. As far as I know, I was the first person to do this, back in the 1960s, when I designed and built fuzz tones and other gadgets into my friends' electric guitars. Today this is a common feature, not only for electric guitars but also for tuners and metronomes and other devices that have a ¼-inch audio output phone jack.

Speaking of batteries, you can quickly test a 9-volt battery by touching both its terminals at once to your tongue. If the battery is fresh and fully charged, you'll get a mild shock that's unpleasant but not too painful. Do this once when you buy a new battery to learn how a new battery feels.

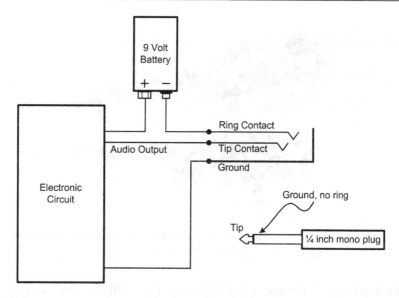

Figure 4.11: When a mono guitar cord is plugged in, the battery's negative terminal is grounded, completing the battery's power connection to the circuit.

There are also ¼-inch plugs available with built-in switches. These are used with guitar cords, with the switch set to short out the audio until the plug is fully inserted into the guitar. This avoids the loud hum that otherwise occurs when a live guitar cord is plugged or unplugged. In this case, the switch is mounted in the outer metal barrel, activated by a small protruding plunger.

It's worth mentioning two variants of the ¼-inch phone plugs and jacks that are used in some studio patch bays: the *long frame*, which is also ¼ inch, and the *bantam TT* (tiny telephone), which is 0.173 inch (4.4 mm) in diameter. Be aware that the tip of a ¼-inch long frame plug is smaller than the usual phone plug, so plugging a regular phone plug into a long frame jack can stretch the jack's tip contact if left in for an extended period of time. Do this only when needed in an emergency.

The last type of low-voltage audio connector we'll consider is the XLR. In the early days of audio these were called Cannon connectors, named for the company that invented them. Cannon called this type of connector its X series, and then later added a latch (the "L" part of the name) so a plug won't pull out by accident. Today, XLR connectors are produced by several manufacturers, such as those in Figure 4.12 made by Neutrik. The most common type of XLR connector has three pins or sockets for plus, minus, and ground. By convention, pin 2 is plus, pin 3 is minus, and pin 1 is ground. But XLR connectors having four, five, and even six contacts are also available.

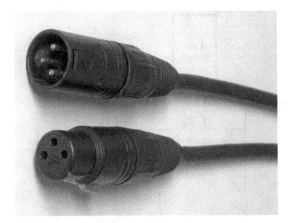

Figure 4.12: XLR connectors are commonly used for both microphone and line-level signals, and most have three pins for plus, minus, and ground.

The standard today for XLR connector wiring is EIA (Electronic Industries Alliance) RS-297-A, which defines pin 2 as plus, or hot. But some older gear treats pin 3 as plus and pin 2 as minus. With most gear it doesn't really matter which pin is plus internally as long as the audio arrives at the same numbered output pins with the same polarity. But it can matter with microphones, especially when different models are in close proximity on a single source such as a drum set.

No discussion of XLR connectors would be complete without a mention of Neutrik's fabulous combination XLR/phone jacks shown in Figure 4.13. Besides accepting standard 3-pin XLR plugs, these connectors also handle ¼-inch phone plugs. The Neutrik website lists 24 different versions, with both mono and stereo phone jacks and many varied switch combinations. The PreSonus FireBOX in Figure 4.13 uses these combo connectors to accept either microphones or electric guitars and other passive instruments through the same input jacks. When a microphone is plugged in, the preamp presents an appropriate input impedance and suitable gain range. But when a phone plug is inserted, switches built into the connector change both the impedance and gain range to suit instruments with passive pickups.

Finally we get to loudspeaker connectors. The banana jacks shown in Figure 4.14 have been a staple for many years because they can handle enough current to pass hundreds of watts. When used with matching banana plugs, their sturdy design and stiff spring contacts ensure a reliable connection that won't easily pull out by accident. Most banana jacks also accept bare wire, which is common, and are secure enough for home stereo installations.

More recently, the speakON connector shown in Figure 4.15, also developed by Neutrik, has become the standard for high-power professional loudspeaker applications. SpeakON connectors can handle very large currents, and they feature a locking mechanism that's reliable enough for professional use.

Figure 4.13: These Neutrik combo XLR connectors also accept ¼-inch phone plugs, making them ideal for small devices like this PreSonus sound card that has little room to spare on its front panel.

Figure 4.14: Banana jacks are commonly used for both professional and consumer loudspeaker connections. Banana jacks are often combined with *binding posts* to accept bare wire. *Photo courtesy of parts-express.com.*

Figure 4.15: SpeakON connectors are ideal for connecting power amplifiers to loudspeakers in professional applications. *Photos courtesy of Neutrik (UK) Ltd.*

Patch Panels

Patch panels—also called patch bays—are the center of every hardware-based recording studio. Originally developed for telephone systems, where calls were routed manually by operators from one phone directly to another, today patch panels allow connecting outboard audio gear in any conceivable combination. The basic premise is for every piece of audio equipment in the studio to have its input and output connected to a patch panel that's centrally located, usually in a nearby rack cabinet. This way, short patch cords can be used to connect the various devices, rather than running long wires across the room to reach a distant equipment rack.

In the old days, patch panels were hardwired to each piece of equipment, which required a lot of soldering for a major installation. When I built a large professional recording studio in the 1970s, it took me several days to solder the wires to connect all the outboard gear, and console inputs and outputs, to five 48-jack patch panels. Back then, pro audio gear often used screw-down terminal strips for input and output connections, so each piece of equipment required soldering one end of a wire to the patch panel, and soldering terminals at the other end to attach to each device's terminal strip. Audio gear has come a long way since then!

Today, modern patch panels have matching pairs of jacks on the front and rear, so no soldering is required unless you want to wire your own cables to the custom lengths needed behind the rack. A typical modern patch panel is shown in Figure 4.16. However, the downside is you have to buy pre-made wires for every input and output of each device you want to connect. And with twice as many connections, there's more chance a loose plug will drop out or cause distortion.

Patch panels are often wired in a configuration known as *normalled*, or optionally *half-normalled*. When jacks are normalled, that means they have a default "normal" connection

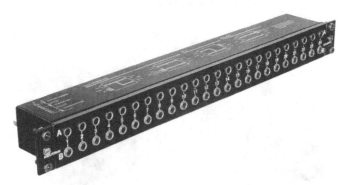

Figure 4.16: This Neutrik patch panel has 24 pairs of stereo ¼-inch phone jacks. Each pair can serve as a left and right stereo input or output, or as an input and output for one balanced or unbalanced mono channel.

even when nothing is plugged in. For example, it's common to patch a console's Reverb Send to an equalizer (EQ) before it goes on to the reverb unit, perhaps to reduce low frequencies that could muddy the sound or just for general tone shaping. This is shown in Figure 4.17. If you want to also insert a compressor before or after the equalizer, you'll simply patch that in manually. But nothing needs to be patched to have only the EQ in the path.

As you can see, the console's Reverb Send goes to the patch panel for ready access, and the switches for both the Send output and EQ input are connected. With nothing plugged into either jack, the Send goes to the equalizer. Likewise, the EQ's output is normalled to the reverb unit's input. Plugging anything into any of the jacks interrupts the normalled connections, letting you freely create other patch arrangements. But what if you want to route the Reverb Send to two places at the same time—say, to the EQ input as well as somewhere else? This is the purpose of the half-normalled switching shown in Figure 4.18.

With half-normalling, the normalled connection is broken only if you plug into a device's input. Plugging into a device's output leaves the normalled path intact, so you can patch that output to another piece of gear at the same time, or perhaps back into a console input. When one output is sent to more than one input, that's called a *mult*, short for multiple destinations. When I owned a professional recording studio, I wired several groups of four jacks together in parallel to be able to send any one device to as many as three destinations. This type of

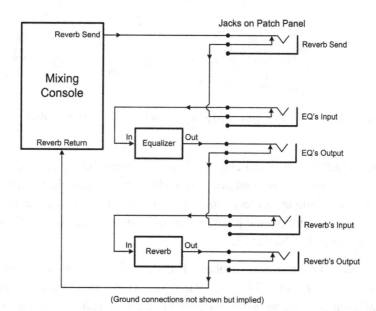

(Ground connections not shown but implied)

Figure 4.17: This drawing shows normalled patch bay connections, where the output of a console goes to an outboard equalizer, then on to a hardware reverb unit. Plugging a patch cord into any of the input or output jacks breaks the normalled connection. The ground wires are omitted for clarity.

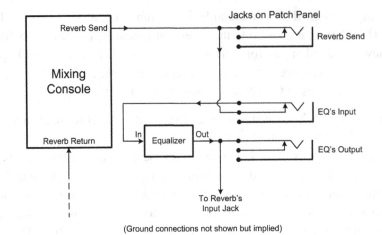

Figure 4.18: With a half-normalled connection, only the input jack switches are used.

Figure 4.19: Stick-on wire labels may be low tech, but they're incredibly useful.

mult does not use a jack's built-in switches. It simply connects the active conductors of three or more jacks together, so you can send one output signal to multiple destinations as needed using patch cords. You could even create line-level *pads* within a patch panel to reduce overly loud signals by wiring resistors to the jacks instead of connecting them directly. Creating resistor pads is explained in Chapter 23.

The last wiring topic I'll address is wire labels. These are simple strips of adhesive tape with preprinted numbers shown in Figure 4.19 that you wrap around the ends of a wire. In a complex installation comprising dozens or even hundreds of cables, it can be a nightmare to track down which wire goes where when something stops working. By applying labels to both ends of every wire during installation, you always know what a wire connects to at the other end. You should also create a printed list of each source and destination with its wire number. Another option is to use write-on labels instead of simple numbers, using a permanent marker to write the name of the gear connected at the other end.

Impedance

The main difference between impedance and simple resistance is that impedance implies a frequency component known as *reactance*. Whereas a resistor has the same resistance at all audio frequencies, the impedance of a capacitor, inductor, loudspeaker, or power amplifier's output varies with frequency. Sometimes this is just what you want, as when using capacitors to create the filters shown in Chapter 1. But just as often, an impedance that changes with frequency is not useful, such as when it harms the frequency response. Note that the term *impedance* is often used even when it doesn't change with frequency. For example, the input impedance of a preamp that doesn't have an input transformer has the same ohms value at all audio frequencies.

As explained earlier, signals from high-impedance outputs such as passive guitar pickups require low-capacitance wire to avoid high-frequency loss. A passive pickup can output only a small amount of current, and that can't charge a wire's inherent capacitance quickly enough to convey high frequencies. Therefore, using long wires having high capacitance attenuates higher frequencies. Wire capacitance is cumulative, so longer wires have higher capacitance. This is why guitar cords benefit from low capacitance and why it's rare to find pre-made cords longer than about 20 feet. A piezo pickup has an extremely high output impedance, so it benefits from even shorter wires. Preamps designed specifically for use with piezo pickups are common, and they're typically placed close to the instrument. The preamp can then drive a much longer wire that goes on to the amplifier.

Line-level inputs typically have an impedance of around 10K (10,000 ohms), though some are as high as 100K, while yet others are as low as 600 ohms. Inputs meant to accept passive guitar and bass pickup are usually 1M (1 million ohms, pronounced "one meg") or even higher. The need for a high input impedance with guitar pickups is related to the need for low-capacitance wire, but it's not exactly the same. Figure 4.20 shows the electrical model of a typical guitar pickup, including its inherent series inductance. (There's also some series resistance, though that's not shown or addressed in this simplified explanation.) When combined with the amplifier's input impedance, the inductance rolls off high frequencies at a rate of 6 dB per octave. Hopefully, the amplifier's input impedance is high enough that the roll-off starts past 20 KHz, but the wire's capacitance adds a second reactive pole, so the rate is potentially 12 dB per octave. The dB per octave slope at which high frequencies are reduced depends on the particular values of the pickup's inductance, the wire's capacitance, and the amplifier's input impedance. So using high-capacitance wire or too low an input impedance, or both, can roll off high frequencies.

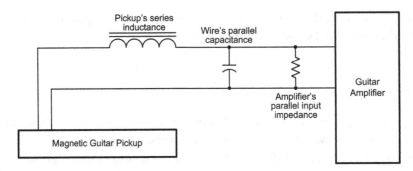

Figure 4.20: A passive guitar pickup has a large inherent inductance because it's basically a coil of wire with many turns. This inductance is effectively in series with the circuit, so it interacts with the capacitance of the wire and with the input impedance of whatever device it's plugged into.

Piezo pickups work best when driving even higher impedances, as well as low capacitance. A cable from a 150-ohm microphone can run to 100 feet or even longer without excessive loss at high frequencies, but a wire connected to a piezo pickup should be 10 feet or less. Unlike a magnetic pickup that has an inherent series inductance, a piezo pickup acts more like it has a series capacitance. So when driving too low an impedance, it's the low frequencies that are reduced. Between their high output impedance that loses high frequencies to wire capacitance and also loses low frequencies with a low load impedance, piezo pickups tend to have an exaggerated midrange that often requires equalization to sound acceptable.

Most modern audio gear has a very low output impedance and fairly high input impedance. Besides reducing the effect of wire capacitance, designing circuits with a low output impedance also allows one device output to feed multiple inputs. Since most audio gear has a high input impedance, you can typically send one output to many inputs without affecting the frequency response or increasing distortion. As mentioned earlier, in the old days professional audio gear had input and output impedances of 600 ohms, based on telephone systems of the day. This arrangement is known as *impedance matching*, and it was done to maximize power transfer between devices. The modern method where a low output impedance drives a high input impedance is better because it reduces noise and potentially improves frequency response, as well as letting one audio device feed several others without degradation. It also wastes less energy because less current is drawn.

Earlier I mentioned wiring a mult from a group of jacks in a patch panel. Using a "Y" splitter to send one output to several inputs is useful and common. But you must never use "Y" wiring to combine two outputs directly together. For that you need a mixer, or at least some resistors to act as a passive mixer. If two low-impedance outputs are connected together, each output acts as a short circuit to the other. At the minimum this will increase distortion, but with a circuit having inadequate protection, this can possibly damage its output stage. The actual output impedance of most audio circuits is less than 1 ohm, but typically a small resistor (10 ohms to 1 K) is wired internally in series with the output connector to protect against a short circuit or improper connection as described here.

The same thing happens if you plug a mono ¼-inch phone plug into a stereo output such as a headphone jack. This shorts out the right channel connected to the ring contact, which may or may not strain the output circuit. Again, well-designed circuits include a small series resistor for protection, but it's possible for a short circuit on one channel to increase distortion on the other because most devices have a single power supply for both channels. It's the same when you connect a balanced output to an unbalanced input with mono phone plugs. If you use stereo phone plugs having a tip and ring at both ends, the ring is simply unconnected at one end and no harm is done. But inserting a mono plug into a balanced output will short out the negative output at the ring. This isn't a problem with older audio gear that has an output transformer, and in fact grounding the transformer's negative output is needed to complete the circuit. But many modern devices use two separate amplifiers to drive the plus and minus outputs. This is also why batteries that power portable devices are never wired in parallel. Unless both batteries have precisely the same voltage, which never happens, each battery shorts out the other battery, quickly draining both and potentially causing damage or even an explosion.

The last impedance issue I'll address is 70-volt loudspeaker systems. These have been around for decades and are used for very large multi-speaker installations, such as stadiums where wires between the power amplifiers and speakers can run to thousands of feet. As explained earlier, when long wires are used to connect loudspeakers, the wires must be thick to avoid voltage loss through the wire's resistance. The formula for this is very simple, where the asterisk (*) represents multiplication:

$$\text{Volts} = \text{amperes} * \text{ohms}$$

In this case, volts is the loss when some amount of current (amperes) flows through some amount of resistance (ohms). With long speaker runs, the resistance can't easily be reduced because that requires using very thick wires, which can get expensive. So 70-volt speaker systems instead use transformers to reduce the amount of current needed to deliver however much power is required to drive the speakers to an adequate volume. The formula for power is equally simple:

$$\text{Watts} = \text{volts} * \text{amperes}$$

As you can see in this formula, a given amount of power can be provided either as a large voltage with a small current, or vice versa. So if the system designer determines that each loudspeaker needs 50 watts to be heard properly over the roar of a cheering crowd, that number of watts can be achieved by sending 20 volts to a typical 8-ohm speaker that in

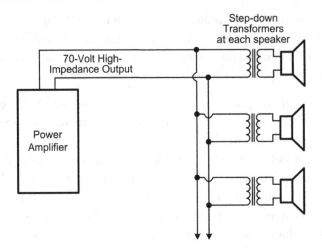

Figure 4.21: A 70-volt speaker system can drive very long wires with little loss because the wire's resistance is low compared to the amount of current each loudspeaker draws.

turn draws about 2.5 amps. But using a 70-volt system as shown in Figure 4.21 reduces the current needed for the same amount of power to only 0.7 amps. When all else is equal, the cost of wire is related to the amount of copper needed to achieve a given thickness, or gauge. So a 70-volt system can yield a substantial savings, even when you factor in the cost of transformers, which are typically inexpensive units meant only for voice frequencies.

By the way, this same principle is used to send AC power over long distances. Sending 120 or 240 volts over hundreds of miles would require extremely thick wires, so for long distances, AC power is sent at 110,000 volts or even higher. Once the power gets from the generator to your town, it passes through transformers that reduce it to a few thousand volts as it travels over utility poles. Then it's reduced again by smaller transformers mounted high up on the poles before entering your home. Large AC power transformers are very expensive, but not as expensive as hundreds of miles of wire thick enough to pass the high current required for lower voltages.

Finally, since my goal for every chapter is to bust at least one audio myth, I'll address the value of using gold in audio connectors. Gold is an excellent conductor, and it's commonly used for circuit board edge connectors and other critical connections. But gold is very expensive, so it's always applied as a thin plating onto another more common metal. Although gold has a fairly low resistance, its real value for electronics is that it doesn't tarnish over time. If you connect a gold-plated plug to a gold-plated socket, the connection will be solid and reliable for many years. Lesser materials that tarnish can not only become intermittent, but the connection points can potentially become diodes or batteries due to oxidization, which in turn creates distortion. Low-level audio signals such as from phono cartridges and analog tape playback heads are especially prone to distortion caused by oxidized connections.

So using gold for audio connectors is a Good Thing. However, gold is most beneficial when used for *both* connectors! If you buy expensive wires with gold-plated RCA connectors, plugging them into an audio device that has tin or nickel connectors loses any advantage of gold. Further, gold plating is very thin—typically measured in microns (1 millionth of a meter, or around 39 millionths of an inch)—so repeatedly plugging and unplugging gold connectors can eventually wear off the plating to expose whatever metal is underneath. Many other, less expensive materials are used to create highly reliable connections, including nickel, brass, and bronze, and these work just as well as gold in most audio applications. Indeed, fancy RCA connectors made from exotic metals are sometimes called "audio jewelry" because they look nice, even if they have little practical value.

Summary

This chapter describes several passive audio sources such as microphones, guitar pickups, and analog tape heads, along with their typical voltage levels. Passive microphones and guitar pickups output very small voltages, requiring a preamplifier to feed devices having line-level inputs. Where most consumer audio equipment uses a nominal level of −10 dBV, professional audio gear operates at a level of +4 dBu. Besides using a higher signal level to overcome noise and hum, most pro audio gear can also drive low-impedance inputs and longer cable lengths without increasing distortion and cross-talk, or losing high frequencies.

You also learned that most audio wiring uses shielded cables to reduce airborne hum and radio interference. Using balanced wiring, with two conductors for each channel, further reduces hum and radio interference, as well as avoiding hum caused by ground loops. Although wire having too much capacitance can roll off high frequencies, this is mostly a problem when long runs are used with passive guitar pickups that have a high output impedance, because guitar pickups don't provide enough current to charge the wire's inherent capacitance quickly enough.

This chapter further explains the various types of connectors used for audio and presents several clever techniques that use the switches built into phone jacks. Further, switches built into some patch panels allow normalled and half-normalled connections, which avoids having to create frequently used connections manually every time with patch cords.

Finally, this chapter explains the basics of impedance. High-impedance magnetic and piezo pickups work best when driving high-impedance inputs to maintain a flat frequency response. Further, most audio gear is designed with a low output impedance and high input impedance, allowing one device output to feed several inputs at once. However, when a system needs to send high-powered audio (or AC power) over long distances, a high impedance is better because larger voltages require less current to transmit the same amount of power.

PART 2

Analog and Digital Recording, Processing, and Methods

This section explains how audio hardware and software devices work, and how to use them. Audio systems are exactly analogous to plumbing, because they both have inputs and outputs, with something flowing through them. The plumbing in a house works in much the same way as audio passing through a mixing console. Where a mixer has a preamp gain trimmer, pan pot, and master volume, the pipes under your sink have a safety shutoff valve in series with the sink's faucet valve. Sliding the faucet handle left and right "pans" the balance between full hot and full cold, or any setting in between. The concept of audio as plumbing applies equally to both analog and digital devices. Of course, a mixing console offers many more ways to route and alter sound, versus plumbing, which can only heat the water and maybe add a filter to remove impurities.

To continue the analogy, voltage is like water pressure, and current is the same as water flow expressed as gallons per minute. The voltage at an AC power outlet is always present, but it's not consumed until you plug something in and turn it on. The pressure at a water valve behaves the same, passing nothing when it's shut off. Then when you open the valve, pressure drives the water down a pipe, just as voltage propels current down a wire. If you can see how the plumbing in a large apartment building is a collection of many identical smaller systems, you'll understand that a complex multi-channel audio mixing console is configured similarly from many individual channels.

Over the years I've noticed that the best mix engineers also take on creative roles, including that of a producer and sometimes even musical arranger. As you'll note in the chapters that follow, mixing music is often about sound design: The same tools and processes are used both for solving specific audio problems and for artistic effect to make recordings sound "better" than reality. So while the main intent of this book is not so much to teach production techniques, I'll explain in detail how the various audio processing tools work, how they're used, and why.

Many of the examples in this section refer to software processors and devices, but the concepts apply equally to analog hardware. Most audio software is modeled after analog hardware anyway, although some software does things that would be very complicated to implement as analog hardware. For example, it's very difficult if not impossible to create after-the-fact noise reduction, linear phase equalizers, and 96 dB per octave filters in analog hardware. Further, a lot of audio hardware these days is really a computer running digital audio software, such as hardware reverb units that are computer-based and do all their processing digitally.

CHAPTER 5

Mixers, Buses, Routing, and Summing

Large mixers are exactly the same as small ones, but with more channels and buses, so at first they can seem complicated. Let's start by examining a single channel from a hypothetical mixing console. Although they're called *mixing* consoles, most also include mic preamps and signal routing appropriate for recording. Figure 5.1 shows the front panel layout of a typical channel, with its corresponding block diagram showing the signal flow presented in Figure 5.2.

A block diagram such as Figure 5.2 shows the general signal flow through an audio device. This is different from a schematic diagram that shows component-level detail including resistors and capacitors, though the switches and variable resistors are as they'd appear in a schematic. In this diagram, arrows show the direction of signal flow. The 48-volt phantom power supply at the upper right in Figure 5.2 isn't part of the audio signal path, but it provides power to the microphone connected to that channel when engaged. Most large consoles include individual switches to turn phantom powering on and off for each channel, though many smaller mixers use a single switch to send power to all of the preamps at once. Phantom power is explained in more detail in Chapter 17.

A microphone is plugged into the XLR input jack on the rear of the console, which then goes to the preamp. The Trim control adjusts the amount of *gain*, or amplification, the preamp provides. Most preamps can accommodate a wide range of input levels, from soft sounds picked up by a low-output microphone, through loud sounds captured by a condenser microphone that has a high output level. The gain of a typical preamp is adjustable over a range of 10 dB to 60 dB, though some microphones such as passive ribbons output very small signals, so preamps may offer 70 dB of gain or even more to accommodate those mics.

Letting users control the preamp's gain over a wide range minimizes noise and distortion. The preamps in some recording consoles from years past couldn't accept very high input levels without distorting, so engineers would sometimes add a *pad* between the microphone and preamp. This is a simple barrel-shaped device with XLR connectors at each end, plus a few resistors inside, to reduce the level from a source such as a loud kick drum when the microphone is placed very close. These days many preamps have a minimum gain low enough to accept very loud mic- or even line-level signals without distorting, or they include a pad that can be engaged when needed. Further, many condenser microphones include a

Figure 5.1: This shows the front panel of a typical mixer channel, with knobs and switches for preamp gain, equalizers, auxiliary sends, and the master fader.

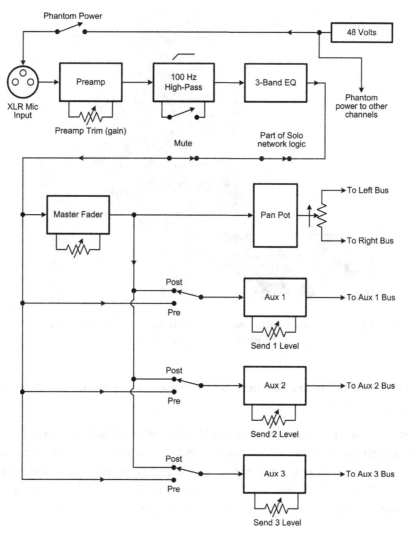

Figure 5.2: This block diagram shows the signal flow within the mixer channel in Figure 5.1.

built-in pad, as shown in Figure 5.3, to avoid overloading their own built-in preamps. So you can use that pad when recording very loud sources.

Continuing on with Figure 5.2, the signal from the preamp goes through a 100 Hz high-pass filter that can be engaged when needed, though I prefer to think of this as a low-cut filter. Either way, reducing low frequencies is very common, so most consoles (and many outboard preamps) routinely include such filters. As you can see in Figure 5.3, it's common for microphones to include a low-cut filter, too. This type of filter is not as flexible as an equalizer because it has a single fixed frequency and a fixed roll-off slope that's often only 6 dB per octave. If you need a different frequency or slope, you'll have to use an equalizer.

Figure 5.3: This Audio-Technica AT4033 microphone has two switches—one for a 10 dB pad and another for a low-cut filter.

But still, simple switched filters like these are often adequate to reduce unwanted rumble or other low-frequency content.

After the low-cut filter, the signal passes through an equalizer. The three-band EQ shown in this hypothetical console channel is typical. While not as flexible or "surgical" as a fully parametric equalizer, it's usually adequate for massaging audio enough to capture a decent recording. The EQ shown here has fixed frequencies for the low and high shelving controls, with a midrange frequency that can be swept over a wide range.

Solo, Mute, and Channel Routing

Next in line are the Solo and Mute switches. The Mute switch is normally closed, so pressing the button on the front panel opens the circuit, which mutes the channel. The Solo switch also mutes the mixer channels, but it's part of a larger network of switches. When engaged, a Solo switch mutes all of the channels *other* than the current channel. So if you think you hear a rattle on the snare drum, for example, solo'ing the snare mic lets you hear only that channel to verify. A solo system is actually quite complex because all of the switches are linked together electronically. Some solo systems disable all Aux sends when activated, though I prefer to hear whatever reverb or echo is active on that track. More elaborate consoles let you solo either way.

Following the Mute switch is the Master Fader, or master volume control for the channel. This is typically a slide control rather than a round knob. A slider lets you make volume changes more precisely—the longer the fader, the easier it is to make fine adjustments. Volume sliders also let you control several adjacent channels at once using different fingers on each hand. After the Master Fader, the signal goes to the pan pot, which sends the single channel to both the left and right outputs in any proportion from full left to full right. By the way, the "pan" in pan pot stands for *panorama*, so the full term (which nobody uses) is panorama potentiometer.

It's worth mentioning that there are two types of large format consoles: those meant for live sound where all of the inputs are sent to left and right main output channels and those meant for multi-track recording. The console shown in Figures 5.1 and 5.2 is the simpler live sound type. A mixing console meant for multi-track recording will have a pan pot on every channel and a bank of switches to let the engineer send that channel to any single track or stereo pair of tracks on a multi-track recorder. A typical Track Assign switch matrix is shown in Figure 5.4.

In Figure 5.4, the left and right channels can be assigned to outputs 1 and 2, or 3 and 4, or 5 and 6, and so forth, as stereo pairs. There are many such arrangements, often using banks of push buttons, but this simplified drawing shows the general idea. When using a console with a multi-track recorder having only 16 or 24 tracks, you may need to record more microphones at once than the number of available tracks. It's not uncommon to use eight microphones or more just for a drum set. In that case you'll premix some of the microphones to fewer tracks when recording. One common setup records the snare and kick drum microphones to their own separate tracks, with all of the other drum mics mixed together in stereo. When a group of microphones is premixed to stereo, the result sent to the recorder is called a *sub-mix*.

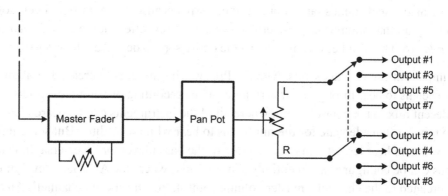

Figure 5.4: This switch matrix lets you send a single channel to any one output or to a pair of outputs for recording the channel as part of a stereo mix. The dashed line indicates that both switch sections are physically connected to change together. For clarity, only 8 outputs are shown here, but much larger switch matrices are common, such as 16, 24, or even 32 outputs.

Of course, when microphones are mixed together while recording, there's no chance to change the balance between them later during mix-down, nor can you change the EQ of some mics without affecting all of the others. Other effects such as reverb or compression must also be applied in the same amount to the entire group.

Buses and Routing

The last section in Figure 5.2 is three Aux Send groups, each having a Pre/Post switch. An Aux output—short for auxiliary output—is an alternate parallel output that's active at the same time as the main output. It's typically used to send some amount of that channel to a reverb or echo unit, which then is added back into the main left and right outputs during mixdown. Most audio effects, such as equalizers and compressors, are patched in series with a channel's signal. It makes little sense to mix a bass or guitar track with an equalized version of itself, as that just dilutes the effect and potentially causes unwanted comb filtering due to slight delays as audio passes through the devices. But reverb and echo *add new content* to a track, so both the original sound and its reverb are typically present together. Chapter 7 explores further the difference between inserting effects onto a track versus adding them to an Aux bus.

Most of the time, when you add reverb to a channel, you'll want the relative amount of reverb to remain constant as you raise or lower the volume for that channel. This is the purpose of the Pre/Post switch. Here, Pre and Post refer to before and after the channel's master volume control. When set to Post, the signal sent to the reverb unit follows the volume setting for that channel. So when you make the acoustic guitar louder, its reverb gets louder, too. But you may want the amount of reverb to stay the same even as the main volume changes—for example, if you want to make an instrument seem to fade into the distance. As you lower the volume, the instrument sounds farther and farther away because the main signal gets softer, and eventually all that remains is its reverb. You can hear this effect at the end of *Pleasant Valley Sunday* by The Monkees from 1967 as the entire song fades away, leaving only reverb.

Another important use for Aux sends, with the Pre switch setting, is to create a monitor mix for the musicians to hear through their earphones while recording or adding overdubs. If you create a decent mix in the control room, where all of the instruments can be heard clearly, that same mix is often adequate for the musicians to hear while recording. But sometimes a drummer wants to hear less of himself and more of the bass player, or vice versa. In that case, an Aux bus can be configured as an entirely separate mix, where the Aux Send level of each track is unrelated to the channel's master volume control. You'll set each channel's Aux Send #1 to approximately the same volume as the main channel volume, but with more or less drums or bass as requested by the performers. A flexible console will let you route any Aux bus output to the monitor speakers, so you can hear the mix you're creating.

The mixing console used for these examples has three separate Aux buses, so up to three different sets of parallel mixes can be created. Or you could use one Aux bus for a reverb unit whose output goes to the main mix you hear in the control room, with the other two Aux buses set up as monitor mixes. Sophisticated mixing consoles allow you to configure many different groupings of channels and buses. Although it's not shown in these simplified block diagrams, when an Aux bus is used for reverb, the output of the reverb unit usually comes back into the console through the same-numbered Aux Return input on the rear of the console. An Aux Return always includes a knob to control the incoming volume level, and most also include a pan pot. However, when used to create a monitor mix, the output of an Aux Bus goes to the power amplifier that drives the studio headphones, rather than to a reverb or echo unit. In that case the corresponding Aux Return is not used, or it could be used as an extra general purpose stereo input.

Most large-format mixers have only mono input channels, but some have stereo inputs that use a single volume fader to control the left and right channels together. In that case, the pan pots are configured to handle stereo sources. Many consoles have both mono and stereo buses, or stereo buses that can be used with mono sources. For example, reverb buses are often set up with a mono send and stereo return. Many reverb units create a stereo effect, generating different left and right side reflection patterns from a mono input source.

Console Automation

Another feature common to sophisticated mixing consoles is *automation* and *scene recall*. A complex mix often requires many fader moves during the course of a tune. For example, it's common to "ride" the lead vocal track to be sure every word is clearly heard. Or parts of a guitar solo may end up a bit too loud, requiring the mix engineer to lower the volume for just a few notes here and there. There are many schemes to incorporate automation into mixing consoles. One popular method uses faders that have small motors inside, as well as sensors that know where the fader is currently positioned. Another method uses a *voltage controlled amplifier* (VCA) rather than a traditional passive slider with an added motor. In that case, a pair of LED lights indicates if the fader's current physical position is louder or softer than its actual volume as set by the VCA.

Scene recall is similar, letting you set up several elaborate mixes and routing schemes, then recall them exactly with a single button push. Modern digital consoles used for live sound and TV production often include this feature. For example, late-night talk shows often have a house band plus a musical guest act with its own instruments. During rehearsal the mix engineer can store one scene for the house band and another for the guest band. It's then simple to switch all of the microphones, and their console settings, from one setup to the other.

When combined with a built-in computer, fader moves that you make while mixing can be recorded by the system, then later replayed automatically. So on one pass through the mix you might ride the vocal level, and then on subsequent playbacks the fader will recreate those fader changes. Next you'll manually control the volume for the guitar track, and so forth until the mix is complete. If the faders contain motors, the sliders will move as the console's computer replays the automation data. As you can imagine, volume faders that contain motors and position sensors are much more expensive than simple passive faders that control the volume.

In the 1970s, I built my own 16-track console, shown in Figure 5.5. This was a huge project that took two years to design and another nine months to build. Back then console automation was new, and rare, and very expensive. So I came up with a clever arrangement that was almost as good as real automation but cost only a few dollars per channel. Rather than try to automate a volume knob with motors, I used two volume controls for each channel, with a switch that selected one or the other. So if the guitar volume needed to be raised during a solo, then lowered again after, I'd set normal and loud levels for each volume control. Then during mixing I'd simply flip the switch when needed, rather than hunt to find the right levels, or mark the fader with grease pencil to identify both positions.

One final point about mixing consoles is that they're set up differently when recording versus when mixing. Some consoles contain two entirely different sections, with one section dedicated to each purpose. That's the design I used for my console in Figure 5.5. But most modern consoles can be switched between recording and mixing modes, a concept developed by David Harrison of Harrison Consoles, and popularized by Grover "Jeep" Harned, who

Figure 5.5: The author built this 16-channel console in the 1970s for a professional recording studio he designed and ran in East Norwalk, Connecticut. Besides this console, Ethan also designed and built much of the outboard gear you see in the rack.

developed his company's MCI consoles in the early 1970s. When a single 24-channel console can handle 24 microphone inputs as well as 24 tape recorder outputs, it can be half the width of a console that uses separate sections for each purpose.

When recording, each input comes from a microphone, or optionally a direct feed from an electric bass or electronic keyboard. The channel outputs are then routed to any combination of recorder tracks, depending on the console's design. Once recording is complete, the console is switched to mixdown mode. At that point, all of the channel inputs come from each recorder track, and the outputs (typically plain stereo) are panned left and right as they're sent on to the power amplifier and monitor speakers. Of course, inputs can be also switched individually to record mode for recording overdubs.

Other Console Features

All recording consoles contain microphone preamps, basic routing, and Aux buses, and many also include a polarity switch for each input channel. But some include a fully parametric EQ on each input channel and Aux bus, which is more flexible than the simple three-band type shown in Figure 5.1. Some high-end mixing consoles even include a compressor on every channel, and optionally on the Aux buses as well. Many large-format consoles also include built-in patch panels that connect key input and output points within the mixer to outboard gear in racks. This is more convenient than having the patch panels in a rack off to the side or behind you.

Almost all consoles include some sort of metering for every channel and bus, and these need to be calibrated to read the same as the VU meters on the connected recording device. This way you can look at the meters in front of you on the console, rather than have to look over to the analog or digital recorder's meters when setting record levels.

Most larger consoles include a built-in microphone and *talk-back system* for communicating with performers. Unlike home studios that usually have only one room, larger studios have a control room where the engineer works, plus a separate acoustically isolated live room where the musicians perform. The recording engineer can easily hear the musicians talk, because one or more microphones are active in the live room. But the performers can't hear the engineer unless there's a microphone in the control room and a loudspeaker out in the live room. A talk-back system adds a button on the console that does two things when pressed: It engages a microphone in the control room (often built into the console), sending it out to loudspeakers in the live room or the earphone monitor mix, and it lowers or mutes the control room monitors to avoid a feedback loop.

Some consoles also include a stereo phase meter to monitor *mono compatibility* to ensure that important mix elements remain audible if a stereo mix is reduced to mono. Most people

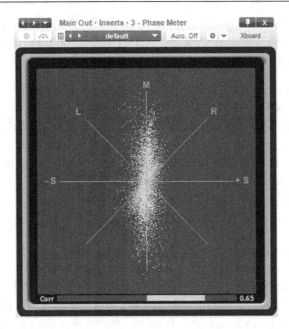

Figure 5.6: A phase correlation meter lets you assess mono compatibility for stereo program material.

listen to music in stereo, but mono compatibility is still important for music that will be heard over AM radio, a mono television, or "music on hold" through a telephone. Figure 5.6 shows a typical software *phase correlation* meter. With this type of display, the shape of the graph pattern indicates mono compatibility. A purely mono source, where both the left and right channels are identical, displays as a vertical pattern similar to what's shown here. The pattern tilts left or right when one channel is louder than the other. But if the pattern expands horizontally, that indicates some of the content common to both channels is out of phase. Such content will therefore become softer or even silent if the stereo channels are summed to mono.

Some consoles include 5.1 surround panning with six or more separate outputs, and this will be described shortly. Finally, some modern digital mixers also serve as a computer "sound card," connecting to a computer via USB or FireWire or a similar connection type. This is very convenient for folks who record to a computer because it combines both elements into a single hardware unit.

Digital Audio Workstation Software and Mixing

Modern audio production software serves as both a multi-track recorder and mixer, modeled to include both functions, and many software versions are even more sophisticated.

Software is certainly more affordable than hardware for a given number of channels and other features. Most modern DAW (digital audio workstation) software includes not only a sophisticated mixing console and multi-track recorder, but also a complete set of effects, including EQ, reverb, compressors, and more. Many DAWs also include software instruments such as synthesizers, electronic drum machines with built-in patterns, and MIDI sample players. Some people consider "DAW" to mean a computer that runs audio software, while others use the term more loosely to mean the entire system, including software, or just the software itself. To my way of thinking, a *workstation* is the complete system of hardware and software, though it's equally proper to consider DAW software and DAW computers separately.

Figure 5.7 shows a flowchart for one channel in Cakewalk's SONAR, the DAW software I use. As you can see, the signal flow is basically the same as a large format mixing console with full automation. It includes faders for Input Trim, Channel Volume and Pan, as many Pre- or Post-fader Aux Sends and Returns as you'll ever need, plus an unlimited number of plug-in effects. Besides automation for all volume and pan levels, SONAR also lets you automate every parameter of every track and bus plug-in effect. So you can turn a reverb on or off at any point in the tune, change EQ frequencies and boost/cut amounts, a flanger effect's sweep rate, and anything else for which a control knob is available on the plug-in. Further, the playback meters can be set pre- or post-fader, with a wide range of dB scales and response times. The meters can also display RMS and peak levels at the same time and optionally hold the peaks as described in Chapter 1. Other DAW programs from other vendors provide similar features.

As you can see, SONAR follows the inline console model described earlier, where most of the controls for each channel are switched between recording and mix-down mode as required. When recording, the input comes from either a physical hardware input—a computer sound card or outboard A/D converter—or from a software synthesizer controlled by a MIDI track. This is shown at the upper right of Figure 5.7. In practice, it's not usually necessary to record the output from a software synthesizer because it creates its sounds in real time when you press Play. In other words, when a synthesizer is inserted onto an audio track, that track plays the synthesizer's audio output. However, it's not a bad idea to record a synthesizer track as a Wave file because, for one reason or another, your next computer may not be able to load and run the same synthesizer in case you want to change something else in the mix.

With regular audio tracks, the source is a Wave file on the hard drive. A Wave file on a track in a DAW program is often called a *clip*, and it might contain only a portion of a larger audio file. Regardless of the source, after passing through all of the volume, pan, and effects stages, and the Aux sends, the output of each channel can be routed to either a stereo or 5.1 surround output bus for playback. The output bus then goes to a physical hardware output, which is either a computer sound card or outboard D/A converter.

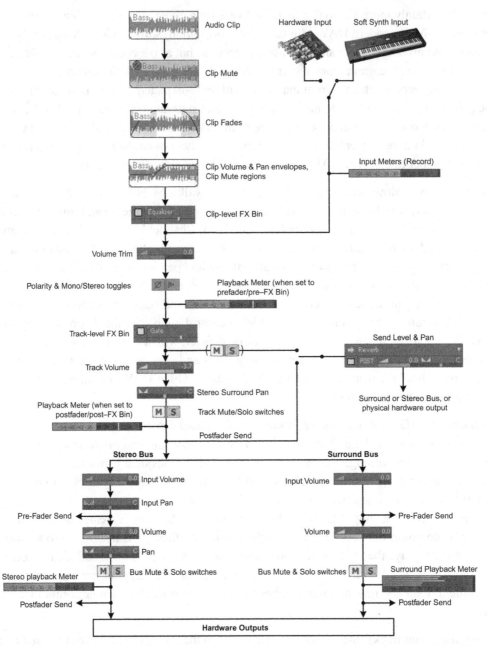

Figure 5.7: This flowchart shows the internal structure of one channel in Cakewalk's SONAR program. As you can see, modern DAW software is functionally equivalent to a complete recording studio.

As with hardware mixers, tracks and channels in modern DAW programs can be either mono or stereo, and all relevant volume and pan controls behave appropriately for either automatically. In SONAR, if you select a single sound card input as the record source, the track is automatically set to mono and records a mono Wave file. If you select a stereo pair of inputs, recording will be in stereo. Newly created tracks default to stereo, but if you import a mono Wave file to that track, it switches to mono automatically. Likewise, importing a stereo Wave file into a mono track automatically changes the track to stereo.

The Pan Law

One final but important aspect of hardware and software mixers is called the *pan law* or *pan rule*. This defines how the volume level of a mono source changes when panned from left to right, through center, in a stereo mix. As you will see, this is a surprisingly complex topic for what might seem like a simple process. Technically, pan law is the premise that says in order to keep the perceived volume constant, the signal level must be reduced by some amount when the pan pot is centered. Pan rule is the specific implementation applied in a mixer or DAW program.

When a mono source is panned fully left or fully right, it emits from one speaker at a level determined by the volume control. But when that same source is panned to the middle of the stereo field by centering the pan pot, it plays through both the left and right speakers. So now the sound is twice as loud. Obviously, it's a nuisance to have to adjust the volume every time you change the panning for a track, so hardware and software mixers automatically lower the volume a little when the pan pot is centered. For positions other than full left or full right, a "curve" is applied to the in-between settings to keep the volume constant no matter where the pan pot is set.

The problem is deciding how much to reduce the volume when the pan pot is centered. Some consoles and DAW programs lower the volume by 3 dB for each side when centered, since that sends half the acoustic power through each loudspeaker. So far so good. But if a stereo mix is summed to mono electrically, as happens with AM radios and other mono playback devices, any instruments and vocals that are panned to the center become twice as loud. That's not so good! If the pan rule instead reduces the level of centered sounds by 6 dB, they may seem soft when played in the control room, but at least the level will not change when summed to mono. Many DAW programs, such as SONAR, let you choose from several pan rules because there's no one best setting. Letting users change the pan rule is more difficult with hardware mixers, so some console manufacturers split the difference, reducing the level by 4.5 dB when centered.

But wait, there's more. Believe it or not, the ideal pan rule also depends on the quality of your monitoring environment. When you play a mono source in a room with no acoustic treatment, you hear a combination of the direct sound from both speakers, plus reflections from nearby room surfaces. Both speakers play the same sound, so their sum is coherent and plays 6 dB louder than just one speaker. But the reflections are probably not coherent due to asymmetry in the room and other factors such as your head not being precisely centered at all times. So when a mono sound common to both the left and right speakers sums acoustically in the air, the combined sound pressure level (SPL) also depends on the strength and coherence of the room reflections.

When the listener is in a reflection-free zone (RFZ), a pan rule of −6 dB does not reduce the volume for centered sounds as much as when a room has untamed reflections. This is similar to the difference between doubling the level of noise versus music, explained in Chapter 2 under "The Stacking Myth." That section showed that music rises by 6 dB when doubled, but noise rises only 3 dB. The same thing happens with reflections in a room. The reflections you hear are likely different on the left and right sides, but the music played by both speakers is the same.

I suspect that pan rule implementation could account for reports of perceived quality differences between various DAW programs. If you import a group of Wave files into a DAW that uses one pan rule, then import the same series of files to another DAW that uses a different pan rule, the mixes will sound different even if all the track volumes and pans are set exactly the same in both programs.

Connecting a Digital Audio Workstation to a Mixer

Modern DAW software contains a complete mixing console with full automation of every parameter, so project studios can usually get by with only a compact "utility" mixer, or even no hardware mixer at all. Most compact mixers offer microphone preamps for recording and a master volume control for the monitor speakers during playback and mixing. Many studios have a few other audio devices around, such as a synthesizer or maybe a set of electronic drums, and those can also be connected to the same small mixer.

One of the most frequent questions I see in audio forums asks how to connect a mixer to a computer-based recording setup to be able to record basic tracks first, then add overdubs later. Many people have a small mixer such as a Mackie or similar. The Mackie manual shows several setups for combining musical instruments and other sources to play them through loudspeakers. But they ignore what may be the most common setup of all: connecting the mixer to a computer DAW. I'll use the Mackie 1402-VLZ3 for the examples that follow, but the concept applies to any mixer that provides *insert points* for each preamp output.

With most DAW setups, it's best to record each instrument and microphone on a separate track. This gives the most flexibility when mixing, letting you change the volume and equalization separately for each sound source, add more or less reverb to just that instrument, and so forth. I prefer doing all the mixing within the DAW program using the software's volume, pan, and plug-in effects, rather than sending individual tracks back out to a hardware mixer. Mixing in a DAW is more flexible because it allows the mix to be automated and recalled exactly. This also ensures that when you export the final mix to a stereo Wave file, it will sound exactly the same as what you heard while mixing, without regard to the hardware mixer's settings. Rendering a mix from software directly to a Wave file also usually goes faster than playing a mix in real time while recording to tape or to another Wave file.

Inputs and Outputs

Most small mixers contain two independent sections: an input section and a mixer section. The input section contains the XLR and ¼-inch input connectors, plus the preamps that raise the input signals to line level suitable for recording. The mixer section then combines all of the preamplified inputs into one stereo mix that can be played through loudspeakers or headphones. It's important to understand the difference between the input channels, whose outputs are available independently, and the combined stereo mix. You'll record each microphone or instrument through the mixer's preamps to a separate input of your sound card, and you *also* need to monitor those same inputs to hear yourself through earphones as you play or sing. Further, when adding tracks as overdubs to a song in progress, you need to hear the tracks that were already recorded as well.

The first step is to route each microphone or instrument from the mixer to a separate input on the sound card. If your sound card has only one stereo input, you can record only one stereo source at a time, or optionally two mono sources. (Of course, you can overdub any number of additional mono or stereo tracks later.) A multi-channel interface lets you simultaneously record as many separate tracks as the interface has inputs. However, it's usually best to play all of the tracks through one stereo output, even if your interface offers more outputs. This way, your mix is controlled entirely by the settings in your DAW program, independently of the volume knobs on the hardware mixer.

Figure 5.8 shows the input section of a Mackie 1402-VLZ3 mixer. You can connect either XLR microphone or ¼-inch instrument cables, which go through the mixer's preamps and then on to the mixer section that combines all of the inputs for monitoring. The channel inserts section is on the mixer's rear panel, as shown in Figure 5.9. That's where you'll connect each of the mixer's preamp outputs to one input of your sound card.

The original purpose of an insert point was to insert an external hardware effects unit, such as a compressor, into one channel of the mixer. With a DAW, you can do that with plug-ins later. But the insert point can also serve as a multi splitter to send the preamplified microphone

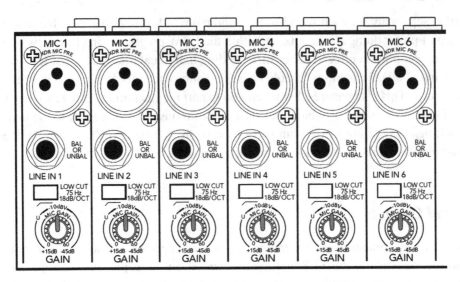

Figure 5.8: The Mackie 1402-VLZ3 mixer has six inputs that accept either XLR microphones or ¼-inch phone plugs. *Drawing courtesy of Mackie.*

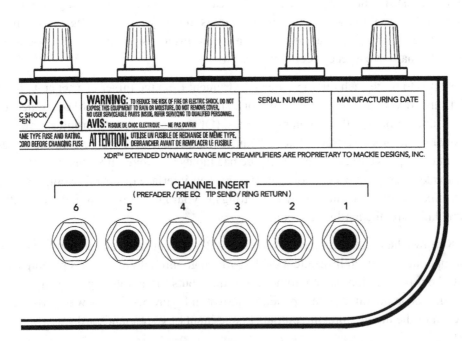

Figure 5.9: The channel inserts on the mixer's rear panel let you send its preamp outputs to a sound card or external interface, without interrupting monitoring within the mixer. *Drawing courtesy of Mackie.*

out to the sound card while also sending it to the rest of the mixer so you can hear yourself while recording. When an output is taken from an insert point this way, it's called a *direct out*, because the signal comes directly from the preamp's output, before passing through the mixer's volume and tone control circuits.

With nothing plugged into an insert jack, the output of the preamp goes through the channel volume, pan, and EQ controls on the front panel and is then combined in the mixer section with all of the other channels. When you insert a ¼-inch plug only partway into the jack, the output of the preamp is sent to the inserted plug and also continues on to the rest of the mixer. This jack switch arrangement is shown in Figure 5.10, and it's the key to recording each input to a separate track.

Unfortunately, these jacks are sometimes too loose to make a reliable connection when the plug is not fully seated. Or the jack switch may wear over time, so you have to wiggle the plug occasionally to get the signal back. The best solution is a phone plug adapter such as the *Cable Up by Vu PF2-PM3-ADPTR* available from Full Compass and other distributors. This inexpensive gadget can be inserted fully into the mixer's insert jack to connect the mixer's tip and ring at the adapter's plug end to the tip (only) at its female end. This retains the connection needed from the preamp to the rest of the mixer while making the Send available to go to your sound card. Of course, you could wire up your own custom insert cables using ¼-inch stereo phone plugs that can be inserted fully, with the tip and ring connected together.

Since the preamp output still goes to the mixer section, you can control how loudly you hear that source through your speakers with that channel's volume control. And since each preamp

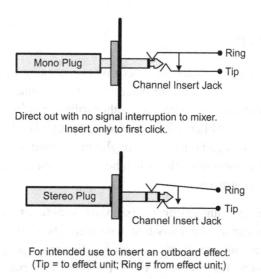

Figure 5.10: The mixer's rear panel insert point lets you tap into a channel to send its preamp output to a sound card, without interrupting the signal flow within the mixer.

output goes directly to the sound card, the recording for that track contains only that one channel. Had you recorded from the mixer's main stereo output, the new recording would include all of the tracks already recorded, as well as any live inputs being recorded.

Figure 5.11 shows how each channel's insert point output goes to one input of the sound card or external interface. The sound card's main stereo output then goes to one of the mixer's stereo inputs so you can hear your DAW program's playback. With this arrangement you control the volume you hear for each track being recorded with its channel volume slider, and set the playback volume of everything else using the stereo channel's slider. In this example, mixer channels 1 through 6 control how loudly you hear each source being recorded, and stereo input pair 13 and 14 controls how loudly you hear the tracks that were already recorded.

Setting Record Levels

Each of the mixer's six mic/line mono input channels has two volume controls: the preamp gain knob (also called Trim) and the channel's main volume slider. Both affect the volume you hear through the loudspeakers, but only the preamp gain changes the level sent to your sound card. Therefore, when recording you'll first set the preamp gain control for a suitable recording level in your software, then adjust the channel output level for a comfortable volume through your loudspeakers or headphones. Since the channel volume and equalizer are not in the path from preamp to sound card, you can freely change them without affecting the recording.

Monitoring with Effects

It's useful and even inspiring to hear a little reverb when recording yourself singing or playing, and also to get a sense of how your performance will sound in the final mix. Many DAW programs let you monitor with reverb and other software effects while recording, but I don't recommend that. One problem with monitoring through DAW software with effects applied is the inherent delay as audio passes through the program and plug-ins, especially with slower computers. Even a small delay when hearing yourself can throw off your timing while singing or playing an instrument.

Another problem is that using plug-in effects taxes the computer more than recording alone. Modern computers can handle many tracks with effects all at once, but with an older computer you may end up with gaps in the recorded audio, or the program might even stop recording. Most small mixers like the 1402-VLZ3 used for these examples have Aux buses to patch in a hardware reverb. Any reverb you apply on an Aux bus affects only the monitoring path and is not recorded. Therefore, you can hear yourself sing (using earphones) with all

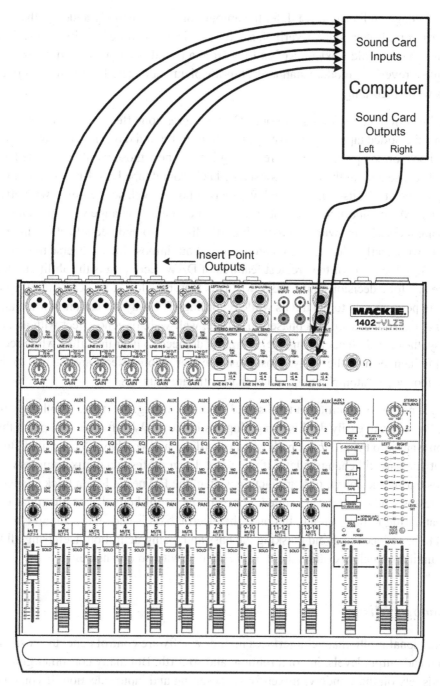

Figure 5.11: Each channel insert goes to one input on the sound card, and the sound card's output comes back into the mixer on a stereo channel for monitoring. *Drawing courtesy of Mackie.*

the glory of a huge auditorium, yet defer the amount of reverb actually added to that track until the final mix. Better, since reverb on an Aux bus is not recorded, an inexpensive unit is adequate. I have an old Lexicon Reflex I got from a friend when he upgraded his studio. It's not a great reverb by modern standards, but it's plenty adequate for adding reverb to earphones while recording.

Likewise, I recommend recording without EQ or compression effects. When recording to analog tape, the added tape hiss is always a problem. In the old days, it was common to add treble boost or compression while recording if you knew those would be needed during mixdown, because adding those later boosts the background tape hiss. But modern digital recording—even at 16 bits—has a very low inherent noise level, so recording with effects is not needed. More important, it's a lot easier to experiment or change your mind later if the tracks are recorded dry with no processing. It's difficult to undo equalization, unless you write down exactly what you did, and it's just about impossible to reverse the effects of compression. Indeed, one of the greatest features of DAW recording is the ability to defer all balance and tone decisions until mixdown. This feature is lost if you commit effects by adding them permanently to the tracks while recording. The only exception is when an effect is integral to the sound, such as a phaser, fuzz tone, wah-wah, or echo effect with an electric guitar or synthesizer. In that case, musicians really do need to hear the effect while recording, because it influences how they play. Most musicians for whom effects are integral to their sound use foot pedal "stomp" boxes rather than software plug-ins.

Some people prefer to commit to a final sound when recording, and there's nothing wrong with that. But you can easily do this using plug-ins in the DAW software. Just patch the effects you want into the *playback* path when recording, and you have "committed" to that sound. But you can still change it later if you want. Even professional mix engineers change their minds during the course of a production. Often, the EQ needed for a track changes as a song progresses, and overdubs are added. For example, a full-bodied acoustic guitar might sound great during initial tracking, but once the electric bass is added, the guitar's low end may need to be reduced to avoid masking the bass and losing definition. Deferring all tone decisions until the final mix offers the most flexibility, with no downside.

The Windows Mixer

Most professional multi-channel interfaces include a software control panel to set sample rates, input and output levels, internal routing, and so forth. But many consumer grade sound cards rely on the Windows mixer for level setting and input selection. If you create music entirely within the computer, or just listen to music and don't record at all, a regular sound card can be perfectly adequate. You launch the Windows mixer shown in Figure 5.12 by double-clicking the small loudspeaker icon at the lower right of the screen. If that's not

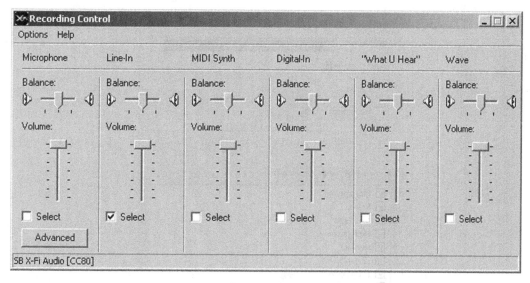

Figure 5.12: The Windows mixer record panel lets you choose which of several inputs to record from.

showing, you can get to it from the Windows control panel, then *Hardware and Sound*. Depending on your version of Windows, the wording may be slightly different.

Once the Windows mixer is showing, click the *Options* menu, then *Properties*. Next, under *Adjust volume for*, select *Recording*. Figure 5.13 shows the Playback source choices, but the Recording screen is very similar and works the same way. If you have a typical consumer sound card, you should select Line-In as the recording source, and be sure to set the input level to maximum.

The Windows mixer Play Control panel in Figure 5.14 adjusts the mix of sources that play through the sound card's Line Output jack. The Record Control panel lets you select only one input from which to record, but playback can be from several sources at the same time. Be aware that the Windows mixer is probably hiding some input and output level controls. Yet those sources can contribute hiss from an unused input, or even add noise generated by a hard drive or video card. For example, you might hear scratching sounds as you move your mouse or as things change on the screen. Therefore, you should select all of the available playback sources in the Properties screen of Figure 5.13 to make them visible. Then mute or turn down all the sources you don't need, such as Auxiliary, Microphone, and so forth. You can hide them again afterward if you don't want to see them.

Most Creative Labs SoundBlaster sound cards have a "What U Hear" input source that records the same mix defined in the Play Control panel. If you have a SoundBlaster card, do not select "What U Hear" when recording, because that also records the tracks you're playing

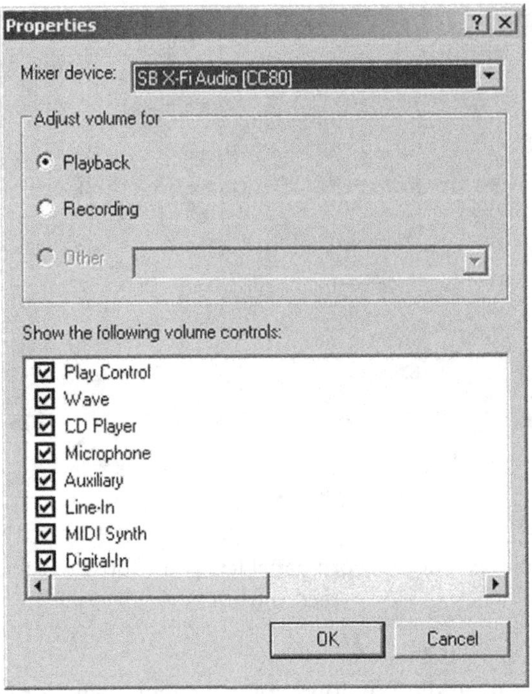

Figure 5.13: The Windows mixer properties panel lets you choose which inputs and outputs to show in the mixer, with separate check boxes for the Record and Playback sections. Note that even when a playback channel's slider isn't showing, that source is still active and can add hiss and other noise.

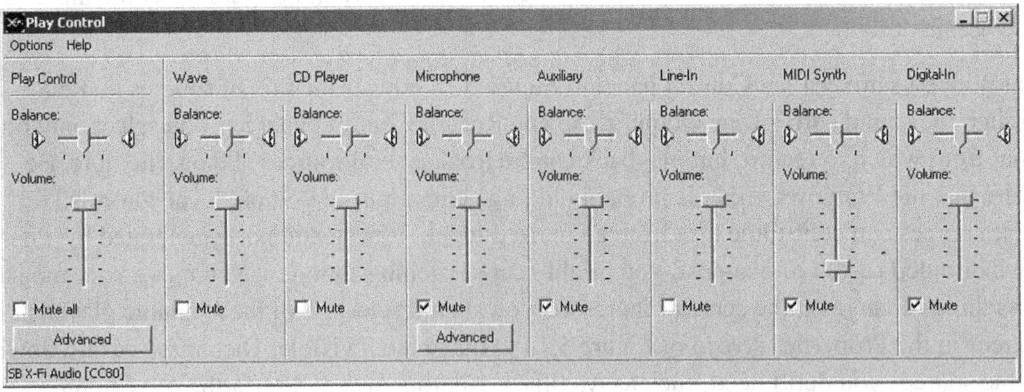

Figure 5.14: The Windows mixer playback panel lets you set the volume level for each available playback source. All of the outputs can play at the same time, though you probably don't want that.

along to. It can also add hiss to the recording because it includes the MIDI synthesizer, CD Audio, and all other playback sources that aren't muted. The main use for "What U Hear" is to capture audio that's streaming from a website when the site doesn't allow you to download it as a file.

Related Digital Audio Workstation Advice

In Figure 5.11, the sound card's main stereo output comes back into the mixer on stereo channels 13–14. This provides a separate volume control for the DAW's output, which can be adjusted independently from the volume of each input being recorded. But you may be using all of the mixer's stereo inputs for other sources like a CD player, smart phone audio, and so forth. Or maybe you have a smaller mixer that has fewer stereo inputs. In that case you can connect the sound card's stereo output to the mixer's second Aux Return, if available, or even to the Tape Input.

Even though the Windows mixer has record volume controls for each input source, it's important to set these to maximum and adjust the record level using the preamp gain setting on your hardware mixer. The same is true for the software control panel that's included with more expensive interfaces. Software volume controls lower the volume digitally *after* the sound card's A/D converter. So if the level from your mixer's preamp is too high and overloads the sound card's input, reducing the software volume control just lowers the recorded volume, yet the signal remains distorted.

People who have used analog tape recorders but are new to digital recording tend to set record levels too high. With open reel tape and cassettes, it's important to record as hot as possible to overcome tape hiss. But analog tape is more forgiving of high levels than digital systems. Analog tape distortion rises gradually as the signal level increases and becomes objectionable only when the recorded level is very high. Digital recorders, on the other hand, are extremely clean right up to the point of gross distortion. Therefore, I recommend aiming for an average record level around −10 dBFS or even lower to avoid distortion ruining a recording of a great performance. The noise floor of 16-bit audio is at least 20 to 30 dB softer than that of the finest professional analog tape recorders, and the inherent noise of 24-bit recording is even lower.

Often when you connect an audio device like a mixer to a computer using analog cables, a ground loop is created between the computer and the mixer that causes hum. Sometimes you can avoid this by plugging both the mixer and computer into the same physical AC power outlet or power strip. If that doesn't solve the problem, a good solution is to place audio isolation transformers in series with every connection between the two devices. High-quality audio transformers can be expensive, but I've had decent results with the EBTECH Hum Eliminator. This device is available in both two- and eight-channel versions, and at less than

$30 per channel, it's reasonably priced for what it is. Other companies offer similar low-cost transformer isolators. However, if the highest audio quality is important, and you're willing to pay upwards of $100 per channel, consider better-quality transformers such as those from Jensen Transformers and other premium manufacturers. Inexpensive transformers can be okay at low signal levels, but often their distortion increases unacceptably at higher levels. Low-quality transformers are also more likely to roll off and distort at the lowest and highest frequencies.

5.1 Surround Sound Basics

Most music is mixed to stereo, but there are many situations where 5.1 surround is useful. For example, music for movie sound tracks is often mixed in surround. Most modern DAW software can create 5.1 surround mixes, including Cakewalk SONAR used for these examples. In fact, surround sound is not limited to 5.1 channels, and some surround systems support 7.1 or even more channels. But let's first consider how surround audio is configured using 5.1 channels.

Figure 5.15 shows a typical surround music playback system having three main loudspeakers in the front, marked L, C, and R, for Left, Center, and Right. The rear surround speakers are LS and RS, for Left Surround and Right Surround. The subwoofer is shown here in the front left corner, but subwoofers are typically placed wherever they yield the flattest bass response. Surround speaker placement is explained more fully in Chapter 19.

In a 5.1 surround system, the left and right main front speakers are also used for regular stereo when listening to CDs and MP3 files, or when watching a concert video recorded in stereo. For true surround material, the center speaker is generally used for dialog, and is often called the *dialog channel* because it anchors the dialog at that location. This is an important concept for both movies and TV shows recorded in surround. In a control room, the engineer sits in the exact center, where the stereo concept of a "phantom image" works very well. But more than one person often watches movies in a home theater, yet only one person can sit in the middle.

Unless you're sitting in the exact center of the room left-to-right, voices panned equally to both the left and right speakers will seem to come from whichever speaker is closer to you. When an actor is in the middle of the screen talking, this usually sounds unnatural. And if the actor walks across the screen while talking, the sound you hear won't track the actor's position on screen unless you're sitting in the center. To solve this problem, the 5.1 surround standard adds a center channel speaker. Early quadraphonic playback from the 1970s included rear surround speakers for ambience, so listeners could feel like they were in a larger virtual space. But it didn't include the all-important center channel that anchors voices or other sounds in the center of the sound field.

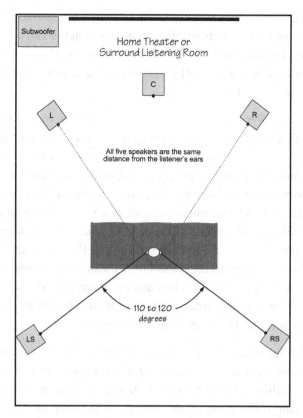

Figure 5.15: A 5.1 surround system comprises five full-range loudspeakers plus a subwoofer.

Surround systems also use *bass management* to route low frequencies to a subwoofer. The ".1" channel contains only low-frequency sound effects such as earthquake rumbles and explosions. The subwoofer channel is called ".1" because its range is limited to bass frequencies only. That channel always goes directly to the subwoofer. But surround receivers also route low frequencies present in the five other channels away from those speakers to the subwoofer. Very low frequencies are not perceived as coming from any particular direction, so having everything below 80 or 100 Hz come from one subwoofer doesn't affect stereo imaging or placement. In a surround system that's set up properly, you should never notice the subwoofer playing or be able to tell where it's located.

Another important advantage of bass management is that it takes much of the load off the five main speakers. Bass frequencies are the most taxing for any loudspeaker to reproduce, so most speakers can play much louder and with less distortion when they don't have to reproduce the lowest two octaves. Further, speakers that don't have to reproduce very low frequencies are physically smaller and generally less expensive than speakers that can play down to 40 Hz or even lower. The standard for bass management specifies a crossover frequency of 80 Hz, but frequencies slightly lower or higher are also used. However, the crossover should never be set too high, or placement and imaging can be affected. This is a problem with surround systems that use too-small satellite speakers with a subwoofer crossover at 200 Hz or higher. You can tell that some of the sound is coming from the subwoofer, which can be distracting.

Professional monitor controllers are available to route 5.1 mixes from a DAW program to surround speakers, including handling bass management, but you can do the same thing with an inexpensive consumer type receiver. My home theater system is based on a Pioneer receiver that is full-featured and sounds excellent, yet was not expensive. However, a receiver used for surround monitoring must have separate analog inputs for all six channels. All consumer receivers accept stereo and multi-channel audio through a digital input, but not all include six separate analog inputs. You'll also need a sound card or external interface with at least six analog outputs. Figure 4.13 in Chapter 4 shows the PreSonus FireBOX interface I use to mix surround music in my living room home theater. The FireBOX uses a FireWire interface to connect to a computer and has two analog inputs plus six separate analog outputs that connect to the receiver's analog inputs. Other computer sound cards with similar features are available using Firewire, USB, or newer interface types.

In order to monitor 5.1 surround mixes, you need to tell your DAW program where to send each surround bus output channel. As mentioned, the PreSonus FireBOX connected to my laptop computer has six discrete outputs, which in turn go to separate analog inputs on my receiver. This basic setup is the same when using a professional monitor controller instead of a receiver. Figure 5.16 shows SONAR's setup screen for assigning surround buses to physical outputs, and this method is typical for other DAW programs having surround capability. Note the check box to monitor with bass management. This lets you hear mixes through your playback system exactly as they'll sound when mastered to a surround format such as Dolby Digital or DTS, when played back from a DVD, Blu-ray, or other multi-channel medium.

Figure 5.17 shows the surround panner SONAR adds to each audio track sent to a surround bus, and other DAW programs use a similar arrangement. Rather than offer left, right, and in-between positions, a surround panner lets you place mono or stereo tracks anywhere within the surround sound field. You can also send some amount of the track to the LFE channel, though that's not usually recommended for music-only productions.

Most consumer receivers offer various "enhancement" modes to create *faux* surround from stereo sources or to enhance surround sources with additional ambience and reverb. It's important to disable such enhancement modes when mixing surround music because the effects you hear are not really present in the mix but are added artificially inside the receiver.

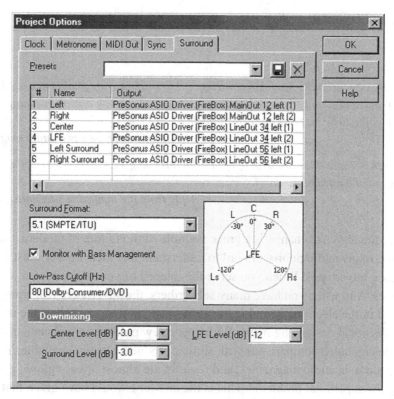

Figure 5.16: The Surround tab under SONAR's Project Options lets you specify which sound card outputs receive each of the 5.1 surround buses for monitoring.

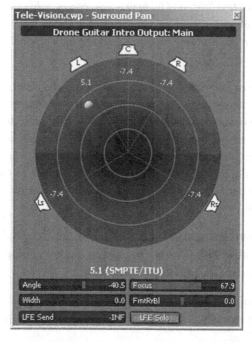

Figure 5.17: The surround panner in SONAR is much more complex than a typical stereo pan pot, letting you place sources anywhere within the surround sound field.

Summing

I'm relatively new to studio production/mixing. . . . I'm finding my final mixes need more separation and space, and I've been reading up on out-of-the-box analog summing. What options are going to be best and most cost-effective?

—Part of a letter to a pro audio magazine

You're in luck because there are a lot of options in this category. Check out summing boxes from [long list of hardware vendors].

—Reply from the magazine's technical editor

In my opinion, the above exchange is a great example of the failure of modern audio journalism. Too many audio publishers fail to understand that their loyalty must be to their readers, not their advertisers. When you serve the interest of your readers, you will sell more magazines. And when you have many subscribers, the advertisers will surely follow. When I worked in the electronics field in the 1970s and 1980s, I read all of the magazines that serve professional design engineers. With every new product announcement or review, these magazines included comparisons with similar products already available. You almost never see that today in audio magazines, and reviews are almost always glowing, criticizing products only for superficial problems. This is the answer I would have given that reader:

Separation and space in a mix are directly related to frequencies in one track masking similar frequencies from instruments in other tracks. This is mostly solved using EQ; knowing what frequencies to adjust, and by how much, comes only with experience and lots of practice. Obviously, reverb and ambience effects influence the spaciousness of a mix, but that's outside the domain of outboard summing boxes, which can only add distortion and alter the frequency response.

You mentioned that you're fairly new to audio production, and this is most likely the real reason your mixes lack space and separation. I suggest you study professional mixes of music you enjoy. Listen carefully to how each instrument and vocal track meshes with all the other tracks. When mixing your projects, if an instrument sounds clear when solo'd but poorly defined in the full mix, try to identify other tracks that, when muted, restore the clarity. I'm sure you'll find that with more experience your mixes will improve. I think it's false hope to expect any hardware device to magically make your mixes come together and sound more professional.

Summing is merely combining sounds, and it's very low tech. Indeed, summing is the simplest audio process of all, adding either voltages in analog equipment, or numbers in a DAW program or digital mixer. Let's take a closer look.

The simplest analog summing mixer is built using only resistors, as shown schematically in Figure 5.18. Only eight inputs are shown in this example, and only one mono output.

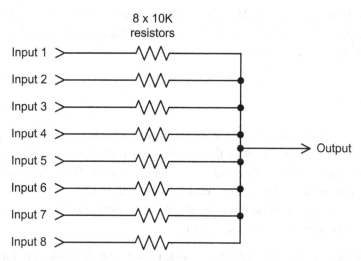

Figure 5.18: A perfectly serviceable analog summing mixer can be built using one 10K resistor for each input channel. Because it uses only resistors, this summing circuit adds virtually no noise or distortion.

But the concept can be expanded to any number of inputs and to include left and right input and output pairs for stereo. Although this mixer contains only resistors, it actually works pretty well. Of course, there's no volume control for each channel, or master volume, nor are there pan pots for stereo. Adding those would require active (powered) circuitry to avoid interaction among the input channels.

With simple passive mixers like this, changing a resistor's value to adjust the volume or pan for one channel affects the volume and panning of all the other channels. But even without volume and pan controls, a passive mixer like this loses some signal level on each channel. The more channels you mix together, the softer each becomes at the output. When mixing 2 channels, each channel is reduced by 6 dB; you lose another 6 dB each time the number of channels is doubled. So mixing 4 channels lowers them all by 12 dB, and 8 channels as shown here loses 18 dB. When mixing 16 channels, each channel is 24 dB softer at the output. Commercial summing boxes typically add an amplifier stage with an appropriate amount of gain at the output to restore the signal levels back to normal.

The most significant thing that happens when you sum tracks is psychoacoustic. A track that sounded clear all by itself may now be masked by another instrument having a similar frequency range. An electric bass track where every note can be distinguished clearly when solo'd might turn into a rumbling mush after you add in a chunky-sounding rhythm guitar or piano. I'm convinced this is the real reason people wrongly accuse "summing" or "stacking" for a lack of clarity in their mixes. The same masking effect happens whether the tracks are mixed with an analog circuit or an equivalent series of sample numbers are added in a DAW program or digital hardware mixer. The loss of clarity occurs in our ears and brain, not the summing device. It seems some people prefer to blame "digital" when they're unable to get instruments to sit well together in a mix.

Gain Staging

Large mixing consoles are complex, allowing flexible routing for audio after it passes through the microphone preamps. It's important that audio signals remain at a reasonable level as they pass through every stage of an analog mixer. If a source is too soft at any point in the chain, noise from the console's circuitry can be heard; if too loud, you risk distortion. The process of keeping signal levels reasonable throughout a mixing console is called *gain staging*, and it's the responsibility of the recording engineer.

Gain staging is always a concern with analog mixers and other hardware, but it matters much less with DAW software. Analog circuits process voltages passing through them, and all circuits add some amount of noise and distortion, especially when signal levels are very low or very high. But modern digital processing uses *floating point calculations* implemented in software, which can handle a huge range of signal levels—more than 1,500 dB. So gain staging within digital software is rarely a problem because it has a minuscule effect on audio quality. Most modern digital software and hardware process audio use 32-bit floating point values, though some use 64 bits for even higher resolution. This is explained more fully in Chapter 8.

Microphone Preamplifiers

These days a lot of fuss is made over microphone preamps, with various magical properties attributed to this model or that. It wasn't so long ago that most recordings were made using whatever preamps were available in the mixing console. Many older recordings made using only the stock preamps in a good-quality console still sound excellent by modern standards. Some people consider mic pres the most important part of the signal path, because any quality loss in a preamp affects everything in the chain that follows. In truth, fidelity can be harmed at any point in the signal path.

With passive dynamic and ribbon microphones, the preamp is indeed the first link in a chain of audio devices. But active microphones have their own preamp inside, so with those microphones the preamp in a mixing console or outboard unit isn't really the first device. Indeed, what really matters with preamps is the same as what matters for every audio device: frequency response, distortion, and noise. Most circuit designers aim to make their preamps *transparent*. If a preamp is transparent, it will add no coloration of its own, which is certainly my preference.

It's not that I don't consider preamps important, because they obviously are. But keep in mind that any two transparent preamps will sound exactly the same—by definition. If the response is within 0.1 dB from 20 Hz to 20 KHz, and the sum of all distortion and noise is 80 dB or more below the signal, a preamp circuit will not alter the sound audibly. A lot of preamps meet that criterion and therefore sound alike by definition. Yet you'll find people who insist all of their preamps sound different anyway. I trust test gear and null tests because they're 100 percent reliable and repeatable versus sighted anecdotal opinions, which are not usually repeatable and are certainly less reliable. However, some preamps color the sound intentionally, and that's a different issue.

Please understand that "specs" are more complicated than may seem from this simplistic explanation. For example, distortion often increases at higher levels. So two preamps may have the same distortion when outputting a −10 dBu signal but be very different at +20 dBu. Frequency response can also change with level or, more accurately, with the amount of gain. So a preamp may be very flat when its trim is set for 20 dB gain, but not so flat with 60 or 70 dB gain. Distortion also increases with gain. In the larger picture, specs *do indeed* tell you everything needed about every circuit, as long as you measure all the parameters at different signal levels. This is especially true for preamps that aim for a clean, uncolored sound. You have to verify that they're transparent at all signal levels and gain settings. But while specs can accurately predict that a preamp is audibly transparent, it's much more difficult to look at the specs for an intentionally colored device and divine how its 4% total harmonic distortion (THD) or other coloration will actually sound when used on various instruments and voices.

Preamp Input Impedance

While we're on the subject of microphones and preamps, it's worth mentioning preamp input impedance. Most microphones have an output impedance of around 150 ohms, letting them drive long cables without high-frequency loss. Most preamps have an input impedance much higher than 150 ohms, which loads the microphone less to avoid losing some of the signal. Mic pres usually have an input impedance of between 1K and 10K, with most around 2K. Ideally, the input impedance of a preamp will be at least 5 to 10 times the microphone's output impedance, to avoid signal loss and possibly increased distortion. If a microphone having an output impedance of 150 ohms is plugged into a preamp whose input impedance is 150 ohms, the microphone will be 6 dB softer than if it were sent into a high-impedance input. Further, active microphones that have a built-in preamp may suffer from higher distortion as they work harder to drive a low impedance.

A recent trend among some preamp makers lets you vary the input impedance yourself, allegedly to fine tune the sound. One popular model lets you choose between 300 and 1,200 ohms. Another has a potentiometer that lets you adjust the input impedance continuously from 100 ohms to 10K. In truth, lowering the input impedance of a preamp mainly rolls off low frequencies and might also increase distortion. What happens as a preamp's input impedance is lowered depends on the microphone's design—whether it's active, how much current the output stage can provide, and whether it has a transformer. This is yet another way that gear vendors try to up-sell us on the value of subtle distortion.

Preamp Noise

One important spec for microphone preamps that really can vary between different models is *equivalent input noise*, abbreviated EIN. All electronic circuits generate some amount of noise. Even a resistor, which is a passive device, generates an amount of noise that can be calculated based on its resistance in ohms and the ambient temperature. A complete explanation of the many different sources and types of circuit noise is beyond the scope of this book, so I'll hit only the high points.

The noise that's always present to affect low-level circuits such as mic pres is called *thermal noise*, or sometimes *Johnson noise* after J. B. Johnson, who discovered it in the 1920s. This type of noise exists at temperatures above absolute zero, rising with increasing temperature, and it's caused by random molecular motion. Assuming a room temperature of 70°F (about 21°C), the theoretical lowest noise possible is around −131 dBu when considering only the audible range. The actual spec for EIN should always include the bandwidth being considered. For example, the low-noise NE5534 op-amp used in some microphone and phonograph preamplifiers has an EIN voltage spec'd as follows:

$$3.5nV * \sqrt{Hz}$$

In other words, the amount of noise is 3.5 nV (nanovolts) times the square root of the bandwidth in Hz. So for the range 20 Hz through 20 KHz:

$$3.5nV * \sqrt{19,980}$$

or

$$3.5*141 = 494 \text{ nV}$$

From this you can estimate the actual signal to noise ratio you'll get from a preamp based on a given microphone output voltage. For example, if a microphone outputs 4.94 millivolts at some SPL, which is reasonable for a low-impedance dynamic microphone, the signal to noise ratio will be 10,000 to 1, or 80 dB.

In the interest of completeness, the approximate noise in nV per root Hz for a resistive source at room temperature is calculated as follows:

$$0.135 * \sqrt{Ohms}$$

A preamp's EIN is typically measured by wiring a 150-ohm metal film resistor across its input terminals, setting the preamp gain to 60 dB, then measuring the noise at the preamp's output. Metal film resistors are used because they are more precise than most other types, and 150 ohms is used because that's the output impedance of most microphones. Given the 60 dB of gain, the EIN of the preamp is simply 60 dB below the noise measured at its output. If you're reviewing preamp specs for a proposed purchase, try to verify that the EIN was in fact measured using a 150-ohm input source. If a manufacturer instead applied a short circuit to the input, that dishonestly biases the test to yield less noise than you'll actually realize in use. However, it's proper to filter out frequencies above and below the audible range (and use A-weighting) to exclude noise that's present but won't be heard. EIN is often spec'd as some number of nanovolts per root Hz:

$$3.5nV \, / \, \sqrt{Hz}$$

However, in this case, the slash (/) means "per" and not "divided by." In fact, you *multiply* the noise times the bandwidth in Hz, as shown in the earlier formulas above. It's also worth mentioning that the EIN for a preamp actually increases at lower gain settings. In practice this isn't a problem because a lower gain setting is used with higher input voltages, which in turn increases the overall signal to noise ratio.

All of that said, transformerless preamps having an EIN within 2 dB of the theoretical limit for a low-noise resistor are common and inexpensive. It's the last dB or so that's elusive and often very expensive to obtain!

Clean and Flat Is Where It's At

My preference is to record most things (though not fuzz guitars) clean and flat. You can add a little grit later to taste when mixing, but you can't take it away if you added too much when recording. To my way of thinking, any preamp that is not clean and flat is colored, and I'd rather add color later when I can hear all of the parts in context. A coloration that sounds good in isolation may sound bad in the mix, and vice versa. So my personal preference is to defer all such artistic choices until making the final mix. If I decide I want the sound of tape or tubes, I'll add that later as an effect.

Not to go off on a rant, but one big problem with vacuum tubes is they're not stable. So their sound changes over time as they age, and eventually they need to be replaced. Few tube circuits are as clean as modern solid state versions, and tubes can also become microphonic. When that happens, the tube resonates, and it sounds like someone is tapping a microphone. This resonance is especially noticeable if the tube is close to a loud source such as a bass amp. Further, tube power amplifiers require a specific amount of DC *bias* voltage to avoid drawing more current than the tube can handle. An amplifier's bias is adjusted using an internal variable resistor, but the optimum bias amount drifts over time as the tube ages and needs to be adjusted occasionally. Setting a tube's bias is not a task most end users are capable of doing correctly. Tubes also have a high output impedance, so most tube-based power amps include an output transformer that further clouds the sound. Finally, with some modern tube gear, the tube is tacked on for marketing purposes only and is operated at less than its optimum power supply voltage.

This point was made exquisitely by Fletcher[1] in a forum post about the value of tube preamps and other tube gear. Fletcher said, "You guys need to understand that the people building 'tube' stuff back in the day were going for the highest possible fidelity attainable. They were going for the lowest distortion possible, they were trying to get the stuff to sound 'neutral.' They *were not* going for the 'toob' sound; they were trying to *get away* from the toob sound."

On this point I completely agree with Fletcher. In the 1950s and 1960s, the electronic and chemical engineers at Ampex and Scully, and 3M and BASF, were aiming for a sound as clean and transparent as possible from analog recorders. They were not aiming for a "tape" sound! This was also true of Rupert Neve and other big-name console designers of the day. The transformers they used were a compromise because they couldn't design circuits to be quiet enough without them. Today, transformers have been replaced with modern op-amps whose inputs are very quiet and are inherently balanced to reject hum. If boosting frequencies with a vintage Neve equalizer distorts and rings due to its inductors, that's a failing of the circuit design and available components, not an intended feature.

> *No listener gives a damn which microphone preamp you used.*
> *—Craig Anderton, audio journalist and magazine editor*

Indeed, the problems I hear with most amateur productions have nothing to do with which preamps were used and everything to do with musical arrangement, EQ choices, and room acoustics. If a mix sounds cluttered with a harsh midrange, it's not because they didn't use vintage preamps. In my opinion, this fascination with the past is misguided. People hear old recordings that sound great and wrongly assume they need the same preamps and compressors and other vintage gear to get that sound. Every day in audio forums I see a dozen new threads with "recording chain" in the title, as a newbie asks what mics and other gear were used to record some favorite song or other. This ignores that the tone of a performance is due mainly to the person playing or singing and the quality of their instrument or voice. I once saw a forum thread asking, "How can I get that Queen layered vocal sound?" I'll tell you how: Capture a super-clean recording of people who can sing like the guys in Queen!

Summary

This chapter explains that audio is similar to household plumbing by showing the layout and signal routing for both large-format and compact mixers, including mute and solo, Aux buses, and automation. A complex system such as a mixing console can be more easily understood by viewing it as a collection of smaller, simpler modules and signal paths. We also covered DAW software, including how to connect a DAW computer to a small-format hardware mixer to add overdubs without also recording previous tracks. One big advantage of the method shown is that you can hear yourself with reverb and EQ or other effects as you record, without including those effects in the recording.

The basics of surround mixing were explained, along with an explanation of summing and gain staging in both analog mixers and digital software and hardware. I couldn't resist including a short rant about the value of vintage mic pres and tube gear. I also explained why I prefer to capture audio sources as cleanly as possible, without distortion or other coloration, because it makes more sense to defer intentional color until mixing, when you can hear all the parts in context.

Note

1 Fletcher is a colorful character who founded Mercenary Audio, a pro audio reseller based in Foxboro, Massachusetts. He's known in audio forums for his strongly worded opinions, often peppered with salty language.

CHAPTER 6

Recording Devices and Methods

Recording Hardware

The two basic types of recording systems are analog and digital. In the context of recording, analog usually refers to old-style tape recorders, but it also includes phonograph records, which are more a means of distribution than a recording medium. However, you'll sometimes hear about an audiophile record label capturing a recording session live in stereo to a record-cutting lathe, which is then used as a master for the pressing plant. Although analog tape is still favored by some recording engineers, considering only the percentage of users, it was surpassed by digital recording many years ago. The main reasons digital recording prevails today are the high costs of both analog recording hardware and blank tape, as well as the superior fidelity and features of modern digital. Compared to editing analog tape with a razor blade, manipulating audio in a digital system is far easier and vastly more powerful. Digital editing also lets you undo anything you don't like or accidentally ruin.

Several different types of digital recording systems are available, including stand-alone hard disk recorders, computer digital audio workstation (DAW) setups, multi-track and two-channel digital audio tape (DAT) recorders, portable recorders that write to solid state memory cards, and stand-alone CD recorders. The most popular of these by far is the computer DAW, for many reasons. Computers powerful enough to record and mix a hundred tracks or more are amazingly affordable these days, and they'll only become less expensive and more powerful in the future. DAW software can run on a regular home or business computer, rather than require custom hardware that's expensive to produce for the relatively few consumers of recording gear. Another big advantage of DAW software is you can upgrade it when new features are added, and you can easily expand a system with purchased (or freeware) plug-ins. With a hard disk recorder or portable device, if you outgrow it, your only recourse is to buy a newer or larger model.

It's easy to prove using the parameters defined in Chapter 2 that modern digital recording is far more accurate than analog tape. Indeed, modern digital beats analog tape and vinyl in every way one could possibly assess fidelity. For example, digital EQ adds less distortion and noise than analog hardware, it's precisely repeatable, and in a stereo EQ the left and right channels always match perfectly. But it's important to point out that perfect fidelity is not everyone's goal. Indeed, recording engineers who favor analog tape prefer it because they *like* the coloration it adds. They're willing to put up with the higher cost, additional maintenance, and poorer editing abilities in exchange for a sound quality they believe is not attainable any other way. And there's nothing wrong with that. However, I believe that digital processing can emulate analog tape coloration convincingly when that effect is desired, for much less cost and inconvenience.

Although this chapter focuses mainly on digital recording, for completeness I won't ignore analog tape because its history is interesting as well as educational from a "how audio works" perspective. The engineering needed to achieve acceptable fidelity with analog tape is clever and elaborate, so let's start there.

Analog Tape Recording

When I was 17 in 1966, I built my first studio in my parents' basement, starting with a Sony quarter-track open reel tape recorder with sound-on-sound capability. This let me record one (mono) track onto ¼-inch tape, then copy that to the other track while also mixing a new source through the microphone or line input. There was only one chance to get the balance correct, and after only a few passes, the tape noise and distortion became quite objectionable. A year later I bought a second Sony stereo tape deck and a four-track record/play head. At $100 for just the tape head, this was a big investment for a teenager in 1967! I mounted the new tape head in the first recorder, replacing the existing record head, and ran wires from the second tape deck to the extra track windings on the new head, thus making a four-track recorder. This let me record and play four tracks all at once or separately, though I used this setup mostly to record and overdub instruments one by one to build a complete performance by myself.

Analog recording uses thin plastic tape, usually 1 or 1.5 mil (0.001 inch) thick, that's been coated with a magnetic material called *slurry*. This is a gooey paste containing tiny particles of iron oxide that's applied to the plastic tape. Each particle can be magnetized separately from the others, which is how audio signals are stored. The amount of magnetization, and how it varies in time, is analogous to the audio signal being recorded—hence the term *analog* tape. The recording head consists of a metal core or *pole piece*, with a coil of wire wound around it. As the tape passes by the record tape head, audio applied to the coil magnetizes the core in direct proportion to the amount of voltage and its polarity. The varying magnetism in the head then transfers to the slurry on the tape in a pattern that changes over time in both amplitude and frequency.

The highest frequency that can be recorded depends on the size of the iron oxide particles, with smaller particles accommodating higher frequencies. The high-frequency limit is also affected by how fast the tape travels as it passes by the head and the size of the gap in the pole piece. If the voltage from a microphone being recorded changes suddenly from one level to another, or from positive to negative, separate particles are needed to capture each new level or polarity. Most professional recorders operate at a tape speed of either 15 or 30 inches per second (IPS), which is sufficient to capture the highest audible frequencies. Consumer tape recorders usually play at either 7-1/2 or 3-3/4 IPS, though speeds as low as 1-7/8 and even 15/16 IPS are used for low-fidelity applications such as cassettes and dictating machines. Magnetic transference also requires close proximity, so the tape must be kept very close to the record head to capture high frequencies. As you can see, particle size, intimate contact between the electromagnet heads and iron oxide, tape speed, and even the head dimensions are all interrelated, and they all conspire to limit the highest frequencies that can be recorded.

Tape Bias

Tape magnetization is not a linear process, so simply applying an audio signal to a tape head will yield a distorted recording. Until the signal to the record head reaches a certain minimum level, the tape retains less than the corresponding amount of magnetization. As the level to the tape head increases, the tape particles retain more of the magnetization, and thus better correspond to the audio signal. This is shown as the *transfer curve* in Figure 6.1. Tape nonlinearity is similar to the nonlinearity caused by crossover distortion described earlier in Chapter 2.

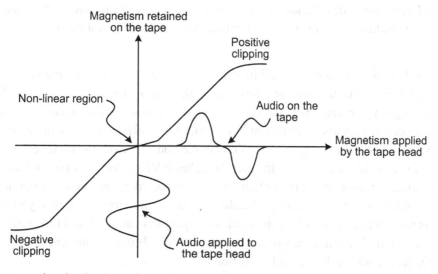

Figure 6.1: At low levels where the audio passes through zero from minus to plus, or vice versa, the tape retains less magnetism than was applied. Then at higher levels it remains fairly linear until the applied magnetism approaches the tape's saturation point. At that point, applying more magnetism again results in less being retained by the tape.

Figure 6.1 labels the "crossover" area near zero as the nonlinear region, but the plus and minus extremes are also nonlinear as the tape's magnetization approaches *saturation*. At that point the tape literally cannot accept any more magnetization. This is similar to a wet sponge that's fully drenched and can't hold any more liquid.

To avoid tape's inherent crossover distortion, analog tape recorders apply a *bias signal* to the record head while recording. The bias signal is a very high frequency—typically 50 KHz or higher. Since the bias frequency is above the audible range, it won't be heard when the tape is played. But it supplies the minimum signal level needed to exceed the magnetization threshold, thus shifting the audio into a range where tape magnetization is more linear. Tape bias also reduces background hiss, and its purity matters, too; a bias oscillator that outputs a low-distortion sine wave yields less noise.

Tape Pre-Emphasis and De-Emphasis

Applying a high-frequency bias signal reduces distortion significantly, but it improves tape's inherently poor signal to noise ratio only slightly. To solve that, clever engineers devised a method called *pre-emphasis* and *de-emphasis*. Pre-emphasis simply boosts the treble when recording, then reduces it by a corresponding amount during playback. Tape hiss is most noticeable at high frequencies, so this is a simple and elegant solution. There are two standards for pre-emphasis: NAB used in North America and CCIR/DIN used in Europe. NAB stands for the National Association of Broadcasters, and CCIR is the group Comité Consultatif International des Radiocommunications. DIN refers to the German standards group Deutsche Industrie-Norm. The two methods are similar but with slightly different EQ curves.

Chapter 1 explained the response time limitation of early-style VU meters and how you must record percussion instruments at a lower level than the meter shows to account for the meter's sluggish response. Because analog tape recorders use pre-emphasis, similar vigilance is required with instruments that have a lot of high-frequency content. Even if an instrument doesn't contain strong transients that come and go before the meter can respond fully, distortion can still occur before the meter reaches 0 VU. This is because the output level meter in a mixing console shows the volume *before* the recorder has applied pre-emphasis, boosting the treble. So when recording a bright sounding tambourine or shaker, you'll typically set the volume so the VU meter reads no higher than about −10. Further, different tape speeds apply different amounts of pre-emphasis. Only through experience will you know how to estimate record levels to avoid distortion.

I'll also mention that analog tape does not have a hard clipping point as do digital audio and most electronic circuits. So here, too, experience—and perhaps a taste for intentional subtle distortion—will be your guide when setting levels using traditional VU meters.

Most analog tape recorders are calibrated such that 0 VU is about 10 to 12 dB below the onset of gross distortion. However, there's no single standard, and users are free to calibrate their machines such that 0 VU corresponds to different levels of tape magnetization, measured in nanowebers per meter (nWb/m). In the 1950s, Ampex defined "standard operating level" such that zero on the VU meter corresponded to a magnetic intensity of 185 nWb/m. Modern tape can accept more magnetization before saturating than earlier formulations, so elevated levels of 250 or 320 nWb/m became the norm. These levels are often stated as a number of dB relative to the earlier standard, such as "plus 3" or "plus 6."

Note that a similar scheme is used with phonograph records to reduce their inherent hiss and scratch noises. In that case, high frequencies are boosted following the RIAA curve when the record master is cut, and de-emphasis built into every phono preamp reverses that boost when a record is played. RIAA stands for Recording Industry Association of America. Phono equalization also reduces low frequencies when recording and raises them when the record is played. However, low frequencies are reduced for a different reason: Pressing LPs with a flat low-frequency response will make the groove too wide, limiting the length of music that can fit on one side. Indeed, vinyl records have many foibles, including requiring mono bass that's equal in both channels.

Sel-Sync

Professional tape recorders contain three tape heads that erase, record, and play, in that order. As the tape moves, it first reaches the erase head, which erases any previous content using a strong signal at the bias frequency. The record head then magnetizes the tape with the audio being recorded. The play head outputs the recorded audio, which will be the audio recorded a moment earlier by the record head if the machine was currently recording. Some consumer tape recorders have only two heads, with one head used both to record and play back and the other to erase the tape. One important reason for separate record and play heads is to verify the result in real time while recording. That is, you listen to the playback while recording rather than the audio sent into the recorder. This also simplifies calibrating the recorder's bias and pre-emphasis: You can adjust those in real time rather than record tones, play them to verify, record again, play again, and so forth.

Because of the physical distance between the record and play heads, there's always a slight delay after audio is recorded before it's played by the play head. When new tracks are recorded as overdubs, adding more tracks to a tune in progress, the time displacement causes the new tracks to be offset in time. That is, the performer hears the prerecorded backing tracks as they come from the play head, but the record head is a few inches earlier. So when you play all of the tracks together, the newer tracks are out of sync, playing slightly later than the original backing tracks.

To avoid this delay, the record head is temporarily used for playback while overdubbing new material. This process is called *Selective Synchronization*, or Sel-Sync for short. The size and shape of a record head is optimized for recording rather than playback, so its frequency response may not be as good as a head optimized for playback. This is not a problem because it affects only what performers hear while recording. When the finished tape is played back as intended using the play head, the response is again normal.

Tape Noise Reduction

To further reduce the background hiss from analog tape, engineers devised a variety of *noise reduction* schemes. The two most popular noise reduction systems were developed by Dolby and dbx, both using a method called *companding*, named for the compression and expansion these devices use. Companding applies a volume compressor to the audio while recording; then on playback the audio passes through an expander that reverses the process. The compressor raises soft passages so they won't be drowned out by tape hiss, and then the expander lowers the volume by a corresponding amount during playback. This restores the audio to its original level while reducing the tape hiss.

Dolby originally offered two different systems: Dolby A for professional recording studios and Dolby B for consumer use with cassette tapes. Dolby A divides the audio into four frequency bands that are compressed separately, and then the bands are combined again before recording to tape. Note that Dolby acts on low-level signals only, raising them to stay above the tape hiss. At playback the process is reversed. The dbx system is broadband, operating over the full range of volume levels, but it adds pre-emphasis to compress high frequencies more than low frequencies. As long as the compressor and expander portions of such systems are calibrated properly, playback expansion exactly reverses the compression applied when recording. But since Dolby A splits the audio into four bands, precise and frequent alignment by studio operators is required. Further, frequency response errors in the recorder or added distortion prevents the expansion from exactly mirroring the compression.

Unfortunately, Dolby and dbx type companding is only half a solution because they don't really reduce the underlying noise level. It just seems that way. Companding merely raises the level of soft music to keep it from being dominated by noise. If a recorder has a signal to noise ratio of 50 dB, companding can't reduce the noise lower than 50 dB below the music. However, while the inherent signal to noise ratio remains the same, for most music the improvement is noticeable and most welcome. I mention this limitation of tape noise reduction because it's related to the examples in Chapter 3 that show the audibility of soft artifacts in the presence of louder music.

Years ago, dbx and Burwen made open-ended systems that reduce tape noise after the fact. The dbx device uses a dynamic low-pass filter whose cutoff frequency changes in response to the music. Burwen's unit was more sophisticated, manipulating separate low-pass and high-pass cutoff frequencies. Such processing would be trivial to implement today as a plug-in, though there are even more effective ways to reduce noise after the fact digitally. Chapter 13 explains software noise reduction in detail.

Tape Pre-Distortion

Unlike electronic circuits that are usually clean up to the point of hard clipping, analog tape distortion creeps up slowly, reducing dynamic range and softening transients. As shown in Figure 6.1, the transfer curve of analog tape slowly flattens as the recorded level increases past the tape's linear region. To combat this, the *tape linearizer* circuit was developed. This clever design was included in recorders made by Nagra, Scully, and MCI, and it reduces distortion by applying an equal but opposite nonlinearity while recording. If calibrated carefully, distortion of 3 percent can be reduced to less than 1 percent for a given signal level. Where tape compresses the highest positive and negative waveform peaks, pre-distortion intentionally *exaggerates* those peaks, yielding less waveform flattening on playback.

Some of these explanations are simplified; an entire book could be devoted just to the various engineering tricks needed to achieve acceptable high fidelity with analog tape. Analog recorders are very "tweaky," and *many* things affect their high- and low-frequency responses, distortion, and noise levels. This brings us to the following.

The Failings of Analog Tape

> *Coloration from analog tape and vinyl records is often revered, but for some reason color added by an A/D/A converter is never acceptable.*

My intent here is not to bash analog recording as much as explain the facts of audio fidelity and convenience. From my perspective, preferring digital recording over analog tape is a no-brainer for many reasons. Analog recorders are expensive, and none are currently manufactured. Unless you can troubleshoot electronics at the circuit level yourself, finding someone knowledgeable to service them is difficult outside of a metropolitan area. Further, finding replacement parts from a dwindling supply is a problem that will only become worse over time. Will any analog recorders even be around to play your archived tapes 20 years from now?

Then there's the high cost of blank tape, which is also becoming difficult to find. Many engineers prefer recording at 30 IPS, which uses twice as much tape as 15 IPS. Tape heads also wear unevenly over time, eventually requiring downtime while they're sent out for expensive relapping, unless you purchase a second set of heads as a backup.

Tape has a relatively high level of background noise, so if you record at too low a level—say, below −15 or −20 VU—the hiss is objectionable. But tape also distorts at higher levels, becoming obnoxious once the level gets much above +3 or so. This requires either carefully watching the recorded levels and adjusting them manually as the recording progresses or using a limiter to do that for you automatically. Of course, if you're recording 24 tracks at once, you need 24 limiters!

To operate optimally, analog tape recorders need constant electrical and mechanical alignment. The heads must be positioned and angled precisely left-right, top-bottom, and front-back, to say nothing of the many adjustments required in the record and playback electronics. If the pre-emphasis and de-emphasis do not match exactly, the frequency response will suffer. Even if they match perfectly within the recorder, they must also match the NAB or CCIR standard. Otherwise, tapes recorded in your studio will sound wrong in other studios, and vice versa. Because the frequency response of analog tape is relatively poor, it's common to record 0 VU tones at several frequencies so other studios can calibrate their machines to match yours, which takes time. Further, the ideal amount of bias applied when recording should be adjusted to match each reel of blank tape. So after recording for 15 to 60 minutes, depending on tape speed, the session comes to a halt before the next reel of blank tape can be used.

Analog tape heads need to be cleaned frequently and occasionally demagnetized. Tape also wears out with repeated use, gradually losing high frequencies, then eventually developing dropouts after the wear is severe enough that the slurry sheds and falls off. An elaborate pop tune can span many tracks and acquire many overdubs during the course of a large production. Every time you play the tape for the performer to practice to or to record along with, the tape wears a little more until eventually its sound is unacceptable. Tape is also fragile, and more than once I've seen an errant recorder snap the tape when its motors or brakes acted too quickly.

When doing overdubs with a tape recorder, you'll often record a part, then rewind to the same place if the performer needs to try again. Getting back to the same place is a time-consuming nuisance for the engineer and an inspiration killer for the artist. MCI developed an auto-locator for their recorders, which helped get to the same place on the tape. But it wasn't entirely accurate, or reliable, and you still had to wait while rewinding.

Overdubs are often done using a method called *punching in*, where portions of an otherwise good performance are recorded over parts deemed unacceptable. For example, if the bass player recorded a good take except for a short section in the middle of the tune, just that one part can be re-recorded, replacing the original performance. So you play the tape, including the parts already recorded successfully on the current track, then press Record a moment before the new part is to be recorded over the old one. But if you press Record too early, you'll erase the tail end of the good portion. And if you don't punch out quickly enough, you'll overwrite the next section that didn't need replacing. Doing tight punch-ins, where only a brief pause separates the existing and replaced passages, is a highly stressful part of any recording session.

In this day of computer DAWs, recording engineers can slice and dice music in numerous ways. If you like the background vocal section in the first chorus, it's easy to copy it to the second chorus rather than require the singers to record the same parts again. This is impossible to do with analog tape directly, though clever engineers would sometimes copy those tracks to another recorder, then copy them back again to the master tape at the new place. The downside is the quality suffers with every copy generation. Indeed, this is another important limitation of analog tape: There's no way to make a backup safety copy whose quality isn't degraded from the original. Editing all of the tracks at once on a multi-track tape is possible using a demagnetized razor blade and splicing tape, but it's risky. One mistake, and your entire project is ruined. There is no Undo with analog tape recorders.

It's possible to synchronize two or more recorders to obtain more than 16 or 24 total tracks using a method called *time-code*. The most common system is *SMPTE*, for Society of Motion Picture and Television Engineers. This method uses a phase modulated sine wave to store data, and when a SMPTE track is played as audio it sounds like an old-fashioned computer modem. SMPTE data identifies the current location on the tape in hours, minutes, seconds, and video frames. The original purpose of time-code was to synchronize audio and video recorders, but it can also synchronize two audio recorders, or a recorder and computer DAW. Because this tone is loud and obnoxious, most engineers recorded SMPTE on an outer track—either Track 1 or Track 24 for a 24-track machine—leaving the adjacent Track 2 or Track 23 empty. Or that track could be used for something like bass that won't be harmed by adding a high-cut EQ to filter out tones that leak through. Time-code is usually recorded at a level around −10 VU to further reduce leaking into adjacent tracks. Indeed, cross-talk is yet another limitation of analog tape recording.

Even when an analog tape recorder is operating optimally, its fidelity is still poor compared to modern digital recording. Besides relatively high levels of distortion and noise, analog tape also suffers from *print through*. This is an echo effect caused by adjacent layers of tape on the reel partially magnetizing each other due to their close proximity. Depending on which way the tape is wound when stored, the louder part of the echo occurs either before or after the original sound. Most engineers store analog tape "tails out" so the echoes come after the original sound, which is usually less noticeable. But it's still there. Tapes that are stored for many years have more print through than tapes stored for only a few days or weeks.

Another failing of analog tape is *flutter*, a rapid speed change that imparts a warbling sound, in addition to long-term speed variations. Flutter is usually low enough that you won't hear it on a professional-quality recorder that's well maintained. But long-term speed changes can make the pitch of music vary between the beginning and end of a reel. However, a different type of flutter, called *scrape flutter*, is audible and disturbing. This occurs when a short section of tape vibrates at a high frequency because there are no supporting rollers between the play and record heads, or some other part of the tape path.

Finally, some types of analog tape have aged badly, most noticeably certain types and batches produced by Ampex in the 1970s. When those tapes have been stored for many years, the binder becomes soft, causing oxide to deposit on heads and tape guides in a gooey mess, and layers of tape on the reel can stick together. Ampex came up with a solution that works well most of the time: baking the tape reel at a temperature of about 120 degrees for a few hours. But this is a risky procedure, and the tape can end up destroyed. Digital recording solves every single one of these problems. In fairness, however, some recording engineers are willing to overlook *all* of these failings in exchange for what they perceive as a sound quality that's more pleasing than digital recording.

Digital Recording

Chapter 8 explores digital audio principles in detail, so this section provides only a brief overview. Digital recording comprises two primary devices: a converter and a storage medium. The first is usually called an A/D/A converter because it converts analog voltages to a series of digital numbers when recording, then does the reverse when playing back. A/D means analog-to-digital, and D/A is digital-to-analog, and most professional outboard converters have both A/D and D/A sections. All computer sound cards also do both, and most today can do both at the same time, which is needed when overdubbing. Sound cards that can record and play at once are known as *full-duplex*. When using a computer to record digital audio, the storage medium is a hard drive inside the computer; an external drive attached through a USB, FireWire, or SATA port; or solid state memory.

When recording, the A/D converter measures the voltage of the incoming audio at regular intervals—44,100 times per second for a 44.1 KHz sample rate—and converts each voltage snapshot to an equivalent number. These numbers are either 16 or 24 bits in size, with more bits yielding a lower noise floor. During playback the D/A section converts the sequence of numbers back to the analog voltages that are eventually sent to your loudspeakers.

Digital recording is sometimes accused of being "sterile" and "cold sounding" by audiophiles, recording engineers, and the audio press. Modern digital audio certainly doesn't add coloration like analog tape, but it's much more accurate. In my opinion, the goal of a recording medium is to record faithfully whatever source you give it. Once you have everything sounding exactly as you want through the console and monitor speakers while recording, when you play back the recording, it should sound exactly the same. This is exactly what modern digital recording does. It may not add a "warm" coloration like analog tape, but it certainly isn't cold. Indeed, competent digital recording has no sound of its own at all.

Table 6.1: Comparison of Audio Fidelity Specs for Three Devices

Model	Frequency Response	Signal to Noise	Distortion
Studer A810 at 30 IPS	40 Hz – 20 KHz +/−1.0 dB	74 dB A-weighted	No spec, level-dependent
SoundBlasterX-Fi	20 Hz – 20 KHz +0/−0.5 dB	>100 dB A-weighted	<0.007%
Delta 66	22 Hz – 20 KHz +0/−0.3 dB	99 dB A-weighted	<0.002%
LavryBlue M AD-824	10 Hz – 20 KHz +/−0.05 dB	113 dB Unweighted	<0.002%

Table 6.1 compares performance specs for a professional Studer analog recorder, plus three digital converters ranging from a $25 SoundBlaster X-Fi consumer sound card through a high-performance LavryBlue model. If transparency is the goal for a recording medium, and I think it should be, even the $25 sound card beats the Studer by a very large margin.

In the Box versus Out of the Box

In this case "the box" refers to a computer. As mentioned earlier, I prefer to do all recording and mixing inside a computer, or In the Box (ITB). I don't even use the Console View in SONAR, which emulates the appearance and controls of a hardware mixer. Everything needed is already available in the Track View, leaving more of the video display available for editing and viewing plug-in settings. But a computer can also be used with external hardware that's Out of the Box, or OTB. In that case a computer runs the DAW program, but the tracks (or subgroups of tracks) are routed through separate D/A converter outputs to an analog mixing console where the actual mixing takes place.

Proponents of OTB mixing believe that analog console summing is superior to the summing math used by DAW software. And again, some people are willing to pay much more for their preferred method. This includes the dollar cost of a mixer, as well as the convenience cost when you have to exactly recreate all the mixer settings each time you work on a previous project. DAW software can save every aspect of a mix, including volume and pan automation changes and automation for every parameter of every plug-in. You can open a project next week or next year, and when you press Play, it will sound exactly the same. You can also render a final mix in one step, which usually happens faster than playing a tune in real time while saving to a Wave file or other recording device.

Using a hardware mixer with outboard effects requires making detailed notes of every setting. Even then it can be difficult to create an identical mix. Many controls on hardware mixers use variable knobs rather than switches that can be set precisely. If you make a note that a pan knob is at 1 o'clock, maybe you'll set it exactly the same and maybe you won't. Electronics also drift over time. Perhaps your converter's input or output level knob shifted by half a dB since you made the last mix. Or maybe you changed the console's input level for one channel three months ago when you had to patch in some odd piece of gear but failed to put it back exactly. When working entirely ITB, recall is complete and precise.

There are many other advantages of working ITB besides perfect recall. One big feature for me is being able to buy a plug-in effect once and use it on as many tracks as needed. Further, plug-ins are perfectly repeatable if you enter the same parameter settings, and the left and right channels of a stereo plug-in always match exactly. Plug-ins also have less noise and distortion than analog outboard gear, limited only by the math precision of the DAW host program. You can also upgrade software more easily and for less cost than hardware, assuming the hardware can be upgraded at all. Even if a hardware company offers an upgrade, it likely must be returned to the factory. Upgrades for software fixes are often free, and new versions are usually cheaper than buying the program all over again as with hardware. Plug-ins never break down, nor do their switches and pots become noisy or intermittent over time. However, if a software company goes out of business and the software doesn't work on your next computer's operating system, you're totally out of luck with no recourse.

Record Levels

It's difficult to determine the optimum recording level when using analog tape because it depends on many factors. Recording at low levels yields more tape hiss, but recording louder increases distortion. And as explained previously, instruments that create strong transients or have a lot of high-frequency content such as cymbals and tambourines can distort even when the VU meter shows a relatively low level. This is due to both the VU meter's slow response time and pre-emphasis that boosts high frequencies inside the recorder but is not reflected by the console's VU meters. So when recording to analog tape, it's common practice to include sufficient *headroom*. This is the difference in decibels between the average and peak volume levels you can record without objectionable distortion.

Digital recording avoids the need for extra headroom just to accommodate the recording medium. An A/D/A converter is perfectly clean right up to 0 dB full scale. Some people believe that digital recording sounds better when recorded at lower levels, such as −18 dBFS or even softer, but this is easy to disprove by measuring as shown in the next section. Table 6.2 compares typical THD + Noise at 1 KHz versus input level for a LavryBlue M · AD-824 converter.

As you can see, this converter adds less distortion at high levels rather than more, as some people believe. This makes perfect sense because a higher input level keeps the signal that much louder than the converter's inherent noise floor. However, this doesn't mean that you should aim to record everything as close to 0 dBFS as possible. Many singers and musicians

Table 6.2: Distortion versus Signal Level

Signal Level	THD+ Noise Level	Equivalent Distortion
−1 dBFS	−98 dBFS	.001%
−3 dBFS	−102 dBFS	.001%
−10 dBFS	−109 dBFS	.001%
−20 dBFS	−112 dBFS	.003%

will be louder when they actually record than when they rehearsed while you set the record level.

Even if you record using "only" 16 bits, the background noise of digital audio is 96 dB below the maximum level. This is more than 20 dB quieter than the best analog recorders. Further, even in a professional studio that's well isolated from outside sounds and has quiet heating and air conditioning, the ambient acoustic noise floor is more often the limiting factor than the digital recording medium. I usually aim for record levels to peak around −10 dBFS. The resulting noise floor of the digital medium is still very low, yet this leaves enough headroom in case an enthusiastic performer gets carried away. If you often record amateur musicians, you could record even lower than −10, or you could add a limiter to the recording chain to be sure the level never exceeds 0 dBFS.

As we have seen, the traditional notion of headroom is irrelevant with modern digital recording, other than to accommodate the performers and the limits of your analog inputs and outputs. At the time I wrote the first edition of this book my home studio was configured as shown in Figure 5.11, with each channel insert of a Mackie mixer going to one input of an M-Audio Delta 66 sound card. The Delta 66 offers three operating levels that can be set independently for the input and output sections: +4 dBu, −10 dBV, and an in-between level that M-Audio calls "consumer." I kept the input sensitivity set to +4 dBu. This requires more output from the mixer's preamps, which in turn puts the recorded signal that much above the mixer's own analog noise floor. At this setting the sound card reaches 0 dBFS with an input level of +5 dBu. This is more than adequate to raise the audio signal level well above the mixer's noise floor, but it never gets anywhere close to the clipping point of either the mixer or sound card.

The Myth of Low Levels

Conventional wisdom says that setting digital record levels to peak around −20 dB below full scale sounds better than recording at levels closer to the 0 dBFS maximum. One popular web article claims that recording at lower levels avoids "recordings that sound 'weak' or 'small' or 'too dense' or 'just not pro enough,'" among other complaints. However, the specs for most sound cards and outboard digital converters show a similar frequency response and distortion amount for a wide range of input levels. So it doesn't make sense that recording at low levels should sound better or even different than recording closer to full scale, assuming the playback levels are eventually matched. Further, as explained in Chapter 8, most modern DAW software uses 32-bit floating point (FP) math, so signal levels within the software should have no effect on sound quality either. If recording at lower volume really does sound better in a given setup—as proven by a proper level-matched blind test—it's more likely the result of clipping due to improper gain staging elsewhere in the analog portion of the signal path.

The benefit of 32-bit FP math is explained in detail in Chapter 8, so I won't elaborate further here. The main point (no pun intended) is that using 32-bit FP math accommodates a range of signal levels exceeding 1,500 dB. Therefore, most plug-in effects sound exactly the same no matter what volume level you send through them. There's no hiss even when levels are very low, and there's no distortion even when signals are hundreds of dB above typical levels. To be clear, some plug-ins *are* affected by signal level, particularly those that add intentional distortion or "vintage analog" effects when overdriven. But most plug-ins accommodate the same huge range of levels as their DAW software host. There should be no audible or measurable difference between mixing a group of tracks that were each recorded near full scale versus mixing tracks that are all 20 dB softer and raising the level later in the chain. As long as the master output volume is adjusted to avoid distortion in the rendered Wave file, both mixes will sound exactly the same.

In the web article mentioned above, the author suggests creating an entire project with each source recorded simultaneously onto two tracks at levels 20 dB apart, then making parallel mixes to prove that recording at lower levels sounds wider, clearer, and generally better overall. But apparently he never actually did that test, or he would have realized he's wrong! The author also claims the improvement in audio quality is more apparent with a larger number of tracks due to the stacking effect, but that's also a myth, disproved in Chapter 2. Of course, it's a good idea to keep levels low when recording a live concert, for safety to avoid distortion if an unexpected loud passage comes along. But that's a different issue, and it's easy to show that the recorded quality *per se* is not improved by recording at lower levels.

To save you the bother, I did this test and the results confirm what should have been obvious. Since it wasn't practical for me to record an entire band in my home studio, which requires dozens of microphones and simultaneous input channels, I did the next best thing: I played each track of an existing song one by one, and re-amped (re-recorded) them using a loudspeaker and microphone to two tracks at the same time with the levels 20 dB apart. (Re-amping is described in more detail in Chapter 7.) Whether re-amping a loudspeaker captures the same sound as a microphone in front of a singer or drum set is irrelevant. The source simply is what it is, and a loudspeaker source is as valid as any other to disprove a myth relating to record levels. For this test I used the 8-track "master" of the old Motown hit *Ain't No Mountain High Enough* that made the rounds a while ago as a group of eight Wave files.

The photo in Figure 3.14 shows the basic setup with my large JBL 4430 loudspeaker, though for this test I used my precision DPA 4090 omni microphone 18 inches in front of the center of the horn. (Skip down to Figure 6.10 to see this microphone pointed at my Fender Sidekick guitar amp.) The output of the mic preamp went through a Y-splitter, then into both line inputs of my Focusrite sound card. I set the levels so that one track of each recording peaked at least above −6 dBFS, with the other 20 dB softer. I didn't change the record levels as each existing track was recorded to a new pair of tracks. I also recorded at 16 bits instead of 24 to make this a worse-case test. Since I had already made a basic mix of these tracks months earlier, I simply solo'd each track as I recorded each pair of re-amped copies. Both groups of recorded tracks were sent to their own output bus whose volumes were set to peak just below 0 dB, then rendered to separate Wave files. I exported only part of the tune to keep the file sizes reasonable. These are online as "levels-mixa.wav" and "levels-mixb.wav" in the section for Chapter 6.

Your mission is to identity which mix was made from the files recorded so they peak near 0 dB, and which mix came from the files that were recorded around −20. If you think you hear a difference, email me from my website ethanwiner.com and I'll send you the results. In all honesty, both mixes sound exactly the same to me, but maybe others have better (and younger) ears than mine. However, when I nulled the two mix files to hear the remaining difference signal, the residual was down around −50 dB, which is extremely soft. I also included the file "levels-nulled.wav" online, to save you the bother of loading both mixes into a DAW to null them yourself.

Recording Methods

Before you hit Record, it's important to verify the quality of whatever you're recording. Listen carefully to all musical sources in the room to identify problems such as rattling drum hardware, squeaky kick drum pedals, mechanical and electrical buzzing from guitar amps, and so forth. It's easy to miss small flaws in the excitement of a session, especially if you're both the engineer and bass player. It's much better to fix problems now rather than curse them (and yourself) later. The same applies to the quality of the source itself. If a drum is badly tuned and sounds lame in the room, it will sound just as lame after it's recorded. If a guitar is out of tune when recording, it will still be out of tune when you play it back later.

When recording others, I suggest that you record every performance, including warm-ups. Some musicians start off great but get worse with each take either from nervousness or exhaustion.

In fact, there's nothing wrong with a little white lie while they're rehearsing. Tell them to take their time and let you know when they're ready for you to record, but record them anyway. It may end up being their best take of the session. Related, when recording inexperienced musicians, I always make a point of telling them there's nothing to be nervous about and that every mistake can be re-recorded or fixed. Often, the best and most successful recording engineers are "people persons" who have a calm demeanor and know how to make their clients feel comfortable.

As mentioned in Chapter 5, adding reverb to performers' earphones is always a good idea because they sound better to themselves, which in turn helps them to play or sing with more confidence. Whether recording analog or digital, ITB or OTB, it's easy to add reverb or EQ or any other effects to a performer's cue mix without recording and committing to those effects. This is not just for amateurs either. I always add reverb to the earphones when I'm recording myself and other professional musicians. It's also important for people to hear themselves at a comfortable volume in their headphones. If a singer's own microphone is too loud, he might sing softly to compensate, which can make him sound timid. But if it's too soft, he might strain or even sing out of tune if he can't hear himself well enough. So always confirm headphone volume with the performer.

Pop bands that have a drummer generally prefer to let the drummer set the tempo, which is as it should be. In classical music a conductor does the same. But some pop music projects are built from the ground up track by track. In that case a *metronome*, or *click track*, is helpful to establish the tempo for all of the overdubs to come. Some programs offer a *tap tempo* feature that lets you click a mouse button repeatedly to establish the pace. The tempo can also be changed over the course of a tune if desired. This usually requires entering tempo data manually into your DAW software, though tempo changes are perhaps more common with classical music than pop tunes. Once the tempo has been set, the DAW software automatically plays a click sound while each new track is being recorded. Most DAWs also let you pick the metronome sounds that play. I generally use high- and low-pitched click sounds, where the higher pitch on beat 1 is louder than the lower-pitched click on beats 2 through 4: *TICK tock tock tock*. Some people prefer to hear open and closed high-hat samples instead of a click sound. Either way, I suggest you keep the metronome volume as soft as possible but still loud enough to easily follow. If clicks are played loudly, a nearby microphone may pick up the sound, especially if the performer is wearing open-back earphones.

My own pop music productions tend toward long and extravagant, with many different sections. I generally write music as I go, making a mock-up of the piece using MIDI samples that will be replaced later with live playing. Once a tune is complete and I'm satisfied with the arrangement, I'll go through the entire piece carefully, deciding on the final tempos for each section, which I enter manually into SONAR's Tempo View. Changes might be small or drastic, depending on the music and context. Even subtle changes of just a few beats per minute (BPM) can add a nice ebb and flow. Tempos can also change over time rather than jump suddenly from one to another. In classical music such tempo changes are called *accelerando* and *rallentando*—progressively speeding up and slowing down, respectively. Click tracks are also used with analog recorders, with the clicks recorded to a dedicated track in real time before any other tracks are added.

Specific Advice on Digital Audio Workstations

In the days of analog multi-track tape, we used *track sheets*—printed paper forms containing a written description of what's on each track, typically stored along with the tape in its cardboard box. DAW software lets you assign meaningful track names that show on the screen, and you can even move tracks up or down to group related tracks logically for convenience. Many DAW programs also let you assign track colors to better identify the tracks in a group. So you could make all the drum tracks orange, acoustic guitars blue, or whatever you prefer.

A recording template is a useful way to begin and organize your sessions. This is simply an empty project having no recorded audio but with all of the inputs and outputs assigned. If you add a few Aux buses with plug-ins such as reverb and echo already patched in, you'll be ready to record an entire band at a moment's notice. I suggest you add everything to the template you could possibly need, including MIDI software synthesizers for an organ, drum machine, and whatever else you might want. You can easily delete any tracks that aren't needed or keep the tracks but hide them from view if your DAW software has that feature. Then each time you start a new project by loading the template, use Save As immediately to save the new project with the proper name. You could also set the template file to Read Only to avoid accidentally overwriting it.

SONAR has a feature called Per-Project Folders, which stores all of the files for a tune under a single folder on the hard drive. Many other DAWs can do the same. I recommend this method of organizing your projects because you'll know exactly what files need to be backed up and where they are on the drive. If the Wave files for a tune are scattered in various locations, you may not remember to include them all in a backup. Also, if you import Wave files from another project or use Wave files from a sample library, I suggest you copy them to the current song's folder. This way, everything needed to open the project is available in case you have to restore the project to another hard drive after a crash or if you bring the project to another studio.

Most DAW software creates an "image" file for every Wave file in the project. This is similar to a GIF image graphics file, and it holds a picture of the waveform your DAW displays while you work on a tune. There's no need to back up these files because the software recreates them as needed. I have SONAR set to store all image files in a dedicated folder just so I won't waste time and disk space backing them up.

Earlier I mentioned a record method called punching in, where you can replace part of an otherwise acceptable performance with a new version by hitting Record on the fly as the tape plays. Most DAW software lets you work that way, but to me punching in is so 1980s. In my opinion, it's much better to record to successive tracks repeatedly. All DAW software lets you define a Start and End region where recording will begin and end automatically. This makes it easy for a musician to get "in the zone" while recording the same part again and again until he's satisfied. The main advantage of this method is there's no risk of accidentally overwriting a good performance. Rather, the software records each successive take to a new track until you press Stop. Some DAWs can record multiple takes—called *layers*, *lanes*, or *virtual tracks*—within a single track. Once you're satisfied that you have a good take in there somewhere, it's easy to edit the tracks or layers to make a single performance from all the various pieces. Editing multiple takes down to a single composite performance is called *comping* and is discussed more fully in Chapter 7.

Copy Protection

My intent with this book is to present the facts about audio more than my personal opinions, but I'll make an exception for what I consider an important issue with all software: copy protection. There are many types of copy protection, and only criminals refuse to acknowledge that software companies deserve to be paid for every copy of their programs that's used. But often the people who suffer most from the inconvenience of copy protection are honest consumers who paid full price for the programs they use.

Copy protection comes in many forms, and it is an attempt by manufacturers to justifiably limit the use of their software to people who actually buy it. The simplest form of protection requires you to enter a serial number when the program is first installed. In practice this protects very little, since anyone can lend the installation disks to a friend along with the serial number. All this does is minimize the chance that someone will upload the program to a web torrent for strangers to retrieve. Since they'd have to include the serial number, that number could be used to identify them—unless they never registered the program in the first place.

A more severe form of copy protection uses a device called a dongle; the most popular system currently is the iLok. Years ago dongles were included for free with the software, but the iLok must be purchased separately, in addition to the software it protects. Old-style dongles plugged into a computer's parallel or serial port, though these days a USB port is standard. If the dongle is not detected, the program refuses to run. USB ports are a big improvement over the old days. I recall years ago visiting a local music store that had many different protected programs installed on its main demo computer. There were half a dozen dongles connected to the computer in a chain, sticking a foot out the back. One day someone bumped into the computer, and it toppled over, snapping off all the dongles. Ouch.

Another protection method, called challenge/response, requires you to phone or email the manufacturer when the program is installed or fill in a form online. After you enter your name and address and the program's serial number, you receive a second code number that's needed along with the main serial number before the software will work. I remember well one Saturday a few years ago when I was having a problem with Sony Vegas Pro, the video editing program I use. I decided to reinstall Vegas to see if the problem would go away. After I reinstalled Vegas, it "phoned home" to verify my ownership. Unfortunately, Sony's website was down. The program that had worked just minutes earlier now refused to open at all. And being a weekend, nobody at Sony was available to help me by telephone. I lost an entire weekend that I could have been working on my project.

It would be difficult to condemn copy protection if it protected publishers without harming legitimate users. Unfortunately, it often does harm legitimate users and rarely thwarts software pirates. Some older protection schemes interfere with disk optimizers, requiring you to uninstall all of the programs each time you defragment your hard disk and then reinstall them all again after. I defragment my system hard drive occasionally, and having to uninstall and reinstall a few dozen programs and plug-ins every time would be a terrible nuisance! Admittedly, this is less of a problem today now that hard disks are cheap and project files are usually kept on a separate drive that can be defragmented independently.

Any copy protection scheme that requires intervention from the publisher has the potential to cause you disaster. Suppose you're working on a project and your hard disk fails. So you go to the local office supply store and buy another, only to learn that you already used up your two allowable installations. Even the seemingly benign method of phoning the vendor for an authorization number is a burden if you're working on a weekend and can't reach them. Or the dongle could simply stop working. You're in the middle of a project with a client paying $150 per hour, but you're totally hosed because even with overnight shipping, the new dongle won't arrive until tomorrow.

The ultimate disaster is when a software vendor goes out of business. In that case you can forget about ever getting a replacement authorization or new challenge/response code. I have *thousands of hours* invested in my music and video programs. This includes not only the time spent creating my audio tracks, MIDI sequences, printed scores, and video edits, but also the time it took me to learn those programs. This is one important reason I chose SONAR: It uses a serial number, plus a challenge/response number you need to obtain only once. Once you have both numbers, they'll work on subsequent installations, even to another computer.

Microphone Types and Methods

Microphones are at the heart of almost every recording. Just as important is the acoustic environment in which you use them. Chapter 17 explains the inner workings of microphones, so this section addresses mainly how to use them. I'll admit up front that I'm not a big fan of dynamic microphones generally, though many pros love them. I prefer condenser microphones that have a flat response, with either a cardioid or omnidirectional pickup pattern as appropriate, to capture a clear sound that's faithful to the source. Few dynamic microphones have a response as flat as condenser models, and even fewer have a response that extends to the highest frequencies. However, you don't need a response out to 20 KHz to capture a good tom or kick drum sound, and any decent dynamic mic is fine for that.

Unlike audio electronic gear that's usually very flat, the frequency response of microphones can vary wildly. Some have an intentional "presence" boost in the high-mid or treble range. All directional microphones also have a *proximity effect*, which boosts low frequencies when the microphone is placed close to a sound source. This is why many cardioid mics include a built-in low-cut filter. Conventional wisdom says you should choose a microphone whose frequency response complements the source you're recording. Again, I'd rather use a flat microphone and add EQ to taste later when mixing, when I can hear all of the parts in their final context. But this is just my opinion, and some pros may disagree. And that's fine.

Microphones are categorized by how they convert acoustic sound waves into electrical voltages and also by their pickup patterns. Dynamic microphones use a thin plastic diaphragm attached to a coil of wire surrounding a permanent magnet. Ribbon microphones are similar and are in fact also considered dynamic because they use electromagnetism to generate a voltage. But ribbon mics use a single thin metal ribbon suspended in the field of a permanent magnet; the ribbon serves as both the diaphragm and the coil. Condenser microphones also have a plastic diaphragm, coated with an extremely thin layer of metal. As sound waves displace the diaphragm, it moves closer or nearer to a fixed metal plate. Together, the diaphragm and fixed plate form a capacitor whose capacitance changes as the diaphragm moves in and out, nearer or farther to the fixed plate. A special preamp built into the microphone converts the varying capacitance to a corresponding electrical voltage. Note that many condenser microphones require a DC *polarizing voltage* to operate. Condenser mics that don't require external polarization are called *electret* condensers, named for the property that lets them hold a permanent DC charge.

A third type of microphone category classifies them by the size of their diaphragms. Most microphones are either large diaphragm or small diaphragm, typically an inch or larger in diameter, or half an inch or smaller. There are also microphones that I call "tiny diaphragm," about ¼ inch or less in diameter. When all else is equal, mics that have a large diaphragm give a better signal to noise ratio because they capture more of the acoustic sound wave and output a higher voltage. Conversely, microphones having a small diaphragm often have a flatter high-frequency response because a diaphragm having a lower mass can vibrate more quickly. They also tend to be flatter because their diaphragm size is smaller than one wavelength at very high frequencies. Small-diaphragm microphones are prized for their flat, extended response, making them a popular choice for acoustic guitars and other instruments having substantial high-frequency content. Note that there's no rule for categorizing microphones by their diaphragm size. The large, small, and tiny size names are merely how I describe them.

As mentioned, active microphones—all condenser types and some newer ribbons—contain a built-in preamp. Therefore, less gain is needed in your mixer or outboard preamp with those types compared to passive dynamic microphones. Some people are concerned about the potential interaction between a microphone and its preamp, but with active microphones, this interaction has already occurred inside the mic. Any further interaction is between one preamp and another.

I'll also mention wireless microphones, which are popular for recording live performances. These are used mostly by singers because it frees them from the constraint of a mic wire and gives animated performers one less thing to worry about tripping over. Wireless guitar and bass transmitters are also available, sending the electrical signal from the instrument to an amplifier elsewhere on the stage. In-ear monitors (IEMs) are equally popular with performers at live shows, and these can be either wired or wireless. Using a wireless microphone or guitar transmitter with a wireless IEM system gives performers a huge degree of freedom. But don't forget to recharge all those batteries before every performance!

Micing Techniques

The first decision you need to make when choosing where to place microphones is whether you want a close-up dry sound or more ambience and room tone. Room tone works best in large rooms—say, at least four or five thousand cubic feet or larger. The room tone in a bedroom is generally very poor, having a small-room sound that's often boxy and hollow due to comb filtering and early echoes from nearby boundaries. In small rooms I suggest placing microphones as close to the source as is practical. You can easily add ambience and reverb later when mixing. An external reverb will have a better sound quality than the real ambience from a small room, even if your only reverb is an inexpensive plug-in.

At the risk of being too obvious, the farther away you place a microphone, the more room tone you'll capture. I remember my first visit to Studio A at the famous Criteria Recording in Florida, being surprised at how reverberant the live room sounded. But the natural reverb has a very neutral quality, thanks to the very large size of the room and the presence of both absorption and diffusion. So engineers there can control the ratio of direct to ambient sound entirely with mic distance when recording. For the rest of us, when in doubt, record closer and dryer because it's easy to add reverb to taste later when you can hear everything in context.

Another basic decision is whether to record as mono or stereo. I know professional recording engineers who always record single instruments using only one microphone, but most record in stereo at least occasionally. For pop music projects, recording each instrument in mono makes a lot of sense. Using two microphones, with each panned fully left and right, can make an instrument sound too large and dominating. That might be just what you want for a solo act where one person plays piano or guitar and sings. And obviously a classical recording of a solo piano or violin benefits from a full-sounding stereo field. But a trumpet or clarinet overdub on a pop tune is often best recorded in mono, then panned and EQ'd to not interfere with the other instruments.

Another option is to record in stereo with two microphones but not pan them fully left and right when mixing. For example, one microphone might be panned fully left, with the other panned to the center or slightly left of center. A string or brass section might be more suitable in stereo to convey the width of the section rather than sound like five players are all standing in the same place, which is unnatural. There are no hard and fast rules for this, and the requirements vary depending on the session. You could always hedge your bets by recording in stereo, then decide whether to use one or both microphones when mixing. Even if you know you'll use only one microphone in the mix, recording in stereo lets you decide later which of the two mic placements sounds better. Ideally, a true stereo recording will use a matched pair of microphones having identical frequency responses. When that's not possible, both mics should at least be the same model.

As mentioned, microphones have different directional properties. There are three basic pickup patterns, with a few minor variations. The most popular pattern is probably *cardioid*, named for its resemblance to a stylized heart shape, as shown in Figure 6.2. Another basic pickup pattern is called *Figure 8*, or *bi-directional*, as shown in Figure 6.3. This pattern is typical for ribbon microphones due to their open design that exposes both sides of the ribbon. Many large diaphragm condenser microphones also offer Figure 8 as one of several pickup patterns.

The third basic pattern is *omnidirectional*, often shortened to omni. In theory, an omni microphone doesn't favor any direction, responding equally to sound arriving from every angle. In practice, most omni mics slightly favor sound from the front at the highest frequencies. Some microphones offer more than one pickup pattern. For example, the Neumann U 87 has a three-position switch to select cardioid, omnidirectional, or Figure 8. Other microphones omit a switch but are sold with optional capsules having different patterns

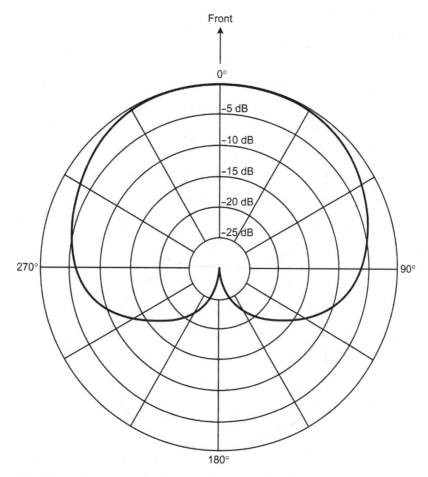

Figure 6.2: The cardioid microphone pickup pattern is named for its heart-like shape when graphed. Sound arriving from the front is typically 5 dB louder than sound arriving from either side and 20 dB or more louder than sound coming from the rear.

that screw onto the mic's body. Ribbon microphones are usually Figure 8, but most non-switchable dynamic and condenser mics are either cardioid or omnidirectional.

One great thing about a Figure 8 pattern is it rejects sounds arriving from both sides completely at all frequencies. The pickup pattern of cardioid microphones generally varies with frequency, having poorer rejection from the rear and sides at the lowest frequencies. This is an important consideration with mic placement because musical instruments and other sounds that are picked up off-axis by a cardioid mic will sound muddy compared to the brighter (flatter) sound of sources captured on-axis. Many professional recording engineers

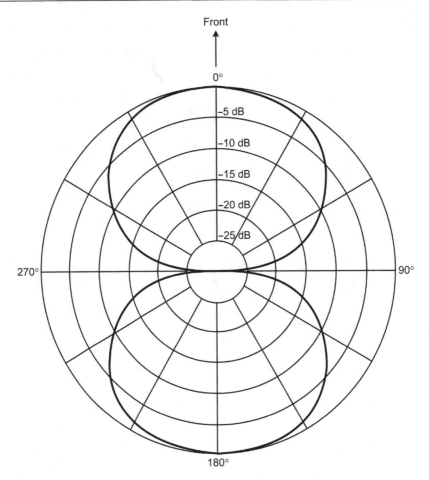

Figure 6.3: A Figure 8 microphone captures sound equally from the front and rear, rejecting completely sound arriving from the sides.

believe it's better to have more leakage that sounds clear rather than less leakage that sounds muffled.

I'll also mention a fourth pattern called *supercardioid*, which is even more directional than cardioid. The supercardioid pattern is common with shotgun microphones, those very long tube-shaped mics attached to a pole that news sound crews use for interviews on the street. The main advantage of a supercardioid pattern is it lets you pinpoint a sound source from farther away, which helps when recording in noisy or reverberant environments.

There are a number of common stereo micing arrangements. One is called X/Y, where a pair of cardioid microphones is placed with their tips adjacent but pointing in opposite directions. Figure 6.4 shows a pair of AKG C-451 cardioid mics set up in an X/Y pattern to capture a

Figure 6.4: This pair of AKG microphones is arranged in an X/Y pattern, with the tips together but pointing left and right.

solo cellist in stereo in my home studio. One big advantage of X/Y placement is it avoids comb filtering due to phase cancellation if the recording is played in mono. No matter which direction a sound arrives from, it reaches both microphones at the same time. A popular X/Y variant used when recording orchestras or other large groups is the Blumlein Pair, named for its inventor Alan Blumlein. In this case, two Figure 8 microphones are used instead of cardioids, and they're angled at exactly 90 degrees. Since the Blumlein method captures sound from both the front and rear, you'd use it when you want to capture more of the room sound.

Another popular stereo arrangement is *spaced omni* or *spaced cardioid*, using two microphones spaced some distance apart. For solo instruments in a studio setting, the microphones can be as close as a few feet, as shown in Figure 6.5. For larger ensembles, a wider spacing is appropriate, as shown in Figure 6.6. You'll sometimes see mics spaced only five or six inches apart to emulate the space between our ears for recordings meant to be heard through earphones.

The strings session in Figure 6.6 was for a recording that emulated an orchestra, so the sound was mic'd to be intentionally large and wide. For this session there were two first violins, two second violins, two violas, and two cellos. Each passage was recorded three times to further give the illusion of a full orchestra. Wide spacing with distant microphones doesn't usually

Figure 6.5: These Audio-Technica AT4033 cardioid microphones are spaced about three feet apart, about two feet in front of the performer.

Figure 6.6: For this recording of an eight-piece string section, the AT4033 microphones were placed farther apart to better convey the width of the section. Only one mic is visible at the top of the photo; the other is at the left outside the frame.

work well in small rooms, but my studio is 33 feet front to back, 18 feet wide, with a ceiling that peaks at 12 feet in the center. You can hear this recording in the example file "string_section.wav."

A variation on the spaced omni method is called the *Decca Tree*, developed in the 1950s by recording engineers at Decca records. This method spaces the left and right omni microphones about six feet apart, but also adds a third omni microphone in the middle about three feet forward of the outer mics. This method is popular for recording classical orchestras and movie sound tracks, with the mic array placed high above the conductor's head. When mixing, the outer mics are panned hard left and right, and the center mic is panned to the middle. A Decca Tree takes advantage of specially chosen omni mics that are somewhat directional at high frequencies. The original version used Neumann M50 mics, and some people claim that this arrangement doesn't work as well with other omni mic models.

The last stereo microphone arrangement I'll mention is called *Mid/Side*, and it's based on a very different principle than the other micing methods. Two microphones are used—one is usually cardioid and the other is always a Figure 8. The cardioid mic points forward, and the Figure 8 is placed adjacent directly above or below, with its null facing forward. With this arrangement, the cardioid mic captures the center information, and the Figure 8 captures both the left and right side ambience only. Individually, the mid mic, when the array is properly placed, will give a good mono picture of the source, and the side mic will sound odd and hollow. When mixed together, the cardioid microphone is panned to the middle, and the Figure 8 is split left and right, with the polarity of one side reversed. Both the microphone placement and electrical mixing setup are shown in Figure 6.7.

The main selling point of Mid/Side micing is that you can adjust the width of the stereo field after the recording is made by raising or lowering the volume of the Mid microphone. Mid/Side micing also avoids phase cancellation when the stereo mix is played in mono. Note that the more common X/Y micing technique also lets you control the width later, simply by panning the mics slightly toward the center rather than fully left and right.

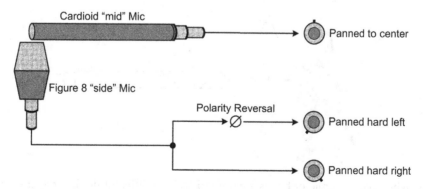

Figure 6.7: Mid/Side stereo recording uses two microphones. One is typically a cardioid type facing toward the sound source, and the other is always a Figure 8 facing away from the source to capture sound from both sides only.

The 3-to-1 Rule

As mentioned earlier, when sounds leak into microphones meant for other performers, the result is excess ambience and echoes. But leakage also causes comb filtering if the delay time is less than about 20 milliseconds and the unwanted sound is fairly loud compared to the desired sound. A general rule for mic placement is to keep unwanted leakage at least 10 dB below the desired sound. Chapter 1 showed the relationship between dB volume levels and acoustic distance, with the level falling off at 6 dB per octave each time the distance is doubled. Obtaining a 10 dB level reduction requires a difference of about three times the distance, as shown in Figure 6.8.

When the volume of the acoustic sources are the same, and the gain of both preamps are equal, this yields peaks of about 2 dB and nulls about 3 dB deep for the unwanted signal as captured by the main microphone. The same happens for intended sound leaking into the other microphone. Leakage doesn't always sound bad, but minimizing comb filtering can only help. Of course, if one source is louder than the other, or the preamps use different

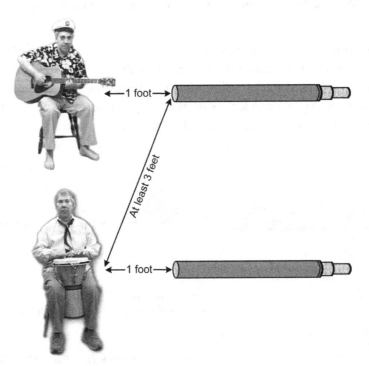

Figure 6.8: When separate microphones are used to capture different performers, the 3-to-1 Rule requires the unwanted source to be at least three times farther from the microphone than the intended source to avoid comb filtering. This assumes that both instruments are sounding at the same acoustic volume and are also mixed at equal levels.

amounts of gain to compensate, the peak and null amounts will vary. The same goes for instruments that are not mixed at the same level for artistic reasons, such as when a rhythm acoustic guitar is mixed at a lower level than a lead acoustic guitar. It should be noted that the 3-to-1 Rule is often quoted but that ratio applies only to omni mics on sources that are the same volume. Since recording setups vary widely, it may be best to forget rules like this and just fix any problems when you hear them.

Microphone Placement

In a large room it almost doesn't matter where you put the performers as long as they're not too close to reflecting walls. But for the small rooms many people record in today, it's difficult to *not* be near a wall or low ceiling. A good general rule is placing the musicians and microphones at least 10 feet away from any reflecting surface. As explained in Chapter 3, reflections arriving within about 20 milliseconds of the direct sound are more audibly damaging than reflections that arrive after 20 milliseconds.

Sound travels at a speed of about 1 foot per millisecond. So after a round trip to a boundary 10 feet away and back, the reflections arrive 20 milliseconds later. Reflections that arrive "late" do not create severe comb filtering, partly because they're softer due to distance. When recording in a small room, the best solution is to put absorbers or diffusers on nearby surfaces, including the ceiling. Chapter 21 explains these treatments in much more detail, so for now I'll focus mainly on practical considerations.

The most important factors affecting the quality of sound captured by a microphone are its distance from the source and its directional pattern and angle to the source. Directional microphones can be aimed toward one instrument and away from others playing at the same time. If you mainly record yourself playing one instrument at a time, you don't have to worry so much about angles and off-axis rejection. However, where the mic is placed, and how close, is still important. All directional mics have a proximity effect that boosts bass frequencies when placed close to the source, though this is easily countered later with EQ when mixing. Or you can use an omnidirectional microphone to avoid proximity effect. Close-micing is common for most pop music, but even orchestra recordings often put a "spot mic" close to a soft instrument, such as a harp.

Ideally, you'll have an assistant who can move microphones around while you listen through speakers in another, totally isolated room, though few home recordists have that luxury. One solution is to put on earphones and listen to how the sound changes as you move the mics around while the performer plays. However, it helps to play the earphones *loudly* to drown out any acoustic leakage into the phones that can influence what you hear and in turn affect your placement choice. You could also make a quick trial recording to confirm everything sounds correct before recording for real. All that said, I admit that I rarely do any of these things. If you have a decent-sounding room and decent microphones, just place the mics a reasonable distance away from the instrument or singer and you'll usually get great results. But what exactly is a "reasonable" distance?

For most sources, the best microphone placement is fairly close to the source. Guitar and bass amps are often mic'd from a few inches away and rarely more than a foot. You might place a second mic farther away in a good-sounding room, but the *main* mic is usually very close. Tambourines and shakers and other hand percussion sound best mic'd closer than a foot away, or you'll get too much room tone if the room is small. I generally try to keep singers a foot away from the mic. Being closer can give a more intimate sound but risks popping Ps and Bs and sibilance on S and T sounds. Many recording engineers use a pop filter with singers to block the air blasts that cause popping, though personally I've never used one. I find it better, and easier, to either put the microphone off to the side or above the singer, pointing down. This way the mic can still be close enough to get a clear, dry sound, but any air blasts from the singer's mouth go forward, away from the mic or underneath it.

A *wind screen* is a common microphone accessory, typically made of open cell foam, which passes sound waves but blocks blasts of air. Wind screens can serve effectively as a pop filter, but their main purpose is to minimize the rumbling sound of wind noise when recording outdoors. Wind screens always affect the high-frequency response at least a little, so they shouldn't be used unless absolutely necessary.

Microphone *shock mounts* are almost always useful and needed. Figure 5.3 shows one of my Audio-Technica 4033 mics with its included shock mount. This is typically a metal ring that holds the microphone, which in turn is supported by rubber bands to decouple the microphone from its stand. Unless your floor is made of solid cement, it most likely flexes and resonates at least a little as people walk or tap their feet. Without a shock mount, low-frequency rumble can travel mechanically from the floor through the stand into the microphone. Worse, the frequencies are so low that many small monitor speakers won't reproduce them. This is a good reason to always check your mixes through earphones, even if briefly just for a spot check.

Drums also benefit from close micing, though it depends on the musical style. A big band sounds most authentic when mic'd from farther away, and even pop recordings sometimes have a stereo pair of mics over the set several feet away. In a large room with a high ceiling, more distant placement can sound great. In a small room with a low ceiling, that's not usually an option. One additional complication with drums is there are so many different sound sources near to each other. So no matter what you do, every microphone picks up at least some of the sound from another nearby drum or cymbal. Using cardioid and Figure 8 microphones and paying attention to their direction and pickup angles helps to capture more of the desired source.

A common technique with snare drums is to place one microphone above the drum, pointing down at the drum head, and a second microphone below the drum, facing up to capture more of the snare sound. When using two mics this way, you must reverse the polarity of one of them to avoid canceling the low end when both mics are mixed together. Usually the polarity of the bottom microphone is reversed to avoid cancellation between the top snare mic and other mics on the kit that face the same way.

It's common to see cardioid microphones suggested for overhead drum mics in a room with a low ceiling to avoid picking up reflections from a low ceiling, with their resultant comb filtering. I've seen this recommended for piano mics, too, when microphones are under the piano's open lid, which also gives strong reflections. But this doesn't really work as intended, as proven by the "comb_filtering" video in Chapter 1. A microphone picks up whatever sound exists at its location. When comb filtering is caused by a reflection, the peaks and nulls occur *acoustically* in the air. So pointing a microphone one way or another is mostly irrelevant, because the sound pressure at that location has already been influenced by the reflections. For this reason, it's common when recording a grand piano to remove the top lid completely, rather than prop it up on the stick.

Some large instruments such as the cello and double bass radiate different frequency ranges in different directions. Likewise, an acoustic guitar sounds very different if a microphone is placed close to the bridge versus close to the sound hole. In both cases, there's no one good place to put a single mic close up. For best results, you need be several feet away to capture all of the frequencies, which is not always possible. When you must use close-micing on these sources, omni microphones often give the best results. Earthworks sells tiny omni microphones for close-micing grand pianos, and DPA offers tiny supercardioid microphones that clip onto the bridge of a violin or string bass, or a trumpet or sax bell. These microphones can sound surprisingly good given their very close proximity.

DI = Direct Injection

Acoustic sources obviously need to be recorded with microphones, but electronic instruments such as synthesizers, electric pianos, and electric basses are often recorded directly from their output jack. The main advantage of recording instruments electrically is that the sound is clearer because there's no contribution from the room. Electric basses benefit further because few bass amps are flat and clean down to the lowest frequencies. Further, most small rooms degrade the sound much more at low frequencies compared to frequencies above a few hundred Hz. Of course, sometimes you want the sound of a bass going through an amplifier. Even then, it's common to record a bass using a microphone as well as record it directly at the same time. If nothing else, this gives more flexibility later when mixing. When an electric instrument is recorded simultaneously through a microphone and directly, you may need to delay the direct sound when mixing to avoid phase cancellation. The farther the microphone was from the bass or keyboard amplifier's loudspeaker, the more you need to delay the direct sound. Rather than add a plug-in delay, it's more common to simply slide one clip or the other in time until the waves align visually.

Any device that has a line-level output can plug directly into a console or sound card input without needing a preamp. This works fine when the instrument is close to whatever you plug it into. But many keyboards have unbalanced outputs, and some line inputs are unbalanced, so hum pickup through the connecting cable is a concern. For an electric piano in the control room, using an unbalanced wire 10 to 20 feet long is usually fine. But if the player is in a separate live room with the other performers or on stage 100 feet away from the mixing console, you'll need a *DI box*. This converts the electric piano or synthesizer output to a low-impedance balanced signal that's sent to a mic input of a console or outboard preamp.

DI boxes are available in two basic types: active and passive. The earliest DI boxes were passive, containing only a transformer. The transformer's high-Z input connects to a ¼-inch phone jack, and its output is typically a male XLR connected to a microphone input through a standard mic cable. It's not practical to build a transformer having a very high-input impedance, at least not as high as 1 MΩ, which is best for an electric bass with a passive pickup. Active direct boxes with very high input impedances first appeared in the 1980s. Some models could even be phantom powered, avoiding the need for batteries.

Every DI box includes a *ground lift switch* to avoid ground loops between a local bass amplifier, if connected, and the distant input it feeds. Some models have a polarity reverse switch, and some include an input pad to accommodate higher-output line-level sources. For synthesizers and electric pianos having a line level output, a passive DI such as the model shown in Figure 6.9 is fine. But for a passive electric bass or an old-style Fender Rhodes electric piano with passive magnetic pickups and no preamp, an active DI box having a very high input impedance is a better choice.

Chapter 4 explained that low-impedance outputs can drive long wires without losing high frequencies due to capacitance. In truth, it's a little more complicated than that. Two factors determine what happens when an output drives a length of wire: A low output impedance minimizes airborne signals such as hum getting through the shield to the signal wires inside because it acts like a short circuit to those signals. But an output can have a low impedance, yet not be able to provide enough current to charge the capacitance of a long wire. In that case the high-frequency response will suffer, and distortion can also increase. This is rarely a problem with power amplifiers, but it could be a concern when an electronic keyboard has to drive a very long wire. So this is another reason a DI box is needed with long wires, even if the keyboard's line output has a very low impedance.

Figure 6.9: These Switchcraft SC800 and SC900 passive DI boxes contain high-quality Jensen transformers that convert a high-impedance source to a low-impedance output suitable for feeding a mic preamp. These models include a pass-through jack in case you want to send the instrument to a keyboard or bass amplifier at the same time. *Photos courtesy of Switchcraft, Inc.*

Additional Recording Considerations

In Chapter 5 I recommend recording without EQ or compression or other effects, unless the effect is integral to a musician's sound and influences her playing. Modern digital recording is extremely clean, so there's no *audio quality* benefit from committing to an effect while recording. Indeed, I'd rather defer all taste decisions until mixdown when I can hear everything in context. In particular, the reverb units built into many guitar amps are generally poor quality. So unless a guitar player requires the particular reverb sound from her amplifier, I prefer she turn off built-in reverb.

It's common to use baffles—often called *gobos*, for go-between—to avoid picking up sounds from instruments other than the intended source a microphone is aimed at. Gobos serve two related purposes: They prevent the sound of other instruments from getting into the microphone, and they can reduce the strength of room reflections reaching the mic. Figure 6.10 shows a RealTraps MiniGobo in a typical application, placed behind a microphone aimed at my Fender guitar amp.

Gobos are never a perfect solution because sound can go over the top or around the sides. Low frequencies are especially difficult to contain with baffles, though they can still help a lot. Many gobos are built with one side absorbing and the other side reflecting, so you can choose which type of surface to put near the performer and microphone. My own preference is for both sides to absorb, because I rarely want a reflecting surface near a microphone. But some players feel more comfortable when hearing some reflected sound, rather than playing

Figure 6.10: This small gobo helps prevent the sound of other instruments in the room from reaching the microphone at the guitar amp.

Figure 6.11: This wheeled gobo is large enough to block much of the sound from an entire drum set.

into a dead wall. Having a choice available allows for experimenting, which is always a good thing when placing microphones.

Most gobos are much larger than the small model shown around my guitar amp. Figure 6.11 shows a RealTraps GoboTrap that's six feet high by four feet wide. This gobo includes large wheels so it's easy to move around, but the wheel base is lower in the center to prevent sound from going under the gobo.

Sound *leakage* from other instruments is not always a bad thing, and it can add a nice "live sound" feel to a studio recording when the instruments are close-mic'd. What matters most is the quality of the leakage—mainly its frequency balance—and how long it takes the delayed sound to arrive at the unintended microphones. I remember learning a valuable lesson when I first started recording professionally years ago. I set up a rock band in my studio's large live room, putting the players far apart from one another, thinking that would reduce the amount of leakage picked up by the mics. But putting people far apart from each other made the inevitable leakage arrive very late. So instead of less added ambience, the sound was dominated by obvious echoes! Sound travels all over an enclosed space, even a large space. Lesson learned: Set up the band with everyone close to one another, as they'd be when performing live. It also helps the band when they can see and hear one another.

Some directional microphones have a non-flat response off-axis, so sound arriving from the rear or sides can sound muddy with less highs. Also, many cardioid microphones are directional at most frequencies but become less directional below a few hundred Hz. I suggest you verify the off-axis response versus frequency for microphones you buy (or rent).

Vocals and spoken narration are almost always recorded with a totally dry sound. A wrap-around baffle called a *carrel* is shown in Figure 6.12. These are used for library study desks and telephone call centers, and they're equally popular with voice-over artists who record

Figure 6.12: The RealTraps Carrel lets voice-over talent capture a clear, dry recording in an overly ambient environment.

at home. Its three large panels wrap all the way around the microphone, creating a very dry environment that also blocks ambient noise from being recorded.

If an artist prefers not to use earphones while overdubbing vocals in the control room, you can play the speakers softly. It's possible to reduce leakage into a microphone further by reversing the polarity of one loudspeaker while recording, assuming your mixer hardware or software offers a way to do that. Leakage will be minimum when the sound picked up from both speakers is the same, which occurs when the microphone is exactly centered left and right. You can adjust the pan pot slightly while playing the track and watching the record level meter to find where rejection is greatest. Rolling off the lowest and highest frequencies playing through the speakers helps further because most cardioid microphones reject less at the frequency extremes. That said, I prefer to avoid all leakage when possible because that gives the most flexibility when mixing. Even if the leakage is soft, it will become louder between words and phrases if the track is later sent through a compressor.

My final mic-related tip is really an admission. I'm the laziest person in the world, and I can't be bothered putting microphones away after every session. But I want to protect them from dust settling on the capsules, which over time can dull high frequencies. So I leave the mics on their stands and place plastic sandwich bags over them when they're not in use.

Advanced Recording Techniques

Low-budget productions often rely on one musician to play an entire section one part at a time, and this is commonly done with string, brass, and woodwind players. If the parts are all different for harmony, having one person play everything can sound convincing. But if your aim is to create the sound of a large unison section with one person playing the same part several times, comb filtering often results, creating an unnatural hollow sound. Fortunately, there are several ways to avoid this.

One approach has the performer play different instruments for each overdub. This is especially effective with stringed instruments because violins and cellos all sound different from one another. Comb filtering relies on two or more sound sources having the same frequencies present in equal levels. So the more each instrument's frequency output varies for a given note, the weaker the comb filtering will be. Few musicians own three or four different instruments of the same type, but even using two helps a lot.

Another approach puts the player in different parts of the room for each take, while leaving the microphones in the same place. The idea is to simulate a real section, where players are typically a few feet apart from one another. If you combine this with the player using different instruments for each performance, the result is identical to recording a real section comprising separate players. A few years ago I recorded a friend playing two unison cello parts for a pop tune. For the first take he played his cello, and for the second he moved his chair three feet to one side and played my cello. The result sounded exactly like two cellists, rather than one part that was double-tracked.

If you're using two or more players to create a larger section comprising multiple parts, you can use a variation on this method where the players exchange parts while recording each pass. For the first recording, one musician plays the lowest part while the other covers a higher harmony. Then for the next recording pass, the performers rotate parts, so unisons are avoided by having different physical instruments, and musicians with slightly different phrasing play the same part. This works equally well when doubling or tripling a small number of background singers to get the sound of a larger group. Instead of singing the same part each time, each person sings a different part on subsequent overdubs.

Vari-Speed

Everyone knows the "chipmunk effect," popularized by the 1958 recording *The Chipmunk Song* by David Seville (real name: Ross Bagdasarian). In fact, this effect is much older than that, going back at least to 1939, when it was used on the voices of the Munchkins in *The Wizard of Oz*. The basic premise is to record a voice or musical instrument at a given tape speed, then increase the speed when you play back the recording. In the old days, most tape recorders had fixed speeds at 2-to-1 multiples. So you could, for example, record at 7.5 inches per second (IPS) and play back at 15 IPS. If you're overdubbing to an existing backing track, the backing plays an octave lower and at half speed while recording. Then when the tape is played back at the original speed, the overdub sounds an octave higher.

Doubling the speed creates a pretty severe effect. For example, some words in the Chipmunks songs can be difficult to understand. Such a severe pitch change also sounds unnatural, though obviously that's the point with the Chipmunks. But much smaller amounts of pitch change can work to great advantage. A friend of mine is an amateur singer, and he often shifts the pitch of his lead vocals up one semitone. That small amount of change makes a huge improvement to the quality of his voice! This also works the other way, playing a backing track at a faster speed while you sing or play. The result when played back is a lower pitch with a corresponding change in timbre. Think of Tony the Tiger, the cartoon mascot for Kellogg's Frosted Flakes; the Jolly Green Giant selling vegetables; or the castle guards, also from *The Wizard of Oz*.

Years ago in the 1970s, I recorded myself singing all four parts to Bach Chorale #8, one of the finest examples of four-part harmony writing you'll ever hear. I'm not much of a singer, and my range is very limited. So to sing the soprano and alto parts, I used Vari-Speed to raise the pitch of my voice; and for the bass part, I lowered the pitch. In those days, the only way to change the tape speed by an amount less than 2-to-1 was to vary the frequency of the 60 Hz AC power that drives the tape recorder's capstan motor. So I bought a 100-watt Bogen PA system power amp and built a sine wave oscillator whose output frequency could be varied between 30 and 120 Hz.

Chapter 4 explained the principle of 70-volt speaker systems, using power amplifiers that have a special output transformer with a 70-volt tap. My Bogen tube amp's transformer included a tap for 70-volt speakers, but it also had a second tap to output 115 volts. I connected my home-made sine wave oscillator to the amplifier's input and wired up a connector to mate with the capstan motor of my Ampex AG-440 four-track tape recorder. I also added a second connector at the power amp's output for a voltmeter so I could set the output to exactly 115 volts to avoid damaging the capstan motor.

Since small pitch changes shift the timbre of a voice or instrument, it can also be used to avoid comb filtering when overdubbing unison parts. Voices and string instruments both contain multiple strong resonances (called *formants*) that increase the level of select frequencies. So shifting the pitch up or down one or two musical half-steps is similar to playing a different violin or using a different singer. Be aware that Vari-Speed also changes the rate of your vibrato. The more you intend to raise the pitch of a performance, the slower your vibrato must be while singing or playing to avoid sounding unnatural. Also be aware that reducing the speed while recording to ultimately raise the pitch requires sustaining notes longer. When I recorded the four-part Bach Chorale mentioned above, I had a heck of a time with the high soprano part. The slower I played the backing track, to better hit the highest notes, the harder it was to *hold* the notes because they had to sustain longer. By the time I reduced the speed enough to hit the highest notes, I'd run out of breath trying to hold them for as long as needed!

Fortunately, having to hack your own Vari-Speed electronically is ancient history, and today's audio software greatly simplifies the process. The computer audio equivalent of Vari-Speed is *sample-rate conversion*. Most audio editing software does this, letting you choose the amount of pitch shift in musical half-steps and cents, rather than have to calculate Hz and IPS tape speeds. When I recorded the audio for my *Tele-Vision* music video in 2007, I recorded myself singing several background vocal tracks using Vari-Speed. My aim was to change the timbre slightly for each performance to sound like several people. I first exported a mix of the backing track and loaded it into Sound Forge. From that I created four more versions, saving each separately—one up a half-step, another up two half-steps, one down a half-step, and another down two half-steps. Then I loaded all five backing tracks into a new SONAR project and solo'd them one by one as I sang along. Next, I loaded each vocal track into Sound Forge and shifted them up or down as needed to get back to the correct key and tempo. After saving all the vocal tracks as new Wave files, I loaded them into the original SONAR project. Whew! Fortunately, some DAW programs can vary playback speed on the fly, in which case all those extra steps aren't needed.

Summary

This chapter covers a lot of territory, including comparing common recording devices such as analog tape recorders and computer DAW software for both fidelity and convenience, as well as cost. Analog tape recorders are complex, and to obtain acceptably high fidelity they require clever engineering, such as tape bias, pre- and de-emphasis, several popular noise reduction schemes, pre-distortion to counter analog tape's high inherent distortion, and Sel-Sync to allow overdubbing.

The basics of digital recording were presented, including sample rate and bit depth. I also busted a few digital audio myths, such as the belief that recording at low levels reduces distortion and why headroom doesn't really apply because digital systems are perfectly clean right up to the onset of gross distortion. I also compared mixing ITB versus OTB, explained how to use a click track, and mentioned the SMPTE system that synchronizes two or more audio or video hardware recorders.

I explained a bit about folder organization with DAW projects, along with the value of setting up a recording template. We also considered why recording overdubs to separate tracks or lanes in a DAW program is superior to the old-school method of punching in, which risks overwriting a previous acceptable recording.

You learned that the quality of the musical source and the acoustics of the room matter more than anything else, and when recording amateurs, it helps to put them at ease. Giving performers a good cue mix that lets them hear everything clearly, hopefully with a touch of reverb, also improves confidence. I couldn't resist injecting my personal feelings about copy protection, though I agree that software piracy is a real problem.

Microphone types and patterns were described in detail, along with typical mic placement and other recording techniques. Several popular methods for stereo micing were shown, including Mid/Side, which lets you adjust the stereo width after the fact when mixing. You also learned that a large room offers more flexibility for mic and instrument placement versus small rooms, where the best choice is to close-mic everything with the performers and microphones far away from reflecting boundaries. In a small room, absorption and diffusion can reduce the strength of reflections, and gobos can minimize leakage in large and small rooms alike. However, a DI box avoids mic placement and room acoustic issues when applicable and also avoids leakage and other problems common to electric bass and keyboard amplifiers.

Finally, you learned some clever tricks to make a single performer sound like a section, using different instruments, different room placement, and even different recording speeds. This led to a bit of history about Vari-Speed and an explanation of how to achieve the same effect using modern audio software.

CHAPTER 7

Mixing Devices and Methods

In the earliest days of recording, "mixing" was performed by placing a group of performers around a single microphone, with some standing closer or farther than others to balance their volumes. If the sax player had a solo, he'd move closer to the microphone, then move back again afterward. All of this was recorded live in mono through one microphone to either a tape recorder or a record-cutting lathe. If anyone flubbed his or her part, the whole group had to start again from the beginning.

As analog recording progressed to multiple microphones, then to multi-track recorders in the 1950s, mixing evolved to the point where a complex mix often required two or more people to control all the volume changes needed over the course of a song. The mixing engineer would set up the basic mix and maybe ride the vocal level, while some of the musicians pitched in to vary other track faders at the direction of the engineer or producer. If only one person was available to do a complex multi-track mix, another option was to mix the song in sections to a stereo recorder, then join the mixed portions together with splicing tape to create the final version.

It's worth mentioning that despite his enormous contribution to recording technology, Les Paul didn't actually invent multi-track recording as is widely believed. Movie studios were placing multiple audio tracks onto optical film as early as the 1930s. When *The Wizard of Oz* was released in 1939, they used separate sound track elements for the orchestra music, spoken dialog, and sound effects. However, Les Paul certainly popularized multi-tracking, and he had a huge hand in refining the technology. His inventiveness brought the process of multi-track recording and overdubbing as we know it today to the forefront of music recording.

This chapter presents an overview of mixing devices and methods, along with relevant tips and concepts. Subsequent chapters explain audio processing in detail, with examples that apply equally to both hardware devices and plug-ins.

Volume Automation

In the 1970s, clever console manufacturers came up with various automation systems to record volume fader moves and replay them later automatically. You enable automation recording for a channel, then raise or lower the level of a vocal or guitar solo while the tune plays. Then on subsequent playbacks, the automation reproduces those level changes, and the mix engineer can perform other volume changes in real time that are captured by the fader automation. Eventually, all of the needed changes are captured and stored, letting the engineer sit back and watch the faders move automatically while the song plays and the mix is captured onto a stereo recorder.

Some automation systems use motors installed into physical faders to move them up and down, though others employ a *voltage controlled amplifier* (VCA) in each channel. When a VCA is used, audio doesn't pass through the fader. Rather, the fader sends a changing DC voltage to the VCA, which in turn varies the volume of the audio passing through it. Further, when a channel's volume is controlled by a voltage, it's then possible for one fader to control the level of several tracks at once. Multiple channels can be raised and lowered in groups, either with the same volume or with some amount of dB offset. So if you have eight drum tracks, each at a different volume, they can be raised and lowered together while maintaining the same relative balance between them.

There are several advantages of mechanical "flying faders," as they are often called. One benefit is that the audio signal goes through a traditional passive volume control, which doesn't add noise or distortion. When you adjust the volume up or down manually, sensors in the fader send "position data" that's recorded for later playback. Then when the automation is replayed, a motor drives the fader up and down, replicating the engineer's moves. Another advantage of moving faders is that you can see the current level for that channel just by looking at the fader's position. Moving faders are still used on large-format studio consoles, but the moving parts make them expensive to build, and they require maintenance to continue working smoothly.

VCA systems are mechanically simpler and therefore less expensive to manufacture, though early versions added a small amount of noise and distortion. With a VCA system, there's no need for the faders to move; each fader merely adjusts the level of a DC control voltage sent to that channel's VCA. The downside is you can't easily determine the current volume for a channel by looking at the fader. If you decide to raise the rhythm guitar by 2 dB, its current playback volume may not match the fader's current physical position. If the track is playing at a level of -15 dB, but the fader happens to be at -20 dB because it was moved when the automation was not in Record mode, "raising" it by 2 dB actually lowers the level by 3 dB to -18.

To solve this problem, a pair of LED lights is placed next to each slider. If the physical slider is currently set louder than the VCA's actual playback volume, the upper LED lights up. If the slider is lower, the lower LED is lit. This way you can adjust the fader position until neither light is on, at which point the fader position and actual playback volume are in sync. Any changes you then record by moving the fader will be as intended.

Editing

Early mono and stereo editing was done either by cutting the tape with a demagnetized razor blade to discard the unwanted portions or by playing a tape while copying only the desired parts to another recorder. When I owned a large professional studio in the early 1980s, a big part of our business was recording educational and training tapes for corporate customers, as well as narration and voice-over work for radio and TV ads. Corporate sessions typically ran for three or four hours, employing one or more voice-over actors. The sessions were eventually edited down to a length of 30 to 60 minutes and duplicated onto cassettes for distribution.

I remember well the tedious burden of editing ¼-inch tape by hand with a razor blade and splicing tape to remove all the gaffs and coughs or to replace a decent section with one deemed even better by the producer/customer. While recording I'd make detailed notes on my copy of the script, marking how many times a passage was repeated in subsequent takes. The talent would read until they goofed, then back up a bit and resume from the start of the sentence or paragraph as needed. By making careful notes while recording, I'd know when editing later how many times a particular section had been read. If my notes said a sentence was read four times, I could quickly skip the first three takes when editing later and remove those from the tape.

You have to be very careful when cutting tape because there's no Undo. Recording narration for big-name clients surely paid well, but it was a huge pain and often boring. A typical session for an hour-long program requires *hundreds* of razor blade edits! And every single cut must be made at the precise point on the tape. The standard method is to grab both tape reels with your hands and rock the tape slowly back and forth, listening for a space between words where the cut would be made. When editing tape this way, you don't hear the words as they normally sound but as a low-pitched roar that's barely intelligible. Unlike modern digital editing, you couldn't see the waveforms to guide you either. And you certainly couldn't edit precisely enough to cut on waveform zero crossings!

Splicing blocks have a long slot that holds the tape steady while you cut and another slot that's perpendicular to guide the razor blade. Figure 7.1 shows a professional quality splicing block made from machined aluminum. To minimize clicks and pops, most splicing blocks include a second slot that's angled to spread the transition over time so one section fades out while the other fades in. This is not unlike cross-fades in a DAW program, where both sections play briefly during the transition. The block in Figure 7.1 has only one angled blade slot, but some models have a second angle that's less severe. Which angle is best depends on the program material and the tape speed. If the gap between words where you'll cut is very short, you can't use a long cross-fade. Stereo material is usually cut perpendicular to the tape so both channels will switch to the new piece of tape together. Today's digital software

Figure 7.1: This splicing block has two razor blade slots to switch from one tape piece to the other either suddenly or gradually over a few milliseconds to avoid clicks and pops.

is vastly easier to use than a splicing block. Editing goes *much* faster, and there are as many levels of Undo as you'll ever need. Life is good. I'll return to audio editing, including specific techniques with video demonstrations, later in this chapter.

Basic Music Mixing Strategies

The basic concepts of mixing music are the same whether you use a separate recorder and mixing console or work entirely within a computer. The general goals are clarity and separation so all the parts sound distinct, and being able to play the mix at realistic volume levels without sounding shrill or tubby. Most mixing engineers also aim to make the music sound better than real—and bigger than life—using various audio processors. These tools include compression, equalization, small- and large-room reverbs, and special effects such as stereo phasers, auto-panners, or echo panned opposite the main sound. Note that many of the same tools are used both for audio correction and for sound design to make things sound more pleasing.

In my experience, anyone can recognize an excellent mix. What separates the enthusiastic listener from a skilled professional mixing engineer is that the engineer has the talent and expertise to know *what to change* to make a mix sound great. A beginner can easily hear that his mix is lacking but may not know the specific steps needed to improve it. By the way, this also applies to video. I spent many hours fiddling with the brightness, contrast, color, tint, and other settings when I first bought a projector for my home theater. It was obvious that the image didn't look great, but I couldn't tell what to change. Eventually I asked a professional video engineer friend to help me. In less than ten minutes, he had the picture looking bright and clear, with perfect color and nothing washed out or lost in a black background.

Most pop music is mixed by adjusting volume levels as well as placing each track somewhere left-to-right in the stereo field with a pan pot. If an instrument or section was recorded in stereo with two microphones onto separate tracks, you'll adjust the panning for both tracks. I usually build a pop mix by starting with the rhythm section—balance and pan the drums, then add the bass, then the rhythm instruments, and finally bring in the lead vocal or main melody if it's an instrumental. Then I'll go back and tweak as needed when I can hear everything in context. I suspect most mix engineers work the same way, though I know that some prefer to start with the vocal and backing chords. Either method works, and you don't have to begin each mix session the same way every time.

Be Organized

Whether you use an analog console or a computer digital audio workstation (DAW), one of the most important skills a recording engineer must develop is being organized. Multiple buses and inserts can quickly get out of hand, and it's easy to end up with tracks in a DAW going through more equalizers or compressors than intended. Perhaps you forgot that an EQ is patched into a Send bus, so you add another EQ to the bus's Return rather than tweak the EQ that's already there. Having fewer devices in the signal path generally yields a cleaner sound.

I also find it useful to keep related tracks next to each other, such as the MIDI track that drives a software synthesizer on an audio track. By keeping them adjacent, you'll never have to hunt through dozens of tracks later to make a change. Likewise, if you record a DAW project that will be mixed by someone else, your organization will be appreciated. For example, if there are two tracks for the bass, one recorded direct and the other through a microphone, it's easy for someone else (or even you) to miss that later when mixing. So you'll wonder why you can still hear the bass even when it's muted. Many DAWs let you link Solo and Mute buttons, so clicking one activates the other, too, which helps when one instrument occupies two or more tracks.

Other aspects of mix organization include giving DAW tracks sensible names, storing session files logically to simplify backing up, and carefully noting everything on each track of a master analog tape reel. Even something as simple as noting if a reel of tape is stored tails out or in will save time and avoid exasperation when you or someone else has to load the tape to make another mix later.

Monitor Volume

When mixing, playback volume greatly affects what you hear and how you apply EQ. Volume level especially affects the balance of bass instruments due to the Fletcher-Munson loudness curves described in Chapter 3. At low levels, the bass may sound weak, but the same mix played loudly might seem bass heavy. A playback volume of 85 dB SPL is standard for the film industry; it's loud enough to hear everything clearly, but not so loud you risk damaging your ears when working for a long period of time. It helps to play your mix at a consistent volume level throughout a session, but it's also useful to listen louder and softer once in a while to avoid some elements being lost at low volume, or overbearing when played loud. If you don't already own an SPL meter, now is the time to buy one. It doesn't have to be a fancy model meant for use by professional acousticians either.

Reference Mixes

Many mix engineers have a few favorite commercial tracks they play as a reference when mixing. While this can help to keep your own mix in perspective, it may not be as useful as you think. The main problem with reference mixes is they work best when their music is in the same key as the tune you're working on. Low frequencies are the most difficult to get right in a mix, even when working in an accurate room. In a typical home studio, peaks and nulls and room resonances dominate what you hear at bass frequencies. But the room frequencies are fixed, and they may or may not align with the key of the music. So if your tune is in the key of A, but the reference mix is in a different key, the resonances in your room will interact differently with each tune. Further, a kick drum's tuning also varies from one song or one CD to another. Unless the kick drum in the reference contains frequencies similar to the kick drum in the song you're mixing, it will be difficult or impossible to dial in the same tone using EQ.

Panning

> *If there's any rule in mixing, it's that there are no rules.*

Well okay, that's not really true. Most of the decisions a mix engineer makes are artistic, and only a few rules apply in art. But there are also technical reasons for doing things one way or another. For example, bass instruments are almost always panned to the center. Bass frequencies contain most of the energy in pop music, so panning the bass and kick drum to the center shares the load equally through both the left and right speakers. This minimizes distortion and lets listeners play your mix at louder volumes. A mono low end is also needed for vinyl records to prevent the needle from jumping out of the groove. However, vinyl mastering engineers routinely sum bass frequencies to mono as part of their processing.

Panning the bass and kick to the center also makes the song's foundation sound more solid. It's the same for the lead vocal, which should be the center of attention. But for other instruments and voices, skillful panning can improve clarity and separation by routing tracks having competing frequencies to different loudspeakers. As explained in earlier chapters, the masking effect makes it difficult to hear one instrument in the presence of another that contains similar frequencies. If a song has two instruments with a similar tone quality, panning one full left and the other full right lets listeners hear both more clearly.

Surround sound offers the mix engineer even more choices for placing competing sources. One of my pet peeves is mix engineers who don't understand the purpose of the center channel. As explained in Chapter 5, when two or more people listen to a surround music mix—or watch a late-night TV talk show broadcast in surround sound—only one person can sit in the middle of the couch. So unless you're the lucky person, anything panned equally to both the left and right speakers will seem to come from whichever speaker is closer. This can be distracting, and it sounds unnatural. In my opinion, surround mixes that pan a lead singer or TV announcer to the left and right channels only, or to all three front speakers equally, are missing the intent and value of a surround system. The center channel is meant to anchor sounds to come from that location. So if a talk show host is in the middle of the TV screen delivering his monologue, his voice should emit mainly from the center speaker. The same goes for the lead singer in a music video.

Panning can also be used to create a nice sense of width, making a mix sound larger than life. For example, it's common to pan doubled rhythm guitars fully left and fully right. With heavy-metal music, the rhythm guitarist often plays the same part twice. This is superior to copying or delaying a single performance, which can sound artificial. Two real performances are never identical, and the slight differences in timing and intonation make the sound wider in a more natural way. For country music it's common to use two hard-panned rhythm acoustic guitars, with one played on a regular guitar and the other on a guitar using *Nashville Tuning*. This tuning replaces the lower four strings of a guitar with thinner strings tuned an octave higher than normal. Or you can use a capo for the second part, playing different inversions of the same chords higher up on the neck. If your song doesn't have two guitars, the same technique works well with a guitar and electric piano, or whatever two "chords" instruments are used in the arrangement.

Getting the Bass Right

One of the most common problems I hear is making the bass instrument too loud. You should be able to play a mix and crank the playback volume up very high, without the speakers breaking up and distorting or having the music turn to mush. One good tip for checking the bass instrument's volume level is to simply mute the track while listening to the mix. The difference between bass and no bass should not be huge but enough to hear that the low end is no longer present or as solid sounding.

Avoid Too Much Reverb

Another common mistake is adding too much reverb. Unless you're aiming for a special effect, reverb should usually be subtle to slight. Again, there's no rule, and convention and tastes change over time. Years ago, in the 1950s, lead vocals were often drenched with reverb. But regardless of the genre, too much reverb can make a mix sound muddy, especially when applied to most of the tracks. You can check for too much reverb the same way you check for too much bass—mute the reverb and confirm that the mix doesn't change drastically. By the way, adding too much reverb is common when mixes are made in a room without acoustic treatment. Untamed early reflections can cloud the sound you hear, making it difficult to tell that excess reverb in the mix is making the sound even cloudier.

There are two basic types of artificial reverb, and I generally use both in my own pop music productions. The first is what usually comes to mind when we think of reverb: a Hall or Plate type preset having a long decay time, reminiscent of a real concert hall, or an old-school hardware plate reverb unit. The other type does not give a huge sound, but it adds a small-room ambience that decays fairly quickly. When added to a track, the player seems to be right there in the room with you. Typical preset names for this type of ambience reverb are Stage or Room.

Besides the usual reason to add reverb—to make an instrument or voice sound larger, or make a choir sound more ethereal or distant—reverb can also be used to hide certain recording gaffs. Years ago, when recording my own cello concerto, I did many separate sessions in my home studio, recording all the various orchestra sections one by one. At each session, everyone wore earphones as they played along with my MIDI backing track. To be sure the players didn't rush or slow down, I played a fairly loud click track through their earphones. Then after each session I'd replace the sampled MIDI parts for that section with the real players I just recorded.

At one of the violin section sessions, I left my own earphones on a chair, not far from one of the microphones. Most of the time the click sound picked up by the microphone was not audible. But one passage had a series of short pizzicato notes with pauses between, and the click was clearly audible in the spaces between notes. It required *hundreds* of tiny edits to duck the volume every time a click sounded with no notes to hide it. And you could hear the sudden drop to total silence as the plucked notes were cut off a little early to hide the click. Applying an ambience type reverb to the track hid those dropouts completely by extending the notes slightly. Had the reverb been applied before editing, the holes would be audible because the reverb would have been silenced, too.

Verify Your Mixes

Besides listening at soft and loud volumes, mixes should also be checked in mono to be sure nothing disappears or takes on a hollow comb filtered sound. Surround mixes should be verified in plain stereo for the same reason. It's also useful to play your mixes through small home or car type loudspeakers. When playing through your main speakers, listen not only for too much bass, or bass distortion, but also for shrillness that hurts your ears at high volumes. Again, as a mix is played louder and louder, it should get clearer and fuller but never muddy or painful. The bass content in a mix should depend mostly on the volume of the main bass instrument and the kick drum. Often one or the other, but not both, carries most of the low-end weight in a mix. If both the bass and kick drum are full sounding, it can be difficult to hear either clearly.

Thin Your Tracks

Many mix engineers roll off low frequencies on all tracks *other* than the bass and kick. Many instruments have more low-frequency content than you want in a mix, especially if they were close-mic'd using a directional microphone that boosts bass due to the proximity effect. Using a low-cut filter to thin out tracks that shouldn't contribute bass energy goes a long way toward achieving clarity in a mix. Instruments might seem painfully thin when solo'd, yet still sound perfectly fine in the context of a complete mix.

Distance and Depth

The two factors that determine front-to-back depth in a mix are high-frequency content and reverb. If you want to bring a track forward to feature it, boost the high end slightly with an equalizer and minimize the amount of reverb. Conversely, to push something farther back in a mix, reduce the highs and add reverb. Adding small room ambience-type reverb also helps make a track more present sounding, though don't overdo it. As with all effects, it's easy to add too much because you get used to the sound as a mix progresses. Again, mute the reverb once in a while to confirm you haven't added so much that it dominates the mix.

Bus versus Insert

All hardware consoles and DAW programs let you apply plug-in effects either individually to one track or on an Aux bus that can be shared by all tracks. As mentioned in Chapter 5, effects that *do something* to the audio are usually inserted onto the track, while effects that *add new content* to the audio typically go on a bus. For example, EQ, wah-wah, chorus, and compression effects all modify the audio passing through them and should therefore be inserted onto a track. It rarely makes sense to mix the dry and processed portions together, and many such effects include a Wet/Dry mix adjustment anyway to vary their strength. On the other hand, reverb and echo add new content—the echoes—and so are better placed on an Aux bus.

Further, reverb plug-ins require a lot of computer calculation to generate the large number of echoes needed to create a realistic reverb effect. So adding separate reverb plug-ins to many tracks is wasteful, and it limits the total number of tracks you can play all at once before the computer bogs down. Moreover, if you use only one or two reverbs for all the tracks, the result will be more coherent and sound like it was recorded in a real room rather than in many different acoustic spaces. Not that there's anything wrong with a combination of acoustic spaces if that's the sound you're after.

Pre and Post, Mute and Solo

As mentioned in Chapter 5, when an Aux Bus Effects Send is set to Post, the Send level follows the track's main volume control. This is usually what you want: As you raise or lower the track volume, the amount of reverb or echo remains the same proportionally. When an Aux Bus is set to Pre, the Send level is independent of the main volume control for that track. So in most cases, when mixing, you'll use Post where the Send level follows the volume control, as well as including any inserted effects such as EQ. One situation where you might want to use Pre is to send a track to an effects bus or subgroup before EQ or compression or other effects are added. Another use for Pre when mixing is to implement *parallel compression*. This special technique mixes a dry signal with a compressed version, and is described more fully in Chapter 9.

Mute and especially Solo are very useful to hear details and find problems. I often solo tracks to listen for and clean up small noises that aren't part of the performance, such as a singer's breath intakes. But don't go overboard and kill every little breath sound, as that can sound unnatural. The amount of trimming needed also depends on how much compression you use. Compressing a voice track always brings up the breath sounds. I also use Solo to listen for thumps or other low-frequency sounds that detract from the overall clarity. If you take the time to do this carefully for every track, you'll often find the end result sounds cleaner.

Room Tone

Even when recording in a very quiet environment, there's always some amount of *room tone* in the background. This is usually a rumble or vent noise from the air conditioning system, but preamp hiss can also add to the background noise. When editing narration and other projects where music will not mask the voice, it's useful to record a bit of room tone you can insert later as needed to hide edits. For example, you may need to insert silence here and there between words to improve the pace and timing of a voice-over narration. If you insert total silence, the sudden loss of background ambience will be obvious. Using a section of room tone sounds more natural, and the edits won't be noticed.

Before the voice-over talent begin reading, ask them to sit quietly for 5 to 10 seconds while you record a section of background silence. If the recorded room tone is too short, you may have to loop (repeat) it a few times in a row, and that could possibly be noticed. Recording at least 5 seconds of silence is also useful if you need to apply software noise reduction to the track later. With this type of processing, the software "learns" what the background silence sounds like, so it can remove only that sound from a music or voice track. Having a longer section of the background room tone helps the software to do a better job. Software noise reduction is explained in Chapter 13.

I used a similar technique when editing a live concert video for a friend. One of his concerts drew a huge crowd, with hundreds of people laughing at the jokes and clapping enthusiastically after every song. But a later concert drew fewer people, and the smaller audience was obvious and less convincing. When editing the later concert video, I copied several different sections of laughing and applause from the earlier video and added them to the audience tracks. Since both concerts were in the same venue with similar microphone placements, it was impossible to tell that the weak applause had been supplemented.

Perception Is Fleeting

> *We've all done stuff that sounds great at one moment, then we listen later and say, 'What was that cheesy sound?' Or vice versa. I'll be doing something at the moment and I'll question whether it works, then I'll listen to it later and note how the performance was really smoking and the sound really worked. Artistic judgment and opinion—it's infinite, and it could vary even in one individual from moment to moment.*
>
> —**Chick Corea, July 2011** Keyboard *magazine interview*

I couldn't agree more with Chick. Understanding that our hearing and perception are fleeting has been a recurring theme in this book, and it affects listeners and professional recording engineers alike. Everyone who mixes music, whether for fun or professionally, has made a mix that sounds stellar at the time but less impressive the next day. Even over the course of one session, perception can change drastically. It's common to raise the playback volume louder and louder to keep the mix sounding clear and full, when what you're really doing is trying to overcome a lousy sound.

I wish I had a great tip that works every time. Sadly, I don't. But try to avoid cranking the playback levels ever higher. Keep the volume loud enough to hear everything clearly, about 85 dB SPL, and occasionally play the mix softer and louder before returning to the normal level. If your mix sounds poor when listening softly, then that's the best volume level to listen at when deciding what needs improving.

Be Creative!

A good mix engineer is part producer, part sound designer. If you can learn to think outside the box, you're halfway there. When I was nearly done mixing the music for my *Cello Rondo* video in 2006, I invited my friend Peter Moshay to listen and offer his advice. Peter is a professional recording and mixing engineer with an amazing client list that includes Mariah Carey, Hall & Oates, Paula Abdul, and Barry Manilow, among other famous names. At one point early in the tune, the lead cello plays a sweeping ascending line that culminates in a high note. One of Peter's suggestions was to increase the reverb during the ascending line, ending up drenched 100 percent in huge reverb by the high note. Change is good, and it adds interest. The included "sonar_rondo" and "sonar_tele-vision" videos show some of the sound design and other editing I did on the mixes for those two music videos.

It's up to the composer and musicians to make musical changes, but a good mix engineer will make, or at least suggest, ideas to vary the audio—for example, adding severe midrange EQ to create a "telephone" effect. But don't overdo it either. An effect is interesting only if it happens once or twice. Once in the 1970s, I sat in on a mixing session at a large professional New York City studio. The tune was typical of that era, with a single lead singer and backing rock band. At one point you could almost see a light bulb go on over the mix engineer's head as he patched an Eventide Harmonizer onto the lead vocal and dialed in a harmony a minor third higher. At a few key places—only—he switched in the Harmonizer. Everyone present agreed that the result sounded very much like Grand Funk Railroad, a popular band at the time.

Many effects are most compelling when they're extreme, but, again, avoid the temptation to overdo it. It might be extreme EQ, or extreme echo, or anything else you can think up. Always be thinking of clever and interesting things to try. Another great example is the flanging effect that happens briefly only a few times in *Killer Queen* by Queen on the line "Dynamite with a laser beam." Or the chugging sounds in *Funkytown* by Lipps, Inc., the cash register sounds in *Money* by Pink Floyd, and the car horns in *Expressway to Your Heart* by the Soul Survivors. Even the bassoon and oboe parts on *I Got You Babe* by Sonny and Cher could be considered a "sound effect" because that instrumentation is not expected in a pop tune. The same applies for wind chimes and synth arpeggios and other ear candy that comes and goes occasionally.

When I was nearly done mixing my pop tune *Lullaby*, I spent several days thinking up sound effects that would add interest to an otherwise slow and placid tune. At one point you can hear the sound of a cat meowing, but I slowed it *way* down to sound ominous and added reverb to push it far back into the distance. Another place I added a *formant filter* plug-in to give a human voice quality to a fuzz guitar line. Elsewhere, I snapped the strings of my acoustic guitar against the fingerboard and added repeating echo with the last repeat building to an extreme runaway echo effect before suddenly dying out. I added several other sound effects—only once each—and at an appropriate level to not draw attention away from the music.

I also used a lot of sound effects in an instrumental tune called *Men At Work*, an original sound track I created to accompany a video tour of my company's factory. I created a number of original samples to give an "industrial" feel to the music and sprinkled them around the song at appropriate places. I sampled a stapler, blasts of air from a pneumatic rivet gun at our factory, large and small saw blades being struck with a screwdriver, and my favorite: the motor that zooms the lens of my digital camera. The motor sample sounded very "small," so I used Vari-Speed pitch shifting to drop the sound nearly an octave. The result sounds much like robot motion effects you'd hear in a movie. In all cases the microphone was very close to the sound source.

If a band is receptive to your input, you might also suggest musical arrangement ideas such as moving a rhythm guitar or piano part up an octave to sound cleaner and clash less with other parts. Or for a full-on banging rock performance with no dynamics, try muting various instruments now and again in the mix so they come and go rather than stay the same throughout the entire tune. Or suggest a key change at a critical point in the song. Not the usual boring half-step up, but maybe shifting to a more distant key for the chorus, then back again for the verse.

In the Box versus Out of the Box—Yes, Again

It's no secret that I'm a huge fan of mixing entirely In the Box. I might be an old man, but I gladly embrace change when it's for the better. To my mind, modern software mixing is vastly superior to the old methods that were necessary during the early years of multi-track recording and mixing. One of the most significant and useful features of DAW mixing is *envelopes* and *nodes*. The first time I used a DAW having a modern implementation of track envelopes, I thought, "This is *so* much better than trying to ride volume levels with a fader." When you're riding levels manually as a song plays, by the time you realize that something is too soft or too loud, the volume should have changed half a second earlier.

A *control surface* is a modern way to emulate the old methods. I understand that some people prefer not to "mix with a mouse," though I personally can't imagine mixing any other way. A control surface interfaces with DAW software via MIDI and provides sliders and other physical controls that adjust volume levels and other parameters in the DAW. All modern DAW software can record automation, so using a control surface lets a $400 program work exactly the same as a $200,000 automated mixing console. But I don't have space on my desk for yet more hardware, and I don't want more things in the way that require touching.

I can add an envelope to a track with just a few mouse clicks, so I don't see how recording automation using a control surface, and having to first specify what parameters to automate, could be easier or more efficient. But everyone has his or her own preference, as it should be.

I once watched a mix session where the engineer used a control surface to adjust the volume levels in ProTools. Due to a Wave file render error, one of the tracks dropped suddenly in level by 20 dB which was clearly visible in the program's waveform view. I watched with amusement as the mix engineer tried in vain repeatedly to counter the instantaneous volume reduction in real time using a physical fader. I had to contain myself not to scream, "Just use your mouse to draw a volume change!" But I understand and respect those who prefer physical controllers to software envelopes. That said, let's move on to the basic operation of DAW software, including some tips and related advice.

Using Digital Audio Workstation Software

The basic premise of DAW software is that each instrument or voice is recorded onto a separate track, which in turn is stored in a Wave file. This paradigm mimics a tape recorder and analog mixing console, but with everything virtualized inside a computer. Modern software includes not only a recorder and mixing console but even virtual outboard gear such as EQ and reverb and other effects. Once a song is complete and sounds as you like, you *render* or *export* or *bounce* the final mix to a new Wave file. All three terms are commonly used, and they all mean the same thing. Most DAW software can render a mix much faster than it takes to play the song in real time, and of course you never have to wait for a tape to rewind.

SONAR's Track View shown in Figure 7.2 is the main display—command central—where you add tracks and adjust their parameters, insert buses, add plug-ins to individual clips or entire tracks, create and edit clip and track envelopes, and so forth. There are other views, including the MIDI Piano Roll, Console View, Tempo Map, and Event List. For now we'll consider only the main Track View.

I have SONAR configured to rewind to where it started when I stop playback. The other option is to function like a tape recorder, where playing resumes at the point you last stopped. This is definitely personal preference, but I find that rewinding automatically makes mixing sessions go much quicker. I can play a section, tweak an EQ or whatever, then play the exact same section again immediately to assess the change.

All of the tracks in a project are numbered sequentially, though you can move tracks up and down to place related tracks adjacent, such as multiple background vocal or drum tracks. When a track is moved up or down, its track number changes automatically to reflect its current position. Some people use a DAW program as if it were a tape recorder. In that case

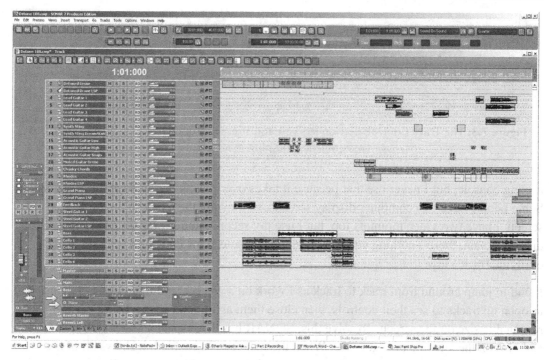

Figure 7.2: SONAR's Track View is where most of the action takes place.

each track contains one Wave file that extends from the beginning to the end of the song. But tracks can also contain one or more *clips*: Wave files or portions of Wave files that are only as long as needed. It makes no sense for a guitar solo that occurs only once in a song to have its Wave file extend for the entire length of the song. That just wastes disk space and taxes the computer more, as silence in the unused portion is mixed with the rest of the tune. You can see many separate clips on the various tracks in Figure 7.2.

All of the tracks in Figure 7.2 are *minimized*, meaning they show only the track name and number, volume, and Mute, Solo, and Record switches in a narrow horizontal strip. When a track is opened, all of the other parameters are also displayed in the Track Controls section shown in Figure 7.3. This includes input and output hardware devices, buses, and inserted plug-in effects or synthesizers. This bass track is opened fully to see all the settings. Recorded input comes from the Left channel of my Delta 66 sound card, and the track's output goes to a bus I set up just for the bass. That bus then goes to the main stereo output bus to be heard and included in a final mix. There are three plug-ins on this track: an EQ, a compressor, then another EQ. If any Aux Send buses had been added to the track, they would show below the Bass Bus.

Every track parameter is available in the Track Controls section, and an entire mix can be done using only these controls. SONAR also offers a Console View that emulates a physical

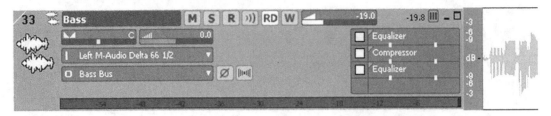

Figure 7.3: The Track Controls section lets you adjust every property of a track, including bus assignments and plug-in effects.

mixing console, but I never use that because it takes over the entire screen and is not needed. Some people have two computer display screens, so they put the Console on the second screen. I have two video monitors too, but I use my second monitor for the MIDI Piano Roll display. Again, there's nothing you can do in the Console View that can't be done just as easily in Track View.

I tend to jump around from track to track as I work on a project, and it gets tiring having to open up tracks to see their controls, then close them again to keep the Track View at a reasonable size on the screen. To solve this, SONAR has the Inspector shown in Figure 7.4 and also on the left side of Figure 7.2. When the Inspector is enabled, it shows all of the parameters for whatever track is currently selected. If the track contains audio, the Inspector shows audio-related controls such as Aux buses and audio plug-ins and software synthesizers. If the track holds MIDI data, controls relevant for MIDI are displayed instead.

SONAR lets you add envelopes that apply to an entire track or to just one clip. Track envelopes can control the track's volume, pan position, mute (on/off), plus any parameter of any plug-in on that track. Figure 7.5 shows a volume envelope on the bass track from Figure 7.3. You can see three places where I added node groups to raise bass notes that were a little too soft and another place I used nodes to mute a note I disliked but didn't want to delete destructively. To create a node, you simply double-click at the appropriate place on the envelope line. Nodes can be slid left and right, as well as up or down. If you click on a line segment between two nodes, both nodes at each end of the line are selected so the nodes go up and down together. You can also select other groups of nodes to adjust many all at once. Nodes are often set to raise or lower the volume by a fixed amount for some duration, as with the middle two node groups. But they can also fade up or down as shown at the left and right.

One limitation with volume envelopes is that they set absolute levels, rather than relative ones. Imagine you spent an hour tweaking a vocal track so every word can be heard clearly, but then you decide later the entire track should be a little louder or softer. It's a nuisance to have to adjust all those envelopes again in dozens of places. The good news is most DAW software offers a second volume control that can scale all of your envelope automation changes up or down. In SONAR this is done using the track's Trim control, and most other programs offer something similar.

Figure 7.4: The Inspector shows all parameters and settings for a track, without having to open up the track. Just click on any track's number, and the Inspector switches to that track.

Slip-Editing and Cross-Fading

Two of the most powerful features of DAW software are *slip-editing* and *cross-fading* between clips. As with most DAW programs, SONAR lets you trim the start and end points of an audio clip and overlap two clips with the volumes automatically cross-fading in and out. Besides splicing together pieces of different takes to create a single best performance,

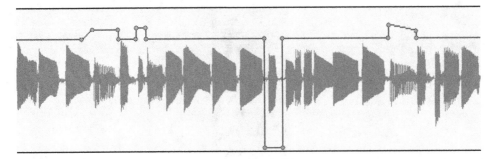

Figure 7.5: Track Clips are displayed as waveforms, and this is where you add envelopes and nodes to adjust the volume, pan, or any parameter of an inserted plug-in. The lines represent the envelope, and the small circles are the nodes.

clip editing can be used to create *stutter edits*, where a single short section of a track or entire mix repeats rapidly for effect.

The video "sonar_envelopes" shows a clip-level volume envelope lowered from its starting point of +0.5 dB gain down to −2.5 dB reduction. Then two nodes are added by double-clicking on the envelope. The trailing portion of the envelope is lowered further to −11.5 dB, and then that portion is slid to the left a bit to start fading the volume earlier. Clips can be faded in or out easily this way, without having to create an envelope or nodes. In this example the fade-out is shifted left to start earlier, and then a fade-in is applied. Finally, the clip is shortened using a method called *slip-editing*. This is a great way to trim the beginning and end of a clip to eliminate any noises on the track before or after a performance. Slip-editing lets you edit the track's Wave file nondestructively, so you can extend the clip again later restoring its original length if needed.

In the "sonar_cross-fade" video, I copy a clip by holding Ctrl while dragging the clip to the right. Then when I slide the copy to the left, overlapping the original clip, SONAR applies a cross-fade automatically, so one clip fades out as the other fades in. Besides the obvious use to cross-fade smoothly between two different sounds, this also helps avoid clicks and pops that might happen when adjacent clips suddenly stop and start. As you can see, the start and end points that control the duration of the cross-fade region are easily changed.

Track Lanes

Most DAWs offer a *loop record* mode that records to the same region repeatedly until you press Stop. A region is defined by Start Time and End Time markers. Each newly recorded take creates either a new track or a new lane within one track. I usually set SONAR to create lanes within one track for simplicity and to save screen space. Once I'm satisfied with one of

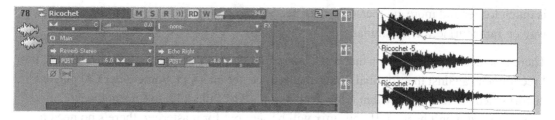

Figure 7.6: Track Lanes let you group related Wave files within a single track.

the takes or I know I have enough good material to create a composite take from all the bits and pieces, it will end up as a single track anyway.

Figure 7.6 shows a track opened up to reveal its three lanes. Note that each lane has Solo and Mute buttons, which simplifies auditioning them after recording. After deleting the bad takes and using slip-editing to arrange the remaining good takes, a single step called Bounce to Clip then combines all the pieces to a new clip with its own new Wave file. All of the original recorded clips can then be safely deleted to reduce clutter and avoid wasting space on the hard drive.

In this example, all three lanes are related and meant to play at once. The top lane is a Wave file of a gunshot ricochet from a sound effects CD. The second lane is the same gunshot sound, but shifted down five musical half-steps in Sound Forge using Vari-Speed type pitch shifting to be fuller sounding. You can see that the clip is also longer because of the pitch change. The bottom lane is the same clip yet again, but shifted down 7 semitones. With all three clips playing at once, the gunshot sounds larger and less wimpy than the original clip playing alone. You can see in Figure 7.6 that this track also goes to a Send bus with an echo that sounds on the right side only. This further animates the sound, which is mono on the CD it came from, making it seem to travel left to right more effectively than panning the track using pan automation.

Normalizing

Normalizing is usually done to a final mix file after it's been rendered. This process raises the volume of the entire track such that the loudest portion is just below the maximum allowable level. There's no technical reason to normalize individual track Wave files, but I sometimes do that for consistency to keep related tracks at roughly the same volume. Depending on the track, I may open a copy of the clip's Wave file in Sound Forge, apply software noise reduction if needed, then normalize the level to −1 dBFS. On large projects having hundreds of audio clips, I often rename the files to shorter versions than SONAR assigned. For example, SONAR typically names files as [Project Name, Track Name, Rec(32).wav], which is much longer and more cumbersome than needed. So I might change that to Fuzz Guitar. wav or Tambourine.wav or similar.

There's no audio quality reason that track Wave files shouldn't be normalized to 0 dBFS, but I recommend against that for final mixes that will be put on a CD. Some older CD players distort when the level gets within a few tenths of a dB of full scale, so I normalize to −1 to be sure that won't happen. If your mix will be sent out for mastering, there's no need to normalize at all because the engineer will handle the final level adjustment.

Editing and Comping

One of the greatest features of modern DAW recording is that editing and mixing are totally nondestructive unless you go out of your way to alter the track's Wave file. With most DAW programs, Undo works only within a single edit session, but nondestructive editing lets you change your mind about any aspect of the mix at any time in the future. If you later notice a cross-fade that's flawed, you can reopen the project and slide the clips around to fix it. If you discover the bass is too loud when hearing your current masterpiece on a friend's expensive hi-fi, you can go back and lower it. Every setting in a DAW project can be changed whenever you want—next week, next month, or next year.

The only times I apply destructive editing to a track's Wave file is to apply software noise reduction or to trim excess when I'm certain I need only a small portion of a much larger file. Otherwise, I use slip-editing to trim tracks to play only the parts I want.

This book is not about SONAR, but most DAW programs are very similar in how they manage editing and *comping*. So I'll show how I comp a single performance from multiple takes in SONAR; the steps you'll use in other programs will be very similar. The concepts are certainly the same. Here, comping means creating a composite performance from one or more separate clips.

Figure 7.7 shows three different recorded takes of a conga drum overdub after trimming them to keep only the best parts. Before the tracks were trimmed, all three clips extended over the full length of the recorded section. I applied cross-fades from one clip to the next manually by sliding each clip's fade-in and fade-out times so they overlap. Once the clips are trimmed as shown, they could all be moved into a single lane. In that case SONAR would cross-fade between them as shown in the "sonar_cross-fade" video. But leaving them in separate lanes makes it easier to adjust timings and cross-fades later if needed.

Another common DAW feature is the ability to loop clips, and this conga overdub is typical of the type of material that is looped. For example, I could combine the three clips spanning eight bars in Figure 7.7 into one clip, then enable looped playback on the result clip.

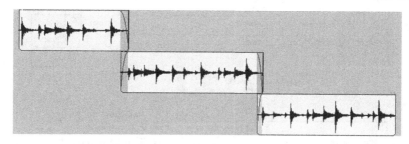

Figure 7.7: Clips in different Track Lanes can be easily trimmed to keep only the best parts of multiple takes. Once comping is complete, you can either combine all the clips to a single new clip or leave them separate as shown here if you might want to change them later.

Once looping is enabled for a clip, it can be repeated as many times as you like using slip-editing. This technique was first popularized by Sony's Acid program, and clips that can be looped this way are called *Acidized* Wave files. Another popular looped format is *REX* files, which stands for Recycle EXport. This is an older format developed by the Swedish company Propellerhead for their ReCycle software. This file type is still used, though probably less so than the more modern Acidized Wave files.

Rendering the Mix

Figure 7.8 shows the main portion of the Bounce dialog in SONAR. When exporting a mix you can include, or not, various aspects of the mix as shown in the check boxes along the right side. You can export as mono or stereo at any supported bit depth and sample rate, and several dither options are available, including none. Note that a stereo mix rendered to a mono Wave file can potentially exceed 0 dBFS (digital zero) if both channels have common content and their levels are near maximum. In that case the result Wave file will be distorted. SONAR simply mixes the two channels together and sends the sum to the mono file.

Who's on First?

One of the most common questions I see asked in audio forums is if it's better to equalize before compressing or vice versa. In many cases you'll have one EQ before compressing and a second EQ after. Let's take a closer look.

If a track has excessive bass content that needs to be filtered, you should do that before the compressor. Otherwise rumbles and footsteps, or just excessive low-frequency energy, will trigger the compressor to lower the volume unnecessarily. If you compress an unfiltered track, the compressor lowers and raises the volume as it attempts to keep the levels even, but those volume changes are not appropriate and just harm the sound. The same applies

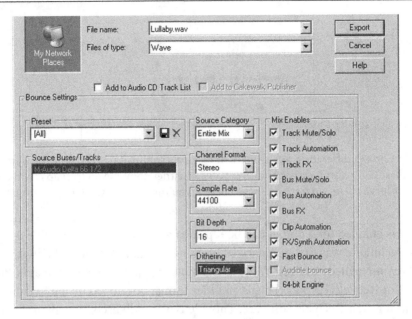

Figure 7.8: The Bounce dialog in SONAR is where you specify what to export, at what bit depth and sample rate, and various other choices. Other DAW programs offer similar options.

for other frequencies that you know will be removed with EQ, such as excess sibilance on a vocal track or a drum resonance you plan to notch out. Therefore, you should do any such *corrective* EQ before compressing.

However, if you *boost desirable* frequencies before compressing, the compression tends to counter that boost. As you apply more and more EQ boost, the compressor keeps lowering the volume, reducing that boost. In fact, this is how many de-essers work: They sense the amount of high-frequency content in the sibilance range, then reduce either the overall volume or just the high frequencies, depending on the particular de-esser's design. So when you're using EQ to change the basic tone of a track, that's best done after compressing. Again, there are few rules with art, and I encourage you to experiment. My intent is merely to explain the logic and theory behind mixing decisions that have a basis in science or that are sensible most of the time.

Figure 7.3 shows the Track Controls for a bass track from one of my projects, with three plug-ins inserted: an EQ, a compressor, then another EQ. The first EQ in the chain applies a gentle 6 dB per octave roll-off below 60 Hz to reduce the overall low-frequency content. This is followed by a compressor having a fairly aggressive 10:1 ratio, which is then followed by the EQ that actually alters the tone of the bass with a slight boost at 175 Hz.

One situation where compressing first usually makes sense is when you intended to insert a severe distortion effect. Distortion tends to bring up the noise floor quite a bit because of the high gain it applies, so compressing after distortion raises the noise even further. Another time you'll want to compress first is with an echo effect whose repeating echoes decay over time. When a compressor follows an echo effect, it can raise the level of the echoes instead of letting them fade away evenly as intended.

Time Alignment

Track clips are often moved forward or back in time. One situation is when micing a bass amp while also recording direct, as mentioned in Chapter 6. In that case you'll record to two adjacent tracks, zoom way in to see the waveform details, then nudge the mic'd track to the left a bit until the waveforms are perfectly aligned. Most DAW software has a *snap* option that slides clips by a fixed amount of one beat or a whole bar. So you'll need to disable this feature in order to slide a clip by a tiny amount. SONAR can either *snap by* whole or partial beats and bars or *snap to* beat or bar boundaries. I find "snap by" more useful because musical parts often start on an upbeat. Moving a clip by bar or beat increments rather than to bar or beat start times preserves the musical timing. For example, if I move a hand-claps clip to start one bar earlier, I want the clip to shift by exactly one bar, even if it started before or after the beat.

You can also slice clips and slide the pieces around to correct timing errors or improve musical phrasing. This is often done as part of the comping process, and this too requires disabling the snap feature so you can move the clips forward or back in time by very small amounts. I've seen people create fake double-tracking by copying a mono clip to another track, shifted slightly later in time to the right, with the two tracks panned left and right. But this is inefficient compared to simply adding a delay effect and panning that to the opposite side.

Editing Music

Music editing applies to both mono and stereo Wave file clips on a DAW track and finished stereo mixes. Most track editing can be done nondestructively using slip-edits and cross-fades described earlier. Hard edits—where the start or end point of a clip turns on or off quickly rather than fading in or out—are usually performed at waveform zero crossings to avoid a click sound. If a clip begins when the wave is not at zero, the sudden jump in level when it starts is equivalent to adding a pulse wave to the audio. In truth, when splicing between two clips, it's not necessary to cut at a zero crossing. What really matters is avoiding a *discontinuity* of the waveform at the splice point.

Figure 7.9 shows a waveform zoomed in enough to see the individual cycles, with the cursor at a zero crossing. Besides zooming way in horizontally, SONAR also lets you zoom the

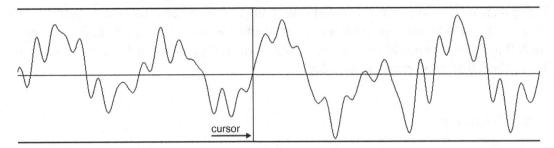

cursor

Figure 7.9: Zooming way in on an audio file lets you find and cut at zero-crossings to avoid clicks and pops.

wave's displayed level vertically to better see the zero crossings on portions of audio that are very soft. To split a clip, you'll put the cursor exactly at the zero crossing, where the waveform passes through the center line, then hit the "S" key or whatever method your software uses. If the clip is a stereo Wave file, it's likely that the left and right channels will not pass through zero at the same point in time. In that case you may also have to apply a fade-in or fade-out, or cross-fade when joining two clips, to avoid a click. Clip fade-ins and fade-outs are also useful to avoid a sudden change in background ambience. Even if a splice doesn't make a click, background hiss or rumble is more noticeable when it starts or stops suddenly.

Another common music editing task is reducing the length of an entire piece or excerpting one short section from a longer song. Modern audio editing software makes this vastly easier than in the old days of splicing blocks. When editing music you'll usually make your cuts on a musical beat and often where a new bar begins. The video "music_editing" shows basic music editing in Sound Forge—in this case, repeating parts of a tune to make a longer "club" version. The voice-over narration describes the editing steps, so there's no need to explain further here.

It's good practice to verify edits using earphones to hear very soft details such as clicks or part of something important being cut off early. Earphones are also useful because they usually have a better response, extending to lower frequencies than most loudspeakers. This helps you to hear footsteps or rumble sounds that might otherwise be missed. But earphones are not usually recommended for mixing music because you can hear everything *too* clearly. This risks making important elements such as the lead vocal too soft in the mix. Mixes made on earphones also tend to get too little reverb, because we hear reverb more clearly when music is played directly into our ears than when it's added to natural room ambience.

Editing Narration

Editing narration is typically more tedious and detailed than editing music, if only because most narration sessions require dozens if not hundreds of separate edits. Every professional engineer has his or her own preferred method of working, so I'll just explain how I do it. Most of the voice-over editing I do is after recording myself narrating educational videos for YouTube or for this book. If I were recording others, I'd have a copy of the script and make detailed notes each time a mistake required a retake. This way I know exactly how many takes to skip before getting to the one I'll actually use. It's difficult to do that when recording myself, so I just read until I make a mistake, pause half a second, then start again. It's not that difficult to sort out later when editing.

After recording, I make a backup copy to another hard drive, then load the original Wave file into Sound Forge. Editing in Sound Forge is destructive, though you could do it with slip-edits in a DAW. The first step is to find coughs, thumps, and other noises that are louder than the voice, then mute or lower them so they're not the loudest part of the file. Then you can normalize the file without those sounds restricting how loud the voice can be made. If noise reduction is warranted, apply that next. Then apply EQ or compression if needed. Noise reduction software requires a consistent background level, so that should always be done before compressing.

Voice recording should be a simple process. If it's done well, you probably won't even need EQ or compression or other processing other than noise reduction if the background noise is objectionable. Of course, there's nothing wrong with using EQ or compression when needed. Most of the voice-overs I recorded for the videos in this chapter were captured by a Zoom H2 portable recorder resting on a small box on my desk about 15 inches from my mouth. I usually record narration with my AT4033 microphone into Sound Forge, but I was already using that program as the object of the video. After recording to the Zoom I normalized the file and edited it in Sound Forge. No EQ or compression was used. I prefer narration in mono, because that keeps the sound solidly focused in the center. I find stereo voice-overs distracting, because the placement keeps changing as the person speaking moves around slightly.

Besides editing out bad takes, coughs, chair squeaks, and other unwanted sounds, you may also need to remove popping "P" and sibilant "S" sounds. Pops and sibilance are easily fixed by highlighting just that part of the word, then reducing the volume 10 to 15 dB. The "voice_ editing" video shows the steps I used to clean up part of a live comedy show I recorded for a friend. Most of the editing techniques shown in this video also apply to single tracks in a multi-track music project, using destructive editing to clean up extraneous noises. Again, earphones are very useful to hear soft noises like page turns and lip smacks that might be missed unless your speakers are playing very loudly.

Re-Amping

Although re-amping (re-amplifying) is a recording process, it's usually done during mix-down. Re-amping was mentioned in Chapter 3 as a way to fairly compare microphones and

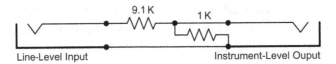

Figure 7.10: This 20 dB pad accepts a line-level input signal and reduces it to instrument level suitable for sending to a guitar or bass amplifier.

preamps, but its main purpose is as an effect when mixing. The name comes from the process where a guitar track that had been recorded direct through a DI is played back into a guitar amplifier. The amplifier is adjusted for the desired tone, then re-recorded with a microphone. This lets you try different amplifiers, different mics and placements, and different acoustic spaces while mixing. Re-amping can also be done using a high-quality speaker and microphone to add genuine room ambience to a MIDI piano or other instrument that was recorded direct. This can be done in the same room as you're mixing if you mute the monitor speakers while recording, but in larger studios it's more common for the loudspeaker to be in a separate live room to capture a more pleasing acoustic character. That also lets you capture a mono source using stereo microphones.

It's equally common to send a bass or synth track through a guitar or bass amp to impart the color of that particular amplifier. I've done this to make sampled guitars sound more convincing. For a dry (or sampled) lead guitar, you can even run the playback signal through a fuzz-tone or other guitar-type effect. Since guitar and bass amps are generally recorded using close-micing, this can be done in the control room for convenience.

Re-recording a clean-sounding track to add ambience is pretty easy: You simply send a line-level output from your mixer to a power amp that drives the loudspeaker, then record the speaker as you would any other "live" source. If you have powered speakers, you can send the line-level audio directly. But sending a line-level signal through a guitar amp requires a resistor pad or a transformer-based direct box. Guitar and bass amps expect a low-level signal from a passive instrument, so sending them a line-level output will cause distortion. Even if you turn the Send volume way down on your mixer to avoid overdriving the guitar amp, the result is likely to be very noisy. Using a pad or DI lowers the line-level output *and* its noise to a level more suitable for an instrument amplifier. A suitable 20 dB pad can be made from two resistors, as shown in Figure 7.10.

Backward Audio

Backward audio is a cool effect, often used with cymbals, reverb, a guitar solo, or even part of an entire mix. The basic premise is to reverse the audio data so it plays from the end to the beginning. Where many musical (and other) sounds start suddenly, then decay to silence over time, backward audio grows and swells to a climax that ends suddenly. SONAR can do this automatically by choosing Audio..Reverse from the Process menu. Sound Forge also offers a Reverse option on its Process menu. Many other DAW and audio editor programs have this feature as well.

In the old days of analog tape, you'd create a backward guitar solo by putting the tape onto the recorder upside down before recording the overdub onto a free track. Since the tape is upside down, the track numbers are reversed, too. So track 1 that had been on top is now track 8, 16, or 24 at the bottom, depending on the recorder. If you're not careful with the reversed track order, you can accidentally overwrite something important on another track! It helps to make a reversed audition mix before the session for the guitarist to learn and practice along with, since the chord changes play in reverse order, too.

To create a backward reverb effect on a single track you'll first make a copy of the track to a separate Wave file. Then reverse the playback direction in an audio editor program and apply reverb as usual. When you reverse the track again putting it back to normal, the track plays as expected but the reverb effect is reversed. So it swells to sudden silence rather than decaying over time. You can also do this with an entire mix, or just a section of the song as a special effect.

Mastering

Earlier I mentioned some of the requirements for preparing audio that will be cut to a vinyl record. For example, bass frequencies must be summed to mono to prevent skipping, and high-frequency pre-emphasis boosts those frequencies quite a lot to overcome the inevitable record scratches and background noise. If a master recording already contains a lot of high-frequency content, there's a risk the record will have noticeable distortion or excess sibilance on vocals and cymbals. Indeed, a loud recording with strong high-frequency content can blow out the expensive vinyl cutting head.

For this reason, a limiter that affects only high frequencies is often part of a record-cutting lathe's signal chain. In one of his *Mix Magazine* columns, Eddie Ciletti opined that part of the appeal of vinyl could be due to the protective high-frequency limiting, which can add a pleasing sheen to the music. A friend of mine is a well-known mastering engineer, and he confirmed that high-frequency limiting is sometimes used even on CD masters for the same effect.

In the early days of mastering, these specialist engineers also had a say in how the songs were sequenced on each side of an LP. Mastering for vinyl requires a trade-off between volume level and music length; the louder you cut the music, the less time will fit on a side. Further, the outer grooves of an LP record have higher fidelity than the inner grooves, simply because the grooves pass by more quickly. For a given rotation speed in RPM, the *linear* speed in inches per second is greater when playing the outer grooves of a record than the inner grooves near the end. So it's common to put louder, brighter tunes at the beginning of a side and softer, mellower songs on the inner grooves where the poorer high-frequency response may not be noticed. It's also best if both sides of an LP are approximately the same length. This was equally important when cassettes were popular to avoid wasting tape if one side was longer than the other.

As you can see, preparing a master tape for cutting to vinyl requires a lot of experience and care. What began as a technical necessity in the early days of recording has evolved into an art form. The most successful early mastering engineers went beyond merely protecting their record-cutting hardware, and some became sought after for their ability to make recordings sound better than the mix-down tapes they received. Besides sequencing songs for better presentation and to more fully utilize the medium, mastering engineers would apply EQ and compression, and sometimes reverb. Today, mastering engineers use all of those tools, and more, to improve the sound of recordings. A good mastering engineer works in an excellent sounding room with a flat response, along with high-quality full-range speakers that play to the lowest frequencies.

Because of the physical limitations of LP records and cutting lathes, the lacquer master used to make the stamped duplicates is sometimes cut at half speed. When a master tape is played at half speed, all of the frequencies are lowered one octave. So 20 KHz is now only 10 KHz, which is easier for the cutter head to handle. While cutting, the lathe spins at $16\frac{2}{3}$ RPM instead of $33\frac{1}{3}$ RPM. The cutter head doesn't need to respond as quickly when creating the grooves, which in turn improves high-frequency response, transient response, and high-frequency headroom.

Save Your Butt

When working on audio and video projects, every few hours I use Save As to save a new version. The first time I save a project, at the very beginning, I name it "[Project Name] 001." Next time I use 002, and so forth. If I work on something for a few weeks or months, I can easily get up to 040 or even higher. While recording and mixing the music for my *Tele-Vision* video, which took the better part of one year, the last SONAR version was 137. And while assembling and editing the video project in Vegas, I got up to version 143.

Audio and video project files are small, especially compared to the recorded data, so this doesn't waste much disk space or take extra time to back up. And being able to go back to where you were yesterday or last week is priceless. At one point while working on *Tele-Vision* in SONAR, before replacing all the MIDI with live audio, I realized part of the MIDI bass track had become corrupted a month earlier. It took about ten minutes to find a previous version where the bass was still correct, and I simply copied that part of the track into the current version.

I don't use the auto-save feature many DAW (and video) editors offer as an option, because I often try things I'm not sure I'll like. I'd rather decide when to save a project, rather than let the program do that automatically every 10 minutes or whatever. I need to know I can Undo, or just close and quit, rather than save my edits. Some programs clear the Undo buffer after every Save, though others let you undo and save again to restore the earlier version. Also, I often call up an existing project for a quick test of something unrelated, rather than start a new project that requires naming it and creating a new folder. In that case I *know* I don't want to save my experiment. And if for some reason I do decide to save it, I can use Save As to create a new project file in a separate folder.

Summary

This chapter explores the early history of mixing and automation to explain how we arrived at the hardware and methods used today. The earliest mixes were balanced acoustically by placing the performers nearer or farther away from a single microphone. Eventually this evolved to using more than one microphone, then to multi-track recording that puts each voice or instrument onto a separate track to defer mixing decisions and allow overdubs. The inherent background noise of modern digital recording is so low that effects such as EQ and compression can also be deferred, without the risk of increasing noise or distortion later.

Early automated consoles were very expensive, but today even entry-level DAW software offers full automation of volume and pan, plus every parameter of every plug-in. Mix changes can be programmed using a control surface having faders and switches or by creating envelopes and nodes that are adjusted manually using a mouse.

In the days of analog tape, the only way to edit music and speech was destructively by literally cutting the tape with a razor blade, then rejoining the pieces with adhesive splicing tape. Today the process is vastly simpler, and there's also Undo, which lets you try an edit without risk if you're not sure it will work.

Mixing is a large and complex subject, but the basic goals for music mixing are clarity and separation, letting the instruments and voices be heard clearly without sounding harsh or tubby at loud volumes. Bass in particular is the most difficult part of a mix to get right, requiring accurate speakers in an accurate room. Muting the bass instrument and overall reverb occasionally helps to keep their level in perspective. Adding low-cut filters to most non-bass tracks is common, and it prevents muddy mixes. Panning instruments intelligently left-right in the stereo field also helps to improve clarity by avoiding competing frequencies coming from the same speaker.

Besides panning instruments and voices left-right to create space, high frequency content and reverb affect front-back depth. Adding treble brings sounds forward in a mix, and small- and large-room reverb settings can help further to define a virtual space. But a good mix engineer brings more to the table than just a good mix. The best mixing engineers contribute creative ideas and sometimes even suggest changes or additions to the musical arrangement.

This chapter also covers the basics of DAW editing, including comping a single performance from multiple takes, using slip-edits and cross-fades, normalizing, as well as explaining the best order of inserted effects such as EQ and compression. This chapter also explains how I organize and name my own projects to simplify backing up. If a picture is worth a thousand words, a video is worth at least a dozen pictures. Therefore, two short videos show the basics of editing music and speech, including the importance of cutting on waveform zero crossings.

Finally, I explained a bit about the history of mastering. What began in the mid-1900s as a necessity to overcome the limitations of the vinyl medium, mastering has evolved into an art form in its own right. A talented mastering engineer can make a good mix sound even better using EQ, compression, and other sound shaping tools.

CHAPTER 8

Digital Audio Basics

Analog audio comprises electrical signals that change over time to represent acoustic sounds. These voltages can be manipulated in various ways, then converted back to sound and played through a loudspeaker. Digital audio takes this one step further, using a series of numbers to represent analog voltages. The process of converting an analog voltage to equivalent numbers is called *digitization*, or *sampling*, and a device that does this is called an *analog to digital converter*, or A/D converter. Once the audio voltages are converted to numbers, those numbers can be manipulated in many useful ways and stored in computer memory or on a hard drive. Eventually the numbers must be converted back to a changing analog voltage in order to hear the audio through a loudspeaker. This job is performed by a *digital to analog converter*, or D/A converter, sometimes abbreviated as DAC.

All computer sound cards contain at least two A/D converters and two D/A converters, with one pair each for recording and playing back the left and right channels. I often use the term *A/D/A converter* for such devices, because most modern converters and sound cards handle both A/D and D/A functions. Many professional converters handle more than two channels, and some are designed to go in an outboard equipment rack rather than inside a computer or connected via USB or FireWire.

Sampling Theory

A number of people contributed to the evolution of modern sampling theory as we know it today, but history has recognized two people as contributing the most. Back in the 1920s, the telegraph was an important method of real-time communication. Scientists back then aimed to increase the bandwidth of those early systems to allow sending more messages at once over the wires. The concept of digital sampling was considered as early as the 1840s, but it wasn't until the 1900s that scientists refined their theories enough to prove them mathematically. In a 1925 article, Harry Nyquist showed that the analog bandwidth of a telegraph line limits the fastest rate at which Morse code pulses could be sent. His subsequent article a few years later clarified further, proving that the fastest pulse rate allowable is equal to half the available bandwidth. Another important pioneer of modern sampling is Claude Shannon, who in 1949 consolidated many of the theories as they're understood and employed today.

Note that the term *sampling* as used here is unrelated to the common practice of sampling portions of a commercial recording or sampling musical instrument notes and phrases for later playback by a MIDI-controlled hardware or software synthesizer. Whereas musical sampling records a "sample" of someone singing or playing an instrument, digital audio sampling as described here takes many very brief individual "snapshot" samples of an analog voltage at regularly timed intervals. Snapshot is an appropriate term for this process because it's very similar to the way a moving picture comprises a sequence of still images. That is, a movie camera captures a new still image at regularly timed intervals. Other similarities between moving pictures and digitized audio will become evident shortly.

Figure 8.1 shows the three stages of audio as it passes through an A/D/A converter operating at a sample rate of 44.1 KHz. The A/D converter samples the input voltage at regular intervals—in this case once every 1/44,100 second—and converts those voltages to equivalent numbers for storage. The D/A converter then converts the numbers back to the original analog voltage for playback. In practice, the numbers would be much larger than those shown here, unless the audio was extremely soft.

Note that the equivalent digitized numbers are not the same as the analog input voltages. Rather, the incoming voltage is scaled by a volume control to span a numeric range from −32,768 through +32,767 (for a 16-bit system). The volume control can be in the A/D converter or the preceding preamp or mixer. If the input volume is set too low, the largest number captured might be only 5,000 instead of 32,767. Likewise, if the volume is set too high, the incoming audio might result in numbers larger than 32,767. If the input signal is small and the loudest parts never create large numbers, the result after digitizing is a poor signal to noise ratio. And if the input voltage exceeds the largest number the converter can accommodate, the result is clipping distortion. By the way, these numbers are a decimal representation of the largest possible binary values that can be stored in a 16-bit word (2^{15} plus one more bit to designate positive or negative).

Quantization

Most A/D/A converters deal with whole numbers only, also called *integers*. If the input voltage falls between two whole numbers when sampled, it's stored as the nearest available whole number. This process is known as *quantization*, though it's also called *rounding* because the voltage is rounded up or down to the nearest available integer value. When the sampled numbers don't exactly match the incoming voltage, the audio waveform is altered slightly, and the result is a small amount of added distortion. In a 16-bit system, the disparity between the actual voltage and the nearest available integer is very small. In most cases the difference won't matter unless the input voltage is extremely small. When recording at 24 bits, the disparity between the input voltage and its stored numeric sample value is even smaller and less consequential.

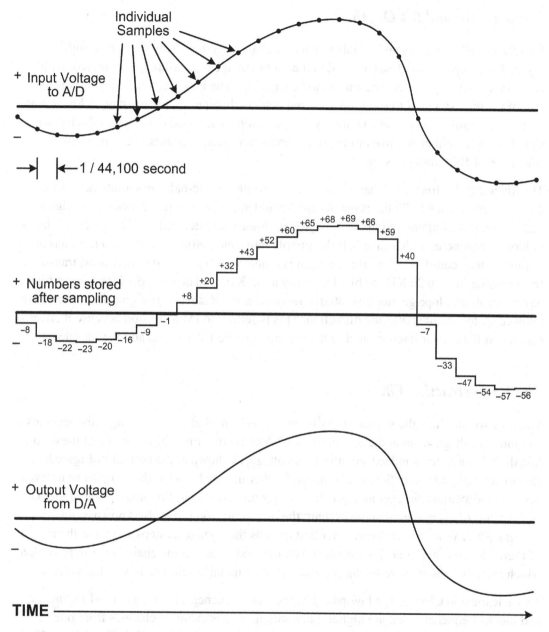

Figure 8.1: When an analog signal enters an A/D converter (top), the converter measures the input voltage at regular intervals. Whatever the voltage is at that moment is captured and held briefly as a number (center) for storage in computer memory or a hard drive. When those numbers are later sent out through the D/A converter (bottom), they're converted back to the original analog voltages.

Sample Rate and Bit Depth

Two factors affect the potential quality of a digital audio system: its *sample rate* and *bit depth*. The sample rate defines how often the input voltage is measured, or sampled, with faster rates allowing higher frequencies to be captured. The bit depth refers to the available size of the numbers used to store the digitized data, with larger numbers giving a lower noise floor. When audio is stored at CD quality, 16-bit numbers are used to represent the incoming signal voltage—either positive or negative—and a new sample passes in or out of the converter 44,100 times per second.

The highest audio frequency that can be accommodated is one-half the sample rate. So in theory, sampling at 44,100 times per second should allow recording frequencies as high as 22,050 Hz. But sampling requires a low-pass *anti-aliasing* filter at the A/D converter's input to block frequencies higher than half the sample rate from getting in. Since no filter has an infinitely steep cutoff slope, a safety margin is required. A typical input filter must transition from passing fully at 20 KHz to blocking fully at 22 KHz and above, which requires a filter having a roll-off slope greater than 80 dB per octave, if not steeper. If higher frequencies are allowed in, the result is *aliasing* distortion. This is similar to IM distortion because it creates sum and difference artifacts related to the incoming audio frequencies and the sample rate.

The Reconstruction Filter

You may wonder how the sequence of discrete-level steps that are stored digitally becomes a continuous voltage again at the output of the D/A converter. First, be assured that the output of a digital converter is indeed a continuous voltage, as shown at the bottom of Figure 8.1. Reports that digital audio "loses information" either in time between the samples or in level between the available integer numbers are incorrect. This assumption misses one of the basic components of every digital audio system: the *reconstruction filter*, also known as an *anti-imaging filter*. When the D/A converter first outputs the voltage as steps shown in the middle of Figure 8.1, the steps really are present. But the next circuit in the chain is a low-pass filter, which smoothes the steps restoring the audio to its original continuously varying voltage.

As we learned in Chapter 1, a low-pass filter passes frequencies below its cutoff frequency and blocks frequencies that are higher. Each sample in this example changes from one value to the next 44,100 times per second, but the low-pass reconstruction filter has a cutoff frequency of 20 KHz. Therefore, the sudden change from one step to the next happens too quickly for the filter to let through. Chapter 4 explained that a capacitor is similar to a battery and can be charged by applying a voltage. Like a battery, it takes a finite amount of time for a capacitor to charge. It also takes time to discharge, or change from one charge level to another. So when a voltage goes through a low-pass filter, sudden step changes are smoothed

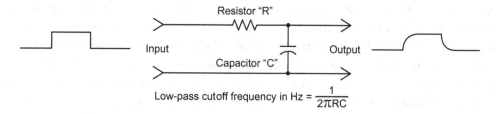

Figure 8.2: The low-pass reconstruction filter inside every D/A converter removes both the vertical and horizontal steps from digitized audio as it plays. The filter at the output of a D/A converter is more complex than the simple one-pole low-pass filter shown here, but the concept is identical.

over and become a continuously varying voltage. This is exactly what's needed at the output of a D/A converter.

The low-pass filter shown schematically in Figure 8.2 smoothes any voltage changes that occur faster than the resistor and capacitor values allow. The resistor limits the amount of current available to charge the capacitor, and the capacitor's value determines how quickly its present voltage can change based on the available current. Capacitor timing works the same whether charging or discharging, so it doesn't matter if the voltage step increases or decreases in level. Either way, the capacitor's voltage can't change faster than the resistor allows.

The filter in Figure 8.2 has a roll-off slope of only 6 dB per octave, but a real reconstruction filter must be *much* steeper. Like an A/D converter's anti-aliasing input filter, this output filter must also transition from passing fully at 20 KHz to blocking fully at 22 KHz and above. The myth that digital audio contains steps or loses data between the samples may have started because audio editor programs show steps on the waveforms when you zoom way in. But the graphic version of a waveform shown by audio software does not include the reconstruction filter, and the displayed images are often reduced to only 8 bits of data (256 total vertical steps) for efficiency and to save disk space. It's easy to prove that digital audio does not contain steps at the output of a D/A converter. One way is to simply look at the waveform on an oscilloscope. Another is to measure the converter's distortion. If steps were present, they would manifest as distortion.

Earlier I mentioned that digital sampling is closely related to moving pictures on film. With digital audio, aliasing distortion occurs if frequencies higher than half the sample rate are allowed into the A/D converter. The same thing happens with visual motion that occurs faster than a moving picture's frame rate can capture. A classic example is when the wagon wheels in an old Western movie seem to go backward. If the wagon is slowing down, you'll often see the wheels go forward, then backward, then forward again until the spokes are no longer moving too quickly for the camera's frame rate to capture. This effect is called *aliasing*, whether it happens with digital audio or motion picture cameras.

The most common sampling rate for digital audio is 44.1 KHz, which supports frequencies as high as 20 KHz. This rate was chosen because early digital audio was stored on video recorders, and 44,100 divides evenly by the number of scan lines used on video recorders in the 1970s. Audio for modern video production typically uses a sample rate of 48 KHz—not so much because the fidelity is deemed better, but because that number can be evenly divided by the common frame rates of 24, 25, and 30 frames per second. Having each video frame correspond to an integer number of audio samples simplifies keeping the audio and video data synchronized. But other sample rates are commonly used for audio, both higher and lower. Voice and other material meant for broadcast can be recorded at a sample rate of 32 KHz to reduce disk space and bandwidth, and 22.05 KHz can be used for even more savings when a 10 KHz bandwidth is sufficient. Many consumer sound cards support even lower sample rates.

Higher rates are also used, mostly by people who believe that ultrasonic frequencies are audible or that a bandwidth higher than is actually needed somehow improves the sound quality for frequencies we can hear. Common high sample rates are 88.2 KHz and 96 KHz, which are exact multiples of 44.1 and 48 KHz. Using integer ratios makes it easier for a *sample rate converter* to reduce the faster rate later when the audio is eventually put onto a CD. Even higher sample rates are sometimes used—192 KHz and 384 KHz—though clearly there's no audible benefit from such overkill, and that just wastes bandwidth and hard drive space. Handling that much data throughput also makes a computer DAW work harder, limiting the total number of tracks you can work with.

Oversampling

Earlier I mentioned that an A/D converter's anti-aliasing input filter must be very steep to pass 20 KHz with no loss while blocking 22 KHz and higher completely. Modern digital converters avoid the need for extremely sharp filters by using a technique known as *oversampling*. Instead of sampling at the desired rate of 44.1 KHz, the converter takes a new snapshot much more often. Sampling at 64 times the desired eventual rate is common, and this is referred to as 64x oversampling. With oversampling, the input filter's roll-off slope doesn't need to be nearly as steep, which in turn avoids problems such as severe ringing at the cutoff frequency. After the A/D converter oversamples through a less aggressive input filter, the higher sample rate is then divided back down using simple integer math as described earlier, before the numbers are sent on to the computer.

Bit Depth

Although 16 bits offers a sufficiently low noise level for almost any audio recording task, many people use 24 bits because, again, they believe the additional bits offer superior fidelity.

Some programs can even record audio data as 32-bit floating point (FP) numbers. But the number of bits used for digital audio doesn't affect fidelity other than establishing the noise floor. The frequency response is not improved, nor is distortion reduced. Claims that recording at 24 bits is cleaner or yields more resolution than 16 bits are simply wrong, other than a potentially lower noise floor. I say *potentially* lower because the background acoustic noise of a microphone in a room is usually 20 to 40 dB louder than the inherent noise floor of 16-bit audio.

To my way of thinking, the main reason to use 24 bits is because it lets you be less careful when setting record levels. The only time I record at 24 bits is for orchestra concerts or other live events. When recording outside of a controlled studio environment, it's better to leave plenty of headroom now rather than be sorry later. In those situations, it's good practice to set your record levels so the peaks never exceed −20 dBFS. Then if something louder comes along, it will be captured cleanly.

Table 8.1 lists the number range and equivalent noise floor for the most commonly used bit depths. Before the advent of MP3-type lossy compression, 8 bits was sometimes used with low fidelity material to save disk space. These days lossy data compression is a much better choice because the files are even smaller and the fidelity is far less compromised. I'll also mention that the −144 dB noise floor of 24-bit audio is purely theoretical, and no converter achieves a noise floor even close to that in practice. The best 24-bit systems achieve a noise floor equivalent to about 21 bits, or −126 dB, and many are closer to −110 dB.

Note that with modern 24-bit converters, the noise floor set by the number of bits is lower than the noise floor of their analog components. So the quantization distortion that results when a sample value is rounded to the nearest number is simply drowned out by the analog noise. In other words, the distortion and noise floor of a 24-bit converter's analog input and output stages dominate rather than the number of bits.

Pulse-Code Modulation versus Direct Stream Digital

The type of digital audio presented so far is called *pulse-code modulation*, abbreviated as PCM. But another method was developed by Sony called *direct stream digital*, or DSD,

Table 8.1: Common Digital Audio Bit Depths

Bit Depth	Numeric Range	Noise Floor
8 bits	−127 through +128	−48 dB
16 bits	−32,768 through +32,767	−96 dB
24 bits	−8,388,608 through +8,388,607	−144 dB
32 bits FP	+/− 3.4 * 2^38	Too low to matter

which is used by SACD players. SACD is short for Super Audio CD, but this format never caught on well for various reasons, including the high cost of players and insufficient interest by publishers to create content. With DSD, instead of periodically converting a varying voltage to equivalent 16- or 24-bit numbers, a technique known as *delta-sigma modulation* outputs only a one or a zero. Which number is generated depends on whether the currently sampled voltage is higher or lower than the previous sampled voltage, respectively. This method requires a very high sample frequency: DSD uses a sample rate of 2.8224 MHz, which is 64 times higher than 44.1 KHz used for CD audio.

In engineering math-speak, the Greek letter delta expresses a change, or difference, and sigma means summation. So delta-sigma implies a sum of differences, which in this case is the stream of one and zero differences that express how the audio changes over time. DSD proponents argue that delta-sigma sampling offers a wider bandwidth (100 KHz for SACD disks) and lower noise floor than standard PCM encoding, and the ultra-high sample rate avoids the need for input and output filters. But the noise of DSD audio rises dramatically at frequencies above 20 KHz, so filters are in fact needed to prevent that noise from polluting the analog output, possibly increasing IM distortion or even damaging your tweeters. The filters must also be steep enough to suppress the noise without affecting the audible band.

This explanation is necessarily simplified, and the actual implementation of modern PCM and DSD involves some pretty heavy math. But this explains the basic principles. The bottom line is that both PCM and DSD systems can sound excellent when executed properly.

Digital Notation

Some people consider digital sample values as binary numbers, but that's not necessarily true. Computers and hard drives do store data in binary form, but the numbers are exactly the same as their decimal counterparts. For example, the decimal number 14 is written in binary as 1110, but both numbers represent the same quantity 14. Binary is just a different way to state the same quantity. Binary notation is common with computers because most memory chips and hard drives can store only one of two voltage states: on or off. So accommodating numbers as large as 256 requires eight separate on-off memory locations, called *bits*.

For efficiency, data pathways and memory chips store data in multiples of 8 bits. This organization is extended to hard drives, live data streams, and most other places digital data is used. This is why audio data is also stored in multiples of 8 bits, such as 16 or 24. You could store audio as 19 bits per sample, but that would leave 5 bits of data unused in each memory location, which is wasteful. It's possible to store an odd number of bits such that some of the bits spill over into adjacent memory locations. But that requires more computer code to implement, as partial samples are split for storage, then gathered up and recombined later, which in turn takes longer to store and retrieve each sample.

Another common computer number notation is *hexadecimal*, often shorted to *hex*. This is similar to a base-2 (binary) system, but the numbers are more manageable by humans than a long string of binary digits. With hex notation, each digit holds a number between 0 and 15, for a total of 16 values. The letters A through F are used to represent values between 10 and 15, with each digit holding four bits of on/off data, as shown in Table 8.2. Computers use both binary and hexadecimal notation because each digit represents a number that's evenly divisible by 2, which corresponds to the way memory chips are organized.

We won't get into computer memory or binary and hex notation too deeply, but a few basics are worth mentioning. As stated, 8 binary bits can store 256 different integer numbers, and 16 bits can store 65,536 different values. These numbers may be considered as any contiguous range, and one common range for 16-bit computer data is 0 through 65,535. But that's not useful for digital audio because audio contains both positive and negative voltages. So digital audio is instead treated as numbers ranging from −32,768 through +32,767. This is the same number of numbers but split in the middle to express both positive and negative values.

When the audio finally comes out of a loudspeaker, the speaker cone is directed to any one of 32,767 different possible forward positions, or one of 32,768 positions when drawn inward. A value of zero leaves the speaker at its normal resting place, neither pushed forward nor pulled into the speaker cabinet. By the way, when a range of numbers is considered to hold only positive values, we call them *unsigned* numbers. If the same number of bits is used to store both positive and negative values, the numbers are considered to be *signed* because a plus or minus sign is part of each value. With signed numbers, the most significant binary bit at the far left has a value of 1 to indicate negative, as shown in Table 8.3.

Table 8.2: Hexadecimal Notation

Decimal	Binary	Hex
0	0000	0
1	0001	1
2	0010	2
3	0011	3
4	0100	4
5	0101	5
6	0110	6
7	0111	7
8	1000	8
9	1001	9
10	1010	A
11	1011	B
12	1100	C
13	1101	D
14	1110	E
15	1111	F

Table 8.3: Signed Numbers

Decimal	Binary
32,767	0111 1111 1111 1111
32,676	0111 1111 1111 1110
32,765	0111 1111 1111 1101
. . .	. . .
3	0000 0000 0000 0011
2	0000 0000 0000 0010
1	0000 0000 0000 0001
0	0000 0000 0000 0000
−1	1111 1111 1111 1111
−2	1111 1111 1111 1110
−3	1111 1111 1111 1101
−4	1111 1111 11111100
. . .	. . .
−32,766	1000 0000 0000 0010
−32,767	1000 0000 0000 0001
−32,768	1000 0000 0000 0000

Sample Rate and Bit Depth Conversion

Sometimes it's necessary to convert audio data from one sample rate or bit depth to another. CD audio requires 44.1 KHz and 16 bits, so a Wave file that was mixed down to any other format must be converted before it can be put onto a CD. Sample rate conversion is also used to apply Vari-Speed type pitch shifting to digital audio. The same process is used by software and hardware music samplers to convert notes recorded at one pitch to other nearby notes as directed by a MIDI keyboard or sequencer program.

Converting between sample rates that are exact multiples, such as 96 KHz and 48 KHz, is simple: The conversion software could simply discard every other sample or repeat every sample. Converting between other integer-related sample rates is similar. For example, to convert 48 KHz down to 32 KHz, you'd discard every third sample, keeping two. And in the other direction you'd repeat every other sample so each pair of samples becomes three. In practice, each sample is not repeated, but rather new samples having a value of 0 are inserted between the existing samples. This reduces the overall signal level by 6 dB, but yields a flatter response than simply repeating samples. The volume loss is then countered by adding 6 dB digitally after the conversion.

Again, this algorithm is used for the simple case of doubling the sample rate, though a similar process can be used for other integer ratios. The actual processing used for modern sample rate conversion is more complicated than described here, requiring interpolation between samples and additional filtering in the digital domain, but this explains the basic logic. However, converting between unrelated sample rates, such as 96 KHz and 44.1 KHz, is more difficult and risks adding aliasing artifacts if done incorrectly.

One method finds the least common denominator sample rate. This is typically a much higher frequency that's evenly divisible by both the source and target sample rates. So the audio is converted to that much higher rate using integer multiplication, then converted back down again using a different integer as the divisor. Perhaps the simplest way to convert between unrelated sample rates is to play the audio from the analog output of one digital system while recording it as analog to a second system at the desired sample rate. This works perfectly well, but it can potentially degrade quality because the audio passes through extra A/D and D/A conversions.

In my opinion, it makes the most sense to record at whatever sample rate the target medium requires. If you know you'll have to throw away half the samples, or risk adding artifacts by using non-integer sample rate conversion, what's the point of recording at 88.2 KHz or 96 KHz? Any perceived advantage of recording at a higher sample rate is lost when the music goes onto a CD anyway, so all you've accomplished are wasting disk space and reducing your total available track count. In fairness, there are some situations where a sample rate higher than 44.1 KHz is useful. For example, if you record an LP, software used later to remove clicks and pops might be better able to separate those noises from the music.

Bit depth conversion is much simpler. To increase bit depth, new bits are added to each sample value at the lowest bit position, then assigned a value of zero. To reduce bit depth, you simply discard the lowest bits, as shown in Tables 8.4 and 8.5. In Table 8.4, a 16-bit sample is expanded to 24 bits by adding eight more lower bits and assigning them a value of zero. In Table 8.5, a 24-bit sample is reduced to 16 bits by discarding the lower eight bits.

Table 8.4: Increasing Bit Depth

Original 16-Bit Value	After Converting to 24 Bits
0010 0100 1110 0110	0010 0100 1110 0110 0000 0000

Table 8.5: Decreasing Bit Depth

Original 24 Bits	After Converting to 16 Bits
0010 0100 1110 0110 0011 1010	0010 0100 1110 0110

Dither and Jitter

When audio bit depth is reduced by discarding the lowest bits, the process is called *truncation*. In this case, the numbers are not even rounded to the nearest value. They're simply truncated to the nearest *lower* value, which again adds a small amount of distortion. As explained in Chapter 3, *dither* avoids truncation distortion and so is applied when reducing the bit-depth of audio data. Dither can also be added in A/D converters when recording at 16 bits. Truncation distortion is not a problem on loud musical passages, but it can affect soft material because the smaller numbers are changed by a larger amount relative to their size. Imagine you need to truncate a series of numbers to the next lower integer value. So 229.8 becomes 229, which is less than half a percent off. But 1.8 gets changed to 1, which is an error of 80 percent!

Dither is a very soft noise, having a volume level equal to the lowest bit. This is about 90 dB below the music when reducing 24-bit data to 16 bits for putting onto a CD. Since all noise is random, adding dither noise assigns a random value to the lowest bit, which is less noticeable than the harmonics that would be added by truncation. In fact, adding dither noise retains some parts of the original, larger bit-depth, even if they fall below the noise floor of the new, smaller bit-depth. Again, truncation distortion created when reducing 24-bit data to 16 bits affects only the lowest bit, which is 90 dB below full scale, so it's not likely to be heard except on very soft material when played back very loudly. But still, dither is included free with most audio editor software, so it only makes sense to use it whenever you reduce the bit-depth.

Another very soft artifact related to digital audio is *jitter*, which is a timing error between sample periods. Ideally, a D/A converter will output its samples at a precise rate of one every 1/44,100 second. But in practice, the time between each sample varies slightly, on the order of 50 to 1,000 picoseconds (trillionths of a second). Timing errors that small aren't perceived as pitch changes. Rather, they add a tiny amount of noise. Like truncation distortion, it's unlikely that anyone could ever hear jitter because it's so much softer than the music, and it is also masked by the music. But it's a real phenomenon, and gear vendors are glad to pretend that jitter is an audible problem, hoping to scare you into buying their latest low-jitter converters and related products.

In order to sample analog audio 44,100 times per second, or output digital data at that rate (or other rates), every digital converter contains an oscillator circuit—called a *clock*—that runs at the desired frequency. Simple oscillators can be made using a few transistors, resistors, and capacitors, but their frequency is neither precise nor stable. Most resistors and capacitors can vary from their stated value by 2 percent or more, and their value also changes as the temperature rises and falls. High-precision components having a tighter tolerance and less temperature drift are available for a much higher cost, but even 0.1 percent variance is unacceptable for a digital audio converter.

The most accurate and stable type of oscillator contains a quartz or ceramic crystal, so those are used in all converters, including budget sound cards. When a crystal is excited by a voltage, it vibrates at a resonant frequency based on its size and mass. Crystals vibrate at frequencies much higher than audio sample rates, so the oscillator actually runs at MHz frequencies, with its output divided down to the desired sample rate. Crystal oscillators are also affected by the surrounding temperature, but much less so than oscillators made from resistors and capacitors. When extreme precision is required, a crystal oscillator is placed inside a tiny oven. As long as the oven's temperature is slightly warmer than the surrounding air, the oven's thermostat can keep the internal temperature constant. But that much stability is not needed for digital audio.

External Clocks

For a simple digital system using only one converter at a time, small variations in the clock frequency don't matter. If a sample arrives a few picoseconds earlier or later than expected, no harm is done. But when digital audio must be combined from several different devices at once, which is common in larger recording and broadcast studios, all of the clocks must be synchronized. When one digital output is connected to one digital input using either shielded wire or an optical cable, the receiving device reads the incoming data stream and locks itself to that timing. But a digital mixer that receives data from several different devices at once can lock to only one device's clock frequency. So even tiny timing variations in the other sources will result in clicks and pops when their mistimed samples are occasionally dropped.

The solution is an *external clock*. This is a dedicated device that does only one thing: It outputs a stable frequency that can be sent to multiple digital devices to ensure they all output their data in lock-step with one another. Most professional converters have a clock input jack for this purpose, plus a switch that tells it to use either its internal clock or the external clock signal coming in through that jack. While an external clock is needed to synchronize multiple digital devices, it's a myth that using an external clock reduces jitter in a single D/A converter as is often claimed.

In truth, when a converter is slaved to an external clock, its jitter can only be made worse, due to the way clock circuits lock to external signals. In a great example of audio journalism at its finest,[1] *Sound On Sound* magazine measured the jitter of four different converters whose prices ranged from affordable to very expensive. The jitter at each converter's output was measured with the converter using its own internal clock, and then again when locked to an external clock. Several different external clocks were used, also ranging greatly in price. Every single converter performed more poorly when slaved to an external clock. So while it's possible that using an external clock might *change* the sound audibly, the only way to know for sure is with a proper blind listening test. Even if the quality does change audibly, it can only be for the worse.

Digital Converter Internals

Figure 8.3 shows a simplified block diagram of an A/D converter. After the audio passes through an anti-aliasing low-pass input filter, the actual sampling is performed by a *Sample and Hold* circuit that's activated repeatedly for each sample. This circuit freezes the input voltage at that moment for the length of the sample period, like taking a snapshot, so the rest of the A/D converter receives a single stable voltage to digitize. In this example only four bits are shown, but the concept is the same for 16- and 24-bit converters. Again, this block diagram is simplified; modern converters use more sophisticated methods to perform the same basic tasks shown here, though this circuit could actually work as shown.

After the Sample and Hold circuit stabilizes the incoming audio to a single non-changing voltage, it's fed to a series of *voltage comparators*. Depending on the analog voltage at its input, the output of a comparator is either fully positive or fully negative to represent a binary one or zero, respectively. As you can see, the input resistors are arranged so that each comparator farther down the chain receives half the voltage (−6 dB) of the one above. Therefore, most of the input voltage goes to the top comparator that represents the most significant bit of data, and subsequent comparators receive lower and lower voltages that represent ever less significant data bits. A comparator is an analog circuit, so other circuits

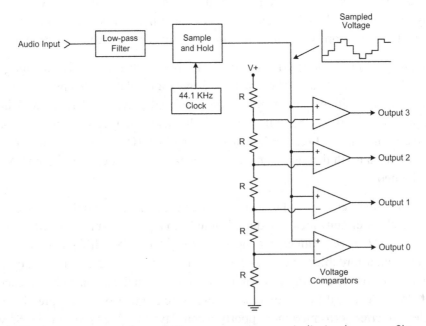

Figure 8.3: The key components in an A/D converter are an anti-aliasing low-pass filter, a Sample and Hold circuit that takes the actual snapshots at a rate dictated by the clock, and a series of voltage comparators that each output a zero or a one, depending on their input voltage.

(not shown) convert the fully positive or negative voltage from each comparator to the On or Off voltages that digital circuits require.

The internal operation of a Sample and Hold circuit is shown in Figure 8.4. Again, this is a simplified version showing only the basic concept. The input voltage is constantly changing, but a switch controlled by the converter's clock closes briefly, charging the capacitor to whatever voltage is present at that moment. The switch then immediately opens to prevent further voltage changes, and the capacitor holds that voltage long enough for the rest of the circuit to convert it to a single binary value. This process repeats continuously at a sample rate set by the clock.

A D/A converter is much less complex than an A/D converter because it simply sums a series of identical binary one or zero voltages using an appropriate *weighting* scheme. Digital bits are either on or off, and many logic circuits consider zero volts as off and 5 volts DC as on. Each bit in digital audio represents a different amount of the analog voltage, but all the bits carry the same DC voltage when on. So the contribution of each bit must be weighted such that the higher-order bits contribute more output voltage than the lower-order bits. In other words, the most significant bit (MSB) dictates half of the analog output voltage, while the least significant bit (LSB) contributes a relatively tiny amount. One way to reduce the contribution of lower bits is by doubling the resistor values for successively lower bits, as shown in Figure 8.5.

The resistor network with doubling values in Figure 8.5 works in theory, but it's difficult to implement in practice. First, the resistor values must be very precise, especially the lower-value resistors that contribute the most to the output voltage. If this 4-bit example were

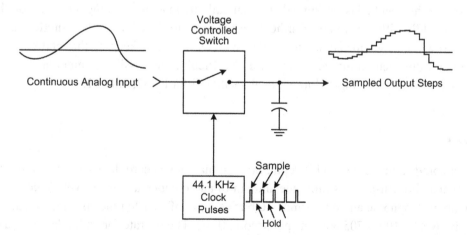

Figure 8.4: A Sample and Hold circuit uses an electronic switch to charge a capacitor to whatever input voltage is currently present, then holds that voltage until the next sample is read.

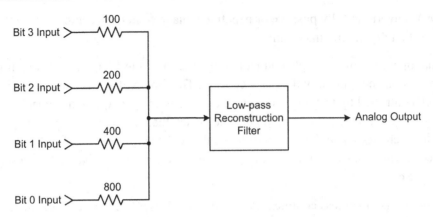

Figure 8.5: Each bit of digitized audio represents half as much signal as the next higher bit, so doubling the resistor values contributes less and less from each lower bit.

expanded to 16 bits, the largest resistor needed for the LSB would be 3,276,800 ohms, which is impractically large compared to 100 ohms used for the MSB. If the value of the 100-ohm resistor was off by even 0.0001 percent, that would affect the output voltage more than the normal contribution from the lowest bit. Again, this type of error results in distortion. Even if you could find resistors as accurate as necessary, temperature changes will affect their values significantly. Such disparate resistor values will also drift differently with temperature, adding even more distortion.

The R/2R *resistor ladder* shown in Figure 8.6 is more accurate and more stable than the doubling resistors circuit in Figure 8.5, and it requires only two different resistor values, with one twice the value of the other. The resistors in an R/2R ladder must also have precise values and be stable with temperature, but less so than for a network of doubling values. Resistor networks like this can be laser-trimmed automatically by machine to a higher precision than if they had wildly different values. Further, if all the resistors are built onto a single substrate, their values will change up and down together with temperature, keeping the relative contribution from each bit the same. It would be difficult or impossible to manufacture a network of resistors on a single chip with values that differ by 32,768 to 1.

Bit-Rate

We've explored sample rate and bit depth, but you may have heard the term *bit-rate* used with digital audio. Bit-rate is simply the amount of audio data that passes every second. CD-quality monophonic audio transmits 16 bits at a rate of 44,100 times per second, so its bit-rate is $16 \times 44,100 = 705,600$ bits per second (bps). The bit-rate for audio data is usually expressed as kilobits per second, or Kbps, so in this case mono audio has a bit-rate of 705.6 Kbps. Stereo audio at CD quality is twice that, or 1,411.2 Kbps.

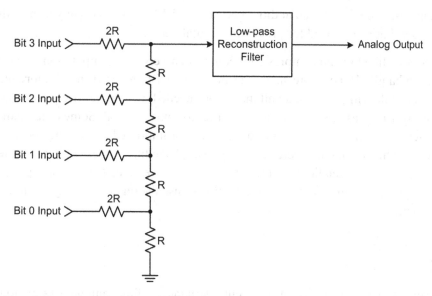

Figure 8.6: An R/2R resistor ladder reduces the contribution from lower bits as required by 6 dB for each step, but uses only two different resistor values.

Bit-rate is more often used to express the quality of an audio file that's been processed using lossy compression, such as an MP3 file. In that case, bit-rate refers to the number of *compressed* bits that pass every second. If a music file is compressed to a bit-rate of 128 Kbps, every second of music occupies 128,000 bits within the MP3 file. One byte contains eight bits, which comes to 16 KB of data being read and decoded per second. So a three-minute song will be 16 KB × 180 seconds = 2,880 KB in size.

More modern MP3-type compression uses variable bit rate (VBR) encoding, which varies the bit-rate at any given moment based on the demands of the music. A slow bass solo can probably be encoded with acceptably high fidelity at a bit-rate as low as 16 Kbps, but when the drummer hits the cymbal as the rest of the band comes back in, you might need 192 Kbps or even higher to capture everything with no audible quality loss. So with VBR encoding you won't know the final file size until after it's been encoded.

Digital Signal Processing

Digital signal processing, often shortened to DSP, refers to the manipulation of audio while it's in the digital domain. Audio plug-ins in your DAW program use DSP to change the frequency response or apply volume compression, or any other task required. "Native" DAW software uses the computer's CPU to do all the calculations, but dedicated DSP chips are used in hardware devices such as PCI "booster" cards, hardware reverb units, surround sound enhancers, and other digital audio processors. A DSP chip is basically a specialized microprocessor that's optimized for processing digital audio.

Simple processes like EQ and compression require relatively few computations, and modern computers can handle 100 or more such plug-ins at once. But reverb is much more intensive, requiring many calculations to create all the echoes needed, with a high-frequency response that falls off over time as the echoes decay in order to sound natural. Some older audio software offers a "quality" choice that trades fidelity for improved processing speed. This is a throwback to when computers were slow so you could audition an effect in real time at low quality, then use a higher quality when rendering the final version. This is not usually needed today, except perhaps with reverb effects, so I always use the highest-quality mode when given an option.

Latency

Electricity may travel at the speed of light, but it still takes a finite amount of time to get digital audio into and out of a computer. The main factor that determines how long this process takes is the converter's *buffer*. This is an area of memory containing a group of digitized samples on their way into the computer when recording or coming out of the computer when playing back. The size of the buffer and the sample rate determine the amount of delay, which is called *latency*.

If a computer had to stop everything else it's doing 44,100 times per second to retrieve each incoming sample after it's been converted, there wouldn't be time for it to do much else. So as each incoming sample is read and converted to digital data, the A/D converter deposits it into an area of buffer memory set aside just for this purpose. When the buffer has filled, the audio program retrieves all of the data in one quick operation, then continues on to update the screen or read a Wave file from a hard drive or whatever else it was doing. The same process is used when playing back audio: The DAW program deposits its outgoing data into a buffer, then when the buffer is full the sound card converts the data back to a changing analog voltage while the computer continues whatever it had been doing. Both operations can happen at the same time because the computer and the sound card each have their own processor and clock.

Most modern sound cards and DAW programs use the ASIO protocol, which lets you control the size of the buffer. ASIO stands for *audio stream input/output*, and it was developed by Steinberg GmbH, which also invented the popular VST plug-in format. Typical buffer sizes range from 64 samples to 2,048 samples. If the buffer is very large, it might take 100 milliseconds or more to fill completely. If you're playing a mix and adjusting a track's volume in real time, hearing the volume change 1/10 second later is not a big problem. But this is an unacceptable amount of delay when playing a software synthesizer, since you'll hear each note some time after pressing the key on your keyboard. With that much delay it's nearly impossible to play in time. But setting the buffer size too small results in dropouts because the computer is unable to keep up with the frequent demands to read from and write to the buffer. So the key is finding the smallest buffer size that doesn't impose audio dropouts or clicks.

Another contributor to latency is the amount of time needed for audio plug-ins to process their effects. As mentioned earlier, some effects require more calculations than others. For example, a volume change requires a single multiplication for each sample value, but an EQ requires many math operations. Reverb effects require even more calculations, creating even longer delays. The total latency in a DAW project depends on whichever plug-in's process takes the longest to complete. Most modern DAWs include a feature called *automatic latency compensation*, which determines the delay imposed by each individual plug-in and makes sure everything gets output at the correct time. But some older programs require you to look up the latency for each plug-in effect or software synthesizer in its owner's manual and enter the numbers manually.

Modern computers can achieve latency delays as small as a few milliseconds, which is plenty fast for playing software instruments in real time. If you consider that 1 millisecond of delay is about the same as you get from one foot of distance acoustically, 10 milliseconds delay is the same as standing ten feet away from your guitar amp. But beware that DAW programs often misstate their latency. I've seen software report its latency as 10 milliseconds, yet when playing a keyboard instrument it was obvious that the delay between pressing a key and hearing the note was much longer.

Floating Point Math

Wave files store each sample as 16 or 24 bits of data, but most modern DAW software uses larger 32-bit *floating point* (FP) math for its internal calculations. Manipulating larger numbers increases the accuracy of the calculations, which in turn reduces noise and distortion. As with analog hardware, any change to the shape of an audio waveform adds distortion or noise. When you send audio through four different pieces of outboard gear, the distortion you get at the end is the sum of distortions added by all four devices.

The same thing happens inside DAW software as your audio passes through many stages of numeric manipulation for volume changes, EQ, compression, and other processing. As a DAW program processes a mix, it reads the 16- or 24-bit sample numbers from each track's Wave file, converts the samples to 32-bit FP format, then does all its calculations on the larger 32-bit numbers. When you play the mix through your sound card or render it to a Wave file, only then are the 32-bit numbers reduced to 16 or 24 bits.

Table 8.6: A 32-Bit Floating Point Number

SEEEEEEE	EMMMMMMM	MMMMMMMM	MMMMMMMM

Table 8.6 shows the internal format of a 32-bit floating point number, which has a dynamic range exceeding 1,500 dB, though most of that range is not needed or used. There are 23 "M" bits to hold the *mantissa*, or basic value, plus eight more "E" bits to hold the *exponent*, or multiplier. The use of a base number plus an exponent is what lets FP numbers hold such a huge range of values. The actual value of an FP number is its mantissa times 2 to the exponent. The "S" bit indicates the number's sign, either 1 or −1, to represent positive or negative, respectively. Therefore:

$$\text{Value} = S*M*2^{\wedge E}$$

Most DAW math adds a tiny amount of distortion and noise, simply because very few operations can be performed with perfect accuracy. For example, increasing the gain of a track by exactly 6.02 dB multiplies the current sample value by two. This yields an even result number having no remainder. Likewise, to reduce the volume by exactly 6.02 dB, you simply multiply the sample times 0.5. That might leave a remainder of 0.5, depending on whether the original sample value was odd or even. But most level changes do not result in a whole number. And that's when the accuracy of each math operation adds a tiny but real amount of distortion and noise.

To learn how much degradation occurs with 32-bit FP math, I created a DAW project that applies 30 sequential gain changes to a 500 Hz sine wave generated in Sound Forge. The sine wave was imported to a track, which in turn was sent to an output bus. I then added 14 more output buses, routing the audio through every bus in turn. Each bus has both an input and an output volume control, so I changed both controls, forcing two math operations per bus.

To avoid the possibility of distortion from one gain change negating the distortion added by other changes, I set each complementary raise/lower pair differently. That is, the first pair lowered, then raised the volume by 3.4 dB, and the next pair reversed the order, raising, then lowering the volume by 5.2 dB. Other gain change pairs varied from 0.1 dB to 6.0 dB. After passing through 15 buses in series, each applying different pairs of volume changes, I exported the result to a new Wave file. The Fast Fourier Transform (FFT) spectrums for both the original and processed sine waves are shown in Figure 8.7.

You can see that most of the added artifacts are noise. A pure frequency would show only a single spike at 500 Hz, but both FFT graphs have many tiny spikes, with most below −120 dB. Distortion added to a 500 Hz tone creates harmonics at 1,000 Hz, 1,500 Hz, 2,000 Hz, and so forth at 500 Hz multiples. But the main difference here is more noise in the processed version, with most of the spikes at frequencies other than those harmonics.

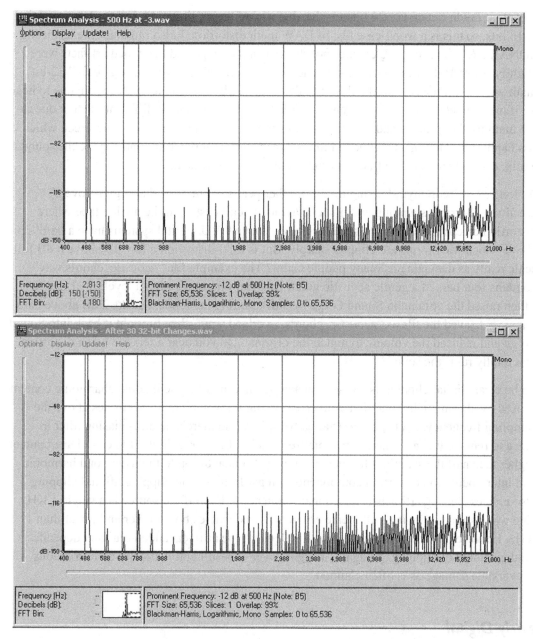

Figure 8.7: These FFT plots show the amount and spectrum of distortion and noise artifacts added by 32-bit DAW processing math. The top graph shows the original 500 Hz sine wave, and the lower graph shows the result after applying 30 gain changes.

Note that I didn't even use 24 bits for the sine wave file, nor did I add dither when exporting at 16 bits, so this is a worst-case test of DAW math distortion. I also ran the same test using SONAR's 64-bit processing engine (not shown), and as expected the added artifacts were slightly softer. But since all of the distortion components added when using "only" 32-bit math with 16-bit files are well below 100 dB, they're already way too soft for anyone to hear. So if mixes made using two different 32-bit DAWs really do sound different, it's not due to the math used to process and sum the tracks—unless the computer code is incorrect, which has happened. Further, as mentioned in Chapter 5, different pan rules can change the sound of a mix, even when all of the track settings are otherwise identical.

As long as we're in myth-busting mode, another persistent myth is that digital "overs" are always horrible sounding and must be avoided at all costs. Unlike analog tape where distortion creeps up gradually as the level rises above 0 VU, digital systems have a hard limit above which waveforms are clipped sharply. But is brief distortion a few dB above 0 dB full scale really as damaging as many people claim? The example file "acoustic_chords.wav" contains four bars of a gentle acoustic guitar part that I normalized to a level of −1 dBFS. I then raised the volume in Sound Forge by 3 dB, putting the loudest peaks 2 dB above hard clipping. I saved the file as "acoustic_chords_overload.wav," then brought it back into Sound Forge and reduced the volume to match the original. Do you hear a terrible harsh distortion that totally ruins the sound?

To be clear, digital clipping is worse than analog clipping because it adds inharmonic content caused by aliasing. Clipping after conversion to digital would create overtones above the sampling frequency, but the audio is already digitized so there's no anti-aliasing filter in place to remove the generated ultrasonic frequencies. Further, a digital file can't have content higher than half the sampling frequency anyway. So in addition to the usual total harmonic and intermodulation distortion components you get from analog clipping, digital clipping also creates aliasing artifacts at additional frequencies. Even if the source is a single 1 KHz sine wave, digital clipping will create artifacts at frequencies both higher and lower than 1 KHz. However, as with analog clipping, the level of the artifacts added digitally depends on the amount of clipping. So again, a slight "over" doesn't automatically do horrendous irreversible damage as is often claimed.

Early Digital

We've all read over and over that "early digital audio" sounded awful. This has been repeated so many times that it's now accepted as true by many. I've even seen people claim that early digital audio sounded terrible while conceding that the fidelity of current converters is almost as good as analog tape and vinyl records!

The 1984 report *The Digital Challenge* on the website of the Boston Audio Society[2] shows that in a blind test, a self-proclaimed audio expert and digital audio hater was unable to tell when digital conversion was inserted into a high quality analog playback chain. Bear in mind this was with the converters available in 1984, which were not nearly as good as what's available today. Another even earlier test was published in the October 1982 issue of *Gramophone* magazine. That report also showed that in blind tests multiple people were unable to tell when a digital audio "bottleneck" was inserted into the playback chain.

As it happens, my wife bought the first "affordable" CD player in 1984, made by Sony. It cost $700 at the time, which would equal more than twice that amount today! She also bought many CDs back then, and she still has them all. So to test the notion that early CDs sounded lousy, I ripped some tracks from three CDs from that era and put short fragments of each on the book's website:

Fleetwood Mac *Dreams*: early-digital-dreams.wav
Pink Floyd *Money*: early-digital-money.wav
Moody Blues *I Am*: early-digital-iam.wav

For fun, I also grabbed a screen-cap of *Dreams* showing how much better music "looked" before the Loudness Wars, shown in Figure 8.8. So if anything, early digital audio probably sounded *better* than the hyper-compressed CDs and MP3s released today. I think a lot of the music was better too, though that's a different discussion. Another factor may be that some early CDs really did sound harsh due to improper mastering. Apparently some CDs were made from master tapes meant for cutting vinyl records, with treble added to counter

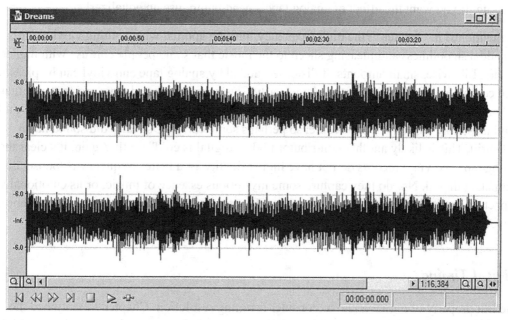

Figure 8.8: As you can see from this waveform, the Fleetwood Mac song *Dreams* has more dynamic range than much of the pop music produced in later years.

the known high frequency loss of LPs. But to my ears, the early CD tracks listed above still sound pretty darn good. What do you think?

Digital Audio Quality

Some people claim that analog recording is "wider" or clearer or more full-sounding than digital audio. However, it's simple to prove by measuring, null tests, and blind listening that modern digital audio is much closer to the original source than either analog tape or vinyl records. To my way of thinking, stereo width is mainly a function of left-right channel differences, which are mostly related to microphone placement and panning. Width can also be increased by adding time-based effects such as delay and reverb. Simply adding stereo hiss at a low level to a mono track, as happens when mixing to analog tape, makes the sound seem wider because the hiss is different left and right. Further, the left-right differences in the minute dropouts that are always present with analog tape can also seem to give a widening effect.

The fullness attributed to analog recording is a frequency response change that can be easily measured and confirmed, and even duplicated with EQ when desired. So the common claim that digital audio is thin-sounding is easy to disprove. Analog tape adds a low-frequency boost called *head bump*. The frequency and amount of boost depends on several factors, mostly the tape speed and record head geometry. A boost of 2 or 3 dB somewhere around 40 to 70 Hz is typical, and this could account for the perception that analog tape is fuller-sounding than digital audio. But the response of modern digital audio is definitely more accurate. If you want the effect of analog tape's head bump, use an equalizer!

As mentioned in Chapter 7, the high-frequency limiting applied when cutting vinyl records can add a smoothness and pleasing sheen to the music that some people confuse with higher fidelity. Likewise, small amounts of distortion added by analog tape and vinyl can be pleasing with certain types of music. Further, some early CDs were duplicated from master tapes that had been optimized for cutting vinyl. But CDs don't suffer from the high-frequency loss that vinyl mastering EQ aims to overcome, so perhaps some early CDs really did sound harsh and shrill. This is likely another contributor to the "digital is cold" myth. Again, it's clear that analog tape and vinyl records do not have higher fidelity, or a better frequency response, than competent digital. Nor do they capture some mysterious essence of music, or its emotional impact, in a way that digital somehow misses. Emotion comes mainly from the performance, as well as the listener's subjective perception and mood.

Digital Timing

The last property of digital audio to consider is timing resolution. Chapter 3 mentioned the flawed research done by Tsutomu Oohashi that attempted to prove people can hear ultrasonic content. Audio engineers have investigated the audibility of ultrasonics for decades, yet no legitimate tests have ever found that people can hear, or otherwise perceive, frequencies much higher than around 20 KHz. Another researcher, Milind Kunchur, thought he found a different way to prove that high sample rates are needed based on our hearing's *temporal resolution*. He claimed that human ears can detect left-right arrival time differences as small as 5–10 microseconds, which is true, but he wrongly concluded that reproducing such small timing offsets requires a sample rate higher than 44.1 KHz. What Dr. Kunchur didn't consider is that bit depth also affects timing resolution, and 44.1 KHz at 16 bits is in fact more than adequate to resolve timing to much finer than anyone can hear. The formula is simple, based on the highest frequency accommodated and the number of bits in use:

$$1 / (2 * \pi * Fmax * 2^\text{Bits})$$

So for CD quality audio with an upper limit of 20 KHz at 16 bits, the resolution is:

$$1 / (2 * \pi * 20,000 * 65,536) = 0.12 \text{ nanoseconds}$$

Even if music is so soft that it occupies only 8 bits (−48 dBFS), the timing resolution is still 31 nanoseconds, or 0.031 microseconds, which is 300 times better than anyone can hear.

Summary

This chapter explains the basics of digital audio hardware and software. Assuming a competent implementation, the fidelity of digital audio is dictated by the sample rate and bit depth, which in turn determine the highest frequency that can be recorded and the background noise floor, respectively. Simplified block diagrams showed how A/D/A converters work, with the incoming voltage captured by a Sample and Hold circuit, then converted by voltage comparators into equivalent 16- or 24-bit numbers that are stored in computer memory.

In order to avoid aliasing artifacts, the input signal must be filtered to prevent frequencies higher than half the sampling rate from getting into the A/D converter. Similarly, the output of a D/A must be filtered to remove the tiny steps that would otherwise be present. Although in theory these filters must have extremely steep slopes, modern A/D converters use oversampling to allow gentler filters, which add fewer artifacts.

When analog audio is sampled, the voltages are scaled by an input volume control to fill the available range of numbers. Samples that fall between two numbers are rounded to the nearest available number, which adds a tiny amount of distortion. The 32-bit floating point math that audio software uses to process digital audio also increases distortion, but even after applying 30 gain changes the distortion added is simply too small to be audible.

You also learned the basics of digital numbering, including binary and hex notation. Although digital audio can be converted by software algorithms to change its sample rate or bit depth, it's usually better to record at the same resolution you'll use to distribute the music. For most projects, CD-quality audio at 44.1 KHz and 16 bits is plenty adequate. However, recording at a lower level using 24 bits for extra headroom makes sense for live concerts or other situations where you're not certain how loud a performance will be.

Several audio myths were busted along the way, including the belief that digital audio contains steps or loses information between the samples, that external clocks can reduce jitter, and that digital overs always sound terrible.

Notes

1 "Does Your Studio Need A Digital Master Clock?" by Hugh Robjohns.
   www.soundonsound.com/sos/jun10/articles/masterclocks.htm
2 "The Digital Challenge: A Report" by Stanley P. Lipshitz.
   www.bostonaudiosociety.org/bas_speaker/abx_testing2.htm

CHAPTER 9

Dynamics Processors

A dynamics processor is a device that automatically varies the volume of audio passing through it. The most common examples are compressors and limiters, which are basically the same, and noise gates and expanders, which are closely related. Where compressors reduce the dynamic range, making volume levels more alike, expanders increase dynamics, making loud parts louder or soft parts even softer. There are also multi-band compressors and expanders that divide the audio into separate frequency bands, then process each band independently.

Compressors and Limiters

A compressor is an automatic level control that reduces the volume when the incoming audio gets too loud. Compressors were originally used to prevent AM radio transmitters from distorting if the announcer got too close to the microphone and to keep the announcer's volume more consistent. Then some creative types discovered that a compressor can sound cool as an effect on voices, individual instrument tracks, and even complete mixes. So a compressor can be used both as a tool to correct uneven volume levels and as an effect to subjectively improve the sound quality. The main thing most people notice when applying compression to a track or mix is that the sound becomes louder and seems more "in your face."

Compressors and limiters were originally implemented in hardware, and most modern DAW programs include one or two plug-in versions. The Sonitus plug-in compressor shown in Figure 9.1 has features typical of both software and hardware compressors, offering the same basic set of controls.

The most fundamental setting for every compressor is its *threshold*—sometimes called the *ceiling*. When the incoming audio is softer than the threshold level, the compressor does nothing and passes the audio with *unity gain*. So if you set the threshold to −12 dB, for example, the compressor reduces the volume only when the input level exceeds that. When the input volume later falls below the threshold, the compressor raises the volume back up again to unity gain with no attenuation.

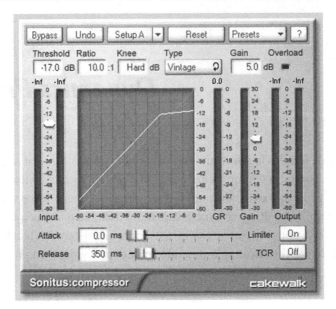

Figure 9.1: A compressor adjusts the volume of an instrument or complete mix automatically to keep the overall level more consistent.

The *compression ratio* dictates how much the volume is reduced, based on how much louder the audio is compared to the threshold. A ratio of 1:1 does nothing, no matter how loud the audio gets. But if the ratio is 2:1, the volume is reduced half as much as the excess. So if the input level is 2 dB above the threshold, the compressor reduces the level by 1 dB, and the output is only 1 dB louder rather than 2 dB louder, as at the input. With a ratio of 10:1, the signal level must be 10 dB above the threshold for the output to increase by 1 dB. When a compressor is used with a high ratio of 10:1 or greater, it's generally considered a limiter, though technically a limiter has an infinite ratio; the output level never exceeds the threshold level. Therefore, a compressor reduces the dynamic range difference between loud and soft parts, whereas a limiter applies a hard limit to the volume to maintain a constant level. Practically speaking, the compression ratio is the only distinction between a compressor and a limiter.

The *gain reduction meter* shows how much the volume is being reduced at any given moment. This meter is abbreviated as GR on the plug-in screen in Figure 9.1. This is the main meter you'll watch when adjusting a compressor, because it shows how much the compressor is doing. The more compression that's applied, the more the sound is affected. Notice that a gain reduction meter displays the opposite of a VU meter. The normal resting position is at zero, and the displayed level goes down toward negative numbers to show how much gain reduction is being applied. Some hardware compressors use a conventional VU meter to display the input or output level, but when switched to display gain reduction, the needle rests at zero and goes backward toward −20 as it reduces the volume.

Internally, a compressor can only lower the volume. So the net result when applying a lot of compression is a softer output. Therefore, the *makeup gain* control lets you raise the compressed audio back up to an acceptable level. The makeup gain is basically an output volume control, and it typically offers a lot of volume boost when needed in order to counter a large amount of gain reduction.

When a compressor lowers the volume and raises it again later, the volume changes occur over some period of time. The *attack time* controls how quickly the volume is reduced once the input exceeds the threshold. If set too slow, a short burst of loud music could get through and possibly cause distortion. So when using a compressor as a tool to prevent overload, you generally want a very fast attack time. But when used on an electric bass to add a little punch, setting the attack time to a moderately slow 30–80 milliseconds lets a short burst of the attack through before the volume is reduced. This adds a little extra definition to each note, while keeping the sustained portion of the note at a consistent level.

The *release time* determines how quickly the gain returns to unity when the input signal is no longer above the threshold. If the release time is too fast, you'll hear the volume as it goes up and down—an effect known as "pumping" or "breathing." Sometimes this sound is desirable for adding presence to vocals, drums, and other instruments, but often it is not wanted. The best setting depends on whether you're using the compressor as a tool to prevent overloading or as an effect to create a pleasing sound or add more sustain to an instrument. If you don't want to hear the compressor work, set the release time fairly long—half a second or longer. If you want the sound of "aggressive" compression, use a shorter release time. Note that as the release time is shortened, distortion increases at low frequencies because the compressor begins to act on individual wave cycles. This is often used by audio engineers as an intentional distortion effect.

Some compressors also have a *knee* setting, which affects only signals that are right around the threshold level. With a "hard knee" setting, signals below the threshold are not compressed at all, and once they exceed the threshold, the gain is reduced by exactly the amount dictated by the compression ratio. A "soft knee" setting works a bit differently. As the signal level approaches the threshold, it's reduced in level slightly, at a lower ratio than specified, and the amount of gain reduction increases gradually until the level crosses the threshold. The compressor does not apply the full ratio until the level is slightly above the threshold. In practice, the difference between a hard and soft knee is subtle.

Besides serving as an automatic volume control to keep levels more consistent, a compressor can also make notes sustain longer. Increasing a note's sustain requires raising the volume of a note as it fades out. That is, making the trailing part of a note louder to counter its natural fadeout makes the note sustain longer. To do this with a compressor, you'll set the threshold low so the volume is reduced most of the time. Then as the note fades out, the compressor reduces the volume less and less, which is the same as raising the volume as the note decays. For example, when you play a note on an electric bass, the compressor immediately reduces the volume by, say, 10 dB because the start of the note exceeds the threshold by 10 dB. You don't hear the volume reducing because it happens so quickly. But as the note fades over time, the compressor applies less gain reduction, making the note sustain longer. You can experiment with the release time to control the strength of the effect. Ideally, the release time will be similar to the note's natural decay time, or at least fast enough to counter the natural decay.

Using a Compressor

First, determine why, or even if, you need to compress. Every track or mix does not need compression! If you compress everything, music becomes less dynamic and can be tiring to listen to. Start by setting the threshold to maximum—too high to do anything—with the attack time to a fast setting such as a few milliseconds and the release time between half a second and one second or so. Then set the ratio to 5:1 or greater. That's the basic setup.

Let's say you want to reduce occasional too loud passages on a vocal track or tame a few overly loud kick drum hits. While the track plays, gradually lower the threshold until the gain reduction meter shows the level is reduced by 2–6 dB for those loud parts. How much gain reduction you want depends on how much too loud those sections are. The lower you set the threshold, the more uniform the drum hits will become.

To add sustain to an electric bass track, use a higher ratio to get more compression generally, maybe 10:1. Then lower the threshold until the gain reduction meter shows 6 to 10 dB reduction or even more most of the time. In all cases, you adjust the threshold to establish the amount of compression, then use the makeup gain to restore the now-softer output level. Besides listening, watching the gain reduction meter is the key to knowing how much you're compressing. The "compression" video lets you hear compression applied in varying amounts to isolated instrument tracks, as well as to a complete mix.

Common Pitfalls

One problem with compressors is they raise the noise floor and exaggerate soft artifacts such as clicks at a poor edit, breath intakes, and acoustic guitar string squeaks. For a given amount of gain reduction applied—based on the incoming volume level, threshold, and compression ratio—that's how much the background noise will be increased whenever the audio is softer than the threshold. In a live sound setting, compression also increases the risk of feedback.

You can switch a compressor in and out of the audio path with its bypass switch, and this is a good way to hear how the sound is being changed. But it's important to match the processed and bypassed volume levels using the compressor's makeup gain to more fairly judge the strength of the effect. Otherwise, you might perceive whichever version is louder as being more solid and full-sounding. Some compressors raise their makeup gain automatically as the threshold is lowered, making the comparison easier.

When compressing stereo audio, whether a stereo recording of a single instrument or a complete mix, the volume of both channels must be raised and lowered by the same amount. Otherwise, if the level of only one channel rises above the threshold and the level is reduced on just that one channel, the image will shift left or right. To avoid this problem, all stereo hardware compressors offer a *link* switch to change both channels equally, using the volume of whichever channel is louder at the moment. Most plug-in compressors accommodate mono or stereo sources automatically and change the volume of both channels together. Stereo hardware units contain two independent mono compressors. Some are dedicated to stereo compression, but most allow you to link the channels for stereo sources when needed or use the two compressors on unrelated mono sources.

Compressing a group of singers or other sources that are already mixed together responds differently than applying compression separately when each musician is on a separate track. When a single audio source contains multiple sounds, whichever sound is loudest will dominate because all the parts are reduced in level together. So if you're recording a group of performers having varying skill levels, it's better to put each on his or her own track. This way you can optimize the compressor settings for each performer, applying more severe compression for the people who are less consistent.

When compressing a complete mix, it's reasonable to assume that large variations in the individual elements have already been dealt with. It's common to apply a few dB of compression to a full mix, sometimes referred to as "glue" that binds the tracks together. This is also known as "bus compression," named for the stereo output bus where it's applied. It's easy to go overboard with compression on a complete mix. Years ago, when AM radio dominated the airways, radio stations would routinely apply heavy compression to every song they played. I recall the song "The Worst That Can Happen," by The Brooklyn Bridge from 1968. Near the end of this tune, the electric bass plays a long, sustained note under a brass fanfare. But the New York City AM stations applied so much compression that all you'd hear for seven seconds was that one bass note. The brass section was barely audible!

Multi-band Compressors

A multi-band compressor is a group of compressors that operate independently on different frequency bands. This lets you tame peaks in one band without affecting the level in other bands.

For example, if you have a mix in which the kick drum is too loud, a single-band compressor would decrease the entire track volume with every beat of the kick making the compression more obvious. By applying compression only to frequencies occupied by the kick, you can control the kick volume without affecting the vocals, horns, or keyboards. A multi-band compressor used this way also minimizes the pumping and breathing effect you get from a normal compressor when the loudest part affects everything else. With a multi-band compressor, the kick drum or bass will not make the mid and high frequencies louder and softer as it works, nor will a loud treble-heavy instrument make bass and midrange content softer.

Radio stations have used multi-band compressors for many years to customize their sound and remain within legal modulation limits. Such compressors help define a unique sound the station managers hope will set them apart from other stations. You define a frequency response curve, and the compressor will enforce that curve even as the music's own frequency balance changes. So if you play a song that's thin-sounding, then play a song with a lot of bass content, the frequency balance of both songs will be more alike. I've also used a multi-band compressor on mic'd bass tracks when the fullness of the part changed too much as different notes were played.

The Sonitus multi-band compressor in Figure 9.2 divides the incoming audio into five separate bands, not unlike a loudspeaker crossover, and its layout and operation are typical of such plug-ins. You control the crossover frequencies by sliding the vertical lines at the

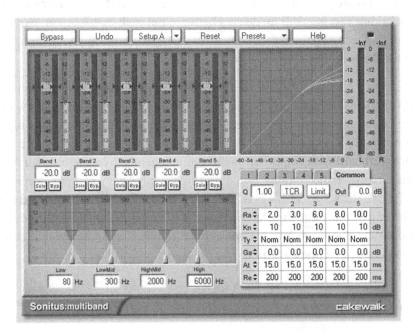

Figure 9.2: A multi-band compressor divides audio into separate frequency bands, then compresses each band separately without regard to the volume levels in other bands.

bottom left of the screen left or right or by typing frequency numbers into the fields below each vertical line. Once you've defined the upper and lower frequency bounds for each band, you can control every parameter of each band to tailor the sound in many ways. All of the compressors are then mixed together again at the output.

As you can imagine, managing five bands of compression can be a handful. To make this easier, the Common display at the lower right of the compressor screen shows all of the parameters currently set for each band. From top to bottom, the abbreviations stand for Ratio, Knee, Type (normal or "vintage"), Makeup Gain, Attack Time, and Release Time. Clicking a number to the left of the Common label displays a screen with sliders and data entry fields to adjust all the parameters for that band. The makeup gain for each band is adjusted by raising or lowering the top portion of each band's "trapezoid" at the lower left.

The threshold controls for each band are at the upper left of the screen, with a data entry field below each to type a dB value manually if you prefer. Note the Solo and Bypass buttons for each band. Solo lets you easily hear the contribution from only the selected band or bands, and Bypass disables gain reduction for a band passing its content unchanged.

Besides its usual job of enforcing a consistent level across multiple frequency bands, a multi-band compressor can also serve as a vocal de-esser, or to reduce popping P's, or to minimize acoustic guitar string squeaks (similar to the de-esser setting). If all of the bands but one are set to Bypass, only the remaining band's content is compressed. A popping P creates a sudden burst of energy below 80 Hz or so, so you'd use only the lowest band and set its parameters appropriately including a high ratio for a large amount of gain reduction when a pop comes along. Sibilance usually appears somewhere in the range between 4 and 8 KHz, so you'd engage only Band 4 to compress that range.

Noise Gates and Expanders

A noise gate minimizes background noise such as hum or tape hiss by lowering the volume when the input audio falls below a threshold you set. This is the opposite of a limiter that attenuates the signal when its level rises *above* the threshold. If a recording has a constant background hum, or perhaps air noise from a heating system, you'd set the threshold to just above the level of the noise. So when a vocalist is not singing, that track is muted automatically. But as soon as she begins to sing, the gate opens and passes both the music and the background noise. Hopefully, the singing will help to mask the background noise.

One problem with gating is that completely muting a sound is usually objectionable. Most gates let you control the amount of attenuation applied, rather than simply switching the audio fully on and off. This yields a more natural sound because the noise isn't suddenly replaced by total silence. The Sonitus gate shown in Figure 9.3 uses the label *Depth* for the

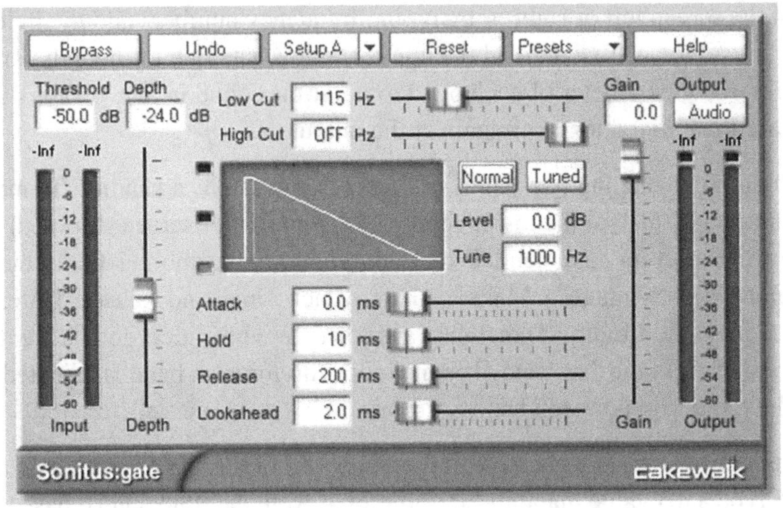

Figure 9.3: A noise gate is the opposite of a compressor. When the incoming audio is softer than the threshold, the gate attenuates the sound.

attenuation amount. Other models might use a related term such as *Gain Reduction* for this same parameter.

The normal state for a gate is closed, which mutes the audio, and then it opens when the input signal rises above the threshold. The *attack time* determines how long it takes for the volume to increase to normal and let the audio pass through. This is generally made as fast as possible to avoid cutting off the first part of the music or voice when the gate opens. But with some types of noise, a very fast attack time can create an audible click when the gate opens. The *release time* determines how quickly the volume is reduced after the signal level drops below the threshold. If this is too slow, you'll hear the background noise as it fades away. But if the release time is too fast, you risk a "chatter" effect when valid sounds near to the threshold make the gate quickly open and close repeatedly. When a noise gate turns off and on frequently, that can also draw attention to the noise.

It's common to use a gate to reduce background noise, plus a compressor to even out volume levels. In that case you'll add the gate first, then the compressor. Otherwise, the compressor will bring up the background noise and keep varying its level, which makes finding appropriate gate settings more difficult.

The Sonitus gate I use adds several other useful features: One is a *hold time* that allows a fast release, but without the risk of chatter; once the gate opens, it stays open for the designated hold time regardless of the input level. This is similar to the ADSR envelope generator in a synthesizer, where the note sustains for a specified duration. When a signal comes and goes quickly, the gate opens, holds for a minimum period of time, then fades out over time based on the release setting. Another useful addition is high- and low-cut filters that affect only the trigger signal but don't filter the actual audio passing through the gate. This could be used to roll off the bass range to avoid opening the gate on a voice microphone when a truck passes by outdoors.

Even the fastest gate takes a finite amount of time to open, which can affect transients such as the attack of a snare drum. To avoid this, some gates have a *look-ahead* feature that sees the audio before it arrives to open the gate in advance. The look-ahead setting delays the main audio by the specified amount but doesn't delay the gate's internal level sensor. So when the sensor's input level rises above the threshold, the gate opens immediately, a millisecond or so before the sound gets to the gate's audio input. This adds a small delay to the audio, so you should be careful to use identical settings when look-ahead gating is used on related tracks, such as stereo overhead mics on a drum set, to avoid timing or phase problems.

Finally, this gate also offers a "punch mode" that can serve as a crude *transient shaper* effect by boosting the volume more than needed each time the gate opens. This adds an extra emphasis, or *punch* as they call it, to an instrument. Transient shapers are described later in this chapter.

Noise Gate Tricks

Besides its intended use to mute background noise when a performer isn't singing or playing, a noise gate can also be used to reduce reverb decay times. If you recorded a track in an overly reverberant room, each time the musician stops singing or playing, the reverb can be heard as it decays over several seconds. By carefully adjusting the threshold and release time, the gate can be directed to close more quickly than the reverb decays on its own. This can make the reverb seem less intense, though the amount and decay time of the reverb isn't changed while the music plays.

A noise gate can also be used as an effect. One classic example is gated reverb, which was very popular during the 1980s. The idea is to create a huge wash of reverb that extends for some duration, then cuts off suddenly to silence. Typically, the output of a reverb unit is fed into a compressor that counters the reverb's normal decay, keeping it louder longer. The compressor's output is then sent to a noise gate that shuts off the audio suddenly when the reverb decays to whatever level is set by the threshold control.

Expanders

An expander is similar to a noise gate, but it operates over a wider range of volume levels, rather than simply turn on and off based on a single threshold level. In essence, an expander is the opposite of a compressor. Expanders can also be used to reduce background noise levels but without the sudden on-off transition a gate applies. For example, if you set the ratio to a fairly gentle 0.9:1, you won't notice the expansion much when the audio plays, but the background hiss will be reduced by 5 to 10 dB during silent parts. The Sonitus compressor plug-in I use also serves as an expander, and some other compressors also perform both tasks. Instead of setting the ratio higher than 1:1 for compression, you use a smaller ratio such as 0.8:1.

When an expander is used to mimic a noise gate, its behavior is called *downward expansion*. That is, the volume level is lowered when the input level falls below the threshold. This is similar to the old dbx and Burwen single-ended noise reduction devices mentioned in Chapter 6. But an expander can also do the opposite: leaving low-level audio as is and raising the volume when the input exceeds the threshold. In that case, the effect is called *upward expansion*, or just expansion. Which type of expansion you get depends on the design of the expander.

If you have a multi-band compressor that also can expand like the Sonitus plug-in, you could do even more sophisticated noise reduction in different frequency bands to lower the noise in bands that don't currently have content. But more modern software noise reduction works even better, so multi-band gates and expanders are less necessary. However, hardware gates are still useful for live sound applications.

But . . .

As useful as compressors are, I use them only rarely. Rather, I use volume envelopes in SONAR as needed to raise individual soft vocal syllables or to lower longer passages that are too loud. Programming volume changes manually is more effective than finding the right compressor settings, which are rarely correct for the entire length of a tune. And, of course, you can change a volume envelope later if needed. Volume envelopes can also be set to raise or lower the volume for a precise length of time, including extremely short periods to control popping P's, and you can draw the attack and release times. Indeed, the main reason to avoid compressors is to prevent pumping and breathing sounds. The only things I routinely compress—always after recording, nondestructively—are acoustic guitar and electric bass if they need a little more sustain or presence as an effect. I might also compress an entire mix if it will benefit using either a normal or multi-band compressor.

Likewise, I avoid using gates, instead muting unwanted audio manually either by splitting track regions and slip-editing the clips, or by adding volume envelopes. When you need to reduce constant background noise, software noise reduction is more effective than a broadband gate, and adding envelopes is often easier and more direct than experimenting to find the best gate settings. Further, when an effect like a gate is patched into an entire track, you have to play it all the way through each time you change a setting to be sure no portion of the track is being harmed.

Dynamics Processor Special Techniques

Dynamics processors can be used to perform some clever tricks in addition to their intended purpose. Some compressors and gates offer a *side-chain input*, which lets them vary the level of one audio stream in response to volume changes in another. When I owned a professional recording studio in the 1970s and 1980s, besides offering recording services to outside clients, we also had our own in-house production company. I recall one jingle session where the music was to be very funky. I created a patch that sent my direct electric bass track through a Kepex, an early gate manufactured by Allison Research, with the side-chain input taken from the kick drum track. I played a steady stream of 1/16 notes on the bass that changed with the backing chords, but only those notes that coincided with the kick drum passed through the gate. The result sounded incredibly tight, with the kick and bass playing together in perfect synchronization. This recording is on the Tunes page of my website ethanwiner.com (search the page for "Kepex" to find it).

My studio also recorded a lot of voice-over sessions, including corporate training tapes. These projects typically employed one or more voice actors, with music playing in the background. After recording and editing the ¼-inch voice tape as needed, a second pass would lay in the music underneath. But you can't just leave the music playing at some level. Usually such a production starts with music at a normal level, and then the music is lowered just before the narration begins. During longer pauses in the speaking, the music comes up a bit, then gets softer just before the announcer resumes.

When I do projects like this today, I use a DAW with the mono voice on one track and the stereo music on another. Then I can easily reduce the music's volume half a second before the voice comes in and raise it gently over half a second or so during the pauses. But in the days of analog tape where you couldn't see the waveforms as we do today in audio software, you had to guess when the voice would reenter.

To automate this process, I devised a patch that used a delay and a compressor with a side-chain input. The music went through the compressor, but its volume was controlled by the speaking voice sent into the side-chain. The voice was delayed half a second through a Lexicon Prime Time into the final stereo mix, but it was not delayed into the compressor's side-chain input. So when the announcer spoke, the music would be lowered immediately *in anticipation* of the voice, then the delayed voice would begin. I used slow attack and decay times to make the volume changes sound more natural, as if a person was gently riding the levels. The basic setup is shown in Figure 9.4. Even today, with a long project requiring many dozens of volume changes, the time to set up a patch like this in a DAW is more than offset

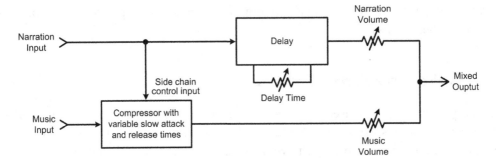

Figure 9.4: The compressor's Attack Time controls how quickly the music drops in volume, and the Delay Time determines how soon before the narration starts the volume drops. The compressor's Release Time controls how quickly the music comes back up when the talking stops, which should be similar to the Delay Time.

by not having to program all the volume changes. And today, with total recall, once you have the setup in a DAW template, you can use it again and again in the future. By the way, when a side-chain input is used this way, the process is called *ducking* the music, and some plug-ins offer a preset already connected this way.

Finally, when compressing an entire mix, you can sometimes achieve a smoother leveling action by using a low compression ratio such as 1.2:1 with a very low threshold, possibly as low as −30 or even −40. This changes the level very gradually, so you won't hear the compressor working.

Other Dynamics Processors

A *transient shaper* lets you increase or decrease the attack portion of an instrument at the start of each new note or chord. This acts on only the first brief portion of the sound, raising or lowering the level for a limited amount of time, so the initial strum of a guitar chord can be brought out more or less than the sustained portion that follows. It works the same way for piano chords and especially percussion instruments. But a transient shaper needs a clearly defined attack. If you use it on a recording of sustained strings or a synthesizer pad that swells slowly, a transient shaper won't be able to do much simply because there are no attack transients to trigger its level changes.

The *tremolo* effect raises and lowers the volume in a steady repeating cycle, and it's been around forever. Many early Fender guitar amplifiers included this effect, such as the Tremolux, which offered controls to vary the depth, or amount of the effect, as well as the speed. Note that many people confuse the term *tremolo*, a cyclic volume change, with *vibrato*, which repeatedly raises and lowers the pitch. A tremolo circuit is fairly simple to design and build into a guitar amp, but vibrato is much more difficult and usually relies on DSP, though it can be done with transistors and tubes. Likewise, a whammy bar—also called a vibrato bar or pitch bend bar—on an electric guitar lowers the pitch when pressed. But this is not tremolo, so calling this a tremolo bar is also incorrect.

An *auto-panner* is similar to a tremolo effect, except it adjusts the volume for the stereo channels symmetrically to pan the sound left and right. Early models simply swept the panning back and forth repeatedly at a fixed rate, but more modern plug-in versions can sync to the tempo of a DAW project. Some even offer a *tap tempo* feature, where you click a button with the mouse along with the beat of the song to set the left-right pan rate. If you need an auto-panning effect for only a short section of a song, it might be simpler to just add a track envelope and draw in the panning manually.

Another type of dynamics processor is the *volume maximizer*, for lack of a more universal term. Like a limiter, this type of processor reduces the level of peaks that exceed a threshold. But it does so without attack and release times. Rather, it searches for the zero-crossing boundaries before and after the peak and lowers the volume for just that overly loud portion of the wave. Brief peaks in a Wave file often prevent normalizing from raising the volume of a track to its full potential. Figure 9.5 shows a portion of a song file whose peak level hovers mostly below −6 dB. But looking at the left channel on top, you can see five short peaks that are louder than −6. So if you normalized this file, those peaks will prevent normalizing from raising the level a full 6 dB.

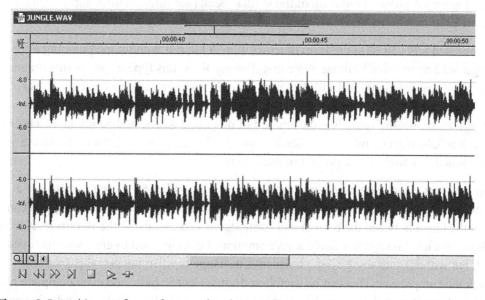

Figure 9.5: In this waveform of a completed mix, a few peaks extend briefly above the −6 dB marker. By reducing only those peaks, normalizing can then raise the volume to a higher level.

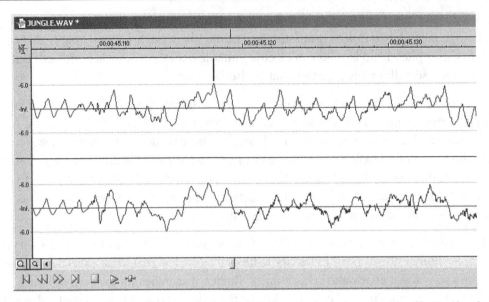

Figure 9.6: After reducing all the peaks to −6 dB, you can then normalize the file to a louder volume. The marker shows the loudest peak near the center of the screen in Figure 9.5.

A volume maximizer reduces the level of only the peaks, so if you reduce the peaks to −6 dB, or lower for an even more aggressive volume increase, you can then normalize the file to be louder. Figure 9.6 shows a close-up of the loudest peak (see marker) after applying the Peak Slammer plug-in with its threshold set to −6 dB.

This approach lets you raise the level of a mix quite a lot, but without the pumping effect you often get with aggressive limiting. Note that it's easy to do this type of processing manually, too, unless there are many peaks to deal with. Simply zoom in on each peak and lower the volume manually. As long as the range you select for processing is bounded by zero crossings, as shown in Figure 9.7, the amount of distortion added will be minimal. There's no free lunch, however, and some distortion is added by the process. Your ears will tell you if you've added too much in the pursuit of more loudness.

Parallel compression is not a separate processor type but rather a different way to use a regular compressor. If you feed a compressor from an auxiliary or subgroup bus, you can control the blend between the normal and compressed versions using the Return level from that bus. This lets you apply a fairly severe amount of compression to get a smooth, even sound, but without squashing all the transients, as happens when the signal passes only through the compressor. This can be useful on drums, electric bass, and even complete mixes. On a bass track it can add punch similar to a long attack time described earlier, letting some of the initial burst through to increase articulation.

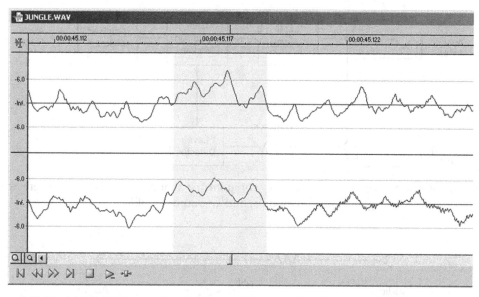

Figure 9.7: The highlighted area shows the same loudest peak from Figure 9.5, in preparation for reducing its volume manually.

Compressor Internals

Common to every compressor are a level detector and gain reduction module, as shown in the block diagram of Figure 9.8. The level detector converts the incoming audio to an equivalent DC voltage, which in turn controls the volume of the *gain reduction module*—basically a voltage-controlled attenuator circuit. The Attack and Decay time controls are part of the level detector's conversion of audio to DC. The Threshold setting is simply a volume control for the level detector, whereby raising the audio volume sent to the detector lowers the threshold. That is, sending more signal into the detector tells it that the audio is too loud at a lower level.

There are a number of ways to create a gain reduction module; the main trade-offs are noise and distortion and the ability to control attack and release times. Early compressors used a light bulb and photoresistor to vary the signal level. The audio is amplified to drive the light bulb—it doesn't even need to be converted to DC—and the photoresistor diverts some of the audio to ground, as shown in Figure 9.9. This method works very well, with very low distortion, but it takes time for a light bulb to turn on and off, so very fast attack and release times are not possible. In fact, most compressors that use a light bulb don't even offer attack and release time settings. You might think that an incandescent bulb can turn on and off quickly, but that's only when the full voltage is applied. When used in a compressor, most of the time the audio is so soft that the bulb barely glows. At those levels the response times are fairly long.

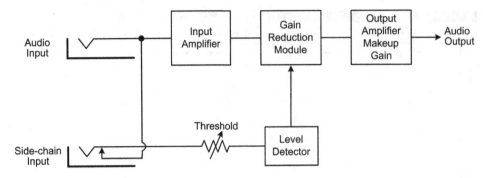

Figure 9.8: A compressor comprises four key components: an input amplifier, a level detector, a gain reduction module, and an output amplifier.

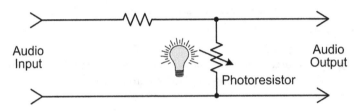

Figure 9.9: When light strikes a photoresistor, its resistance is lowered, shunting some of the audio signal to ground.

To get around the relatively long on and off times of incandescent bulbs, some early compressors used electroluminescent lights. These green, flat panels, also used as night-lights that glow in the dark, respond much more quickly than incandescent bulbs. With either bulb type this design is commonly called an "optical" or "opto" compressor. Another type of gain reduction module uses a field effect transistor (FET) in place of the photoresistor, with a DC voltage controlling the FET's resistance. A more sophisticated type of gain reduction module is the voltage controlled amplifier (VCA). This is the same device that's often used for console automation as described in Chapter 7, where a DC voltage controls the volume level.

Time Constants

Compressor and gate attack and decay times are based on an electrical concept known as *time constants*. Everything in nature takes some time to occur; nothing happens immediately in zero seconds. If you charge a capacitor through a resistor, it begins to charge quickly, and then the voltage increases more and more slowly as it approaches the maximum. Since it takes a relatively long time to reach the eventual destination, electrical engineers consider charge and discharge times based on the mathematical constant *e*, the base value for natural logarithms whose value is approximately 2.718. The same concept applies to attack and decay times in audio gear, where resistors and capacitors are used to control those parameters. A hardware compressor typically uses a fixed capacitor value, so the knobs for attack and release times control variable resistors (also called *potentiometers*).

If a limiter is set for an attack time of one second, and the audio input level rises above the threshold enough for the gain to be reduced by 10 dB, it won't be reduced by the full 10 dB in that one second. Rather, it is reduced by about 6 dB, and the full 10 dB reduction isn't reached until some time later. The same applies for release times. If the release time is also one second, once the input level falls below the threshold, the volume is raised about 6 dB over that one-second period, and full level isn't restored until later. Figure 9.10 shows the relationship between actual voltages and time constants. As you can see, the stated time is shorter than the total time required to reach the final value.

For the sake of completeness, the charge and discharge times of a simple resistor/capacitor network are based on the constant *e* as follows:

$$e = 2.718$$
$$\text{discharge time} = 1/e = 0.368 \text{ of final value}$$
$$\text{charge time} = 1 - (1/e) = 0.632 \text{ of final value}$$

To illustrate this concept in more practical terms, I created a 1 KHz sine wave at three volume levels, then repeated that block and processed the copy with a compressor having a high ratio. Figure 9.11 shows the Wave file containing both the original and processed versions. As you can see, the volume in the second block isn't reduced immediately after it jumps up to a level of −1, so a brief burst gets through before the level is reduced to −8 dB. Then, after the input level drops down to −20, it takes some amount of time for the volume to be restored. Figure 9.12 shows the compressor screen with 100 ms attack and 500 ms release time settings. The ratio is set high at 20:1 to apply hard limiting, making it easier to understand the expected gain changes.

Figure 9.13 shows a close-up of the attack portion to see how long the compressor took to reduce the volume. Figure 9.14 shows a close-up of the release portion, and again the volume is not restored fully until after the specified release time.

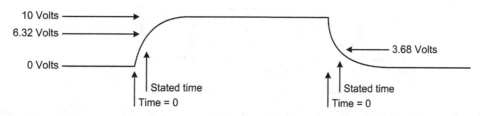

Figure 9.10: The actual attack and decay times in electrical circuits are longer than their stated values.

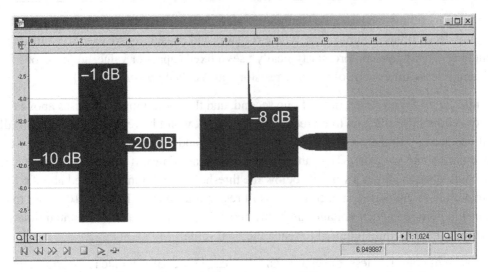

Figure 9.11: This Wave file contains a 1 KHz tone at three different volume levels for two seconds each. Then that block of three tones was repeated and compressed, to see how the actual attack and decay times relate to the specified times.

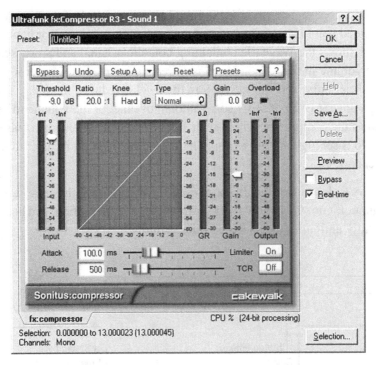

Figure 9.12: The compressor used for this test has an attack time of 100 milliseconds and a release time of 500 milliseconds.

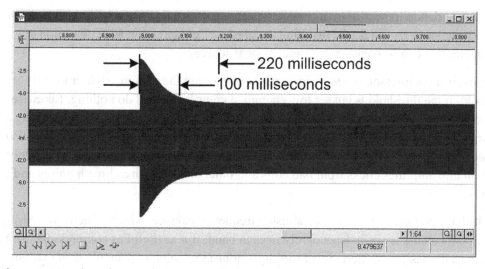

Figure 9.13: When the attack time is set for 100 milliseconds, the volume is reduced to only 37 percent in that period. The full amount of gain reduction isn't reached until more than twice that much time.

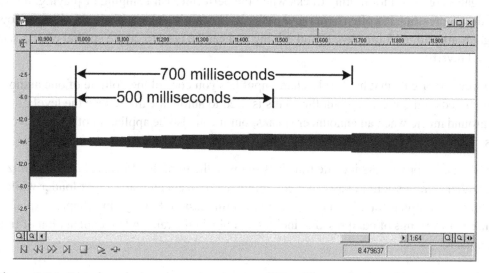

Figure 9.14: Even though the release time is set to 500 milliseconds, the volume is restored to only 63 percent during that time.

Summary

This chapter explains how compressors came to be invented and describes the most common parameters used by hardware and software versions. A compressor can be used as a tool to make volume levels more consistent, or as an effect to add power and fullness to a single track or complete mix. It can also be used to counter an instrument's normal decay to make it sound more powerful by increasing its apparent sustain. When both the attack and release times are very quick, the resulting "aggressive" compression adds distortion at low frequencies that can be pleasing on some musical sources.

When using a compressor, the most important controls are the compression ratio and threshold. If the threshold is not set low enough, a compressor will do nothing. Likewise, the ratio must be greater than 1:1 in order for the volume to be reduced. When adjusting a compressor, you can verify how hard it's working with the gain reduction meter. You also learned some common pitfalls of compression: Compression always raises the noise floor, which in turn amplifies clicks from bad edits and other sounds such as breath noises and chair squeaks.

A multi-band compressor lets you establish a frequency balance, which it then enforces by varying the volume levels independently in each band. But as useful as compressors can be, sometimes it's easier and more direct to simply adjust the volume manually using volume envelopes in your DAW software. This is especially true if the corrections are "surgical" in nature, such as controlling a very brief event.

Noise gates are useful for muting tracks when the performer isn't singing or playing, though as with compressors, it's often better and easier to just apply volume changes with envelopes. But gates are still useful for live sound mixing and can be used to create special effects such as gated reverb.

A compressor or gate that has a side-chain input lets you control the volume of one audio source based on the volume of another. This is used to automatically reduce the level of background music when an announcer speaks, but it can also be applied in other creative ways.

Other dynamics processors include transient shapers, the tremolo effect, volume maximizers, and parallel compression. In particular, a volume maximizer can make a mix louder while avoiding the pumping sound you often get with compression. Finally, this chapter explains the internal workings of compressors, including a block diagram and an explanation of time constants.

CHAPTER 10

Frequency Processors

The equalizer—often shortened to EQ—is the most common type of frequency processor. An equalizer can take many forms, from as simple as the bass and treble knobs on your home or car stereo to multi-band and parametric EQ sections on a mixing console or a graphic equalizer in a PA system. Equalizers were first developed for use with telephone systems to help counter high-frequency losses due to the inherent capacitance of very long wires. Today, EQ is an important tool that's used to solve many specific problems, as well as being a key ingredient in a mix engineer's creative palette.

Classical composers, professional orchestrators, and big band arrangers use their knowledge of instrument frequencies and overtones to create tonal textures and to prevent instruments from competing with one another for sonic space, which would lose clarity. Often mix engineers must use EQ in a similar manner to compensate for poor musical arranging, ensuring that instruments and voices will be clearly heard. But EQ is just as often used to enhance the sound of individual instruments and mixes to add sparkle or fullness, helping to make the music sound even better than reality.

Equalizer Types

There are four basic types of equalizers. The simplest is the standard bass and treble shelving controls found on home hi-fi receivers, with boost and cut frequency ranges similar to those shown in Figure 10.1. The most commonly used circuit for bass and treble tone controls was developed in the 1950s by P. J. Baxandall, and that design remains popular to this day. Some receivers and amplifiers add a midrange control (often called *presence*) to boost or cut the middle range at whatever center frequency and Q are deemed appropriate by the manufacturer.

A *graphic equalizer* divides the audible spectrum into five or more frequency bands and allows adjusting each band independently via a boost/cut control. Common arrangements are 5 bands (each two octaves wide), 10 bands (one octave wide), 16 bands (⅔ octaves), and 27 to 32 bands (⅓ octave). This type of EQ is called graphic because adjusting the front-panel sliders shows an approximate display of the resulting frequency response. So instead of broad adjustments for treble, bass, and maybe midrange, a graphic EQ has independent control over

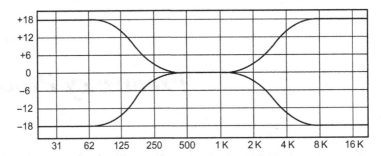

Figure 10.1: Treble and bass shelving controls boost or cut a broad range of frequencies, starting with minimal boost or cut in the midrange, and extending to the frequency extremes. This graph simultaneously shows the maximum boost and cut available at both high and low frequencies to convey the range of frequencies affected.

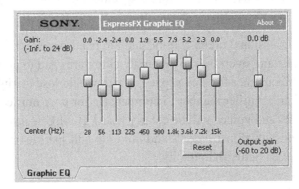

Figure 10.2: Graphic equalizers offer one boost or cut control for each available band.

the low bass, mid-bass, high bass, low midrange, and so forth. Third-octave graphic EQs are often used with PA systems to improve the overall frequency balance, though graphic equalizers are not often used in studios on individual tracks, or a mix bus, or when mastering. However, experimenting with a 10-band unit such as the plug-in shown in Figure 10.2 is a good way to become familiar with the sound of the various frequency ranges.

Console equalizers typically have three or four bands, with switches or continuous knobs to vary the frequency and boost/cut amounts. Often the highest- and lowest-frequency bands can also be switched between peaking and shelving. Professional consoles usually employ switches to select between fixed frequencies, and some also use switches for the boost and cut amounts in 1 dB steps. While not as flexible as continuous controls, switches let you note a frequency and recall it exactly at a later session.

The console equalizer in Figure 10.3 shows a typical model with rotary switches to set the frequencies in repeatable steps and continuously variable resistors to select any boost

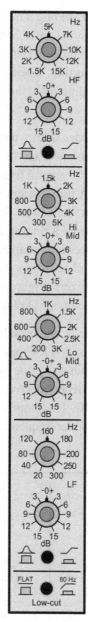

Figure 10.3: This typical console EQ has four bands with rotary switches to vary the frequencies and potentiometers to control the boost and cut amounts. The low and high bands can be switched between shelving and peaking, and a low-cut filter is also offered.

or cut amount between −15 and +15 dB. Building a hardware EQ with rotary switches costs much more than using variable resistors—also called *potentiometers*, or simply *pots*. A potentiometer is a relatively simple component, but a high-quality rotary switch having 10 or more positions is very expensive. Further, when a switch is used, expensive precision resistors are needed for each available setting, plus the labor to solder all those resistors to the switch, which adds yet more to the cost.

The most powerful equalizer type is the *parametric*, named because it lets you vary every EQ parameter—frequency, boost or cut amount, and bandwidth or Q. Many console EQs are "semi-parametric," providing a selection of several fixed frequencies and either no bandwidth control or a simple wide/narrow switch. A fully parametric equalizer allows adjustment of all parameters continuously rather than in fixed steps. This makes it ideal for "surgical" correction, such as cutting a single tone from a ringing snare drum with minimal change at surrounding frequencies or boosting a specific fundamental frequency to increase fullness. The downside is it's more difficult to recall settings exactly later, though this, of course, doesn't apply with software plug-in equalizers. Figure 10.4 shows a typical single-channel hardware parametric equalizer, and the Sonitus EQ plug-in I use is in Figure 10.5.

The Sonitus EQ offers six independent bands, and each band can be switched between peaking, high- or low-frequency shelving, high-pass, and low-pass. The high-pass and low-pass filters roll off at 6 dB per octave, so you can obtain any slope by activating more than one band. For example, to roll off low frequencies at 12 dB per octave starting at 100 Hz, you'd set Band 1 and Band 2 to high-pass, using the same 100 Hz frequency for both bands. You can also control the steepness of the slope around the crossover frequency using the Q setting.

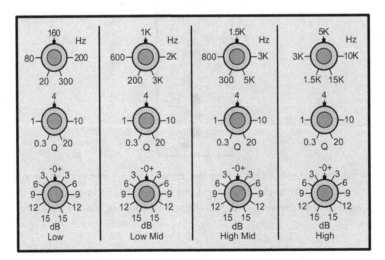

Figure 10.4: The parametric EQ is the most powerful type because you can choose any frequency, bandwidth, and boost/cut amount for every band.

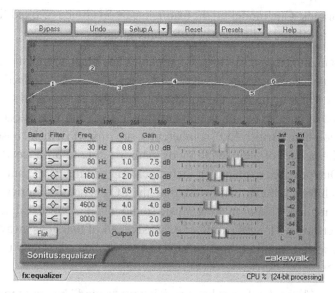

Figure 10.5: A plug-in parametric EQ has the same features as a hardware model, but it can also save and recall every setting.

There's also a hybrid equalizer called *paragraphic*. This type has multiple bands like a graphic EQ, but you can control the frequency of each band. Some let you vary the Q as well. In that case, there's no real difference between a paragraphic EQ and a parametric EQ having the same number of frequency bands.

Finally, there are equalizers that can learn the sound of one audio source and change another source automatically to have a similar frequency response. The first such automatic equalizer I'm aware of was Steinberg's FreeFilter plug-in, which analyzes the frequency balance of a source file, then applies EQ in third-octave bands to the file being EQ'd to match. Har-Bal (Harmonic Balance) is a more recent product, and it's even more sophisticated because it analyzes the audio at a higher resolution and also considers the difference between peak and average volume levels in the source and destination. Both products let you limit the amount of EQ applied to avoid adding an unreasonable amount of boost or cut if frequencies present in one source don't exist at all in the other.

All Equalizers (Should) Sound the Same

Years ago, before parametric equalizers were common in even entry-level setups, most console equalizers offered a limited number of fixed frequencies that were selected via switches. The Q was also fixed at whatever the designer felt was appropriate and "musical" sounding. So in those days there were audible and measurable differences in the sound of various equalizer brands. One model might feature a certain choice of fixed frequencies at a given Q, while others offered a different set of frequencies.

Console equalizers manufactured by British companies have had a reputation for better sound quality than other nationalities, though in my view this is silly. Indeed, many veteran recording engineers will tell you that the equalizers in British consoles all sound different; some love the sound of an SSL equalizer but hate the Trident, and vice versa. To my mind, this refutes the notion that "British EQ" is inherently superior. If British console equalizers are so different, what aspect of their sound binds them together to produce a commonality called "British"? One possible explanation is that peaking equalizers in some British recording consoles have a broader bandwidth than the EQs in some American consoles, so more boost can be applied without making music sound unnatural or nasal. But now that parametric equalizers are no longer exotic or expensive, all of that goes away. So whatever one might describe as the sound of "British EQ" can be duplicated exactly using any fully parametric equalizer.

In my opinion, an equalizer should not have an inherent "sound" beyond the frequency response changes it applies. Different EQ circuits might sound different as they approach clipping, but sensible engineers don't normally operate at those levels. Some older equalizer designs use inductors, which can ring and add audible distortion. However, most modern equalizer designs use op-amps and capacitors because of these problems with real inductors, to say nothing of their high cost and susceptibility to picking up airborne hum.

Finally, I can't resist dismissing the claim by some EQ manufacturers that their products sound better because they include an "air band" at some ultrasonic frequency. In truth, what you're really hearing is a change within the *audible* band due to the EQ's finite bandwidth. Unless the bandwidth is extremely narrow, which it obviously isn't, applying a boost at 25 KHz also boosts to some extent the audible frequencies below 20 KHz.

Digital Equalizers

Ultimately, skeptics and believers have the same goal: to separate fact from fiction. The main difference between these two camps is what evidence they deem acceptable.

Contrary to what you might read in magazine ads, many parametric EQ plug-ins are exactly the same and use the same standard "textbook" algorithms. Magazine reviews proclaiming that some model plug-in is more "musical sounding" or "adds sweeter highs" than other models further proves the points made in Chapter 3 about the frailty of our hearing perception. However, some digital equalizers add intentional distortion or other noises to emulate analog designs, and those might sound different unless you disable the extra effects.

In 2009, the online audio forums exploded after someone posted their results comparing many EQ plug-ins ranging from freeware VSTs to very expensive brands. The tests showed that most of the EQs nulled against each other to silence, as long as extra effects such as distortion or "vintage mode" were disabled. As you know, when two signals null to silence, or very nearly so, then by definition they must sound identical. However, some equalizers that are otherwise identical may not sound the same, or null against each other, because there's no industry standard for how the Q of a peaking boost and cut is defined. Chapter 1 explained that the bandwidth of a filter is defined by the −3 dB points, but that's for plain filters. Unlike a band-pass filter that continues to roll off frequencies above and below its cutoff, an equalizer has a flat response at frequencies other than those you boost or cut. And equalizers can boost or cut less than 3 dB, so in that case there's no −3 dB point to reference. In the comparison of many plug-in EQs, the author noted in his report that simply typing the same numbers into different equalizers didn't always give the same results. He also had to vary the Q values to achieve the smallest null residuals.

The same applies to many hardware equalizers: There are only so many ways to build a parametric EQ. Design engineers tend to use and reuse the most elegant and efficient designs. Indeed, while analog circuit design (and computer programming) are both based on science, there's also an artistic component. An elegant hardware design accomplishes the most using the fewest number of parts, while drawing the least amount of current from the power supply. For software, the goal is smaller code that's also efficient to minimize taxing the CPU. The best analog and digital designs are often in college textbooks or application guides published by component manufacturers, and any enterprising individual can freely copy and manufacture a commercial product based on those designs. This is not to say that all hardware equalizers use identical circuitry, but many models are similar.

Another common audio myth claims that good equalizers require less boost or cut to achieve a given effect, compared to "poorer" EQ designs. This simply is not true. How much the sound of a track or mix is changed depends entirely on the amount of boost or cut applied, and the Q or bandwidth. As explained, plug-in equalizers may interpret the Q you specify differently from one brand and model to another. But whatever the Q really is, *that* is what determines how audible a given amount of boost or cut will be. The notion that linear phase equalizers achieve more of an effect with less boost and cut than standard minimum phase types is also unfounded. Indeed, how audio responds to EQ depends entirely on the frequencies it contains. I'll have more to say about linear phase equalizers shortly.

EQ Techniques

> *At the same time we said we wanted to try for a Grammy, we also said we didn't want to use any EQ. That lasted about eight hours.*
>
> **—Ken Caillat, engineer on Fleetwood Mac's Rumours *album, interviewed in the*
> October 1975 *issue of* Recording Engineer/Producer *magazine*

The art of equalization is identifying what should be changed in order to improve the sound. Anyone can tell when a track or complete mix sounds bad, but it takes talent and practice to know what EQ settings to apply. As with compressors, it's important to match volume levels when comparing the sound of EQ active versus bypassed. Most equalizers include both a volume control and a bypass switch, making such comparisons easier. It's difficult to explain in words how to best use an equalizer, but I can convey the basic concepts, and the "equalization" demo video that accompanies this book lets you hear equalizers at work.

You should avoid the urge to boost low frequencies too much because small speakers and TV sets can't reproduce such content and will just distort when the music is played loudly. The same applies for extreme high frequencies. If your mix depends on a substantial boost at very low or very high frequencies to sound right to you, it's probably best to change something else. Often, it's better to identify what's harming the sound and remove it rather than look for frequencies that sound good when boosted. For example, cutting excessive low frequencies better reveals the important midrange frequencies that define the character of an instrument or voice.

Standard mixing practice adds a low-cut filter to thin out tracks that do not benefit from low frequencies. This avoids mud generally and reduces frequency clashes with the bass or kick drum due to masking. Cutting lows or mids is often better than boosting highs to achieve clarity. However, adding an overall high-frequency boost can be useful on some types of instruments, such as an acoustic guitar or snare drum. This is usually done with a broad peaking boost centered around 4 KHz or higher, depending on the instrument. A shelving boost can work, too, but that always adds maximum boost at the highest frequencies, which increases preamp hiss and other noise at frequencies that may not need to be boosted, or may not even be present in the source.

The same applies for overall low-frequency boost to add fullness. It's easy to add too much, and you risk raising muddy frequencies and other low-level subsonic content. Further, if your loudspeakers don't extend low enough, you're not even hearing everything in the track or mix. So always verify a final mix all the way through with earphones, unless your speakers can play to below 40 Hz. Adding too much low end is a common problem with mixes made on the popular Yamaha NS-10 loudspeakers. These speakers are 3 dB down at 80 Hz, and their response below that falls off quickly at 12 dB per octave. So if you boost 40 Hz to increase fullness while listening through NS-10s, by the time you hear the fullness you want, you've added 12 dB more boost than you should have. Some engineers watch the speaker cone of the NS-10 to spot excess low-frequency content, though using full-range speakers seems more sensible to me.

As mentioned in Chapter 7, boosting high frequencies helps to bring a track forward in the mix, and reducing them pushes the sound farther back. This mimics the absorption of air that reduces high frequencies more than lows with distance. For example, an outdoor rock concert heard from far away consists mostly of low frequencies. So when combined with an appropriate amount of artificial reverb, you can effectively make instruments seem closer or farther away. A lead vocal should usually be up front and present sounding, while background vocals are often farther back.

The "equalization" video lets you hear EQ applied to a variety of sources including several isolated tracks, plus a complete mix of a tune by my friend Ed Dzubak, a professional composer who specializes in soundtracks for TV shows and movies.

Boosting versus Cutting

The instinctive way to use a parametric equalizer is to set it for some amount of boost, then twiddle the frequency until the instrument or mix sounds better. But often what's wrong with a track is frequency content that sounds bad. A better way to improve many tracks is to set the EQ to boost, then sweep the frequency to find the range that sounds worst. Once you find that frequency, set the boost back to zero, wait a moment so your ears get used to the original timbre, then cut that range until it sounds better.
  —***Ethan Winer, from a letter to*** Electronic Musician *magazine, December 1993*

One of the most frequently asked questions I see in audio forums is whether it's better to boost or to cut. Replies often cite phase shift as a reason why cutting is better, but Chapters 1 through 3 explained why that's not true. All "normal" equalizers use phase shift, and the amount of phase shift added when using typical settings is not audible. So the real answer to which is better is *both*—or neither. Whether boosting or cutting is better depends on the particular problem you're trying to solve. If this wasn't true, then all equalizers would offer only boost, or only cut.

With midrange EQ, a low Q (wide bandwidth) lets you make a large change in the sound quality with less boost or cut, and without making the track sound nasal or otherwise affected. A high Q boost always adds resonance and ringing. Now, this might be useful to, for example, bring out the low tone of a snare drum by zeroing in on that one frequency while applying a fair amount of boost. When I recorded the audio for my *Tele-Vision* music video, one of the tracks was me playing a Djembe drum. I'm not much of a drummer, and my touch isn't very good. So I didn't get enough of the sustained ringing tone that Djembes are known for. While mixing the track, I swept a parametric EQ with a high Q to find the fundamental pitch of the drum, then boosted that frequency a few dB. This created a nice, round sound, as the dull thud of my inept hand slaps became more like pure ringing tones. You can hear this in the "equalization" demo video.

Often when something doesn't sound bright enough, the real culprit is a nasal or boxy sounding midrange. When you find and reduce that midrange resonance by sweeping a boost to find what frequencies sound worst, the sound then becomes brighter and clearer, and often fuller by comparison. Today, this is a common technique—often called "surgical" EQ—where a narrow cut is applied to reduce a bad-sounding resonance or other artifact. Indeed, identifying and eliminating offensive resonances is one of the most important skills mixing and mastering engineers must learn.

So it's not that cutting is generally preferred to boosting, or vice versa, but rather identifying the real problem, and often this is a low-end buildup due to having many tracks that were close-mic'd. Cutting boxy frequencies usually improves the sound more than boosting highs and lows, though not all cuts need a narrow bandwidth. For example, a medium bandwidth cut around 300 Hz often works well to improve the sound of a tom drum. This is also demonstrated in the "equalization" video. But sometimes a broad peaking boost, or a shelving boost, is exactly what's needed. I wish I could give you a list of simple rules to follow, but it just doesn't work that way. However, I can tell you that cutting resonance is best done with a narrow high Q setting, while boosting usually sounds better using a low Q or with a shelving curve.

Common EQ Frequencies

Table 10.1 lists some common instruments with frequencies at which boost or cut can be applied to cure various problems or to improve the perceived sound quality. The indicated frequencies are necessarily approximate because every instrument (and microphone placement) sounds different, and different styles of music call for different treatments. These equalization frequencies for various instruments are merely suggested starting points. The Comments column gives cautions or observations based on experience. These should be taken as guidelines rather than prescriptions because every situation is different, and mixing engineers often have different sonic goals. Additional advice includes the following:

- Your memory is shorter than you think; return to a flat setting now and then to remind yourself where you began.
- Make side-by-side comparisons against commercial releases of similar types of music to help judge the overall blend and tonality.
- You can alter the sound of an instrument only so much without losing its identity.
- Every instrument can't be full, deep, bright, sparkly, and so forth all at once. Leave room for contrast.
- Take a break once in a while. Critical listening tends to numb your senses, especially if you listen for long periods of time at loud volumes. The sound quality of a mix may seem very different to you the next day.
- Don't be afraid to experiment or to try extreme settings when required.

Table 10.1: Boost or Cut Frequencies for Different Instruments

Instrument	Cutting	Boosting	Comments
Human voice	Scratchy at 2 KHz, nasal at 1 KHz, popping P's below 80 Hz	Crisp at 6–8 KHz, clarity above 3 KHz, body at 200–400 Hz	Aim for a thinner sound when blending many voices, especially if the backing track is full.
Piano	Tinny at 1–2 KHz, boomy at 300 Hz	Presence at 5 KHz, bottom at 100 Hz	Don't add too much bottom when mixing with a full rhythm section.
Electric Guitar	Muddy below 150 Hz, boxy at 600 Hz	Clarity at 3 KHz, bottom at 150 Hz	There are no rules with an electric guitar!
Acoustic Guitar	Tinny at 2–3 KHz, boomy at 200 Hz	Sparkle above 5 KHz, full at 125 Hz	Keep the microphone away from the sound hole.
Electric Bass	Thin at 1 KHz, boomy at 125 Hz	Growl at 600 Hz, full below 100 Hz	Sound varies greatly depending on the type of bass and brand of strings used.
Acoustic Bass	Hollow at 600 Hz, boomy at 200 Hz	Slap at 2–5 KHz, bottom below 125 Hz	Cutting nasty resonances usually works better than boosting.
Snare Drum	Annoying at 1 KHz	Crisp above 2 KHz, full at 150–250 Hz, deep at 80 Hz	Also try adjusting the tightness of the snare wires.
Kick Drum	Floppy at 600 Hz, boomy below 80 Hz	Slap at 1–3 KHz, bottom at 60–125 Hz	For most pop music, remove the front head, then put a heavy blanket inside resting against the rear head.
Toms	Boxy at 300 Hz	Snap at 2–4 KHz, bottom at 80–200 Hz	Tuning the head tension makes a huge difference too!
Cymbals, Bells, Tambourines, etc	Harsh at 2–4 KHz, annoying at 1 KHz	Sparkle above 5 KHz	(Analog tape only:) Record at conservative levels to leave headroom for transients.
Brass and Strings	Scratchy at 3 KHz, honky at 1 KHz, muddy below 120 Hz	Hot at 8–12 KHz, clarity above 2 KHz, strings are lush at 400–600 Hz	Resist the temptation to add too much high end on strings.

Mixes that Sound Great Loud

> *THIS RECORD WAS MADE TO BE PLAYED LOUD.*
>
> **—Liner note on the 1970 album** Climbing **by the band**
> **Mountain featuring Leslie West**

This is a standard test for me when mixing: If I turn it up really loud, does it still sound good? The first recording I ever heard that sounded fantastic when played very loud was *The Yes Album* by the band Yes in 1971. Not that it sounded bad at lower volumes! But for me, that was a breakthrough record that set new standards for mixing pop music and showed how amazing a recording could sound.

One factor is the bass level. I don't mean the bass instrument, but the amount of content below around 100 Hz. You simply have to make the bass and kick a bit on the thin side to be able to play a mix loudly without sounding tubby or distorting the speakers. Then when the music is played really loud, the fullness kicks in, as per Fletcher-Munson. If you listen to the original 1978 recording of *The War of the Worlds* by Jeff Wayne, the tonality is surprisingly thin, yet it sounds fantastic anyway. You also have to decide if the bass or the kick drum will provide most of the fullness for the bottom end. There's no firm rule, of course, but often with a good mix, the kick is thin-sounding while the bass is full, or the bass is on the thin side and the kick is full with more thump and less click.

Another factor is the harshness range around 2 to 4 KHz. A good mix will be very controlled in this frequency range, again letting you turn up the volume without sounding piercing or fatiguing. I watch a lot of concert DVDs, and the ones that sound best to me are never harsh in that frequency range. You should be able to play a mix at realistic concert levels—say, 100 dB SPL—and it should sound clear and full but never hurt your ears. You should also be able to play it at 75 dB SPL without losing low midrange "warmth" due to a less than ideal arrangement or limited instrumentation. A five-piece rock band sounds great playing live at full stage volume, but a well-made recording of that band played at a living room level sounds thin. Many successful recordings include parts you don't necessarily hear but that still fill out the sound at low volume.

Complementary EQ

Complementary EQ is an important technique that lets you single out an instrument or voice to be heard clearly above everything else in a mix. The basic idea is to boost a range of frequencies for the track you want to feature and cut that same range by the same amount from other tracks that have energy in that frequency range. This is sometimes referred to as "carving out space" for the track you want to feature. A typical complementary boost and cut amount is 2 or 3 dB for each. Obviously, using too much boost and cut will affect the overall tonality. But modest amounts can be applied with little change to the sound, while still making the featured track stand out clearly.

When doing this with spoken voice over a music bed, you'll boost a midrange frequency band for the voice and cut the same range by the same amount on the backing music. Depending on the character of the voice and music, experiment with frequencies between 600 Hz and 2 KHz, with a Q somewhere around 1 or 2. To feature one instrument or voice in a complete mix, it's easiest to set up a bus for all the tracks that will be cut. So you'll send everything but the lead singer to a bus and add an EQ to that bus. Then EQ the vocal track with 2 dB of boost at 1.5 KHz or whatever, and apply an opposite cut to the bus EQ.

This also works very well with bass tracks. In that case you'd boost and cut somewhere between 100 and 600 Hz, depending on the type of bass tone you want to bring out. Again, there's no set formula because every song and every mix is different. But the concept is powerful, and it works very well. As mentioned in Chapter 9, I used to do a lot of advertising voice-over recording. With complementary EQ, the music can be made loud enough to be satisfying, yet every word spoken can be understood clearly without straining. Ad agencies like that.

Mid/Side Equalization

Another useful EQ technique is *Mid/Side processing*, which is directly related to the Mid/Side microphone technique described in Chapter 6. A left and right stereo track is dissected into its equivalent middle and side components. These are equalized separately, then combined back to left and right stereo channels. This is a common technique used by mastering engineers to bring out a lead vocal when they don't have access to the original multi-track recording. To bring out a lead vocal that's panned to the center of a mix, you'd equalize just the mid component where the lead vocal is more prominent. You could also use complementary EQ to cut the same midrange frequencies from the side portion if needed.

Since Mid/Side EQ lets you process the center and side information separately, you can also more easily EQ the bass and kick drum on a finished mix, with less change to everything else. Likewise, boosting high frequencies on the side channels can make a mix seem wider, because only the left and right sides are brightened. The concept of processing mid and side components separately dates back to vinyl disk mastering where it was often necessary to compress the center material in a mix after centering the bass. The Fairchild 670 mastering limiter worked in this manner. Extending the concept to equalization evolved as a contemporary mastering tool. Often adding 1–3 dB low-end boost only to the sides adds a nice warmth to a mix without adding mud to the kick and bass which are centered. Although Mid/Side equalization is mainly a mastering tool, you may find it useful on individual stereo tracks—for example, to adjust the apparent width of a piano.

Extreme EQ

> *A mixer's job is very simply to do whatever it takes to make a mix sound great. For me, there are no arbitrary lines not to cross.*
>
> **—Legendary mix engineer Charles Dye**

Conventional recording wisdom says you should always aim to capture the best sound quality at the source, usually by placing the microphones optimally, versus "Fix it in the mix." Of course, nobody will argue with that. Poor sound due to comb filtering is especially difficult to counter later with EQ because there are so many peaks and nulls to contend with. But mix engineers don't usually have the luxury of re-recording, and often you have to work with tracks as they are.

Some projects simply require extreme measures. When I was mixing my *Cello Rondo* music video, I had to turn 37 tracks of the same instrument into a complete pop tune with bright-sounding percussion, a full-sounding bass track, and everything in between. Because some of the effects needed were so extreme, I show the plug-in settings at the end of the video so viewers can see what was done. Not just extreme EQ, but also extreme compression to turn a cello's fast-decaying pizzicato into chords that sustain more like an electric guitar. My cello's fundamental resonance is around 95 Hz, and many of the tracks required severely cutting that frequency with EQ to avoid a muddy mess. The "sonar_rondo" video shows many of the plug-in settings I used.

I recall a recording session I did for a pop band where extreme EQ was needed for a ham-fisted piano player who pounded out his entire part in the muddy octave below middle C. But even good piano players can benefit from extreme EQ. Listen to the acoustic piano on some of the harder-rocking early Elton John recordings, and you'll notice the piano is quite thin-sounding. This is often needed in pop music to prevent the piano from conflicting with the bass and drums.

Don't be afraid to use extreme EQ when needed; again, much of music mixing is also sound design. It seems to me that for a lot of pop music, getting good snare and kick drum sounds is at least 50 percent sound design via EQ and sometimes compression. As they say, if it sounds good, it *is* good. In his educational DVD *Mix It Like a Record*, Charles Dye takes a perfectly competent multi-track recording of a pop tune and turns it into a masterpiece using only plug-ins—a *lot* of them! Some of the EQ he applied could be considered over the top, but what really matters is the result.

Linear Phase Equalizers

As mentioned in Chapter 1, most equalizers rely on phase shift, which is benign and necessary. Whether implemented in hardware or software, standard equalizers use a design known as *minimum phase*, where an original signal is combined with a copy of itself after applying phase shift. In this case, "minimum phase" describes the minimum amount of phase shift needed to achieve the required response. Equalizers simply won't work without phase shift. But many people indict phase shift as evil anyway. In a misguided effort to avoid phase shift, or perhaps for marketing reasons because so many people believe this myth, software developers created *linear phase* equalizers. These delay the audio the same amount for all frequencies, rather than selectively by frequency via phase shift as with regular equalizers.

You'll recall from Chapter 1 that comb filter effects can be created using either phase shift or time delay. A flanger effect uses time delay, while phaser effects and some types of stereo synthesizers instead use phase shift. Equalizers can likewise be designed using either method to alter the frequency response. A minimum phase filter emulates phase shift that occurs naturally, and any ringing that results from boosting frequencies occurs after the signal. If you strike a resonant object such as a drum, ringing caused by its resonance occurs after you strike it, then decays over time. But linear phase filters use pure delay instead of phase shift, and the delay as implemented causes ringing to occur *before* audio events. This phenomenon is known as *pre-ringing*, and it's audibly more damaging than regular ringing because the ringing is not masked by the event itself. This is especially a problem with transient sounds like a snare drum or hand claps.

The file "impulse_ringing.wav" in Figures 3.5 and 3.6 from Chapter 3 shows an impulse wave alone, then after passing through a conventional EQ that applied 18 dB of boost with a Q of 6, and again with a Q of 24. To show the effects of pre-ringing and let you hear what it sounds like, I created the "impulse_ringing_lp.wav" example using the LP64 linear phase equalizer bundled with SONAR. Figure 10.6 shows the wave file with the original and EQ'd impulses, and a close-up of a processed portion is in Figure 10.7. This particular equalizer limits the Q to 20, so it's not a perfect comparison with the impulse wave in Chapter 3 that uses a Q of 24. Then again, you can see in Figure 10.6 that there's little difference between a Q of 6 and a Q of 20 with this linear phase equalizer anyway.

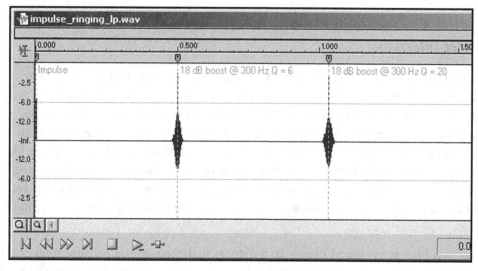

Figure 10.6: This file contains the same impulse three times in a row. The second version used a linear phase equalizer to apply 18 dB of boost at 300 Hz with a Q of 6, and the third is the same impulse but EQ'd with a Q of 20.

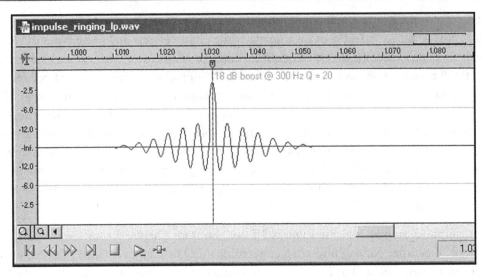

Figure 10.7: Zooming in to see the wave cycles more clearly shows that a linear phase EQ adds ringing both after *and* before the impulse.

When you play the example file in Figure 10.6, you'll hear that it sounds very different from the version in Chapter 3 that applied conventional minimum phase EQ. To my ears, this version sounds more affected, and not in a good way. Nor is there much difference in the waveform—both visually and audibly—whether the Q is set to 6 or 20.

Equalizer Internals

Figure 8.2 from Chapter 8 shows how it takes time for a capacitor to charge when fed a current whose amount is limited by a resistor. An inductor, or coil of wire, is exactly the opposite of a capacitor: It charges immediately, then dissipates over time. This simple behavior is the basis for all audio filters. As explained in Chapter 1, equalizers are built from a combination of high-pass, band-pass, and low-pass filters. Good inductors are large, expensive, and susceptible to picking up hum, so most modern circuits use only capacitors. As you can see in Figure 10.8, a capacitor can create either a low- or high-frequency roll-off. Adding active circuitry (usually op-amps) offers even more flexibility without resorting to inductors.

The low-pass filter at the top of Figure 10.8 reduces high frequencies because it takes time for the capacitor to charge, so rapid voltage changes don't pass through as readily. The high-pass filter at the bottom is exactly the opposite: High frequencies easily get through because the voltage keeps changing so the capacitor never gets a chance to charge fully and stabilize.

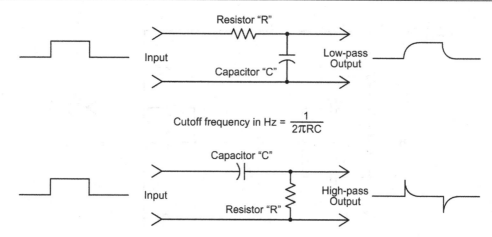

Figure 10.8: Both low-pass and high-pass filters can be created with only one resistor and one capacitor.

Note that the same turnover frequency formula applies to both filter types. Here, the symbol π is Pi, which is approximately 3.1416. Further, replacing the capacitors with inductors reverses the filter types, turning the top filter into a high-pass and the bottom into a low-pass. Then the formula to find the cutoff frequency becomes:

$$Cutoff\ frequency = \frac{R}{2\pi L}$$

The passive version of the Baxandall treble and bass tone control circuit shown in Figure 10.9 incorporates both low-pass and high-pass filters. In practice, this circuit would be modified to add an op-amp active stage, and that was an important part of Baxandall's contribution. He is credited with incorporating the symmetrical boost/cut controls within the feedback path of an active gain stage. A passive equalizer that allows boost must reduce the overall volume level when set to flat. How much loss is incurred depends on the amount of available boost. An active implementation eliminates that and also avoids other problems such as the frequency response being affected by the input impedance of the next circuit in the chain.

Looking at Figure 10.9, if the Bass knob is set to full boost, capacitor C1 is shorted out, creating a low-pass filter similar to the top of Figure 10.8. In this case, resistor R1 combines with capacitor C2 to reduce frequencies above the midrange. Resistor R2 limits the amount of bass boost relative to midrange and higher frequencies. The Treble control at the right behaves exactly the opposite: When boosted, the capacitor C3 becomes the "C" at the bottom of Figure 10.8, and the Treble potentiometer serves as the "R" in that same figure. In practice, C1 and C2 will have larger capacitance values than C3 and C4 because C1 and C2 affect lower frequencies.

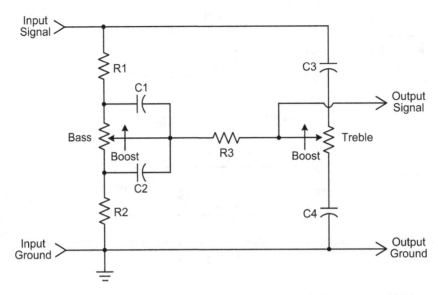

Figure 10.9: A Baxandall tone control is more complex than simple low-pass and high-pass filters, but the same basic concepts apply.

Other Frequency Processors

Most of this chapter has been about equalizers, but there are several other types of frequency processors used by musicians and in recording studios. These go by names suggestive of their functions, but at their heart they're all variable filters, just like equalizers. One effect that's very popular with electric guitar players is the *wah-wah pedal*. This is simply a very high-Q band-pass filter whose frequency is swept up and down with a variable resistor controlled by the foot pedal. When you step on the rear of the pedal, the center frequency shifts lower, and stepping near the front sweeps the frequency higher, as shown in Figure 10.10.

Wah-wah effects are also available as plug-ins. Most wah plug-ins offer three basic modes of operation: cyclical, with the frequency sweeping up and down automatically at a rate you choose; tempo-matched, so the repetitive sweeping follows the tempo of the song; and triggered, where each upward sweep is initiated by a transient in the source track. The triggered mode is probably the most common and useful, because it sounds more like a human player is controlling the sweep frequency manually. When using auto-trigger, most wah plug-ins also let you control the up and down sweep times.

Chapter 9 clarified the common confusion between tremolo and *vibrato*. Where tremolo is a cyclical volume change, vibrato—also known as *Frequency Modulation*—repeatedly raises and lowers the pitch. To create vibrato in a plug-in, the audio samples are loaded into a memory buffer as usual, but the clock that controls the output data stream sweeps continuously between faster and slower rates.

Another useful frequency processor is the *formant filter*, which emulates the human vocal tract to add a vocal quality to synthesizers and other instruments. When you mouth vowels such as "eee eye oh," the resonances of different areas inside your mouth create simultaneous high-Q acoustic band-pass filters. This is one of the ways we recognize people by their voices. The fundamental pitch of speech depends on the tension of the vocal chords, but just as important is the complex filtering that occurs acoustically inside the mouth. Where a wah-wah effect comprises a single narrow band-pass filter, a formant filter applies three or more high-Q filters that can be tuned independently. This is shown in Figure 10.11.

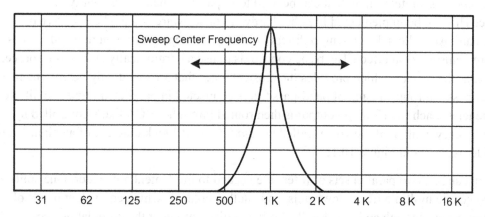

Figure 10.10: The wah-wah effect is created by sweeping a high-Q band-pass filter up and down.

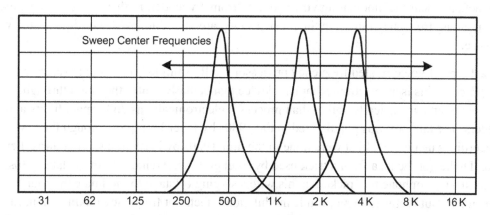

Figure 10.11: A formant filter comprises three or more high-Q band-pass filters that emulate the vocal cavities inside a human mouth.

The relationship between center frequencies and how the amplitude of each frequency band changes over time is what makes each person's voice unique. The audio example file "formant_filter.wav" plays a few seconds of a single droning synthesizer note, with a formant filter set to jump randomly to different frequencies. You'll notice that it sounds sort of like a babbling robot because of the human voice quality. Carefully controlling the filter frequencies enhances text-to-speech synthesizers, such as used by Stephen Hawking.

Vocal tuning, or pitch correction, is common in pop music, and sophisticated formant filtering has improved the effect considerably compared to early plug-ins. Small changes in the fundamental pitch of a voice can pass unnoticed, but a change large enough to create a harmony part benefits from applying formant filtering similar to that of the original voice. This reduces the "chipmunk effect" substantially, giving a less processed sound.

The *vocoder* is another effect that can be used to impart a human voice quality onto music. This can be implemented in either hardware or software using a bank of five or more band-pass filters. Having more filters offers higher frequency resolution and creates a more realistic vocal effect. The basic concept is to use a harmonically rich music source, known as the *carrier*, which provides the musical notes that are eventually heard. The music carrier passes through a bank of band-pass filters connected in parallel, all tuned to different frequencies. Each filter's audio output is then routed through a VCA that's controlled by the frequency content of the voice, called the *modulator*. A block diagram of an eight-band vocoder is shown in Figure 10.12.

As you can see, two parallel sets of filters are needed to implement a vocoder: One splits the voice into multiple frequency bands, and another controls which equivalent bands of the music pass through and how loudly. So if at a given moment the modulating speech has some amount of energy at 100 Hz, 350 Hz, and 1.2 KHz, any content in the music at those frequencies passes through to the output at the same relative volume levels. You never actually hear the modulating voice coming from a vocoder. Rather, what you hear are corresponding frequency ranges in the music as they are varied in level over time in step with the voice.

The last type of voice-related frequency processor we'll examine is commonly referred to as a *talk box*. This is an electromechanical device that actually routes the audio through your own mouth acoustically, rather than process it electronically as do formant filters and vocoders. The talk box was popularized in the 1980s by Peter Frampton, though the first hit recordings using this effect were done in the mid-1960s by Nashville steel guitarist Pete Drake. Drake got the idea from a trick used by steel guitarist Alvino Rey in the late 1930s, who used a microphone that picked up his wife's singing offstage to modulate the sound of the guitar. But that was an amplitude modulation effect, not frequency modification. In the 1960s, radio stations often used a talk box in their own station ID jingles, and lately it's becoming popular again.

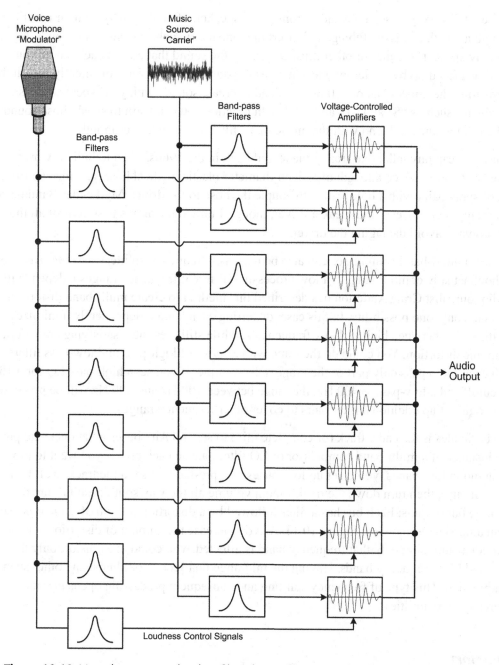

Voice
Microphone
"Modulator"

Music
Source
"Carrier"

Band-pass
Filters

Voltage-Controlled
Amplifiers

Band-pass
Filters

Band-pass
Filters

Audio
Output

Loudness Control Signals

Figure 10.12: Vocoders use two banks of band-pass filters: One bank evaluates frequencies present in the modulating source, usually a spoken voice, and the other lets the same frequencies from the music "carrier" pass through at varying amplitudes.

To use a talk box, you send the audio from a loudspeaker, often driven by a guitar amplifier, through a length of plastic tubing, and insert the tube a few inches inside your mouth. As you play an electric guitar or other musical source, the sound that comes back out of your mouth is picked up by a microphone. Changes in your vocal cavity as you silently mouth the words filter the music. The only time you need to create sounds with your vocal cords is for consonants such as "S" and "T" and "B." With practice you can learn to speak those sounds at the right volume to blend with the music that's filtered through your mouth.

Today you can buy talk boxes ready-made, but back in the 1960s, I had to build my own. I used a PA horn driver, a length of ½-inch-diameter plastic medical hose, and the cap from a can of spray paint with a hole drilled to couple the hose to the driver. My Fender Bandmaster guitar amplifier powered the speaker driver, though I turned the amp's bass control all the way down to avoid damaging the driver.

Finally, a multi-band compressor can also be used as a frequency splitter—or *crossover*—but without actually compressing, to allow processing different frequency ranges independently. Audio journalist Craig Anderton has described this method to create multi-band distortion, but that's only one possibility. In this case, distorting each band independently minimizes IM distortion between high and low frequencies, while still creating a satisfying amount of harmonic distortion. You could do the same with a series of high-pass and low-pass filters, with each set to pass only part of the range. For example, if you set a high-pass EQ for a 100 Hz cutoff and a low-pass to 400 Hz, the range between 100 Hz and 400 Hz will be processed. You could set up additional filter pairs to cover other frequency ranges.

The basic idea is to send a track (or complete mix) to several Aux buses all at once and put one instance of a multi-band compressor or EQ filter pair on each bus. If you want to process the audio in four bands, you'd set up four buses, use pre-fader to send the track to all four buses at once, then turn down the track's main volume all the way so it doesn't dilute the separate bus outputs. Each bus has a filter followed by a distortion plug-in of some type, such as an amp-sim, as shown in Figure 10.13. As you increase the amount of distortion for each individual amp-sim, only that frequency band is affected. You could even distort only the low- and high-frequency bands, leaving the midrange unaffected, or almost any other such combination. This type of frequency splitting and subsequent processing opens up many interesting possibilities!

Summary

This chapter covers common frequency processors, mostly equalizers, but also several methods for imparting a human voice quality onto musical instruments. The simplest type of equalizer is the basic treble and bass tone control, and most modern implementations use a circuit developed in the 1950s by P. J. Baxandall. Graphic equalizers expand on that by

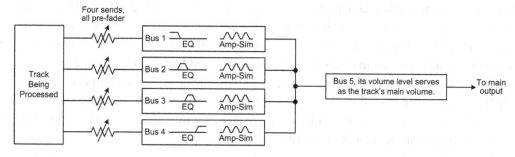

Figure 10.13: When audio is split into multiple-frequency bands, each band can be processed independently. In this example a track is sent to four different Aux buses at once, each with an EQ filter to pass only one band, in turn followed by an amp-sim or other device that adds distortion. The four buses then go on to another bus that serves as the main volume control for the track.

offering more bands, giving even more control over the tonal balance. The two most common graphic equalizer types offer either 10 bands or about 30 bands. While graphic equalizers are useful for balancing PA systems, they're not as popular for studio use because their center frequencies and bandwidth can't be varied.

Console equalizers generally have only three or four frequency bands, but each band can be adjusted over a wide range of frequencies. Equalizers in high-end consoles often use rotary switches to select the frequencies and boost/cut amounts. Equalizers that use switches cost more to manufacture, but their settings can be recalled exactly. The downside is that using switches limits the available frequencies to whatever the manufacturer chose.

The most flexible equalizer type is the parametric, which lets users pick any arbitrary frequency and boost/cut amount. A parametric EQ also lets you adjust the bandwidth, or Q, to affect a narrow or wide range of frequencies and everything in between. By being able to hone in on a very narrow frequency range, you can exaggerate, or minimize, resonances that may not align with the fixed frequencies of other equalizer types.

This chapter also explains a number of EQ techniques, such as improving overall clarity by thinning tracks that shouldn't contribute low frequencies to the mix. Other important EQ techniques include sweeping a parametric equalizer with boost to more easily find frequencies that sound bad and should then be cut. Parametric equalizers are especially useful to "carve out" competing frequencies to avoid clashes between instruments. Complementary EQ takes the concept a step further, letting you feature one track to be heard clearly above the others, without having to make the featured track substantially louder. A few simple schematics showed how equalizers work internally, using resistors and capacitors to alter the frequency response.

We also busted a few EQ myths along the way, including the common claim that linear phase equalizers sound better than conventional minimum phase types. In fact, the opposite is true, because linear phase equalizers add pre-ringing, which is more noticeable and more objectionable than normal ringing that occurs after the fact. Another common myth is that cutting is better than boosting because it avoids phase shift. In truth, you'll either cut or boost based on what actually benefits the source. You also learned that most equalizers work more or less the same, no matter what advertisements and magazine reviews might claim. Further, every frequency curve has a finite bandwidth. So claims of superiority for equalizers that have an "air band" letting you boost ultrasonic frequencies are unfounded; what you're really hearing are changes at frequencies that are audible.

Finally we looked at other popular frequency processors including wah-wah pedals, formant filters, vocoders, and talk boxes. All of these use band-pass filters to change the sound, though a talk box is potentially the most realistic way to emulate the human voice because it uses the acoustic resonances inside your mouth rather than emulate them with electronic circuits.

CHAPTER 11

Time Domain Processors

Echo

Everyone knows the "HELLO, Hello, hello" echo effect. In the early days of audio recording, this was created using a tape recorder having separate heads for recording and playback. A tape recorder with separate record and play heads can play back while recording, and the playback is delayed slightly from the original sound letting you mix the two together to get a single echo. The delay time depends on the tape speed, as well as the distance between the record and play heads. A faster speed, or closer spacing, creates a shorter delay. A single echo having a short delay time was a common effect on early pop music in the 1950s; you can hear it on recordings such as *Great Balls of Fire* by Jerry Lee Lewis from 1957. This is often called *slap-back* echo because it imitates the sound of an acoustic reflection from a nearby room surface.

Creating multiple repeating echoes requires *feedback*, where the output from the tape recorder's play head is mixed back into its input along with the original source. As long as the amount of signal fed back to the input is softer than the original sound, the echo will eventually decay to silence, or, more accurately, into a background of distortion and tape hiss. If the signal level fed back into the input is above unity gain, then the echoes build over time and eventually become so loud the sound becomes horribly distorted. This is called *runaway*, and it can be a useful effect if used sparingly.

Eventually, manufacturers offered stand-alone tape echo units; the most popular early model was the Echoplex, introduced in 1959. I owned one of these in the 1960s, and I recall well the hassle of replacing the special lubricated tape that frequently wore out. When I owned a large professional recording studio in the 1970s, one of our most prized outboard effects was a Lexicon Prime Time, the first commercially successful digital delay unit. Today many stand-alone hardware echo effect units are available, including inexpensive "stomp box" pedal models, and most use digital technology to be reliable and affordable and have high quality.

With old tape-based echo units, the sound would become more grungy with each repeat. Each repeat also lost high frequencies, as the additional generations took their toll on the response. However, losing high-end as the echoes decay isn't necessarily harmful, because the same thing happens when sound decays in a room. So this by-product more closely emulates the sound of echoes that occur in nature. Many modern hardware echo boxes and software plug-ins can do the same, letting you roll off highs and sometimes lows, too, in the feedback loop to sound more natural and more like older tape units.

Different delay times give a different type of effect. A very short single delay adds presence, by simulating the sound of an acoustic echo from a nearby surface. Beware that when used to create a stereo effect by panning a source and its delay to opposite sides, comb filtering will result if the mix is played in mono. The cancellations will be most severe if the volume of the original and delayed signals are similar. The "echo_demo.wav" file plays the same short spoken fragment four times: First is the plain voice, then again with 30 milliseconds of mono delay applied to add a presence effect, then with 15 milliseconds of delay panned opposite the original to add width, and finally with the original and a very short (1.7 ms) delay summed to mono. The last example clearly shows the hollow sound of comb filtering.

You can add width and depth to a track using short single-echoes mixed in at a low level, maybe 10 to 15 dB below the main signal. I've had good results panning instruments slightly off to one side of a stereo mix, then panning a short echo to the far left or right side. When done carefully, this can make a track sound larger than life, as if it was recorded in a much larger space. Sound travels about one foot for every millisecond (ms), so to make an echo sound like it's coming from a wall 25 feet away, you'll set the delay to about 50 ms—that is, 25 ms for the sound to reach that far wall, then another 25 ms to get back to your ears. Again, the key is to mix the echo about 10 dB softer than the main track to not overwhelm and seem too obvious. Unless, of course, that's the sound you're aiming for. The "echo_space.wav" file first plays a sequence of timpani strikes dry, panned 20 percent to the left. Then you hear them again with 60 ms of delay mixed in to the right side about 10 dB softer. These timpani samples are already fairly reverberant and large-sounding, but the 60 ms delay further increases their perceived size.

By using echoes that are timed to match the tempo of the music, you can play live with yourself in harmony, as shown in the audio demo "echo_harmony.wav." You can add this type of echo after the fact, as was done here, but it's easier to "play to the effect" if you hear it while recording yourself. Hardware "loop samplers" can capture long sections of music live, then repeat them under control of a foot switch to create backings to play along with. A cellist friend of mine uses an original Lexicon JamMan loop sampler to create elaborate backings on the fly, playing along with them live. He'll start with percussive taps on the cello body or strings to establish a basic rhythm, then add a bass line over that, then play chords or arpeggios. All three continue to repeat while he plays lead melodies live along with the backing. Les Paul did this, too, years ago, using a tape recorder hidden backstage.

The Sonitus Delay plug-in in Figure 11.1 can be inserted onto a mono or stereo track, or added to a bus if more than one track requires echo with the same settings. Like many

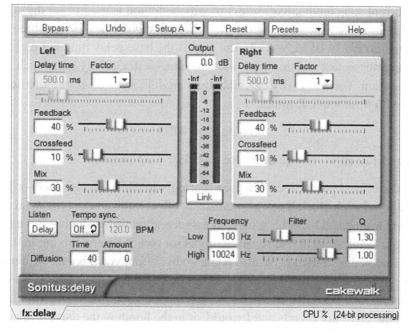

Figure 11.1: The Sonitus Delay plug-in offers a number of useful features including high- and low-cut filters, cross-channel feedback, and the ability to sync automatically to the tempo of a song in a DAW host that supports *tempo sync.*

plug-in delay effects, the Sonitus accepts either a mono or stereo input and offers separate adjustments for the left and right output channels. A mono input can therefore be made wider-sounding by using different delay times left and right. Or the Mix on one side can be set to zero for no echo, but set to 100 percent on the other side.

Notice the Crossfeed setting for each channel. This is similar to the Feedback amount, but it feeds some amount of echo back to the input of the *opposite* channel, which also increases stereo width. If both the Crossfeed and Feedback are set to a high value, the effect goes into runaway, as subsequent echoes become louder and louder. This plug-in also includes adjustable high- and low-cut filters for the delayed portion of the audio to prevent the sound of the echoes from becoming muddy or overly bright.

The Link switch joins the controls for both channels, so changing any parameter for one channel changes the other equally. The Low and High Frequency filters in the lower right affect only the delayed output to reduce excess low end in the echoes or to create the popular "telephone echo" effect where the echo has only midrange with no highs or lows.

Finally, the delay time can be dialed in manually as some number of milliseconds, set to a specific tempo in *beats per minute* (BPM), or synchronized to the host DAW so the timing follows the current song tempo. Regardless of which method is used to set the delay time, the Factor amount lets you scale that from 1/32 through eight times the specified time. So if the song's tempo is 120 BPM, you can use that as the base delay time but make the left channel one-quarter as fast and the right side two times faster, or any other such musically related combination. You can easily calculate the length in seconds of one-quarter note at a given tempo with this simple formula:

$$Length = 60/BPM$$

So at a tempo of 120 BPM, one-quarter note = 60/120 = 0.5 seconds.

Echo is a terrific effect because it can be used sparingly to add a hint of space, or in extreme amounts for a psychedelic feel. It can also create an effect known as *automatic double tracking*, where a single added echo sounds like a second person is singing. An early recording studio I built in 1970 had separate rooms for recording and mixing in the upstairs of a large barn owned by a friend. We also had a 10 by 7–foot vocal booth, with microphone and earphone jacks connected to the control room. Besides recording ourselves and the occasional paying rock band, the studio was also a fun hangout. I remember fondly many late nights when a bunch of my friends and I would sit in the booth in the dark. A microphone was on a short stand on the floor of the booth, connected to a repeating echo patch from a three-head tape recorder in the control room. We'd all put on earphones, making noises and blabbering on for hours, listening to ourselves through the extreme echo. (Hey, it was the 1970s!)

Reverb

Reverb is basically many echoes all sounding at once. This effect is denser than regular echo, so you don't hear the individual repeats. When recording with microphones in a reverberant space, you can control the amount of reverb picked up by varying the distance between the microphone and sound source. When placed very close, the microphone picks up mostly the direct sound with little room tone. If the microphone is farther away, then you get more of the room's sound. The distance between the source and mic where the level of direct and reverberant sound is equal is known as the *critical distance*. Modern recording methods often place microphones close to the source, well forward of the critical distance, then add reverb as an effect later during mixdown when it can be heard in context to decide how much to add.

The earliest reverb effect used for audio recording was an actual room containing only a loudspeaker and one or two microphones. According to audio manufacturer Universal Audio, Bill Putnam Sr. was the first person to add artificial room reverb to a pop music recording in 1947, using the sound of a bathroom on the tune "Peg o' My Heart," by The Harmonicats. To get the best sound quality, a live reverb room should be highly reflective and also avoid favoring single resonant frequencies. Getting a sufficiently long reverb time is often done by painting the walls and ceiling with shellac, but sometimes ceramic tiles are used for all of the room surfaces.

The advantage of a live room is that the reverb sounds natural because it is, after all, a real room. The best live reverb chambers have angled walls to avoid the "boing" sound of repetitive flutter echo and unrelated dimensions that minimize a buildup of energy at some low frequencies more than others. The downside of a live reverb chamber is the physical space required and the cost of construction because the room must be well isolated to prevent outside sounds from being picked up by the microphones.

In 1957, the German audio company EMT released their model 140 *plate reverb* unit. This is a large wooden box containing a thin steel plate stretched very tightly and suspended on springs inside a metal frame. A single loudspeaker driver is attached near the center of the plate, and contact microphones at either end return a stereo signal. The sound quality of this early unit was remarkable, and its price tag was equally remarkable. When I built my large professional recording studio in the 1970s, we paid more than $7,000 for our EMT 140. Adjusted for inflation, that's about $20,000 today! Being a mechanical device with speakers and microphones, it too was susceptible to picking up outside sounds unless it was isolated. We put our EMT plate reverb in an unused storage closet near the studio offices, far away from the recording and mixing rooms.

One very cool feature of the EMT plate is being able to vary its decay time. A large absorbing pad is placed very close to the steel plate, with a hand crank to adjust the spacing. As the pad is moved closer to the vibrating plate, the decay time becomes shorter. An optional motor attached instead of the hand crank lets you adjust the spacing remotely. A later model, the EMT 240, was much smaller and lighter than the original 140, taking advantage of the higher mass and density of gold foil versus steel. An EMT 240 is only two by two feet, versus four by eight feet for the earlier 140, and it's much less sensitive to ambient sound. You won't be surprised to learn that the EMT 240 was also incredibly expensive. The cost of audio gear has certainly improved over the years.

Another mechanical reverb type is the *spring reverb*, which transmits vibration through long metal springs similar to screen door springs to create the effect. A miniature loudspeaker voice coil at one end vibrates the springs, and an electromagnetic pickup at the other end converts the vibrations back to an audio signal. This design was first used on Hammond electric organs and was quickly adapted to guitar amplifiers. Some units used springs of different lengths to get a more varied sound quality, and some immersed the springs in oil. Similar to having two pickup transducers on a plate reverb, using two or more springs can also create different left and right stereo outputs. Like a live room and plate reverb, a spring reverb must also be isolated from ambient sound and mechanical vibration. Indeed, every guitar player knows the horrible—and very loud!—sound that results from accidentally jarring a guitar amplifier that has a built-in spring reverb unit.

The typical small spring reverbs in guitar amps have a poor sound quality, but all spring units don't sound bad. The first professional reverb unit I owned, in the mid 1970s, was made by AKG and cost $2,000 at the time. Rather than long coil springs, it used metal rods that were twisted by a torsion arrangement. It wasn't as good as the EMT plate I eventually bought, but it was vastly better than the cheap reverb units built into guitar amps.

EMT also created the first digital reverb unit in 1976, which sold for the staggering price of $15,000. Since it was entirely electronic, the EMT 250 was the first reverb unit that didn't require acoustic isolation. But it was susceptible to static electricity. I remember my visit to the 1977 AES show in New York City, where I hoped to hear a demo. When my friends and I arrived at the distributor's hotel suite, the unit had just died moments before after someone walked across the carpet and touched it. And the distributor had only that one demo unit. D'oh!

Modern personal computers are powerful enough to process reverb digitally with a reasonably high quality, so as of this writing, dedicated hardware units are becoming less necessary. Unlike equalizers, compressors, and most other digital processes, reverb has always been the last frontier for achieving acceptable quality with plug-ins. A digital reverb algorithm demands hundreds or even thousands of calculations per second to create all the echoes needed, and it must also roll off high frequencies as the echoes decay to better emulate what happens acoustically in real rooms. Because of this complexity, all digital reverbs are definitely not the same. In fact, there are two fundamentally different approaches used to create digital reverb.

The most direct way to create reverb digitally is called *algorithmic*: A computer algorithm generates the required echoes using math to calculate the delay, level, and frequency response of each individual echo. Algorithm is just a fancy term for a logical method to solve a problem. The other method is called *convolution*: Instead of calculating the echoes one reflection at a time, a convolution reverb superimposes an *impulse* that was recorded in a real room onto the audio being processed. The process is simple but very CPU-intensive.

Each digital audio sample of the original sound is multiplied by *all* of the samples of the impulse's reverb as they decay over time. Let's say the recording of an impulse's reverb was done at a sample rate of 44.1 KHz and extends for five seconds. This means there are $44,100 * 5 = 220,500$ total impulse samples. The first sample of the music is multiplied by each of the impulse samples, one after the other, with the result numbers placed into a sequential memory buffer. The next music sample is multiplied by the remaining impulse samples, then added to the sample numbers currently in the buffer. This process repeats for the entire duration of the music.

In theory, the type of impulse you'd record would be a very brief click sound that contains every audible frequency at once, similar to the impulse in Figure 3.5 from Chapter 3. A recording of such an impulse contains all the information needed about the properties of the room at the location the microphone was placed. But in practice, real acoustic impulses are not ideal for seeding convolution reverbs.

The main problem with recording a pure impulse is achieving a high enough signal to noise ratio. Realistic reverb must be clean until it decays by 60 dB or even more, so an impulse recorded in a church or concert hall requires that the background noise level of the venue be at least that soft, too. Early acoustic impulses were created by popping a balloon or shooting a starter pistol. A better, more modern method records a slow sine wave sweep that includes the entire frequency range of interest. The slower the sweep progresses, the higher the signal to noise ratio will be. Then digital signal processing (DSP) converts the sweep into an equivalent impulse using a process called *deconvolution*. This same technique is used by most modern room measuring software to convert a recording of a swept sine wave into an impulse that the software requires to assess the room's properties.

Thankfully, developers of convolution reverbs have done the dirty work for us; all we have to do is record a sweep, and the software does everything else. Figure 11.2 shows the Impulse Recovery portion of Sonic Foundry's Acoustic Mirror convolution reverb. The installation CD for this plug-in includes several Wave files containing sweeps you'll use to record

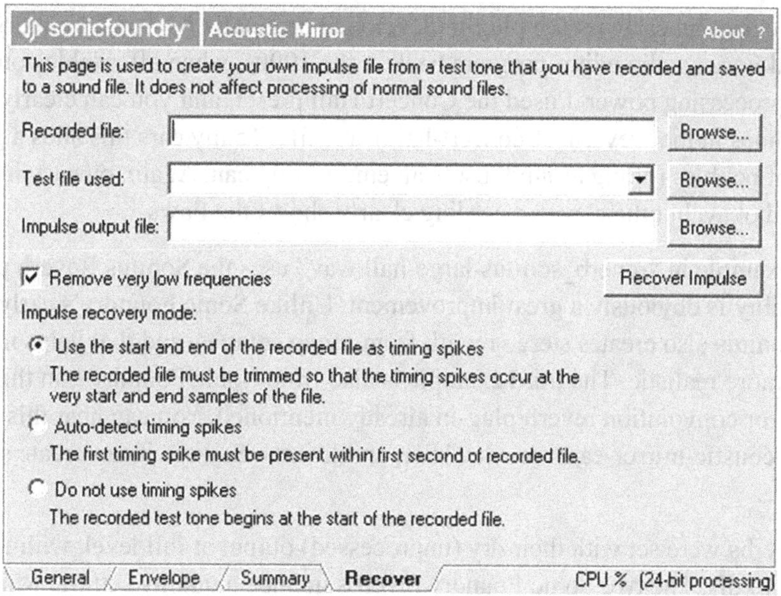

Figure 11.2: The Acoustic Mirror Impulse Recovery processor lets you record your own convolution impulses.

and process your own impulse files, without needing a balloon or starter pistol. Note that convolution is not limited to reverb and can be used for other digital effects such as applying the sound character of one audio source onto another.

Most reverb effects, whether hardware or software, accept a mono input and output two different-sounding left and right signals to add a realistic sounding three-dimensional space to mono sources. Lexicon and other companies make surround sound digital reverbs that spread the effect across five or even more outputs. Some hardware reverbs have left and right inputs, but often those are mixed together before creating the stereo reverb effect. However, some reverb units have two separate processors for the left and right channels. Even when the input channels are summed, dual inputs are needed to maintain stereo when the device is used "inline" with the dry and reverb signals balanced internally. If you really need separate left and right channel reverbs, with independent settings, this is easy enough to rig up in a DAW by adding two reverb plug-ins panned hard left and right on separate stereo buses.

Because reverb is such a demanding process, and some digital models are decidedly inferior to others, I created a Wave file to help audition reverb units. Even a cheap reverb unit can sound acceptable on sustained sources such as a string section or vocal. But when applied to percussive sounds such as hand claps, the reverb's flaws are more clearly revealed. The audio file "reverb_test.wav" contains a recording of sampled claves struck several times in a row. The brief sound is very percussive, so it's ideal for assessing reverb quality. I also used this file to generate three additional audio demos so you can hear some typical reverb plug-ins.

The first demo uses an early reverb plug-in developed originally by Sonic Foundry to include in their Sound Forge audio editor program back in the 1990s, when affordable computers had very limited processing power. I used the Concert Hall preset, and you can clearly hear the individual echoes in the "reverb_sf-concert-hall.wav" file. To my ears this adds a gritty sound reminiscent of pebbles rolling around inside an empty soup can. Again, on sustained sources it might sound okay, but the claves recording clearly shows the flaws.

The second example in "reverb_sonitus-large-hall.wav" uses the Sonitus Reverb plug-in, and the sound quality is obviously a great improvement. Unlike Sonic Foundry's early digital reverb, the Sonitus also creates stereo reverb from mono sources, and that helps make the effect sound more realistic. The third example is also from Sonic Foundry—in this case, their Acoustic Mirror convolution reverb plug-in already mentioned. You can hear this reverb in the "reverb_acoustic-mirror-eastman.wav" file, and like the Sonitus it can create stereo from a mono source.

All of the reverbs were set with their dry (unprocessed) output at full level, with the reverb mixed in at −10 dB. The two Sonic Foundry reverbs include a quality setting, which is another throwback to the days of less powerful computers. Of course, I used the highest Quality setting, since my modern computer can easily process the effect in real time. For the

Sonitus reverb demo, I used the Large Hall preset. The Acoustic Mirror demo uses an impulse recorded at the University of Wisconsin–Madison, taken 30 feet from the stage in the Eastman Organ Recital Hall. This is just one of *many* impulses that are included with Acoustic Mirror.

Modern reverb plug-ins offer many ways to control the sound, beyond just adding reverb. The Sonitus Reverb shown in Figure 11.3 has features similar to those of other high-quality plug-in reverbs, so I'll use that for my explanations. From top to bottom: The Mute button at the upper left of the display lets you hear only the reverb as it decays after being excited by a sound. Many DAW programs will continue the sound from reverb and echo effects after you press Stop so you can better hear the effects in isolation. But if your DAW doesn't do this, the Mute button stops further processing while letting the reverb output continue. The volume slider to the right of this Mute button controls the overall input level to the plug-in, affecting both the dry and processed sounds.

Next are high- and low-cut filters in series with the reverb's input. These are used to reduce bass energy that could make the reverb sound muddy or excessive high end, which would sound harsh or unnatural when processed through reverb. Note that these filters affect only

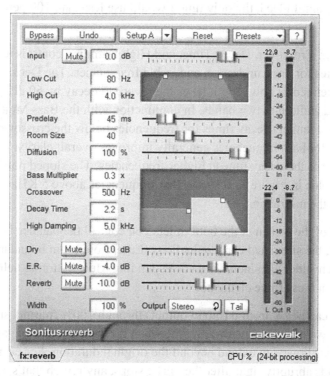

Figure 11.3: The Sonitus Reverb provides many controls to tailor the sound.

the sound going into the reverb processor. They do not change the frequency response of direct sound that might be mixed into the output.

The Predelay setting delays the sound going into the reverb processor by up to 250 milliseconds. This is useful when emulating the sound of a large space like a concert hall. When you hear music live in an auditorium or other large space, the first sound you hear is the direct sound from the stage. Moments later, sounds that bounced off the walls and ceiling arrive at your ears, and over time those reflections continue to bounce around the room, becoming more and more dense. So in a real room, the onset of reverberation is always delayed some amount of time, as dictated by the room's size. Predelay does the same to more closely emulate a real room.

The Room Size has no specific time or distance values; it merely scales the virtual length of the internal delay paths that create reverberation. This is usually adjusted to match the reverb's overall decay time. Specifying a small room size sounds more natural when the decay time (farther down the screen) is set fairly short, while a larger room setting sounds better and more natural when used with longer decay times.

The Diffusion setting controls the density of the individual echoes. If you prefer to avoid the gravelly sound of older reverbs, where each separate echo can be distinguished, set the Diffusion to maximum. Indeed, the only time I would use less than 100 percent diffusion is for special effect.

The Bass Multiplier, Crossover, Decay Time, and High Damping controls let you establish different decay times for bass, midrange, and treble frequencies. The Decay Time sets the basic RT60 time, defined as how long it takes the reverb to decay by 60 dB. The Crossover splits the processed audio into two bands. In conjunction with the Bass Multiplier, this allows relatively shorter (or longer) decay times for frequencies below the crossover point. Using a shorter decay time for bass frequencies generally improves overall clarity. High Damping is similar, but it shortens the decay time at higher frequencies. The sloped portion at the right of the decay time graphic in Figure 11.3 shows that frequencies above 5 KHz are set to decay more quickly than the midrange.

The mixer portion of the screen at the bottom lets you control the balance between the original dry sound, the simulated early reflections (E.R.) from room boundaries, and the overall reverb. Each slider also includes a corresponding Mute button to help you audition the balance between these three elements.

The width control can be set from 0 (mono output) through 100% (normal stereo), to 200% (adds exaggerated width to the reverb). Finally, when the Tail button is engaged, the reverb added to a source is allowed to extend beyond the original duration. For example, if you have a Wave file that ends abruptly right after the music stops, any reverb that's added will also stop at the end of the file. But when Tail is engaged, the Wave file is actually made longer to accommodate the added final reverb decay.

As explained in Chapter 5, effects that *do something* to the audio are usually inserted onto a track, while effects that *add new content* should go on a bus. Reverb and echo effects add new content—the echoes—so these are usually placed onto a bus. In that case, the reverb volume should be set near 0 dB, and the output should be 100 percent "wet," since the dry signal is already present in the main mix. Reverb plug-ins can also be patched into individual tracks; in that case you'll use the three level controls to adjust the amount of reverb relative to the dry sound.

In Chapter 7 I mentioned that I often add two buses to my DAW mixes, with each applying a different type of reverb. To add the sound of a performer being right there in the room with you, I use a Stage type reverb preset. For the larger sound expected from "normal" reverb, I'll use a Plate or Hall type preset. The specific settings I use in many of my pop tunes for both reverb types are shown in Table 11.1. These are just suggested starting points! I encourage you to experiment and develop your own personalized settings. Note that in all cases the Dry output is muted because these settings are meant for reverb that's placed on a bus.

Finally, I'll offer a simple tip that can help to improve the sound of a budget digital reverb: If the reverb you use is not dense enough, and you can hear the individual echoes, try using *two* reverb plug-ins with the various parameters set a little differently. This gives twice the echo density, making grainy-sounding reverbs a little smoother and good reverbs even better. If your reverb plug-ins don't generate a satisfying stereo image, pan the returns for each reverb fully left and right. When panned hard left and right, the difference in each reverb's character and frequency content over time can add a nice widening effect.

Table 11.1: Stage and Normal Reverb Settings

Parameter	Basic Reverb	Ambience
Low Cut	75 Hz	55 Hz
High Cut	4 KHz	11 KHz
Predelay	50 ms	35 ms
Room Size	50	20
Diffusion	100%	100%
Bass Multiplier	0.3	0.7
Crossover	350 Hz	800 Hz
Decay Time	1.9 seconds	0.4 seconds
High Damping	4.0 KHz	9.5 KHz
Early Reflections (E.R.)	−10.5 dB	0.0 dB
Reverb	−3.0 dB	0.0 dB
Width	150%	200%

Phasers and Flangers

The first time I heard the flanging effect was on the 1959 tune *The Big Hurt* by Toni Fisher. Flanging was featured on many subsequent pop music recordings, including *Itchycoo Park* from 1967 by the Small Faces, *Bold as Love* by Jimi Hendrix in 1967, and *Sky Pilot* in 1968 by Eric Burdon and The Animals, among others. Today it's a very common effect. Before plug-ins, the flanging effect was created by playing the same music on two different tape recorders at once. You'd press Play to start both playbacks together, then lightly drag your hand against one of the tape reel flanges (side plates) to slow it down. Figure 11.4 shows an even better method that avoids the need for synchronization by using two tape recorders set up as an inline effect.

Chapter 1 showed how phase shift and time delay are used to create phaser and flanger effects, respectively. As explained there, phaser effects use phase shift, and flangers use time delay, but that's not what creates their characteristic sound. Rather, what you hear is the resulting comb filtered *frequency response*.

Phaser and flanger effects often sweep the comb filter up and down at a steady rate adjustable from very slow to very fast. When the sweep speed is very slow, this effect is called *chorus* because the constantly changing response sounds like two people are singing. Sweeping a comb filter can also create a Leslie rotating speaker effect. If the effect generates a stereo output from a mono source, that can enhance the sound further. Some plug-ins can even synchronize the filter sweep rate to a song's tempo.

The Sonitus plug-in I use in Figure 11.5 wraps both effect types into one unit called Modulator, so you can select a flanger, one of several phaser types, or tremolo. The Modulator plug-in also outputs different signals left and right to create a stereo effect from a mono source.

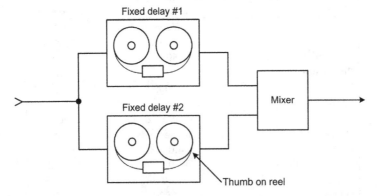

Figure 11.4: The flanging effect is created by sending audio through two recorders at once, then slowing the speed of one recorder slightly by dragging your thumb lightly against the reel.

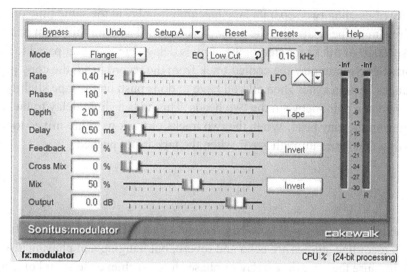

Figure 11.5: The Sonitus Modulator plug-in bundles a flanger, tremolo, and several types of phaser effects into a single plug-in. All can be swept either manually or in sync with the music.

When the phase shift or delay time is swept up and down repeatedly, the Rate or Speed control sets the sweep speed. Most flangers and phasers also include a Depth or Mix slider to set the strength of the effect. Unlike most plug-ins where 100 percent gives the strongest effect, the Sonitus Modulator has a maximum effect when the Mix slider is set to 50 percent. This blends the source and shifted signals at equal volumes, creating the deepest nulls and largest peaks.

Many flangers and phasers include a feedback control to make the effect even stronger. Not unlike feedback on an echo effect, this sends part of the output audio back into the input to be processed again. And also like an echo effect, applying too much feedback risks runaway, where the sound becomes louder and louder and highly resonant. That might be exactly what you want. This particular plug-in also offers polarity Invert switches for the main Mix and Feedback paths to create a different series of peak and null frequencies.

One of the failings of many digital flangers is an inability to "pass through zero" time, needed to set the lowest peak and null frequencies above a few KHz. When two tape recorders are used to create flanging as in Figure 11.4, the timing difference is zero when both machines are in sync. But most plug-in flangers add a minimum delay time, which limits their sweep range. The Sonitus Modulator's Tape button adds a small delay to the incoming audio, so both the original and delayed version have the same minimum delay time, enabling the sound of real tape flanging passing through zero time.

Back in the 1960s, I came up with a cool way to create flanging using a single tape recorder having separate record and playback heads and Sel-Sync capability. This avoids the complexity of needing two separate tape machines to play at the same time or varying the speed of one machine relative to the other. If you record a mono mix from one track to another or use a pair of tracks for stereo, the copy is delayed in time compared to the original. If you then use Sel-Sync to play the copied track from the record head, the copy plays earlier putting the tracks back in sync. Then all you have to do is push on the tape slightly with your finger or a pencil at a point between the two heads. This adds a slight delay to the original track playing from the play head. When the two tracks or track pairs are mixed together, the result is flanging that "goes through zero" when the tape is not pushed.

Summary

This chapter explains the basics of time-based audio effects. The most common time-based effects used in recording studios are echo and reverb, but flanging and phasing also fall into this category because they use time delay or phase shift to achieve the sound of a comb filtered frequency response.

Echo is useful to add a subtle sense of space to tracks when the delays are short and soft, but it can also be used in extreme amounts. The longer you set the delay time, the larger the perceived space becomes. Older analog tape echo units are prized for their vintage sound, but I prefer modern digital models because their sound quality doesn't degrade quickly when feedback is used to add many repeats. Modern plug-in echo effects are very versatile, and most can be set to automatically synchronize their delay time to the tempo of a DAW project. Many plug-ins also let you cross-feed the echoes between the left and right channels to add a nice sense of width.

Reverb is equally ubiquitous, and it's probably safe to say that reverb is used in one form or another on more recordings than not. Early reverb was created either by placing speakers and microphones in a real room or with mechanical devices such as springs and steel plates under tension. Unfortunately, mechanical reverb units are susceptible to picking up external noise and vibration. Modern digital reverb avoids that, though achieving a truly realistic effect is complex and often expensive. Most reverbs, whether mechanical or electronic, can synthesize a stereo effect from mono sources by creating different reverb sounds for the left and right outputs.

The two basic types of digital reverb are algorithmic and convolution. The first calculates all the echoes using computer code, and the second applies the characteristics of an impulse file recorded in a real room onto the audio being processed. Early reverb plug-ins often sounded grainy, but newer models take advantage of the increased processing power of modern computers. A good reverb unit lets you adjust decay times separately for low frequencies, and many include other features such as a low-cut filter to further reduce muddy reverberant bass. Reverb is useful not only for pop music, but classical music also benefits when too little natural reverb was picked up by the microphones.

Finally, flanger and phaser effects are explained, including the various parameters available on most models. In the old days, flanging required varying the speed of a tape recorder, but modern plug-ins wrap this into a single effect that's simple to manage. Most flanger and phaser effects include an automatic sweep mode, and most plug-in versions can be set to sweep at a rate related to the tempo of a DAW project. The strongest effect is created when the original and delayed versions are mixed equally, though feedback can make the effect even stronger.

CHAPTER 12

Pitch and Time Manipulation Processors

Chapter 6 described what Vari-Speed is and how it works, implemented originally by changing the speed of an analog tape recorder's capstan motor while it records or plays back. Today this effect is much easier to achieve using digital signal processing built into audio editor software and dedicated plug-ins.

There are two basic pitch-shifting methods: One alters the pitch and timing together, and the other changes them independently. Changing pitch and timing together is the simpler method, and it also degrades the sound less. This is the way Vari-Speed works, where lowering the pitch of music makes it play longer, and vice versa. Large changes in tape speed can affect the high-frequency response, though this doesn't matter much when Vari-Speed is used for extreme effect such as "chipmunks" or creating eerie sounds at half speed or slower. The other type of pitch shifting, where the pitch and timing are adjusted independently, is more complex and often creates clicks and gurgling artifacts as a by-product.

Pitch Shifting Basics

Implementing Vari-Speed type pitch shifting digitally is relatively straightforward: The audio is simply played back at a faster or slower sample rate than when it was recorded. Internally, digital Vari-Speed is implemented via sample rate conversion, as explained in Chapter 8. You can effectively speed up digital audio by dropping samples at regular intervals or slow it down by repeating samples. To shift a track up by a musical fifth, you'd discard every third sample, keeping two. And to drop the frequency by an octave, you'd repeat every sample once. When the audio is later filtered, which is needed after sample rate conversion, the dropped or repeated samples become part of the smoothed waveform, avoiding artifacts.

The other type of pitch shifting adjusts the pitch while preserving the duration. It can also be used the other way to change the length of audio while preserving the pitch. Changing only the pitch is needed when "tuning" vocals or other tracks that are flat or sharp. If the length is also changed, the remainder of the track will then be slightly ahead or behind in time. Changing pitch and timing independently is much more difficult to implement than Vari-Speed. This too uses resampling to alter the tuning, but it must also repeat or delete longer portions of the audio. For example, if you raise the pitch of a track by a musical fifth, as above, the length of the audio is reduced by one-third. So small chunks of audio must be repeated to recreate the original length. Likewise, lowering the pitch requires that small portions of the audio be repeatedly deleted.

The first commercial digital pitch shifter that didn't change the timing was the Eventide H910 Harmonizer, introduced in 1974. Back then people used it mainly as an effect, rather than to correct out of tune singers as is common today. As mentioned in Chapter 7, a Harmonizer can, of course, be used to create harmonies. There's no easy way to automate parameters on a hardware device as with modern plug-ins, so to correct individual notes, you have to adjust the Pitch knob manually while the mix plays. Or you could process a copy onto another track, then mute the original track in those places. This is obviously tedious compared to the luxury we enjoy today with DAW software.

One common use for a Harmonizer is to simulate the effect of double-tracked vocals. In the 1960s, it was common for singers to sing the same part twice onto different tracks to enhance the vocal, making it more rich-sounding. This isn't a great vocal effect for intimate ballads, but it's common with faster pop tunes. Double-tracking is also done with guitars and other instruments, not just voices. One difficulty with double-tracking is the singer or musician has to perform exactly the same way twice. Small timing errors that might pass unnoticed on a single track become much more obvious with two tracks in unison.

A Harmonizer offers a much easier way to create the effect of double-tracking from a single performance. The original and shifted versions are mixed together, with both typically panned to the center. This is called *automatic double-tracking*, or ADT for short. You can also use a Harmonizer to add a widening effect by panning the original and shifted versions to opposite sides by some amount. Either way, the Harmonizer is typically set to raise (or lower) the pitch by 1 or 2 percent. A larger shift amount adds more of the effect, but eventually it sounds out of tune. One big advantage of using a Harmonizer effect for ADT rather than simple delay is to avoid the hollow sound of static comb filtering. Since the effective delay through a Harmonizer is constantly changing, the sound becomes more full rather than thin and hollow. Reducing treble frequencies a little on the Harmonizer's output keeps the blended sound even smoother and less affected.

Figure 12.1 shows the basic concept of pitch shifting without changing duration. In the example at left that raises the pitch, the wave is first converted to a higher sample rate, which also shortens its duration. Then the last two cycles are repeated to restore the original length. Of course, a different amount of pitch shift requires repeating a different number of cycles. Lowering the pitch is similar, as shown at right in Figure 12.1. After converting the wave to a lower sample rate, the last two cycles are deleted to restore the original length. Time-stretching without changing the pitch is essentially the same: The wave is resampled up or down to obtain the desired new length, then the pitch is shifted to compensate.

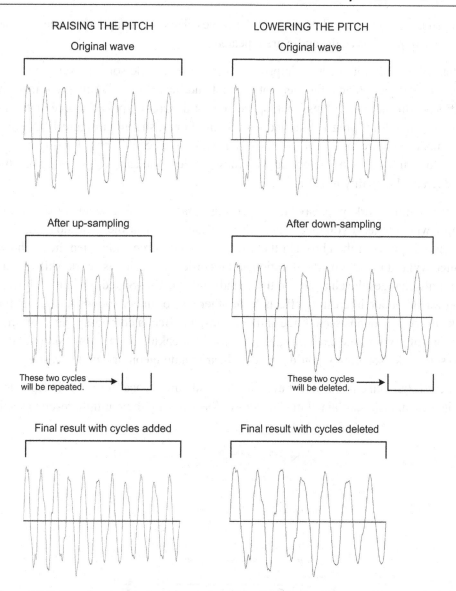

Figure 12.1: To raise the pitch without also changing the duration, as shown at the left, the wave is up-sampled to shift the frequencies higher, then the last two cycles are repeated to restore the original length. Lowering the pitch instead down-samples the wave to lower the frequencies, then the last two cycles are deleted.

When short sections of audio—say, a few dozen milliseconds—are either repeated or deleted, small glitches are added to the sound. These glitches occur repeatedly as small sequential chunks of the audio are processed, which adds a gurgling quality. Applying very small amounts of pitch change or time stretch usually doesn't sound too bad. But once the amounts exceed 5 or 10 percent, the degradation is often audible. Shifting the pitch upward, or making the audio shorter, deletes small portions of the waves. This is usually less noticeable than the opposite, where portions of the audio are repeated.

DSP pitch shifting algorithms have improved over the years, and some newer programs and plug-ins can shift more than a few percent without audible artifacts. Many programs also offer different shifting algorithms, each optimized for different types of source material. For example, the Pitch Shift plug-in bundled with Sound Forge, shown in Figure 12.2, offers 19 different modes. Six are meant for sustained music, 3 for speech, 7 more for solo instruments, and 3 just for drums. By experimenting with these modes, you can probably find one that doesn't degrade the quality too much.

The audio example "cork_pop.wav" is a recording I made of a cork being yanked from a half-empty wine bottle. If you listen carefully, you can even hear the wine sloshing around after the cork is pulled. I then brought that file into Sound Forge and varied the pitch over a wide range without preserving the duration and recorded the result as "cork_shifted.wav." This certainly changes the character of the sound! Indeed, Vari-Speed type pitch shifting is a common way to create interesting effects from otherwise ordinary sounds. Lowering the pitch can make everyday sounds seem huge and ominous, an effect a composer friend of mine uses often. In one of my favorite examples, he processed a creaking door spring, slowing it *way* down to sound like something you'd expect to hear in a dungeon or haunted castle.

Some pitch shifters take this concept even further and adjust vocal formants independently of the pitch change. As explained in Chapter 10, formants are the multiple resonances that

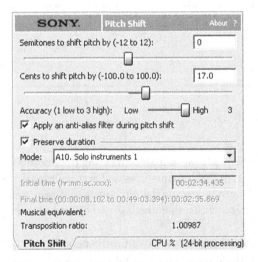

Figure 12.2: The Pitch Shift plug-in can change the pitch either with or without preserving the duration. When Preserve Duration is enabled, the available pitch shift ranges from one octave down to one octave up. But when pitch shifting is allowed to change the length, the available range is increased to plus or minus 50 semitones, or more than four octaves in each direction.

form inside our mouths as we create various vowel sounds. So when the pitch of speech or singing is shifted, the frequencies of those resonance are also shifted. A pitch shifter that can manipulate formants adjusts the resonances independently. You can even use such software to shift only the formants, leaving the basic pitch and timing alone. This might be used to make a male singer sound like a female, and vice versa, or make someone sound younger or older.

Both of the pitch shift methods described so far change the pitch by multiplying the frequencies to be higher or lower. Multiplying retains the musical relationship between notes and between the harmonics within a single note. So if a track or mix is shifted up an octave, all of the note frequencies and their harmonics are doubled, and the music remains in tune with itself. But a third type of pitch shifter instead adds a constant frequency offset. This is called *bode* shifting, or *sideband modulation*, or sometimes *heterodyning*. This type of frequency shifting is used in radio receivers, but less often for audio.

If you apply sideband modulation to shift an A-440 note up a whole step to the B at 494 Hz, the fundamental pitch is shifted by 54 Hz as expected, but all the harmonics are also shifted by the same amount. So the new B note's second harmonic that would have become 988 Hz is now only 930 Hz, which is nearer to an A# note. Other notes and their harmonics are also shifted by the same fixed offset, which puts them out of tune. For this reason, sideband modulation is not often used to process music. However, it has other audio uses, such as disguising voices as in spy movies or creating inharmonic bell-like sound effects from other audio sources.

Auto-Tune and Melodyne

One problem with correcting the pitch of singers or other musicians is they're not usually consistently out of tune. If a singer was always 30 cents flat through the entire song, fixing that would be trivial. But most singers who need correction vary some notes more than others, and, worse, often the pitch changes over the course of a single held note. It's very difficult to apply pitch correction to a moving target.

Auto-Tune, from the company Antares, was the first contemporary hardware device—soon followed by a plug-in version—that could automatically track and correct pitch as it changes over time. If a singer swoops up into a note, or warbles, or goes flat only at the end, Auto-Tune can make the pitch consistent. It can also add vibrato to the corrected note. You can even enter a musical scale containing valid notes for the song's key, and Auto-Tune will coerce pitches to only those notes. This avoids shifting the pitch to a wrong note when a singer is so flat or sharp that his or her pitch is closer to the next higher or lower note than to the intended note. So if a song is in the key of C, you'll tell Auto-Tune to allow only notes in a C scale. You can also use a MIDI keyboard to enter the specific note you want at a given moment, regardless of what was actually sung or played. Auto-Tune then applies whatever amount of pitch change is needed to obtain that note.

An additional parameter lets you control how quickly the pitch correction responds. For example, if the correction speed is set to slow, normal intended vibrato will not be "flattened out," while long-term deviations will be corrected. But when the correction speed is set to the fastest setting, changes in pitch can happen more quickly than would be possible for a human to sing. The result is a vocal quality reminiscent of a synthesizer, and this effect was used famously on the song *Believe* by Cher. This use of Auto-Tune has become so ubiquitous that it's commonly referred to as the *Cher effect*. Coupled with MIDI keyboard input, this can make a singer sound even more like a synthesizer. It can also be used to impart a musical pitch to non-pitched sounds such as normal speech. You can hear that effect applied with great success in the very funny video series *Auto-Tune the News* on YouTube.

There's no doubt that the development of pitch correction was a remarkable feat. But one important shortcoming of Auto-Tune is that it's monophonic and can detect and correct only one pitch at a time. That's great for singers and solo instruments, but if one string is out of tune on a rhythm guitar track, any correction applied to fix the wrong note also affects all the other notes. Celemony Melodyne, another pitch-shifting program, solves this by breaking down a polyphonic source into individual notes that it displays in a grid. You can then change the one note that's out of tune without affecting other notes that don't need correction. It can also fix the pitch of several sour notes independently so they're all in tune, plus related tricks such as changing a major chord to minor. Since you have access to individual notes on the grid, you can even change the entire chord. In case it's not obvious, this is an amazing technical achievement for audio software!

Acidized Wave Files

Besides correcting out-of-tune singers, independent pitch shift and time stretch are used with *Acidized* Wave files in audio looping programs. The first program for constructing music by looping pre-recorded musical fragments was Acid, and it's still popular. Acid looks much like a regular DAW program, and in fact it is, but it also lets you combine disparate musical clips to create a complete song without actually performing or recording anything yourself. Hundreds of third-party Acid libraries are available in every musical style imaginable, and some other DAW programs can import and manipulate Acidized Wave files. You can piece together all of these elements to create a complete song much faster than actually performing all the parts. If the prerecorded files you've chosen aren't in the right key or tempo for your composition, Acid will pitch-shift and time-stretch them automatically.

To create an Acid project, you first establish a tempo for the tune. As you import the various prerecorded loops onto different tracks, the software adjusts their timing automatically to match the tempo you chose. You can also change the pitch of each loop, so a bass part in the key of A can be played back in the key of C for four bars, then in the key of F for the next two bars, and so forth. As with all pitch/time-shifting processes, if the current tempo or key is very different from the original recordings, artifacts may result, but for the most part this process works very well. It's a terrific way for singers to quickly create backing tracks for their original songs without having to hire musicians or a recording studio. It's also a great way to create songs to practice or jam with, and for karaoke.

Acidized files store extra data in the header section of the Wave file. This is a non-audio portion of the file meant for copyright and other text information about the file. In this case, acidizing a file stores data that tells programs like Acid where the musical transients are located to help them change the pitch and time with less degradation. Another aspect of Acid-specific data is whether the loop's musical key should change along with chord changes in the song. If you add an organ backing to an Acid project, the organ parts need to change tempo and key along with the song. But a drum loop should ignore chord changes and only follow the tempo. Otherwise the pitch of the drums and cymbals will shift, too.

Summary

This chapter explains the basics of pitch shifting and time stretching and shows how they're implemented. Vari-Speed type pitch shifting is a simple process, using resampling to change the pitch and duration together. Changing the pitch and duration independently is much more complex because individual wave cycles must be repeated or deleted. When the amount of change is large, glitches in the audio may result. But modern implementations are very good, given the considerable manipulation applied.

Pitch shifting is useful for fixing a singer's poor intonation, and it's an equally valuable tool for sound designers. When used with extreme settings, it can make a singer sound like a synthesizer. Modern implementations can adjust vocal formants independently from the pitch and duration, and Melodyne goes even further by letting you tune or even change individual notes within an entire chord.

Finally, Acidized Wave files are used to create complete songs and backings, using loop libraries that contain short musical phrases that are strung together. The beauty of Acid, and programs like it, is that the loops can be played back in a key and tempo other than when they were originally recorded.

CHAPTER 13

Other Audio Processors

Tape-Sims and Amp-Sims

Tape simulators and guitar amp simulators usually take the form of digital plug-ins that can be added to tracks or a bus in DAW software. There are also hardware guitar amp simulators that use either digital processing or analog circuits to sound like natural amplifier distortion when overdriven. Even hardware tape simulators are available, though the vast majority are plug-ins. The main difference between a tape-sim and an amp-sim is the amount and character (frequency spectrum) of the distortion that's added and the labeling on the controls. Guitar players often require huge amounts of distortion, where distortion expected from an analog tape recorder is usually subtle.

My first experience with a tape simulator was the original Magneto plug-in from Steinberg. It worked pretty well—certainly much better than recording my DAW mixes onto cassettes and back again, as I had done a few times previously. The Ferox tape-sim in Figure 13.1 is typical of more modern plug-in versions. Some models let you choose the virtual tape speed and pre-emphasis type, and other settings can be used to vary the quality of the added distortion using analog tape nomenclature. With Ferox, you instead enter the parameters directly.

The audio example "tape-sim.wav" plays five bars of music written by my friend Ed Dzubak, first as he mixed it, then with the Ferox tape-sim applied. I added only a small amount of the effect, but you can easily hear the added crunch on the soft tambourine tap near the end of bar 2. To make this even clearer, at the end of the file, I repeated the tambourine part four times without, then again with, Ferox enabled.

Even though I own a nice Fender SideKick guitar amp, I still sometimes use an amp-sim. When I recorded the rhythm guitar parts for my *Tele-Vision* music video, I didn't know how crunchy I'd want the sound to be in the final mix. It's difficult to have too much distortion on a lead electric guitar, but with a rhythm guitar, clarity can suffer if you add too much. So I recorded through my Fender amp with only a hint of distortion, then added more with an amp-sim during mixdown when I could hear everything in context. The Alien Connections ReValver plug-in bundled with SONAR Producer shown in Figure 13.2 is typical, and it includes several different modules you can patch in any order to alter in sound in various ways.

Figure 13.1: The Ferox tape simulator imitates the various types of audio degradation that analog tape recorders are known and prized for. *Image courtesy of ToneBoosters*.

Figure 13.2: The Alien Connections ReValver guitar amp simulator offers overdrive, EQ, reverb, auto-wah, and several amplifier and speaker types.

The first hardware amp-sim I'm aware of was the original SansAmp from Tech 21, introduced in 1989, with newer models still in production. This is basically a fuzz tone effect for guitar players, but it can add subtle amounts of distortion as well as extreme fuzz. Line 6 is another popular maker of hardware amp-sims, and their POD line of digital effects includes simulations of many different guitar amplifier types and sounds.

In the audio file "amp-sim.wav," you can hear a short recording of two clean guitar chords plain, then with the ReValver amp-sim plug-in engaged. This example shows just one of the many amplifier character types that can be added. For lead guitars—or anything else—you can add some very extreme fuzz effects.

Other Distortion Effects

Another type of distortion effect is *bit-depth reduction*, sometimes called *bit crushing*. These are always in the form of plug-ins that work in a DAW host or other audio editor program. With this type of distortion, you specify the number of output data bits. Bit-reduction effects are used mainly for intentionally low-fi productions, and in this application the bit-reduced audio is not dithered, adding truncation distortion as well as reducing resolution.

I used the freeware Decimate plug-in to let viewers of my *AES Audio Myths* YouTube video hear how the audio quality of music degrades as the bit-depth is reduced to below 16 bits. The video demo "other_effects" contains part of a cello track I recorded for a friend's pop tune to understand the relationship between what you hear and the number of bits used at that moment as shown on the screen. It starts at 16 bits, then transitions smoothly down to only 3 bits by the end. Cellos are rich in harmonics, so on this track the effect becomes most noticeable when the bit-depth is reduced to below 8 bits.

The *Aphex Aural Exciter* was already described in Chapter 3, and I explained there how to emulate the effect using EQ and distortion plug-ins. It's mentioned here again only because it adds a subtle trebly distortion to give the impression of added clarity, so it, too, falls under the category of distortion effects.

Software Noise Reduction

Chapter 9 showed how to use a noise gate to reduce hiss and other constant background noises, but software noise reduction is more sophisticated and much more effective. With this type of processing, you highlight a section of the Wave file containing only the noise, and then the software "learns" that sound to know what to remove. This process is most applicable to noise that's continuous throughout the recording such as tape or preamp hiss, hum and buzz, and air conditioning rumble. Software noise reduction uses a series of *many* parallel noise gates, each operating over a very narrow range of frequencies. The more individual gates that are used, the better this process works. Note that this is unrelated to companding noise reduction such as Dolby and dbx that compress and expand the volume levels recorded to analog tape. Software noise reduction works after the fact to remove noise that's already present in a Wave file.

As mentioned in Chapter 9, a problem with conventional gates is they process all frequencies at once. If a bit of bass leaks into the microphone pointing at a tambourine, the gate will open even if the tambourine is not playing at that moment. The same happens if high frequencies leak into a microphone meant to pick up an instrument containing mostly low frequencies. By splitting the audio into different bands, each gate opens only when frequencies in that band are present, leaving all the other bands muted. So if the noise is mostly trebly hiss, and the audio contains mostly low frequencies, the higher-frequency gates never open. When set for a high enough resolution—meaning many narrow bands—noise reduction software can effectively remove 60 Hz hum and its harmonics, but without touching the nearby musical notes at A# and B. Since hundreds or even thousands of bands are typically employed, the threshold for each band is set automatically by the software to be just above the noise present in that band. For this reason, software noise reduction is sometimes called *adaptive noise reduction*, because the software adapts itself to the noise.

Figure 13.3 shows the main screen of the original Sonic Foundry Noise Reduction plug-in, which I still use because it works very well. Other, newer software works in a similar fashion. Although this software is contained in a plug-in, the process is fairly CPU-intensive, so it makes sense to use it with a destructive audio editor that applies the noise reduction permanently to a Wave file. The first step is to highlight a section of the file that contains only the noise to be removed. Then you call up the plug-in, check the box labeled Capture noiseprint, and click Preview. It's best to highlight at least a few seconds of the noise if possible. With a longer noise sample, the software can better set the thresholds for each band. Be sure that only pure noise is highlighted and the section doesn't contain the decaying sound of music. If any musical notes are present in the sampled noise, their frequencies will be reduced along with the noise.

When you click Preview, you'll hear the background noise sample play once without noise reduction, and then it repeats continuously with the noise reduction applied while you adjust the various settings. This lets you hear right away how much the noise is being reduced. After you stop playback, select the entire file using the Selection button at the bottom right of the screen. This tells the plug-in to process the entire file, not just the noise-only portion you highlighted. When you finally click OK, the noise reduction is applied. Long files take a while to process, especially if the Fast Fourier Transform (FFT) size is set to a large value. Making the FFT size larger divides the audio into more individual bands, which reduces the noise more effectively. Using fewer bands processes the audio faster, but using more bands does a better job. I don't mind waiting, since what really matters is how well it works.

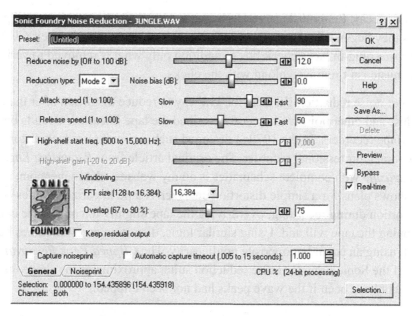

Figure 13.3: The Sonic Foundry Noise Reduction plug-in comprises a large number of noise gates, each controlling a different narrow range of frequencies. The software analyzes the noise, then sets all of the gate parameters automatically.

As with pitch and time manipulation plug-ins, several different modes are available so you can find one that works best with your particular program material and noise character. Note that one of the settings is how much to reduce the noise. This is equivalent to the Attenuation, Depth, or Gain Reduction setting on a noise gate. Applying a large amount of noise reduction often leaves audible artifacts. So when the noise to be removed is extreme, two passes that each reduce the noise by 10 dB usually gives better results than a single pass with 20 dB of reduction. Applying a lot of noise reduction sounds much like low bit-rate lossy MP3 type compression, because the technologies are very similar. Both remove content in specific frequency bands when it's below a certain threshold.

You can also tell the plug-in to audition only what it will remove, to assess how damaging the process will be. This is the check box labeled Keep residual output in Figure 13.3. You'll do that when using Preview to set the other parameters to ensure that you're not removing too much of the music along with the noise. If you hear bits of the music playing, you may be reducing the noise too aggressively.

Another type of noise reduction software is called *click and pop removal*, and this is designed to remove scratch sounds on vinyl records. If you think about it, it would seem very difficult for software to identify clicks and pops without being falsely triggered by transient sounds that occur naturally in music. The key is the software looks for a fast rise time *and* a subsequent fast decay. Real music never decays immediately, but most clicks and pops end as quickly as they begin. Like Sonic Foundry's Noise Reduction plug-in, their *Click and Crackle Remover* plug-in has a check box to audition only what is removed to verify that too much of the music isn't removed along with the clicks.

The last type of noise reduction I'll describe is a way to reduce distortion after the fact, rather than hiss and other noises. Chapter 6 described the tape linearizer circuits built into some analog tape recorders. In the 1970s I designed such a circuit to reduce distortion on my Otari two-track professional recorder. The original article from *Recording Engineer/ Producer Magazine* with complete schematics is on my website ethanwiner.com, and the article also shows plans for a simple distortion analyzer. A tape linearizer applies equal but opposite distortion during recording, overdriving the tape slightly to counter the compression and soft-clipping the tape will add. Using similar logic, distortion can sometimes be reduced after the fact, using an equal but opposite nonlinearity. The *Clipped Peak Restoration* plug-in, also part of the Sonic Foundry noise reduction suite, approximates the shape the original waveform might have been if the wave peaks had not been clipped.

Other Processors

One of the most useful freeware plug-ins I use is GPan from GSonic, shown in Figure 13.4. This simple plug-in lets you control the volume and panning separately for each channel of a stereo track. I use this all the time on tracks and buses to narrow a too wide stereo recording or to keep only one channel of a stereo track and pan it to the center or to mix both channels to mono at the same or different volume levels.

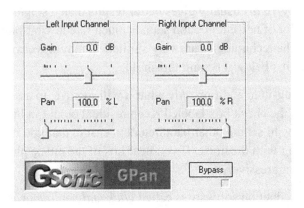

Figure 13.4: The GSonic GPan plug-in lets you control the volume and panning for each channel of a stereo track independently. Simple but highly useful!

Chapter 1 explained one way to create a *stereo synthesizer*, using phase shift to create opposite peak and null frequencies in the left and right channels. But some designs are more sophisticated, and less fake-sounding, than simple comb filtering that differs left and right. By applying different amounts of phase shift to the left and right channels, sounds can seem to come from locations wider than the speakers, or even from other places in the room, including overhead. This popular effect can also make stereo material sound wider than usual. Several companies offer widening plug-ins that accept either mono or stereo sources.

Vocal Removal

Another "effect" type is the *vocal remover*, sometimes called a *center channel eliminator*. I put "effect" in quotes because this is more of a process than an effect, though plug-ins are available to do this automatically. Since a lead vocal is usually panned to the center of a stereo mix, its level can be reduced by subtracting one channel from the other. This is similar to the Mid/Side processing described in Chapters 6 and 10. The basic procedure is to reverse the polarity of one channel, then combine that with the other channel at an equal volume. Any content common to both channels will be canceled, leaving only those parts of the stereo mix that are different on the left and right sides. Unfortunately, most vocal removal methods reduce a stereo mix to mono because the two channels are combined to one. However, you could synthesize a stereo effect as explained previously. Chapter 19 explains a different method using Dolby ProLogic available in most consumer receivers.

It's impossible to completely remove a vocal or reduce its level without affecting other elements in the mix. Even though most vocals are placed equally in the left and right channels, stereo reverb is often added to vocal tracks. So even if you could completely remove the vocal, some or all of the reverb is likely to remain. If you plan to record yourself or someone else singing over the resultant track, the new vocal can have its own reverb added, and you may be able to mix the new voice and reverb loud enough to mask the ghost reverb from the original track. Another problem is that vocals are not the only thing panned to the center of a mix. Usually, the bass and kick drum are also in the middle, so those are also canceled. You can avoid this by using EQ to roll off the bass on one channel before combining it with the other. If one channel has less low-frequency content than the other, bass instruments will not completely cancel, though their tonality will be affected.

I've used Sound Forge to remove vocals in Wave files destructively, but processing the left and right channels on separate tracks of a DAW is usually faster. This way you can more easily adjust the channel levels while the song plays to achieve the most complete cancellation. If your DAW doesn't offer a direct way to split one stereo track to two separate mono tracks, a plug-in such as GPan described earlier is needed. You'll put the same music on two tracks, then send the left channel (only) of one track panned to the center and do the same for the right channel on the other track. Then reverse the polarity of one track while adjusting its volume as the song plays, listening for the most complete cancellation. Finally, insert a low-cut EQ filter onto either track, and adjust the cutoff frequency to bring back the bass and kick drum.

Ring Modulators

The last audio effect I'll describe is the *ring modulator*, which applies amplitude modulation (AM) to audio. This is a popular feature in many synthesizers for creating outer space–type sounds and other interesting effects, but it can be implemented as a plug-in and applied to any audio source. Amplitude modulation is a cyclical variation in amplitude, or volume. The simplest type of AM is tremolo, as described in Chapter 9. Tremolo rates typically vary from less than 1 Hz up to 10 Hz or maybe 20 Hz. But when amplitude modulation becomes fast enough, it transitions from sounding like tremolo into audible sum and difference frequencies, also called *side bands*. In fact, there's no difference between tremolo and ring modulation other than the speed of the modulation. By the way, the term "ring" derives from the circle arrangement of diodes used to implement this in analog circuitry, similar to the full-wave diode bridge in Figure 23.15 in Chapter 23.

When the volume of one frequency is modulated by another, sum and difference frequencies result. But unlike IM distortion that *adds* sum and difference components, the output of a ring modulator contains *only* the sum and difference frequencies. Even a basic tremolo does this, though it's difficult to tell by listening. If you pass a 500 Hz sine wave through a tremolo running at 3 Hz, it sounds like the 500 Hz tone varies in volume at that rate. In truth, the output contains 497 Hz and 503 Hz. Musicians know this effect as *beat frequencies*, and they use the repeated pulsing of two strings playing together to help tune their instruments. When the pulsing slows to a stop, both strings are then at the same pitch, as shown in the video "beat_tones."

To prove that tremolo generates sum and difference frequencies, and vice versa, I created the audio demo "beat_frequencies.wav" from two similar frequencies. I started by generating a 200 Hz sine wave in Sound Forge lasting two seconds. Then I mixed in a second sine wave at 203 Hz at the same volume. The result sounds like a 201.5 Hz tone whose volume goes up and down at a 3 Hz rate via tremolo. The two processes are very similar, whether you create the effect by mixing two sine waves together or modulate one sine wave's volume with another. So even though this demo sounds like tremolo, an FFT of the wave file shows that it contains the original two frequencies.

Most ring modulator plug-ins use a sine wave for the modulating frequency, though some offer other waveforms such as triangle, sawtooth, square, or even noise. In that case, the sum and difference frequencies generated are more complex because of the additional modulating frequencies, which add more overtones giving a richer sound. Note that frequency modulation (FM) also changes audibly with the modulation speed, where the sound transitions from vibrato at slow rates to many added frequencies as the speed increases. Even with two pure sine waves, FM creates a much more complex series of overtones than AM. This is the basis for FM synthesis and is described in more detail in Chapter 14.

To show the effect of ring modulation at a fast rate, which is how it's normally used, the second section of the "other_effects" video applies AM using the freeware Bode Frequency Shifter plug-in from Christian-W Budde. I programmed a track envelope in SONAR to sweep the modulating frequency continuously upward from 0.01 Hz to 3.5 KHz, first on a clean electric guitar track, then again on a triangle recording that repeats four times. You can hear the effect begin as stereo tremolo, varying the volume differently left and right. Notice that the fourth time the triangle plays, the pitch seems to go down even though the ring modulation continues toward a higher frequency. This is the same as digital aliasing, which, as explained in earlier chapters, also creates sum and difference frequencies. In this case, as the modulating speed increases, the *difference* between the modulating frequency and frequencies present in the triangle becomes smaller. So the result is audible difference frequencies that become lower as the modulation rate increases.

Summary

This chapter explains miscellaneous audio effects, including tape-sims, amp-sims, and bit-depth reduction. Although software noise reduction is not an audio effect, it's a valuable tool for audio engineers. Unlike a conventional noise gate, this process applies hundreds or even thousands of gates simultaneously, with each controlling a different very narrow range of frequencies. Digital audio software can also remove clicks and pops from LP records and even restore peaks that had been clipped to reduce distortion after the fact. Finally, this chapter demonstrates that the same amplitude modulation used to create a tremolo effect can also generate interesting sounding sum and difference frequency aliasing effects when the modulating frequency is very fast.

CHAPTER 14

Synthesizers

A synthesizer is a device that creates sounds electronically, rather than acoustically using mechanical parts as with a piano, clarinet, or electric guitar. The sound is literally synthesized, and electronic sound shaping is often available as well. The first synthesizer was designed in 1874 by electrical engineer Elisha Gray. Gray is also credited with inventing the telephone, though Alexander Graham Bell eventually won the patent and resulting fame. Gray's synthesizer was constructed from a series of metal reeds that were excited by electromagnets and controlled by a two-octave keyboard, but the first practical synthesizer I'm aware of was the Theremin, invented by Léon Theremin and patented in 1928.

Several modern versions of the Theremin are available from Moog Music, the company started by another important synthesizer pioneer, Robert Moog. One of the current Moog models is the Etherwave Plus, shown in Figure 14.1. Unlike most modern synthesizers, the Theremin plays only one note at a time, and it has no keyboard. Rather, moving your hand closer or farther away from an antenna changes the pitch of the note, and proximity to a second antenna controls the volume in a similar fashion. By the way, synthesizers that can play only one note at a time are called *monophonic*, versus *polyphonic* synthesizers that can play two or more notes at once. These are sometimes called *monosynths* or *polysynths* for short.

Controlling the pitch by waving your hand around lets you slide the pitch smoothly from one note to another, a musical effect called *portamento*. Using antennas also lets you easily add vibrato that can be varied both in frequency and intensity just by moving your hand. However, the Theremin is not an easy instrument to play because there are no "anchor points" for the notes. Unlike a guitar or piano with fixed frets or keys, the Theremin is more like a violin or cello with no frets. The original Theremin was built with tube circuits, though modern versions are of course solid state.

Analog versus Digital Synthesizers

Early commercial keyboard synthesizers were analog designs using transistor circuits for the audio oscillators that create the musical notes and for the various modulators and filters that process the sounds from the oscillators. These days, most hardware synthesizers are digital, even when they're meant to mimic the features and sounds of earlier analog models. Modern digital synthesizers go far beyond simply playing back recorded snippets of analog synths.

Figure 14.1: The Etherwave Plus from Moog Music is a modern recreation of the original Theremin from the 1920s. *Photo courtesy of Moog Music, Inc.*

Most use digital technology to generate the same types of sounds produced by analog synthesizers. In this case the synthesizer is really a computer running DSP software, with a traditional synthesizer appearance and user interface. Many offer additional capabilities that are not possible at all using analog techniques. There are many reasons for preferring digital technology to recreate the sound of analog synthesizers: The pitch doesn't drift out of tune as the circuits warm up, there's less unwanted distortion and noise, and *patches*—the arrangements of sound sources and their modifiers—can be stored and recalled exactly.

Another type of analog-style digital synthesizer is programmed entirely in software that runs either as a stand-alone computer program or as a plug-in. You can also buy sample libraries that contain recordings of analog synthesizers to be played back on a keyboard. In my opinion, using samples of analog synthesizers misses the point because the original parameters can't be varied as the notes play. One of the joys of analog synthesis is being able to change the filter frequency, volume, and other parameters in real time to make the music sound more interesting. With samples of an analog synthesizer you can add additional processing, but you can't vary the underlying parameters that defined the original sound character.

Additive versus Subtractive Synthesis

There are two basic types of synthesis: *additive* and *subtractive*. An additive synthesizer creates complex sounds by adding together individual components—often sine waves that determine the strength of each individual harmonic (see Figure 14.2). A subtractive synthesizer instead starts with a complex waveform containing many overtones, followed by

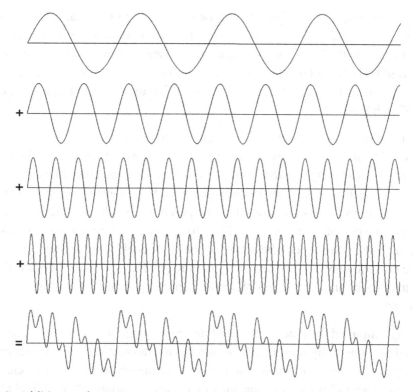

Figure 14.2: Additive synthesis creates a complex sound by adding different frequencies together. Varying the volume of each frequency affects the tone color, also called *timbre*, as mentioned in Chapter 3.

a low-pass or other filter type that selectively removes frequencies. Both methods have merit, though additive synthesis is more complicated because many individual components must be synthesized all at once and mixed together. With subtractive synthesis, a single oscillator can create an audio wave containing many harmonics, and a single filter can process that complex sound in many ways.

The Hammond organ was arguably the first commercial additive synthesizer, though acoustic pipe organs have been around for many centuries. A Hammond organ contains a long spinning rod with 91 adjacent metal gear-like "tone wheels" driven by an electric motor. Each gear spins near a magnetic pickup that generates an electrical pulse as each tooth crosses its magnetic field. The rod spins at a constant rate, and each tone wheel has a different number of teeth that excite its own pickup. So each wheel generates a different frequency, depending on how many gear teeth pass by the pickup each second.

Nine harmonically related frequencies are available at once—each with its own volume control, called a *drawbar*—using an elaborate switching mechanism attached to each key of the organ's keyboard. The drawbars determine the volume of each harmonic, letting the player create a large variety of timbres. You can adjust the drawbars while playing to vary the tone in real time. As with all organs, the sound starts when the key is first pressed and stops when it's released. Hammond organs have a feature called *percussion* that mimics the sound of a pipe organ. When enabled, it plays a short burst of an upper harmonic, adding emphasis to the start of each note.

The first commercially viable all-in-one subtractive synthesizer was the Minimoog, introduced in 1970, and it plays only one note at a time. However, inventor Robert Moog sold *modular synthesizers* as early as 1964. Unlike modern self-contained synthesizers, a modular synthesizer is a collection of separate modules meant to be connected together in various configurations using patch cords—hence the term *patch* to describe a particular sound. So you might route the output of an audio oscillator to a voltage controlled amplifier (VCA) module to control the volume, and the output of the VCA could go to a filter module to adjust the harmonic content. Few synthesizers today require patch cords, though the term "patch" is still used to describe an arrangement of synthesizer components and settings that creates a particular sound.

The Minimoog simplified the process of sound design enormously by pre-connecting the various modules in musically useful ways. Instead of having to patch together several modules every time you want to create a sound, you can simply flip a switch or change a potentiometer setting. Other monophonic analog synthesizers soon followed, including the ARP 2500 and 2600, and models from Buchla & Associates, E-MU Systems, Sequential Circuits, and Electronic Music Labs (EML), among others. A modern version of the Minimoog is shown in Figure 14.3.

Figure 14.3: The Minimoog was the first commercially successful subtractive synthesizer, bringing the joys of analog synthesis to the masses. *Photo courtesy of Moog Music, Inc.*

Voltage Control

Another important concept pioneered by Moog was the use of *voltage control* to manipulate the various modules. Using a DC voltage to control the frequency and volume of an oscillator, and the frequency of a filter, allows these properties to be automated. Unlike an organ that requires a complex switching system to control many individual oscillators or tone wheels simultaneously, the keyboard of a voltage controlled synthesizer needs only to send out a single voltage corresponding to the note that's pressed. When an oscillator accepts a control voltage to set its pitch, it's called a *voltage controlled oscillator*, or VCO. Further, the output of one oscillator can control other aspects of the sound, such as driving a VCA to modulate the volume. Or it can add vibrato to another oscillator by combining its output voltage with the voltage from the keyboard, or vary a note's timbre by manipulating a *voltage controlled filter* (VCF).

Music sounds more interesting when the note qualities change over time, and being able to automate sound parameters makes possible many interesting performance variations. For example, a real violinist doesn't usually play every note at full volume with extreme vibrato for the entire note duration. Rather, a good player may start a note softly with a mellow tone, then slowly make the note louder and brighter while adding vibrato, and then increase the vibrato intensity and speed. A synthesizer whose tone can be varied by control voltages can be programmed to perform similar changes over time, adding warmth and a human quality to musical notes created entirely by a machine. Of course, you can also turn the knobs while a note plays to change these parameters, and good synthesizer players do this routinely.

Sound Generators

Oscillators are the heart of every analog synthesizer, and they create the sounds you hear. An oscillator in a synthesizer is the electronic equivalent of strings on a guitar or piano. But where a string vibrates the air to create sound directly, an oscillator creates a varying voltage that eventually goes to a loudspeaker. Either way, the end result is sound vibration in the air you can hear.

The five basic waveforms used by most synthesizers are sine, triangle, sawtooth, square, and pulse. These were shown in Figure 1.23 and are repeated here in Figure 14.4. As you learned in Chapter 1, fast movement at a wave's rising or falling edges creates harmonic overtones. The steeper the wave transition, the faster the speaker moves, and in turn the more high-frequency content produced.

A sine wave is not very interesting to hear, but it's perfect for creating the eerie sound of a Theremin. Most patches created with subtractive synthesis use a sawtooth, square, or pulse wave shape. These are all rich in harmonics, and each has a characteristic sound. Sawtooth

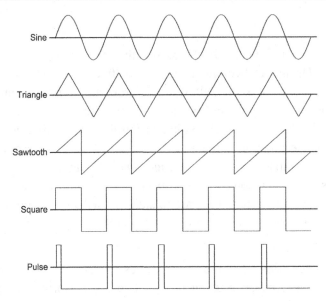

Figure 14.4: Most oscillators that generate musical pitches offer several different wave shapes, each having a characteristic sound quality, or timbre. The same wave shapes are also used for oscillators that modulate the volume, pitch, and filter frequency.

and pulse waves contain both even and odd harmonics, where square and triangle waves contain only odd harmonics. The classic sound Keith Emerson used in the song *Lucky Man* from 1970 is based on a square wave.

One important way to make a synthesizer sound more interesting is to use two or more oscillators playing in unison but set slightly out of tune with each other. This is not unlike the difference between a solo violin and a violin section in an orchestra. No two violin players are ever *perfectly* in tune, and the slight differences in pitch and timing enhance the sound.

Modulators

When one oscillator of an analog-style synthesizer varies the pitch of another oscillator, the controlling oscillator usually operates at a low frequency to add vibrato, while the main (audio) oscillator runs at higher frequencies to create the musical notes. Many analog synthesizers contain dedicated low-frequency oscillators intended only to modulate the frequency of audio oscillators, or vary other sound attributes. As you might expect, these are called *low-frequency oscillators*, or LFOs for short. Most LFOs can be switched between the various wave shapes shown in Figure 14.4, and some can also output noise to create a random modulation.

An LFO creates a tremolo effect when used to modulate the volume, a vibrato effect when varying the pitch, or a wah-wah effect when applied to a filter. If the modulating wave shape is sine or triangle, the pitch (or volume) glides up and down. A sawtooth wave instead gives a swooping effect when modulating the pitch, and a square or pulse wave switches between two notes. When sent to a VCA for tremolo, square and pulse waves turn the volume on and off, or switch between two volume levels.

One important way synthesizer players add interest is by varying the sound quality of notes over their duration. The simplest parameter to vary is volume. This is handled by another common modulator, the *ADSR*, shown in Figure 14.5. An ADSR is an *envelope generator*, which is a fancy way of saying it adjusts the volume or some other aspect of the sound over time. ADSR stands for Attack, Decay, Sustain, and Release. Some synthesizers have additional parameters, such as an initial delay before the ADSR sequence begins, or more than one attack and decay segment. Note that the Attack, Decay, and Release parameters control the *length* of each event, while the Sustain knob controls the sustained *volume level*. The solid portion in Figure 14.5 shows an audio waveform zoomed out too far to see the individual cycles. In other words, this shows a volume envelope of the sound.

The *Attack* time controls how long it takes for the volume to fade up initially. Once the volume reaches maximum, it then decays at a rate set by the *Decay* setting. But rather than always decay back to zero volume, it instead fades to the *Sustain* volume level. This lets you create a short note burst for articulation, then settle to a lower volume for the duration of the note. Only after you release the key on the keyboard does the volume go back to zero at a rate determined by the *Release* setting.

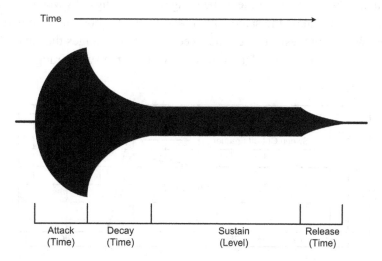

Figure 14.5: The ADSR is an envelope generator that controls the volume of a note over time, or the cutoff frequency of a filter, or any other property that can be voltage-controlled.

An ADSR lets you create very short staccato notes as used in the instrumental hit *Popcorn* from 1972 by Gershon Kingsley. For the lead synth sound from that song, the attack is very fast, and the decay is almost as fast. In this case the note should not sustain, so the sustain level is set to zero. By extending the decay time, notes can be made to sound somewhat like a guitar, then more like a piano. Notes can also continue after you release the key by setting a longer release time. Early synthesizers played all notes at the same volume, like an organ, with the same attack and sustain levels used throughout a passage. But when players can control the volume of every note—for example, by striking the key harder or softer—the ADSR levels are relative to the performed volume of the note.

An ADSR is often used to change the volume over time, but it can also control a filter to vary a note's timbre as it sustains. When an ADSR controls a filter to change the tone quality, an ADSR volume envelope can also be used, or it can be disabled. Many synth sounds sweep the filter frequency either up or down only, though it can of course sweep both ways. In truth, an ADSR always sweeps the filter both up and down, but one direction may be so fast that you don't hear the sweep. Many synth patches use one of those two basic ADSR filter settings, where the attack is short with a longer decay, or vice versa.

Filters

Classic Moog-type synthesizers use a low-pass filter with a fairly steep slope of 24 dB per octave, with the cutoff frequency controlled by an ADSR (see Figure 14.6). This type of filter is offered in many analog synthesizers. The filter's *Initial Frequency* knob sets its static cutoff frequency before an ADSR or LFO is applied. The filter's ADSR then sweeps the frequency up from there and back down over time, depending on its settings. As when using an ADSR to vary the volume, you can do the same with a filter to create short staccato notes or long sustained tones. With a synthesizer filter, the sweep direction changes the basic character of the sound. The filter's Gain control determines the width of the sweep range.

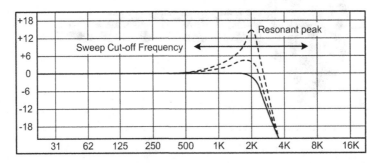

Figure 14.6: The classic analog synthesizer filter is a low-pass at 24 dB per octave, with an adjustable resonant peak at the cutoff frequency.

Another important filter parameter is *resonance*, described in Chapter 1. When the resonance (or Q) of a low-pass filter is increased, a narrow peak forms at the cutoff frequency. This emphasizes that frequency and brings out single harmonics one at a time as it sweeps up or down. Many analog-style synthesizers let you increase the resonance to the point of self-oscillation, not unlike echo effects that let you increase the feedback above unity gain. Setting the resonance very high, just short of outright feedback, imparts a unique whistling character to the sound.

The last filter parameter we'll consider is *keyboard tracking*. Normally, a synthesizer filter is either left at a static position or swept automatically by its own ADSR as each note is played. Most analog synths also have an option for the filter to track the keyboard, so playing higher notes moves the initial filter frequency higher in addition to any sweeping effect. For example, if you have a resonant filter boost that brings out the third harmonic of a note, you'll want the filter to track the notes as you play up and down the keyboard. So whatever note you play, the filter emphasizes that same-numbered harmonic. Keyboard tracking also makes possible sound effects such as wind noise that can be varied in pitch or even played as a melody. To do that you'll use white or pink noise as the sound source, then increase the filter resonance, and have the filter track the keyboard without influence from an ADSR.

MIDI Keyboards

Most hardware synthesizers include a piano-like keyboard to play the musical notes, though there are also smaller desktop and rack-mount sound modules containing only the sound-generating electronics. Modern hardware synthesizers that include a keyboard also send MIDI data as you play to control other synths or sound modules, or to be recorded for later playback. Dedicated keyboard controllers that contain no sounds of their own are also available. A keyboard controller usually includes many knobs to manipulate various functions, and these are typically used as a master controller for several synthesizers in a larger system. Software synthesizers running within a digital audio workstation (DAW) program can also be controlled by a keyboard controller or a conventional keyboard synthesizer with MIDI output.

Modern synthesizers send MIDI data each time a key is pressed or released. In that case, the data for each note-on and note-off message includes the MIDI note number (0–127), the channel number (0–15), and the note-on volume, called *velocity* (0–127). MIDI note data and other parameters are described more fully in Chapter 15. Old-school analog synthesizer keyboards instead output DC voltages that correspond to which note was struck, as well as trigger signals when notes are pressed and released.

Most MIDI keyboards are *velocity sensitive*, sometimes called *touch sensitive*, and their output data include how loudly you play each note. The standard MIDI term for how hard you strike a key is *velocity*, and that's one factor that determines the volume of the notes you play. The other factor is the overall volume setting, which is another standard MIDI data type. Internally, most MIDI keyboards contain two switches for every key: One engages when you first push down on the key, then another senses when the key is fully depressed. By measuring how long it took for the key to reach the end of its travel, the keyboard's electronics can calculate how hard you struck the key and translate that into a MIDI velocity value.

In addition to transmitting note-on velocity to indicate how loudly a note was played, some MIDI keyboards include a feature called *aftertouch*. After pressing a key to trigger a note, you can then press the key even harder to generate aftertouch data. Some synthesizers let you program aftertouch to add a varying amount of vibrato or to make the note louder or brighter or vary other aspects of the note's quality while the note sustains. One of the coolest features of the Yamaha SY77 synthesizer I owned in the 1990s was the harmonica patch, which used aftertouch to shift the pitch downward. While holding a note, pressing the key harder mimics a harmonica player's reed-bending technique. Most MIDI keyboards that respond to aftertouch do not distinguish which key is pressed harder, sending a single data value that affects all notes currently sounding. This is called *monophonic aftertouch*. But some keyboards support *polyphonic aftertouch*, where each note of a chord can be modified independently, depending on how much harder individual keys are pressed.

Beyond Presets

Modern synthesizers are very sophisticated, and most include tons of great preset patches programmed by professional sound designers. Sadly, many musicians never venture beyond the presets and don't bother to learn what all the knobs and buttons really do. But understanding how a synthesizer works internally to make your own patches is very rewarding, and it's also a lot of fun!

Figure 14.7 shows the block diagram of a basic old-school analog synthesizer that uses control voltages instead of MIDI data. It includes two audio VCOs, a mixer to combine their audio outputs, a VCA to vary the volume automatically, and a VCF that changes its filter cutoff frequency to shape the tone color. There are also two ADSR envelope generators, one each for the VCA and VCF, and an LFO to add either vibrato, tremolo, or both.

Note that there are two distinct signal paths: One is the audio you hear, and the other routes the control voltages that trigger the ADSRs and otherwise modulate the sound. Every time a key is pressed, two different signals are generated. One is a DC voltage that corresponds to which note was pressed to set the pitch of the audio VCOs. Each key sends a unique voltage; a low note sends a small voltage, and a higher note sends a larger voltage. In addition,

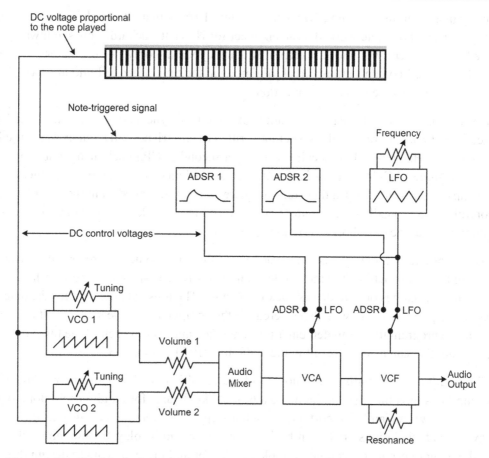

Figure 14.7: This block diagram shows the basic organization of a pre-MIDI monophonic subtractive analog synthesizer.

pressing a key sends a separate note-on trigger signal telling the ADSRs to begin a new sequence of rising, falling, and sustaining control voltages. When the key is finally released, a note-off signal tells the ADSR to initiate the Release phase and fade its output voltage down to zero.

Also note the switches in this example synthesizer that let you choose whether the VCA and VCF are controlled by the ADSR or the LFO. When either switch is set to ADSR, the VCA or VCF is controlled by the ADSR's programmed voltage, which in turn varies the volume or sweeps the filter over time. Setting either to LFO instead applies tremolo or vibrato, respectively. In practice, many synthesizers use volume knobs to continuously adjust the contribution from both the ADSR and the LFO. This lets you have both modulators active at once in varying proportions.

I built my first home-made analog synthesizer around 1970 with a lot of help from my friend Leo Taylor, who at the time worked as an engineer for Hewlett-Packard. This synth was modeled loosely after a Minimoog, except it had no keyboard—only two oscillators, a VCF, and a VCA. Since I couldn't afford a real Minimoog at the time, I had to build my own. In this case, *poverty* was the mother of invention.

In what was surely one of the earliest attempts by anyone to sync a computer with an analog tape recorder, Leo and I recorded a short tune written by my friend Phil Cramer. We started by recording 60 Hz hum—Hz was called cycles per second or CPS back then—onto one track of my Ampex AG-440 4-track half-inch professional recorder. That hum was then squared up (overdriven to clip and become a square wave) and played from the recorder into the control port of a small H-P computer Leo borrowed from work. (In those days a "small" computer was the size of a short refrigerator!)

Leo wrote a program in binary machine code to read the 60 Hz sync tone recorded on Track 1, entering it one instruction at a time on the computer's 16 front-panel toggle switches. There were no diskettes or hard drives back then either. The musical note data for the tune, a modern three-part invention, was also entered via the computer's toggle switches. It took us 11½ hours to program the computer, enter the song data, and record each of the three tracks one at a time. You can hear the result in the audio file "1970_synth.mp3."

In 1974, I built another synthesizer, shown in Figures 14.8 and 14.9. This one was much more ambitious than the first and included a 61-note keyboard, four audio oscillators, two VCFs and two VCAs, an LFO, portamento, a ten-step sequencer, and circuitry to split the keyboard at any point so both synthesizer "halves" could be played simultaneously for a whopping two-note polyphony. It took Leo Taylor and me two years to design this synthesizer, and I spent another nine months building it in the evenings. Leo was the real brains, though I learned a lot about electronics over those two years. By the end of this project, I had learned enough to design all the keyboard and LFO control circuits myself. Fourteen separate plug-in circuit cards are used to hold each module, and each card was hand-wired manually using crimped wiring posts and 24-gauge buss wire with Teflon sleeving.

Also during this period, Leo and I designed what may have been the first working guitar synthesizer based on a pitch-to-voltage converter. I tried to sell the design to Electronic Music Labs (EML), a synthesizer manufacturer near me in Connecticut, but they were too short-sighted to see the potential market. Today, guitar controllers for synthesizers are common!

The best way to learn about analog synthesizers is to see and hear at the same time. The video "analog_synthesizers" pulls together all of the concepts explained so far, and it also shows the waveforms on a software oscilloscope to better relate visually to what you hear.

Figures 14.8 and 14.9: The author built this two-note analog synthesizer in the 1970s. It contains all of the standard analog synth modules, plus a sequencer to play repeating arpeggio type patterns.

Alternate Controllers

In addition to a built-in keyboard for playing notes, many modern synthesizers and keyboard controllers provide a *pitch bend controller*. This usually takes the form of a knurled wheel that protrudes slightly from the left side of the keyboard, letting you glide the pitch between nearby notes, create vibrato manually, or otherwise add expression to a performance. When you rock the wheel forward, the note's pitch goes higher, and pulling it back lowers the pitch. Some synthesizers use a joystick or lever instead of a wheel. Either way, when you let go of the wheel or lever, a spring returns it to its center resting position, restoring the normal pitch. The standard pitch bend range is plus/minus two semitones, or one musical whole step up or down, though most synthesizers let you change the range through a setup menu or via MIDI data.

Being able to set the bend range as much as one or even two octaves in either direction is common, though when the bend range is very large, the pitch is more difficult to control because small movements have a larger effect.

Another common built-in controller is the *modulation wheel*, or *mod wheel* for short. Unlike the pitch bend wheel, a mod wheel stays where you leave it rather than returning to zero when you let go. The most common use is for adding vibrato, but most synthesizers can use this wheel to control other parameters, such as a filter's initial cutoff frequency. Again, assigning the mod wheel to control a different parameter is done in the synth's setup menu or by sending MIDI data.

Most keyboard synthesizers also have a jack that accepts a foot pedal controller. This is typically used to hold notes after you release the keys, like the *sustain pedal* on an acoustic piano. My Yamaha PFp-100 piano synthesizer includes three pedal jacks—one for a sustain pedal and two others for the *sostenuto* and *soft* pedals that behave the same as their counterparts on a real grand piano. An additional jack accepts a continuously variable foot controller to change the volume in real time while you play. Many synths let you reassign these pedals to control other aspects of the sound if desired, and these pedals also transmit MIDI control data.

Some synthesizers include a built-in *ribbon controller*. This is a touch-sensitive strip that's usually programmed to vary the pitch, but instead of rocking a wheel or joystick, you simply glide your finger along the ribbon. If the ribbon is at least a foot long, it's easier to adjust the pitch in fine amounts than with a wheel. It can also be used to add vibrato in a more natural way than using a pitch bend wheel, especially for musicians who play stringed instruments and are used to creating vibrato by rocking a finger back and forth. You can also play melodies by sliding a finger along the ribbon.

Besides the traditional keyboard type we all know and love, other controllers are available for guitar and woodwind players. Most *guitar controllers* use a special pickup to convert the notes played on an electric guitar to MIDI note data, and many also respond to bending the strings by sending MIDI pitch bend data. Converting the analog signal from a guitar pickup to digital MIDI note data is surprisingly difficult for several reasons. First, if more than one note is playing at the same time, it's difficult for the sensing circuits to sort out which pitch is the main one, even if the other notes are softer. To solve this, the pickups on many guitar-to-MIDI converters have six different sections, with a separate electrical output for each string. But even then, a pickup directly under one string may still receive some signal from a different nearby string.

Another obstacle to converting a guitar's electrical output to MIDI reliably is due to the harmonics of a vibrating string changing slightly in pitch over time. A vibrating string produces a complex waveform that varies, depending on how hard and where along the string's length it's plucked. When a string vibrates, it stretches slightly at each back and forth cycle of the fundamental frequency. So when a string is at a far excursion for its fundamental frequency, the pitch of each harmonic rises due to the increased string tension. This causes the waveform to "roll around" over time, rather than present a clean zero crossing to the detection circuit. Unless extra circuitry is added to correct this, the resulting MIDI output will jump repeatedly between the correct note and a false note an octave or two higher.

The short demo video "vibrating_strings" uses a software oscilloscope to show and hear three different wave types: a static sawtooth, a plucked note on an electric bass, and an electric guitar. An oscilloscope usually triggers the start of each horizontal sweep when the wave rises up from zero. This is needed to present a stable waveform display. The sawtooth wave is static, so the oscilloscope can easily lock onto its pitch, and likewise a frequency measuring circuit can easily read the time span between zero crossings. But the harmonics of the bass and guitar notes vary as the notes sustain, and the additional zero crossings make pitch detection very difficult. The electric guitar wave has even more harmonics than the bass, and you can see the waveform dancing wildly. When I built the guitar-to-MIDI converter mentioned earlier, I added a very steep voltage-controlled low-pass filter to remove *all* the harmonics before the circuit that measures the frequency.

Another alternate MIDI input device is the *breath controller*, which you blow into like a saxophone. Unlike a keyboard or guitar controller, a wind controller is not used to play notes. Rather, it sends volume or other MIDI sound-modifying data while a note sustains. Musicians who are fluent with woodwind and brass instruments can use this effectively to vary the volume while they play notes on the keyboard. Some synth patches also respond to a breath controller by varying the timbre. This helps to better mimic a real oboe or trumpet, where blowing harder makes the note louder and also brighter sounding.

One of the most popular alternate MIDI controllers is the *drum pad*. These range from a simple pad you tap with your hand through elaborate setups that look and play just like a real drum set. Modern MIDI drum sets include a full complement of drum types, as well as cymbals and a hi-hat that's controlled by a conventional foot pedal. Most MIDI drum sets are also touch sensitive, so the data they send varies by how hard you strike the pads. Better models also trigger different sounds when you strike a cymbal near the center or hit the snare drum on the rim. Some "pad" controllers instead present one or more rows of plastic push buttons you press with fingers rather than hit with drum sticks.

The last alternate input device isn't really a hardware controller but rather uses the *external audio input* jack available on some synthesizers. This lets you process a guitar or voice—or any other audio source—through the synth's built-in filters and other processors. Of course, you can't apply LFO vibrato to external audio, but you could apply a filter or volume envelope using an ADSR that's triggered each time you play a note on the keyboard.

Samplers

All of the synthesis methods described so far create and modify audio via oscillators, filters, and other electronic circuits. This truly is synthesis, because every aspect of the sound and its character is created manually from scratch. Another popular type of synthesizer is the *sampler*, which instead plays back recorded snippets, or samples, of real acoustic instruments and other sounds. This is sometimes called *wavetable synthesis*, because the sound source is a series of PCM sample numbers representing a digital audio recording. In this case, the "table" is a block of memory locations set aside to hold the wave data. Note the distinction between short recorded samples of musical instruments played by a human versus the rapid stream of samples that comprise digital audio. Both are called "samples," but music and instrument samples are typically several seconds long, where most digital audio samples are 1/44,100 second long.

The earliest samplers were hardware devices that could record sounds from a microphone or line input, store them in memory, and manipulate them in various ways with the playback controlled by a MIDI keyboard. The Synclavier, introduced in 1975, was the first commercial digital sampler I'm aware of, but at $200,000 (and up), it was far too expensive for most musicians to afford. Every hardware sampler is a digital device containing a computer with an operating system of some sort that runs digital audio software. There's no fundamental difference between stand-alone hardware samplers and sampler programs that run on a personal computer.

Early samplers had very limited resources compared to modern types. For example, the E-mu Emulator, first sold in 1981, offered only 128 kilobytes (!) of memory, with a sampling rate of 27.7 KHz at eight bits. This can store only a few seconds of low-quality mono music. To squeeze the most out of these early systems, a concept known as *looping* was devised, which repeats a portion of the recorded note for as long as the key is held. Without looping, every sample Wave file you record would have to extend for as long as the longest note you ever intend to play. The most important part of a musical note is the first half-second or so because it contains the attack portion that defines much of an instrument's basic sound character. Figure 14.10 shows a sampled electric bass note that's been set up in Sound Forge to allow looping within a sampler.

When this Wave file is played in a sampler that recognizes *embedded loop points*, the entire file is played, and then the trailing portion bounded by the Sustaining Loop markers repeats continuously while the synthesizer's key remains pressed. When the key is released, the sustained portion continues to loop until the Release portion of the ADSR completes. Figure 14.11 shows a close-up of just the looped portion.

The start and end loop points are usually placed at zero crossings to avoid clicks. In truth, any arbitrary points on the waveform could be used as long as they both match exactly to

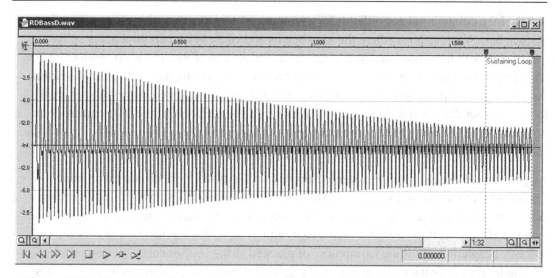

Figure 14.10: This sampled bass note contains a looped region that repeats the last part of the Wave file continuously for as long as the note needs to sound.

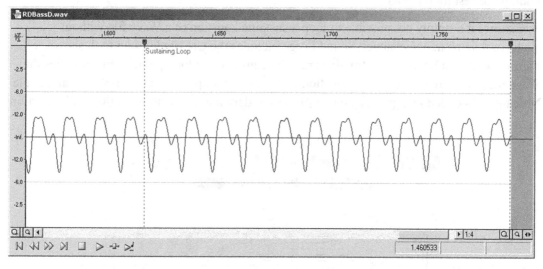

Figure 14.11: This close-up shows the looped portion of the bass note in Figure 14.10. When playback reaches the small upward cycle just before the end of the Wave file, the sampler resumes playing at the beginning of the looped region. This short section of the file continues to repeat for as long as the key is held.

allow seamless repeating. The "sample_looping" video shows this Wave file being played in Sound Forge with loop playback mode enabled. You can see and hear the note as it plays the main portion of the file, then remains in the looped portion until I pressed Stop. Sound Forge includes a set of tools for auditioning and fine-tuning samples to find the best loop points.

Setting loop points is not too difficult for a simple waveform such as this mellow sounding electric bass. You establish where the loop begins and ends, being careful that the splice makes a smooth transition back to the earlier part of the waveform. However, with complex waveforms, such as a string or horn section, or a vocal group singing "Aah," the waveform doesn't repeat predictably within a single cycle. Further, with a stereo Wave file, the ideal loop points for the left and right channels are likely to be different. If the loop ending point does not segue smoothly back to the starting boundary, you'll hear a repeated thump or click each time the loop repeats.

As you can see, looping complex stereo samples well is much more difficult than merely finding zero crossings on a waveform. Many instrument notes decay over time, so you can't just have a soft section repeat back to a louder part. Nor does repeating a single cycle work very well for complex instruments and groups; this sounds unnatural because it doesn't capture the "rolling harmonics" of plucked string instruments. I've done a lot of sample looping over the years, and it's not uncommon to spend 20 minutes finding the best loop point for a single difficult stereo sample. However, I've had good success using an inexpensive program called Zero-X Seamless Looper, which greatly simplifies the process of finding the best loop points.

Sometimes it's simply not possible to loop a sustained portion of a Wave file without clicks or thumps in one channel. Sound Forge includes the Crossfade Loop tool, shown in Figure 14.12, that applies a destructive cross-fade around the loop points. This modifies the Wave file and "forces" a smooth transition when perfect loop points are simply not attainable. Note that embedded loop points are part of the standard *metadata* stored in the header portion

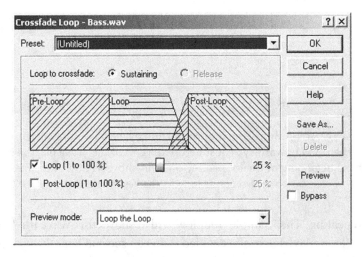

Figure 14.12: The Crossfade Loop tool in Sound Forge lets you force clean loop points by applying a cross-fade directly to the Wave file.

of Wave files, so when those files are imported into Vienna or any other sample creation program, the loop points will be honored.

Also note in Figure 14.12 the Post-Loop section at the right of the screen. I mentioned earlier that a sampler plays the initial portion of the recording when each note's playback begins, and then the looped portion repeats for as long as the note is held, continuing until the ADSR's Release time is exhausted. But looped sample files can also incorporate a third section of the recorded sample, which is played instead of the looped portion during the Release portion of the ADSR envelope. This is useful to add realism to a sampled piano, for example, because a real piano makes a soft mechanical "clunk" sound as the key falls back into place. It's also useful for other instruments such as woodwinds and brass; a note that's ended naturally by a musician sounds different from a sustaining note faded out artificially with a volume control.

Creating a complete high-quality sampled instrument set usually requires a large number of different Wave file recordings. The original "standard" for creating sample sets records every third note the instrument is capable of playing, and then the sampler applies resampling-type pitch shifting to create the in-between notes. If you're sampling a violin whose lowest note is a G at 196 Hz, you'd record that G note, then the Bb a minor third above, then the C# above that, and so forth up to the highest note you want to include. Today, most sample playback is done on computers having gigabytes of memory, and even more gigabytes of hard drive space, so many modern commercial sample libraries include recordings of every note in the instrument's range.

Further, a high-quality sample set requires recording the same notes played at different volume levels. A bassoon played softly has a very different sound quality than when played loudly where the overtones are much more prominent. So you can't just use a single loud recording, then lower the volume to create soft notes. Many sample sets are recorded with the performer playing at two different volume levels—loud and soft—but some include many more in-between levels. This, too, increases the amount of memory and hard drive space needed for a complete sampled instrument. However, with clever programming, a single loud sample can sometimes work. The sampler simply lowers the frequency of a low-pass filter at softer volumes to reduce the harmonic content. The softer you play, the lower the cutoff frequency is set to. This method actually works pretty well for sampled pianos and other percussion instruments, and is a good compromise when available memory is limited.

Software Synthesizers and Samplers

As mentioned, hardware and software samplers are more alike than different, though digital implementations of analog-type synthesizers are very different internally than their analog counterparts, even if outwardly they look and respond the same. Software synthesizers operate in one of two basic modes: as either a stand-alone computer program you play live using a MIDI keyboard or as a plug-in that receives data from a MIDI track in a DAW program. Many software synths provide two versions, so you can use them either way.

A stand-alone program makes sense for playing live concerts, using a laptop computer as the "hardware" platform. But a plug-in version is more practical when using a DAW to create complete productions by recording MIDI data. This way you can record the notes as you would when recording any other type of instrument overdub but with the added benefit of being able to edit wrong notes and make other changes after recording. Plug-in synthesizers can also be run through audio plug-ins in the DAW, such as EQ, compression, reverb, and echo. Not only does it have the ability to change notes and patch sounds, but a plug-in synth also lets you record and edit controller data afterward, such as pitch bend and filter frequency. Of course, a stand-alone synthesizer can be recorded into a DAW program as audio, though that loses much of the flexibility MIDI offers.

Sample Libraries

As mentioned, most commercial sample libraries contain a collection of Wave file recordings of various notes played by an instrument or section or sung performances for sampled solo voices and choirs. These samples are often recorded at two or more volume levels, and they usually contain a looped portion that repeats indefinitely for as long as the key is held. However, some sample libraries don't use looping, with the Wave files instead sustaining for 8 to 10 seconds or even longer. There are many different sample library formats, including AKAI, SoundFont, GigaSampler, Kontact, Roland, Kurzweil, and others. Describing the internal details of every sample format is beyond the intent of this book, so I'll give an overview of the process using the popular SoundFonts format I'm most familiar with. Please understand that the exact same concepts apply to all sample formats.

I group sampled instruments into two broad categories: percussive instruments and sustained instruments. To my way of thinking, percussive instruments are those that create a sound with a single strike that decays naturally over time. This includes drums and bells but also the piano, which, in fact, is officially classified as a percussion instrument. A guitar or plucked bass could also be considered percussive in this context because once a note is struck, the sound eventually decays on its own. Sustained instruments include the violin and other bowed instruments, and clarinets and trombones and other blown instruments. However, some instruments fall into both groups, such as the tambourine, triangle, and maracas. These can be struck once for a percussion effect or shaken repeatedly to sustain indefinitely. The same applies to the tremolo plucking style on a mandolin or a snare drum or timpani roll.

I divide musical instruments into these two categories because it affects how their samples are programmed inside a sample library. Purely percussive instruments do not need to be looped and, if sampled well, are mostly indistinguishable from a "real" recording of a live performance. However, piano notes interact differently when played live versus sampled. When played live with the sustain pedal pressed, a struck piano string excites other strings that were not struck. But by and large, sampled pianos can sound very realistic. And while a continuously shaken tambourine or triangle can be considered a sustained sound, the repeated striking is not usually difficult to loop.

Sampled sustained instruments usually sound less realistic than sampled percussive instruments. To my ears, nothing sounds worse than a sampled saxophone solo. Short sax and brass note "stabs" can often sound acceptable, but sustained passages on a solo clarinet or cello almost always sound fake to an experienced listener. How realistic a sampled performance sounds depends on the type of musical passage being played, the quality of the samples and how well they were looped, and how well the sample reacts to MIDI expression controls. Real violas and flutes can change both their volume and timbre while notes sustain, and it's very difficult to program such changes realistically when using samples. Then again, background violins playing sustained whole notes can usually sound acceptable if their basic recorded tonality is pleasing.

Although samples and synthetic recreations of expressive instruments usually sound pretty poor, you can often get acceptable results by using one recording of a real instrument, along with one or two sampled or synthesized versions. It's best if the real instrument is a little louder in the mix than the others to help further hide the synthetic elements. I've heard sampled brass sections sound pretty good when one real trumpet or sax was prominent in the mix.

Creating Sample Libraries

Figure 14.13 shows the main screen of the Vienna SoundFont Studio, a program included for free with most SoundBlaster sound cards. Years ago, early SoundBlaster cards had a reputation for mediocre sound quality, but modern versions are capable of very high fidelity. The Vienna program requires a SoundBlaster to be present in the computer, but SoundFont files are a universal format usable by most modern software samplers. So once you create a SoundFont bank, it can be played back by DAW software through whatever sound card is attached to your system. Further, SoundFonts are based on standard Wave files, so their fidelity is limited only by quality of the sample recordings they're created from. I have a SoundBlaster card in my computer, though I use it only to create and edit SoundFonts.

Vienna SoundFont Studio shown in Figure 14.13 is a comprehensive sample management program that lets you import and loop Wave files, organize them into banks and presets (patches), specify reverb and other standard MIDI effects, and define split points based on both a range of note pitches and a range of note-on velocities. The MIDI standard allows up to 128 sound banks, each containing up to 128 patches, with all 16,384 patches available for

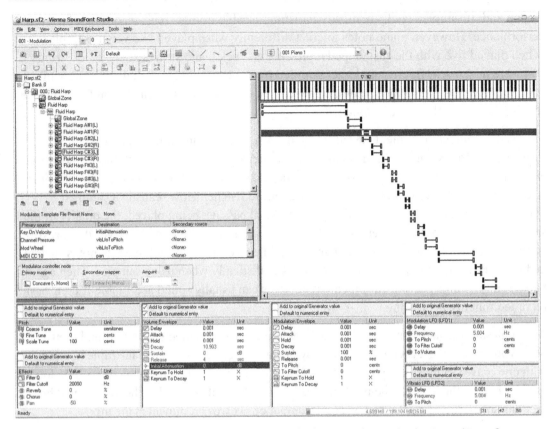

Figure 14.13: The Vienna program that creates and edits sample sets in the SoundFont format is very comprehensive. Besides letting you organize complex groups of instruments into banks and presets, the samples can be programmed to respond to the keyboard and other controllers in every way that's supported by the MIDI standard.

playback at once. Of course, nobody needs that many patches in a single sample set. Further, the amount of memory in your computer will surely limit you to fewer banks and patches, depending on the size of the Wave files used.

The main reason to allow so many banks and patches is for organization. For example, Bank 0 could contain three or four different French horn sample sets, Bank 1 might hold a few different violin section patches, and so forth. But even that type of organization is wasteful of memory, because you'll load every instrument type even if a tune needs only one or two. I organize my own SoundFont collection by instrument categories. Cellos.sf2 contains seven different cellos, Flutes.sf2 has six flutes, and so forth. This way I can load only the instruments I need, then try different versions while the song plays to decide which sounds best for that particular song. I often use two different versions to create a unison section, rather than have three parts all play the same violin patch. This sounds more natural, like different musicians playing different instruments. It also helps to avoid comb filtering when the same note is played by several instruments at once, as is common with string sections.

Creating a custom sample set is reasonably straightforward, if tedious. The biggest hurdle is understanding how the banks and patches are organized. The main Vienna screen in Figure 14.13 is divided into several sections. The upper left area displays a tree view of the currently loaded SoundFont. In this file, Bank 0 contains five different concert harp patches, but only Patch 000 (Fluid Harp) is visible. This patch is opened to show the keyboard zones, or key ranges, that will trigger each Wave file sample. The key ranges for each sample are set in the upper-right portion, though the key range display can be switched to instead show the velocity switch points. This is needed when multiple samples are used to play the same note, with different samples triggered depending on how hard you strike the key. The lower section of the screen is where the various sound modifying parameters are defined, such as ADSR values, filter Q, reverb, LFO rate and intensity, and so forth.

Key and Velocity Switching

As mentioned earlier, ideally a separate sample will be recorded for every note the sampled instrument can play. But that requires a huge amount of memory to store all those samples, not to mention the time and effort to record so many files. Therefore, it's common to record samples at selected intervals—perhaps every two to four half-steps. The samples are then pitch-shifted up or down by the sampler during playback to produce the in-between notes. So instead of recording all 47 notes on a modern concert harp, the sample recording of middle C could serve the range from two notes below middle C through two notes above, and so forth. This particular sampled harp uses 14 separate stereo Wave files, with most used to play a range of only three or four adjacent notes.

If you try to cover too broad a range with a single sample, the notes at each extreme can sound unnatural. This is especially true with instruments that have an inherent body resonance, such as violins and acoustic guitars. The primary pitch of the in-between notes are raised and lowered during playback, but the resonant frequencies of the instrument's body are also shifted. So a low trumpet note might sound more like it was played on a trombone, or a high violin note could sound as if the violin is only five inches long. Another problem when sampling notes at too few intervals is the sudden change in timbre that results when you play a scale as it crosses a sample boundary point. Two adjacent notes on a real harp or cello usually sound similar, so playing an ascending scale flows smoothly from one note to the next and doesn't suddenly change tonality. But a note that's been shifted up in pitch by several whole steps sounds very different from another note sampled an octave away that's now shifted down several whole steps. In that case, playing one note after the other yields a large change in tonality and sounds fake.

MIDI sample sets can contain as many or as few Wave files as you'd like for each patch. You define which samples will play for single notes or note ranges and also which will play at different volumes of the same notes or note ranges, depending on the MIDI key velocity. Most MIDI data ranges from 0 through 127, for a total range of 128 values. So note-on velocities can range from 0 (silence) through 127 (maximum loudness). However, some systems consider the range to be 1 through 128. Again, most musical instruments have a different quality when played loudly versus softly, so you might use one sample of a snare drum that was played softly for note velocities of 0 through 70 and then switch to another sample of the same drum struck harder for higher velocities.

You can also set an *exclusive class* for note groups so that playing one note automatically mutes another in the same group. The classic example is when programming hi-hat samples. When a closed hi-hat sample is triggered, it should immediately turn off the open hi-hat sample if that's currently sounding. Otherwise, you'd have both samples playing at once, which sounds phony. To program one or more samples to mute all the others that are related, assign them the same class number. The SoundFont format supports up to 127 exclusive class groups, though it's doubtful you'd ever need more than two or three.

I prefer to loop Wave files in Sound Forge because of its superior tools and its destructive cross-fade option when finding clean loop points is not otherwise possible. But plain looping without cross-fades can be set up entirely in Vienna, as shown in Figure 14.14.

Figure 14.15 shows the same information as the bottom portion of Figure 14.14 but zoomed in to better see the splice details. As you can see in this close-up, sample number 16,753 has just descended through zero, and sample number 11,847 (earlier in the Wave file) resumes at the same level and continues negative. If you click the Up and Down buttons next to the Local Loop End and Start values, the displays scroll horizontally helping you find the ideal boundaries. You can click the Play Loop button at any time to hear the looped portion of the Wave file played back with the current loop points, using the Key Number field to specify which MIDI note triggers the sample.

Sampler Bank Architecture

Sample banks are organized in a tree-like structure, as shown at the upper left of the screen in Figure 14.13. The basic building block is the samples that are imported as Wave files, and the banks and patches you call up in a DAW or sequencer program are built from those Wave files. To create a new sample set, you'll use File.. New, then browse to find and import the Wave files you recorded and optionally already looped. You can import a single Wave file or an entire group of files at once, and Vienna automatically creates a new Bank 0 with a new Patch 000 that contains all of those files. You can then specify the range of notes each sample will respond to, as well as the *root note* for that sample. For example, middle C is MIDI note

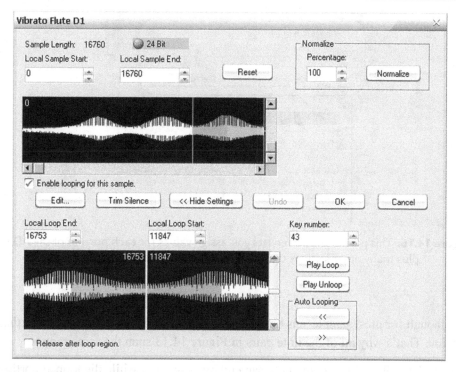

Figure 14.14: Vienna's Looping screen shows an overview of the Wave file at the top, as well as the transition between the loop end (left) and start (right) at the bottom.

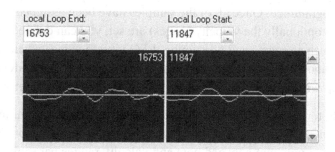

Figure 14.15: This close-up of the loop points from Figure 14.14 is zoomed in to show the individual cycles and samples, to help find the best transition.

number 60, and the A note 1½ octaves lower is 45. So if you have a sample of an A bass note at 110 Hz and want it to play the range from G below through B above, you'll specify the root note as 45, then slide the markers in the upper right of the screen in Figure 14.13 to span only that range. The harp sample highlighted in Figure 14.13 is a C# note, and it's set to play from the B below C# to the D above. Stereo Wave files are treated as separate left and right

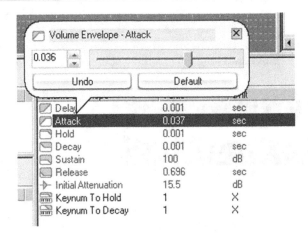

Figure 14.16: This part of the screen lets you assign values for each portion of the ADSR, plus many other settings that determine how a sample is to be played.

samples, though for most sample sets both channels will use identical settings other than their pan position. That's why all of the note pairs in Figure 14.13 span the same ranges.

For samples that will be looped, the note can end by fading out while the looped portion repeats or set to resume playing after the looped region to reproduce the natural sound of the note stopping. Of course, this assumes the sample recording includes the note's natural decay when the musician stopped and that the Wave file was set to play the end portion when the loop points were programmed. Once all of the samples have been assigned to a patch and the note ranges (and optionally the velocity ranges) are set, you can define a large number of effects parameters such as coarse and fine tuning, LFO vibrato values when the mod wheel is used, and envelope and filter ADSR settings. Figure 14.16 shows the ADSR section of Vienna and the little pop-up window that appears when you click to edit a parameter.

Again, most sample formats include features that are the same as or similar to SoundFonts, though some of the nomenclature might be different. But the basic concepts of looping Wave files and assigning root note numbers and note and velocity ranges per sample are the exactly same, as is organizing samples into patches and banks accessed by name in your DAW program.

FM Synthesis

FM synthesis was invented by John Chowning at Stanford University in the 1960s. The patent was later licensed to Yamaha, which produced the DX7, the first commercially successful FM synthesizer. Today, FM synthesis is a popular staple available on many hardware and software synthesizers. The harmonics of the square, pulse, and sawtooth waves used by an analog synthesizer mimic the overtone series of acoustic instruments and most other sounds that occur in nature. FM synthesis can generate conventional harmonics, but it can also create inharmonic overtones, yielding a sound reminiscent of bells and gongs. Coupled with ADSRs to vary the volume, or the frequency of a traditional analog type filter, FM synthesizers can create many unusual and musically interesting sounds.

Earlier I mentioned that when one oscillator controls the frequency of another, the controlling oscillator typically runs at a low frequency to create vibrato. Figure 14.17 shows a 200 Hz sine wave with frequency modulation applied at a slow rate to create vibrato. You can see the wave cycles compress and expand as the wave's frequency changes over time.

When the modulating frequency rises above 20 Hz or so, passing into the audible range, the sum and difference side bands added to the carrier are perceived as changes in timbre rather than as vibrato. In truth, these side bands are always created, even at slow modulating frequencies, but increasing the vibrato rate progressively raises their volume, making them more audible. Figure 14.18 shows the same 200 Hz sine wave but with a modulation frequency of 100 Hz. Here you can see that the basic shape of the waveform has changed, which, of course, affects its tone quality. So it's no longer a pure tone containing only 200 Hz.

In FM-speak, the oscillator that creates the main tone you hear is called the *carrier*, while the oscillator that applies the vibrato is a *modulator*. These oscillators are usually called *operators* when describing how a patch is constructed, and an arrangement of oscillators that modulate one another in various ways is called an *algorithm*. When an ADSR is applied to the modulator's volume, the tone color changes over time in a way reminiscent of sweeping a filter. The "fm_synthesis" video shows simple vibrato added to a 200 Hz sine wave, with the vibrato frequency increased from 1 Hz up into the audible range. Then the amount of

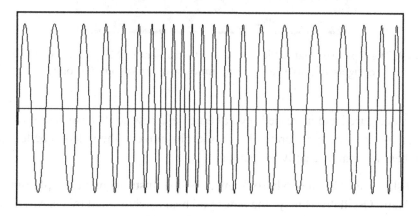

Figure 14.17: Applying FM at a slow rate creates vibrato, which repeatedly sweeps the frequency higher and lower.

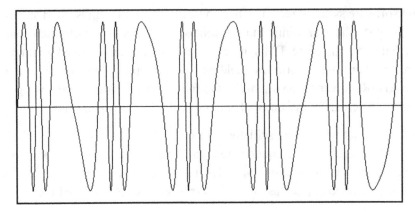

Figure 14.18: When FM is applied at a fast rate, frequency shifts occur within single cycles of the carrier, creating interesting tone colors.

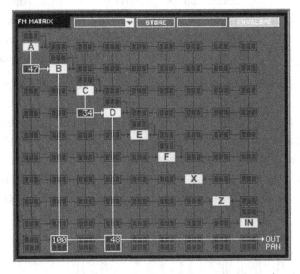

Figure 14.19: The Operator Matrix View of Native Instrument's FM7 plug-in synthesizer lets you configure multiple carriers and modulators that sound at once.

modulation is decreased and increased several times in a row so you can hear how that affects the tone color. You can see the vibrato plug-in's knobs move in the video to better relate what you see and hear to what's actually happening.

Figure 14.19 shows the Operator Matrix View of Native Instrument's FM7 plug-in synthesizer. Here, Oscillator B plays the notes you hear as controlled by the MIDI keyboard, while Oscillator A modulates the frequency of Oscillator B. The modulation depth (amount) ranges from zero to 100 and is set to 47 in this patch. At the same time, Oscillator D also plays notes controlled by the keyboard, with Oscillator C serving as a modulator with a bit less depth. Audio oscillators B and D are mixed with 100 percent of B to 48 percent of A to create the complete sound.

Having two sets of oscillators creates a sound that's even more complex than a single oscillator pair, because the ratio of oscillator frequencies can be different for the two pairs. Different ADSR-type envelopes can be applied to all four oscillators to change the overall volume as usual and also to change the modulation amount over time. Further, modulators can be integer-related to the carrier or not. An A note at 220 Hz modulated by an A an octave below at 110 Hz sounds very different from an A-220 modulated by a C note at 130.8 Hz. A modulator oscillator can also follow the keyboard to rise and fall in pitch as different notes are played, or it can remain at a constant frequency.

FM synthesis is a very deep subject, and operator matrices are often much more complex than the simple algorithm shown in Figure 14.19. Oscillators can modulate their own frequency, and one modulating oscillator can modulate another, which in turn modulates the audio oscillator that creates the musical notes. The number of possible combinations is truly staggering. The best way to learn more about how FM algorithms affect the sound quality is to explore the factory presets in whatever FM synthesizer you happen to own. But I'll also mention synth expert Jim Aiken's online article[1] that gives an excellent overview of some common algorithms and includes audio examples.

Physical Modeling

Physical modeling is a form of additive synthesis, but instead of adding and filtering multiple waveforms, modeled sounds are derived entirely through mathematical calculations. The first commercial physical modeling synthesizer I'm aware of was Yamaha's VL1 in 1994, which was soon followed by the more affordable VL70M in 1996. Rather than simply sum a number of static sine waves, physical modeling uses complex equations that mimic the sound sources, body resonances, and other attributes of acoustic instruments. I remember hearing the VL1 when it was demonstrated at the 1995 New York AES show. What struck me most was the realism of the trumpet patch as sustained notes transitioned from one pitch to another.

Unlike samples that can play only one static note followed by another, physical modeling more realistically creates the sounds that real instruments make *in between* successive notes. When a trumpet or clarinet plays a legato passage, the player's breath continues uninterrupted for the duration of the passage, even as the notes change. But sampled notes start with a new breath. Further, real instruments often create subtle sounds as the notes transition from one to the other. Sampled strings behave similarly. Real violinists often play several notes in a row without reversing the bow direction, and they sometimes slide from one note to another note—an effect called *glissando*. But with string samples, each new note begins with a new bow stroke as the old note ends. This is acceptable for staccato passages, where each note is short and clearly defined. But for flowing passages, samples of instruments such as the trombone and cello almost always sound artificial because real musicians don't play that way.

Physical modeling circumvents the limitations of sampled acoustic instruments by calculating a mathematical representation of the vibration of a physical string or drum head and how it interacts with the instrument's sound chamber and other sources of resonance. An accurate physical model of a flute will have a different harmonic structure when overblown, just like a real flute, and a plucked string will go slightly sharp and add a "splat" sound when played very hard. As you can imagine, physical modeling is a complex process that requires many extensive computations. In my opinion, physical modeling is the final frontier of electronic synthesis—at least the type of synthesis that aims to recreate the sound of real musical instruments.

Granular Synthesis

Granular synthesis joins together short sound fragments into a longer whole. The basic concept is a sequence of many very short sounds that are strung together in various ways to create new tone qualities. Granular synthesis is not unlike *musique concréte*, a twentieth-century composition style that was often created from a montage of disparate sounds by splicing together short pieces of analog recording tape. But with modern digital implementations of granular synthesis, the sounds can be divided into even smaller segments, called *grains*. These typically range in length from 1 to 50 milliseconds. Granular synthesis often employs sampled music as the sound source, but any sounds can be used, including static square waves and the like, or even sounds of nature. The sounds can be played one after the other or morphed from one to another, or several sounds can be played at once.

Prerendering

Many people use MIDI synthesizers in their DAW audio projects, or software samplers to augment or substitute for real instruments they may not know how to play. Many of my own tunes include software synthesizers, and I use samplers all the time for drums, keyboards, and woodwinds.

One important attribute of software synthesizers is how much computing power they require. The number of notes playing simultaneously at any moment usually affects this as well. The common term for synthesizers and other audio processes that require a lot of computer resources is *CPU intensive*. Here, CPU refers to the *central processing unit*, the heart and brains of every personal computer. A software synthesizer that plays a simple square wave with only volume changes over time is not very CPU intensive, compared to a complex patch having two sweeping filters, three ADSR envelopes, plus five Wave files samples that cross-fade from one to the other as each note sustains. If you have a large number of tracks, each with a complex patch playing several notes at once, at some point the computer's CPU will not be able to perform all the calculations quickly enough to play the song in real time.

As computers have become more powerful, overtaxing the CPU is less of a problem today. Then again, it's a truism that software increases in complexity to take advantage of more powerful hardware. The solution for synthesizers that are highly CPU intensive is to pre-render the audio to a new Wave file. Most DAWs can do this automatically. In SONAR this process is called *freezing* a synth track, and there are many available options. Freezing a synthesizer track is not difficult even if your DAW doesn't offer an automatic method. You solo the synth track, then export the audio to a new Wave file. It doesn't matter if the computer can play that synthesizer's audio in real time. The render takes as long as it takes to complete. Then you load the resulting Wave file to a new audio track and mute or disable the synthesizer's MIDI and audio tracks so they don't require further computation.

Algorithmic Composition

Algorithmic composition isn't a synthesis device or method, but it's used exclusively with synthesizers, so it fits well in this chapter. The simplest form of computer-generated performance is the arpeggiator. Early versions were designed as plug-ins for MIDI sequencer programs to create arpeggios and other simple musical patterns, without the performer having to play all the notes. The idea is you'll sustain a single chord, and the arpeggiator plays the notes of that chord individually up or down over some number of octaves. The "arpeggiator.wav" demo file plays a few bars from one of my pop tunes as I recorded it through a Fender Rhodes sample set, then again after adding the Arpeggiator MIDI plug-in shown in Figure 14.20.

Another, more advanced type of "composing" MIDI plug-in is the drum pattern generator, based on the programmable drum machines that became popular in the 1970s. These range from simple types useful for little more than a metronome, through programs that play sophisticated patterns using high-quality sampled drum sets. But algorithmic composition can go far beyond simple one-instrument musical patterns. Indeed, the concept of using computers to compose music has been around for many years.

Beyond arpeggios and drum patterns, programs are available that create complete MIDI backing tracks for songs in many different musical styles. This is a useful way for songwriters to record their original tunes without having to hire musicians or a recording studio, or to get inspiration from a polished accompaniment while trying out different lyrics. It's also great for singers to make their own karaoke backing tracks or for musicians who just want to create music to play along with for fun.

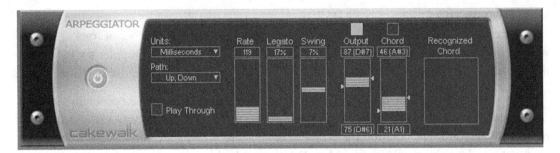

Figure 14.20: The Cakewalk Arpeggiator offers a number of ways to create arpeggiated note patterns. It reads the chord to know what notes to play, then generates new notes at the tempo, and within the note range, you specify. The Legato control determines how long the notes play for, ranging from choppy to very smooth. It can even add a swing feel to the patterns.

One of the most popular *auto-arranger* programs is Band-in-a-Box (BIAB) from PG Music, available for both Windows and Mac computers. With this type of program, you enter the chords for your song, which can change as often as every eighth note, then choose a song style. BIAB offers literally hundreds of styles to choose from, which the program then uses to generate bass, drum, keyboard, guitar, and other instrument parts. Once all of the chords have been entered, you can try different style patterns, and you can also overdub melodies as either MIDI or Wave files.

Another popular auto-arranger program is SoundTrek's Jammer, for Windows only, shown in Figure 14.21. Jammer includes fewer styles than BIAB, but it offers more ways to customize and control the music it generates. Where BIAB is mostly pattern-based, applying set patterns onto the chord changes you enter, Jammer uses intelligent algorithms to generate original performances. To my ears, the music Jammer creates seems a little more hip, though Band-in-a-Box has a loyal following, too.

Many keyboard hardware synthesizers also offer auto-arranger features, from inexpensive Casio models to high-end workstations from Korg and Roland. With these keyboards you play single notes in the bass range with your left hand, and the keyboard generates music in various styles while you play melodies with your right hand. If you play only one note, the keyboard assumes a major chord at that note for the music it creates. But you can play other chord types with your left hand to specify minor chords, seventh chords, and so forth. The music and sound quality from some of these keyboards can be very good!

Notation Software

Another popular type of music-making computer program is notation software, used mainly by composers and musicians who need to create printed music for performers to play. All

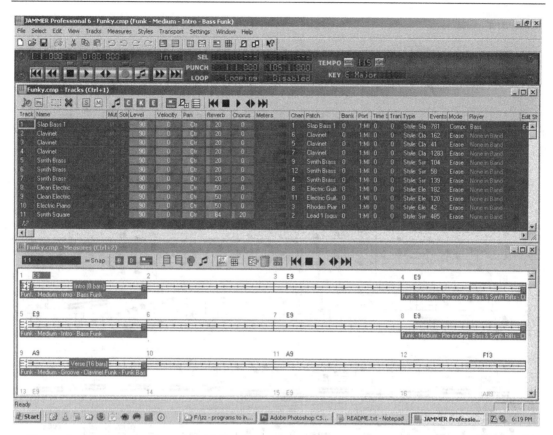

Figure 14.21: SoundTrek's Jammer is a complete song creation environment, and the music it generates can be surprisingly realistic.

notation programs can import standard MIDI files, so you can start a project in your DAW as a MIDI sequence playing virtual instruments, then import the MIDI tracks into notation software to publish printed sheet music. Modern notation software offers many of the features of a full MIDI sequencer, such as changing instrument sounds to experiment while composing, varying the tempo, or trying a part an octave higher or lower. You can also change the printed (only) key automatically for trumpets, French horns, and other transposing instruments.

Orchestra and big band music is often written as a full score using software samplers to play the various parts, and then the notation software extracts each instrument part to a separate file for more refined formatting and printing. The flexibility offered by these programs is impressive, and you can spend a lot of time making things look as perfect as you want. It took me about a month working full time to typeset the score and 26 individual orchestra parts for my cello concerto, and Figure 14.22 shows one page of the solo cello part. As you

Figure 14.22: Modern notation software can create a full orchestra score with publication quality, then extract all the individual parts to separate files automatically.

can see, all of the standard music notation conventions and symbols are supported, including dynamic markings, fingerings, clef types, and natural and artificial harmonics, and every element can be sized and placed precisely on the page.

Summary

This chapter covers a lot of territory, including detailed explanations of popular synthesis methods. Along the way, I show the internal details of analog synthesizers, including sound generators, modulators, and filters. A comparison of additive and subtractive synthesis methods is presented, as well as an explanation of digital synthesis types including FM, physical modeling, and granular synthesis. In addition, you learned about Bob Moog's clever use of DC voltages to control oscillator and filter frequencies, and other synthesizer parameters.

This chapter also includes an in-depth explanation of MIDI keyboards and alternate control devices such as guitar and breath controllers. Samplers are also covered, including an overview of sample libraries and the way they employ key and velocity switching to increase realism. Software and hardware auto-accompaniment products are described, along with a brief overview of notation software.

Note

1   "Fee, Fi, Fo, FM: Explore the World of FM Synthesis" by Jim Aiken.
http://archive.oreilly.com/pub/a/oreilly/digitalmedia/2006/04/12/fm-synthesis-tutorial.html

CHAPTER 15

MIDI Basics

MIDI Internal Details

MIDI stands for Musical Instrument Digital Interface. It was developed by a consortium of musical product manufacturers to create a common language and protocol for musical software and hardware to communicate with each other. Before MIDI, the keyboard of one synthesizer brand could not send note data to another brand of synthesizer, nor was there a standard way to connect sound modules to the early computers of the day. I recall a session in 1982 when composer Jay Chattaway recorded the music for the movie *Vigilante* at my studio. The music was hard hitting and created entirely with early hardware synthesizers and samplers. For the week of sessions Jay brought to my studio a very elaborate synthesizer setup, including a custom computer that played everything at once. It was a very complex and expensive system that took many hours just to connect and get working.

MIDI changed all that. The original MIDI Specification 1.0 was published in 1982, and while it still carries the 1.0 designation, there have since been many additions and refinements. It's amazing how well this standard has held up for so long. You can connect a 1983 synthesizer to a modern digital audio workstation (DAW), and it will play music. Indeed, the endurance of MIDI is a testament to the power of standards. Even though the founding developers of the MIDI Manufacturers Association (MMA) competed directly for sales of musical instrument products, they were able to agree on a standard that benefited them all. If only politicians would work together as amicably toward the common good.

As old as MIDI may be, it's still a valuable tool because it lets composers experiment for hours on end without paying musicians. MIDI data is much smaller than the audio it creates because most commands are 2- or 3-byte instructions. Compare this to CD-quality stereo audio that occupies 176,400 bytes per second. When I write a piece of music, I always make a MIDI mockup in SONAR first, even for pieces that will eventually be played by a full orchestra (heck—*especially* if they will be played by a real orchestra). This way I know exactly how all the parts will sound and fit together, and avoid being embarrassed on the first day of rehearsal. Imagine how much more productive Bach and Haydn might have been if they had access to the creative tools we enjoy today.

MIDI is also a great tool for practicing your instrument or just for making music to play along with for fun. Google will find MIDI files for almost any popular or classical music you want. You can load a MIDI file into your DAW program, mute the track containing your instrument, and play along. Further, a MIDI backing band never makes a mistake or plays out of tune, and it never complains about your taste in music either.

MIDI Hardware

The original MIDI communications spec was a hardware protocol that established the data format and connector types. A five-pin female DIN connector is used on the MIDI hardware, with corresponding male connectors at each end of the connecting cables. (DIN stands for Deutsche Industrie-Normen, a German standards organization.) Although MIDI was originally intended as a way for disparate products to communicate with one another, it's since been adapted by other industries. For example, MIDI is used to control concert and stage lighting systems and to synchronize digital and analog tape recorders to computer DAW programs via SMPTE as described in Chapter 6. MIDI is also used to send performance data from a hardware control surface to manipulate volume and panning and plug-in parameters in DAW software.

Figure 15.1 shows the standard arrangement of three MIDI jacks present on most keyboard synthesizers. The MIDI In jack accepts data from a computer or other controlling device to play the keyboard's own built-in sounds, and the MIDI Out sends data as you press keys or move the mod wheel, and so forth. The MIDI Thru port echoes whatever is received at the MIDI In jack to pass along data meant for other synthesizers. For example, a complex live performance synth rig might contain four or even more keyboard synths and sound modules, each playing a different set of sounds. So a master keyboard could send all the data to the first module in the chain, which passes that data on to all the others using their Thru connectors.

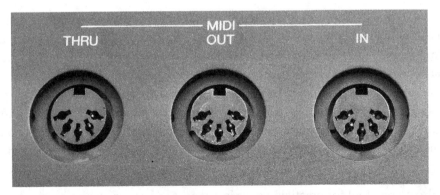

Figure 15.1: MIDI hardware uses a 5-pin DIN connector. Most MIDI devices include three ports for input, output, and pass-through (Thru) for chaining multiple devices.

MIDI Channels and Data

A single MIDI wire can send performance and other data to more than one synthesizer, or it can play several different voices at once within a single synth. In order to properly route data on a single wire to different synthesizers or voices, most MIDI data includes a *channel number*. This number ranges between 0 and 15, representing channels 1 through 16, with the channel number embedded within the command data. For example, a note-on message to play middle C (note 60) with a velocity of 85 on Channel 3 sends three bytes of data one after the other, as shown in Table 15.1. MIDI channel numbers are zero-based, so the 2 in 92 means the note will play through Channel 3.

There are many types of MIDI data—note-on, note-off, pitch bend, sustain pedal up or down, and so forth—and most comprise three bytes as shown in Table 15.1. The first byte is treated as two half-bytes, sometimes called "nybbles" (or "nibbles"), each holding a range of 0 through 15 (decimal, or 0-F Hex). The higher (left-most) nybble contains the command, and the lower nybble holds the channel number the message is intended for. The next two bytes contain the actual data, which usually ranges from 0 through 127 (7F Hex) for each byte.

Some MIDI messages contain only two bytes when the data portion can be expressed in one byte. For example, aftertouch requires only a single byte, so the first byte is the command and channel number, and the second byte is the aftertouch data value. Likewise, a program change that selects a different voice for playback requires only two bytes. The program change command and channel number occupy one byte, and the new voice number is the second byte. Some MIDI messages require two bytes for each piece of data to express values larger than 127. The pitch bend wheel is one example, because 127 steps can't express enough in-between values to give adequate pitch resolution. So pitch bend data is sent as the command with channel number in one byte, followed by the lower value data byte, then the higher value data byte.

Every note sent by a MIDI keyboard includes a channel number, which usually defaults to Channel 1 unless you program the keyboard otherwise. If one keyboard is to play more than one voice, or control more than one physical sound module, you'll tell it to transmit on different channels when you want to play the other voices or access other sound modules.

Table 15.1: The MIDI Note-On Command

Binary	Hex	Explanation
1001 0011	92	9 = Note-on command, 2 = Channel 3
0011 1100	3C	3C Hex = 60 Decimal = Note number
0101 0101	55	55 Hex = 85 Decimal = Velocity

Some keyboards have a split feature that sends bass notes below, say, middle C out through one channel and higher notes out another. Likewise, each receiving device must be set to respond to the specific channels to avoid accidentally playing more than one voice or sound module at a time. Of course, if each synthesizer you'll play contains its own keyboard, this complication goes away.

Receiving devices can also be set to omni mode, in which case they'll respond to all incoming data no matter what channels it arrives on. This is common when recording to a MIDI track in a DAW program because it avoids those embarrassing head-scratching moments when you press a key and don't understand why you hear nothing. In omni mode, the incoming channel is ignored, so the receiving synthesizer will respond no matter which channel the sending device specifies. However, omni mode may not be honored when recording or playing a drum set through MIDI because many MIDI drum synths respond only to Channel 10.

Note that the term "channel" can be misleading, compared to TV and radio channels that are sent over different frequencies all at once. With MIDI data, the same wires or internal computer data paths are used for all channels. A header portion of the data identifies the channel number so the receiving device or program knows whether to honor or ignore that data. If you're familiar with data networking, the channel is similar to an IP address, only shorter.

MIDI Data Transmission

The original hardware MIDI standard uses a serial protocol, similar to the serial ports on older personal computers used to connect modems and early printers. Newer devices use USB, which is also a type of serial communication. Serial connections send data sequentially, one bit after the other, through a single pair of wires. So if you play a chord with eight notes at once, the notes are not sent all at the same time, nor are they received at the same time by the connected equipment.

Early (before USB) MIDI hardware uses a speed of 31,250 bits per second—called the *baud rate*—which sends about three bytes of data per millisecond. But most MIDI messages comprise at least three bytes each, so the data for that eight-note chord will be spread out over a span of about eight milliseconds by the time it arrives at the receiving synthesizer or sound module. For a solo piano performance, small delays like this are usually acceptable. But when sending MIDI data for a complete piece of music over a single MIDI cable, the accumulated delays for all the voices can be objectionable.

Imagine what happens on the downbeat of a typical pop tune as a new section begins. It's not uncommon for 20 or more notes to play all at once—the kick drum, a cymbal crash, 5 to

10 piano notes, a bass note, another five to 10 organ notes, and maybe all six notes of a full guitar chord. Now, that 8-millisecond time span has expanded to 20 or even 30 milliseconds, and that will surely be noticed. The effect of hearing notes staggered over time is sometimes called flamming, after the "flam" drum technique where two hits are played in rapid succession.

Today, with software synthesizers and samplers that run within a DAW, MIDI messages are passed around inside the computer as fast as the computer can process them, so serial hardware delays are no longer a problem. But you still may occasionally want to move time-critical MIDI data between hardware synthesizers or between a hardware synth and a computer. I learned a great trick in the 1990s, when I wanted to copy all the songs in my Yamaha SY77 synthesizer to a MIDI sequencer program to edit the music more efficiently on my computer.

The SY77 is an all-in-one workstation that includes a 16-voice synthesizer using both samples and FM synthesis, several audio effects, plus a built-in sequencer to create and edit songs that play all 16 voices at once. Rather than play each song in real time on the SY77 while capturing its MIDI output on the computer, I set the SY77's tempo to the slowest allowed. I think that was 20 beats per minute (BPM). So the data for a song whose tempo is 120 BPM is sent from the synthesizer at a rate equivalent to six times faster than normal, thus reducing greatly the time offset between notes when played back on a computer at the correct tempo.

General MIDI

As valuable as MIDI was in the early days, the original 1.0 version didn't include standardized voice names and their corresponding program numbers. The computer could tell a synthesizer to switch to Patch number 14, but one model synth might play a grand piano, while the same patch number was a tuba on another brand. To solve this, in 1991 the MIDI spec was expanded to add General MIDI (GM), which defines a large number of voice names and their equivalent patch numbers. The standard General MIDI voice names and patch numbers are listed in Table 15.2. Note that some instrument makers consider these numbers as ranging from 1 to 128. So the tenth patch in the list will always be a glockenspiel, but it might be listed as Patch 10 instead of 9.

Of course, how different sound modules respond to the same note data varies, as do the basic sound qualities of each instrument. So a purely MIDI composition that sounds perfect when played through one synthesizer or sound module might sound very different when played through another brand or model, even if the intended instruments are correct. A soft drum hit might now be too soft or too loud, and a piano that sounded bright and clear when played at

Table 15.2: Standard MIDI Patch Definitions

Number	Patch Name	Number	Patch Name
0	Acoustic grand piano	64	Soprano sax
1	Bright acoustic piano	65	Alto sax
2	Electric grand piano	66	Tenor sax
3	Honkytonk piano	67	Baritone sax
4	Electric piano 1	68	Oboe
5	Electric piano 2	69	English horn
6	Harpsichord	70	Bassoon
7	Clavichord	71	Clarinet
8	Celesta	72	Piccolo
9	Glockenspiel	73	Flute
10	Music box	74	Recorder
11	Vibraphone	75	Pan flute
12	Marimba	76	Blown bottle
13	Xylophone	77	Shakuhachi
14	Tubular bells	78	Whistle
15	Dulcimer	79	Ocarina
16	Drawbar organ	80	Lead 1 (square)
17	Percussive organ	81	Lead 2 (sawtooth)
18	Rock organ	82	Lead 3 (calliope)
19	Church organ	83	Lead 4 (chiff)
20	Reed organ	84	Lead 5 (charang)
21	Accordion	85	Lead 6 (voice)
22	Harmonica	86	Lead 7 (fifths)
23	Accordion	87	Lead 8 (bass and lead)
24	Acoustic guitar (nylon)	88	Pad 1 (new age)
25	Acoustic guitar (steel)	89	Pad 2 (warm)
26	Electric guitar (jazz)	90	Pad 3 (polysynth)
27	Electric guitar (clean)	91	Pad 4 (choir)
28	Electric guitar (muted)	92	Pad 5 (bowed)
29	Overdriven guitar	93	Pad 6 (metallic)
30	Distortion guitar	94	Pad 7 (halo)
31	Guitar harmonics	95	Pad 8 (sweep)
32	Acoustic bass	96	FX 1 (rain)
33	Electric bass (finger)	97	FX 2 (soundtrack)
34	Electric bass (pick)	98	FX 3 (crystal)
35	Fretless bass	99	FX 4 (atmosphere)
36	Slap bass 1	100	FX 5 (brightness)
37	Slap bass 2	101	FX 6 (goblins)
38	Synth bass 1	102	FX 7 (echoes)
39	Synth bass 2	103	FX 8 (sci-fi)
40	Violin	104	Sitar
41	Viola	105	Banjo
42	Cello	106	Shamisen
43	Contrabass	107	Koto
44	Tremolo strings	108	Kalimba

Number	Patch Name	Number	Patch Name
45	Pizzicato strings	109	Bagpipe
46	Orchestral harp	110	Fiddle
47	Timpani	111	Shanai
48	String ensemble 1	112	Tinkle bell
49	String ensemble 2	113	Agogo
50	Synth strings 1	114	Steel drums
51	Synth strings 2	115	Woodblock
52	Choir Aaahs	116	Taiko drum
53	Voice Oohs	117	Melodic tom
54	Synth voice	118	Synth drum
55	Orchestra hit	119	Reverse cymbal
56	Trumpet	120	Guitar fret noise
57	Trombone	121	Breath noise
58	Tuba	122	Seashore
59	Muted trumpet	123	Bird tweet
60	French horn	124	Telephone ring
61	Brass section	125	Helicopter
62	Synth brass 1	126	Applause
63	Synth brass 2	127	Gun shot

a medium velocity through one sound module might now sound muffled or brash at the same velocity on another sound module.

Fortunately, many modern samplers contain several different-sounding instruments within each type and offer more than one type of drum set to choose from. So Patch 000 in Bank 0 will play the default grand piano, while the same patches in Banks 1 and 2 have pianos that sound different. Indeed, this is yet another terrific feature of MIDI: Unlike audio Wave files, MIDI data is easy to edit. All modern MIDI-capable DAW software lets you select a range of notes and either set new velocities or scale the existing velocities by some percentage. Scaling is useful to retain the variance within notes, where a musical phrase might get louder, then softer again. Or you could set all of the notes in a passage to the same velocity, not unlike applying limiting to an audio file.

In addition to defining standard patch names and numbers, General MIDI also established a standard set of note names and numbers for all of the sounds in a drum set. Table 15.3 shows the standard GM assignments that specify which drum sounds will play when those notes are struck on the keyboard or sent from a sequencer program via MIDI. This also helps to ensure compatibility between products from different vendors. As with GM voices, the basic tonal character of a given drum set, and how it responds to different velocities, can vary quite a lot. Further, some drum sets are programmed to cut off the sound when you release the key, while others continue sounds to completion even if the note length played is very short. But with GM, at least you know you'll get the correct instrument sound, if not the exact timbre or responsiveness.

Table 15.3: Standard GM Drum Assignments

Note Number	Drum Sound	Note Number	Drum Sound
35	Bass drum 2	59	Ride cymbal 2
36	Bass drum 1	60	High bongo
37	Snare side stick	61	Low bongo
38	Snare drum 1	62	Muted high conga
39	Hand clap	63	Open high conga
40	Snare drum 2	64	Low conga
41	Low tom 2	65	High timbale
42	Closed hi-hat	66	Low timbale
43	Low tom 1	67	High agogo
44	Pedal hi-hat	68	Low agogo
45	Mid tom 2	69	Cabasa
46	Open hi-hat	70	Maracas
47	Mid tom 1	71	Short whistle
48	High tom 2	72	Long whistle
49	Crash cymbal	73	Short guiro
50	High tom	74	Long guiro
51	Ride cymbal 1	75	Claves
52	Chinese cymbal	76	High wood block
53	Ride bell	77	Low wood block
54	Tambourine	78	Mute cuica
55	Splash cymbal	79	Open cuica
56	Cowbell	80	Muted triangle
57	Crash cymbal 2	81	Open triangle
58	Vibraslap	82	Shaker

In addition to standards for voice names and numbers and drum sounds, MIDI also defines a standard set of continuous controller (CC) numbers. The standard continuous controllers are shown in Table 15.4, though not all are truly continuous. For example, the sustain pedal, CC #64, can be only On or Off. So any value of 64 or greater is considered as pushing the pedal down, and any value below 64 releases the pedal. These standard controllers are recognized by most modern software and hardware synthesizers, and some synthesizers recognize additional CC commands specific to features of that particular model.

Standard MIDI Files

Besides all the various data types, the MIDI standard also defines how MIDI computer files are organized. There are two basic types of MIDI file: Type 0 and Type 1. Type 0 MIDI files are an older format not used much anymore, though you'll occasionally find Type 0 files on the Internet when looking for songs to play along with. A Type 0 MIDI file doesn't distinguish between instrument tracks, lumping all of the data together onto a single track.

Table 15.4: Standard MIDI Continuous Controller Assignments

CC#	Name
0	Bank select
1	Modulation wheel
2	Breath controller
4	Foot controller
5	Portamento time
6	Data entry
7	Main volume
10	Pan
11	Expression
32–63	LSB for controllers 0–31 when two bytes are needed
64	Sustain pedal
65	Portamento
66	Sostenuto pedal
67	Soft pedal
98	Non-registered parameter number LSB
99	Non-registered parameter number MSB
100	Registered parameter number LSB
101	Registered parameter number MSB
102–119	Undefined
121	Reset all controllers
122	Local control
123	All notes Off
124	Omni mode Off
125	Omni mode On
126	Mono mode On
127	Poly mode On

In a Type 0 file, the only thing that distinguishes the data for one instrument from another is the channel number. Fortunately, most modern MIDI software expands Type 0 files to multiple tracks automatically when you open them.

Type 1 MIDI files can contain multiple tracks, and that's the preferred format when saving a song as a MIDI file. Regardless, both MIDI file types contain every note and its channel number, velocity, and also a time stamp that specifies when the notes are to start and stop. Other MIDI data can also be embedded, such as on-the-fly program (voice) changes, continuous controller values, song tempo, and so forth. Those also have a time stamp to specify when the data should be sent.

MIDI Clock Resolution

Many DAW sequencer programs let you specify the time resolution of MIDI data, specified as some number of *pulses per quarter note*, abbreviated PPQ. These pulses are sometimes called *clock ticks*. Either way, the PPQ resolution is usually set somewhere in the Options menu of the software and typically ranges from 96 PPQ through 960 PPQ. Since this MIDI time resolution is related to the length of a quarter note, which in turn depends on the current song tempo, it's not an absolute number of milliseconds. But it can be viewed as a form of quantization because notes you play between the clock pulses are moved in time to align with the nearest pulse. To put into perspective the amount of time resolution you can expect, Table 15.5 lists the equivalent number of milliseconds per clock pulse for different PPQ values at the common song tempo of 120 beats per minute.

Older hardware synthesizers often use a resolution of 96 PPQ, which at 5 ms is accurate enough for most applications. I use 240 PPQ for my MIDI projects because the divisions are easy to remember when I have to enter or edit note lengths and start times manually, and 2 ms is more than enough time resolution. To my thinking, using 960 PPQ is akin to recording audio at a sample rate of 192 KHz; it might seem like it should be more accurate, but in truth the improved time resolution is not likely audible. Even good musicians are unable to play with a timing accuracy better than about 20 to 30 milliseconds.[1] As a fun exercise to put this into proper perspective, try tapping your finger along with a 50 Hz square wave, and see how accurately you can hit every tenth cycle. Table 15.6 shows the duration in MIDI clock pulses of the most common note lengths at a resolution of 240 PPQ. If you ever enter or edit MIDI note data manually, it's handy to know the number of clock pulses for the standard note

Table 15.5: Pulse Spacing at 120 Beats per Minute

PPQ	Time between Pulses
96	5.2 ms
240	2.1 ms
480	1.0 ms
960	0.5 ms

Table 15.6: Duration of Common Note Lengths at 240 PPQ

Note Length	Number of Pulses
Whole note	960
Half note	480
Half note triplet	320
Quarter note	240
Quarter note triplet	160
Eighth note	120
Eighth note triplet	80
Sixteenth note	60
Sixteenth note triplet	40

lengths. Even at "only" 240 PPQ, time increments can be as fine as 1/60 of a 1/16th note. It seems unlikely to me that any type of music truly requires a higher resolution.

MIDI Minutiae

The Bank Select command is used when selecting patches on synthesizers that contain more than one bank for sounds. You won't usually need to deal with this data directly, but you may have to choose from among several available bank select methods in your software, depending on the brand of synthesizer you are controlling.

Most DAW and MIDI sequencer programs send an All Notes Off command out all 16 channels every time you press Stop to cut off any notes that might be still sounding. But sometimes you may get "stuck notes" anyway, so it pays to learn where the All Notes Off button is located in your MIDI software and hardware.

Besides the standard continuous controllers, General MIDI also defines *registered parameter numbers* (RPNs) and *non-registered parameter numbers* (NRPNs). There are a few standard registered parameters, such as the data sequence that adjusts the pitch bend range to other than the usual plus/minus two musical half-steps. NRPNs are used to control other aspects of a synthesizer's behavior that are unique to a certain brand or model and thus don't fall under the standard CC definitions. The specific sequence of data needed to enable a nonstandard feature on a given synthesizer should be listed in the owner's manual.

Earlier I mentioned that some MIDI data requires two bytes in order to express a sufficiently large range of values such as pitch bend. In that case, two bytes are sent one after the other, with the lower byte value first. When two bytes are treated as a single larger value, they're called the *least significant byte* (LSB) and *most significant byte* (MSB).

You can buy hardware devices that split one MIDI signal to two or more outputs or that merge two or more inputs into a single MIDI data stream. A MIDI splitter is easy to build; it simply echoes the incoming data through multiple outputs. But a MIDI merger is much more complex because it has to interpret all the incoming messages and combine them sequentially without splitting up data that spans multiple bytes. For example, if one input receives a three-byte note-on command at the same time another input receives a two-byte command to switch from a sax to a flute voice, the merger must output one complete command before beginning to send the other. Otherwise, the data would be scrambled with a command and channel number from one device, followed by data associated with another command from a different device.

Finally, MIDI allows transferring blocks of data of any size for any purpose using a method called *Sysex*, which stands for System Exclusive. Sysex is useful because it can store and recall patch information for older pre-General MIDI synthesizers that use a proprietary format. I've also used Sysex many times to back up custom settings for MIDI devices to avoid having to enter them all over again if the internal battery fails.

Playing and Editing MIDI

Most MIDI-capable DAW programs work more or less the same, or at least have the same basic feature set. Again, I'll use SONAR for my examples because it's a full-featured program that I'm most familiar with. A program that records and edits MIDI data is often called a *sequencer*, though the lines have become blurred over the years now that many DAW programs can handle MIDI tracks as well as Wave file audio. Chapter 7 showed how audio clips can be looped, split, and slip-edited, and MIDI data can be manipulated in the same way. You can't apply cross-fades to MIDI data, but you can manipulate it in many more ways than audio files.

Because MIDI is data rather than actual audio, it's easy to change note pitches, adjust their length and velocity, and make subtle changes in phrasing by varying their start times. There are also MIDI plug-in effects that work in a similar way as audio effects. The MIDI arpeggiator shown in Chapter 14 operates as a plug-in, but there are also compressors that work by varying MIDI note-on velocity, echo and transpose (pitch shift) effects, and even chord analyzers that tell you what chord and its inversion is playing at a given moment.

MIDI tracks in SONAR include a Key Offset field to transpose notes higher or lower, without actually changing the track data. This is useful when writing music for transposing instruments such as the clarinet and French horn. You can write and edit your music in the natural key for that instrument, yet have it play at normal concert pitch. But most MIDI sequencer software also offers destructive editing to change note data permanently. For example, you can quantize notes to start at uniform 1/8 or 1/16 note boundaries. This can often improve musical timing, and it's especially helpful for musicians who are not expert players. Then again, real music always varies at least a little, so quantizing can equally rob a performance of its human qualities. Most sequencer programs let you specify the amount of quantization to apply to bring the start times of errant notes nearer to the closest time boundary, rather than forcing them to an exactly uniform start time. Many sequencers also offer a *humanizing* feature that intentionally moves note start times away from "the grid" to sound less robotic.

Another important MIDI feature is being able to record a performance at half speed or even slower, which again is useful for people who are not accomplished players. If you set up the metronome in your software to click every eighth note instead of every quarter note, it's easier to keep an even tempo at very slow speeds. I'm not much of a keyboard player, but I have a good sense of timing and dynamics control. So I often play solos at full speed on the MIDI keyboard, paying attention to phrasing and how hard I hit the keys, but without worrying about the actual notes I play. After stabbing at a passage "with feeling," I can go back later and fix all the wrong notes. I find that this results in a more musical performance than step-entering notes one by one with a mouse or playing very slowly, which can lose the context and feel of the music.

Figure 15.2 shows the Piano Roll view in SONAR, and other software brands offer a similar type of screen for entering and editing MIDI note data. Every aspect of a note can be edited, including its start time, length, and velocity. Controller data can also be entered and edited in the areas at the bottom of the window. The screen can be zoomed horizontally and vertically to see as much or as little detail as needed.

When you need to see and work with an even finer level of detail, the Event List shown in Figure 15.3 displays every piece of MIDI data contained in the sequence. This includes not just musical notes and controllers, but also tempo changes and RPN and NRPN data. If your synthesizer requires a particular sequence of bytes to enable a special feature, this is where you'll enter that data.

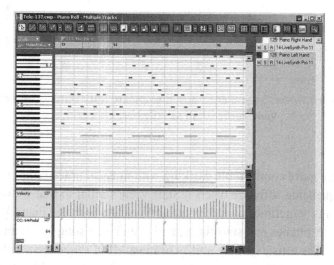

Figure 15.2: The Piano Roll window lets you enter and edit every property of MIDI notes and controller data.

Trk	HMSF	MBT	Ch	Kind	Data		
43	00:01:01:15	31:04:000	1	Note	B 2	110	1:000
43	00:01:02:00	32:01:000	1	Note	E 3	90	2:220
43	00:01:03:15	32:04:000	1	Note	E 3	68	120
43	00:01:03:22	32:04:120	1	Note	F#3	68	124
43	00:01:04:00	33:01:003	1	Note	G#3	67	3:117
43	00:01:06:00	34:01:000	1	Note	B 3	74	3:000
43	00:01:07:22	34:04:105	1	Note	Db4	77	136
43	00:01:08:00	35:01:000	1	Note	G#3	53	3:000
43	00:01:09:23	35:04:120	1	Note	G#3	77	60
43	00:01:10:00	36:01:000	1	Note	A 3	77	3:000
43	00:01:11:23	36:04:120	1	Note	A 3	93	45
43	00:01:11:26	36:04:180	1	Note	A 3	86	45
43	00:01:12:00	37:01:000	1	Note	B 3	74	3:000
43	00:01:13:23	37:04:120	1	Note	B 2	66	120
43	00:01:14:00	38:01:000	1	Note	B 3	75	7:000
43	00:01:17:15	39:04:000	1	Note	B 2	57	1:000
43	00:01:18:00	40:01:000	1	Note	E 3	80	12:000

Figure 15.3: The Event List view offers even more detailed editing and data entry, showing every aspect of a MIDI project, including tempo changes, pitch bends, and other non-note parameters.

Figure 15.4: Most sequencers let you enter and manipulate MIDI data as musical notes in a Staff View, rather than as little bars on a grid.

Some musicians are more comfortable working with musical notes rather than computer data, and most sequencing software offers a Staff View similar to that shown in Figure 15.4. This type of screen is sometimes called Notation View. As with the Piano Roll view, you can edit notes to new pitches and start times, and even insert song lyrics and common dynamics symbols. The music display ability of most MIDI sequencers falls short of dedicated notation software, but many sequencer programs are capable of creating perfectly usable printed music and lead sheets. Some even include standard guitar symbols for all the popular chord types.

The "midi_editing" video shows an overview of MIDI editing basics, using the solo cello and piano section from my Tele-Vision music video as a demo. The piano track is entirely MIDI, which I created partly by playing notes on a keyboard and partly by entering and copying notes one at a time with a mouse. But the cello is my live performance; it's not a sampled cello!

Summary

This chapter covers MIDI internal details in depth, including hardware protocols, and data formats and their use of channels. The standard General MIDI instruments and drum notes are listed, along with miscellaneous MIDI tidbits such as MIDI file types, non-registered parameters, and using Sysex to back up custom settings on MIDI hardware. Finally, a short video tutorial shows the basics of editing MIDI notes in a sequencer program.

Note

1 *Movement-Related Feedback and Temporal Accuracy in Clarinet Performance*, by Caroline Palmer, Erik Koopmans, Janeen D. Loehr, and Christine Carter. McGill University, 2009.

PART 3

Video Production

With so many bands and solo performers using videos to promote their music, video production has become an important skill for musicians and audio producers to acquire. While this section can't explain video cameras and editing techniques as deeply as a dedicated book, I'll cover the basics for creating music videos and concert videos. I'll use as examples my one-person music videos A Cello Rondo and Tele-Vision, a song from a live concert I produced for a friend's band, as well as a cello concerto with full orchestra that I worked on as a camera operator and advisor. It will help if you watch those first:

> Ethan Winer—*A Cello Rondo* music video:
> www.youtube.com/watch?v=ve4cBOnSU9Q
> Ethan Winer—*Tele-Vision* music video:
> www.youtube.com/watch?v=cWMNw-rM5xk
> Rob Carlson—*Folk Music in the Nude* music video:
> www.youtube.com/watch?v=eg0LY8kBO08
> Allison Eldredge—*Dvorak Cello Concerto* fragment:
> www.youtube.com/watch?v=Fgw54_uGDDg

Because video editing is as much visual as intellectual, I created three tutorial videos to better show the principles. The video "vegas_basics" gives an overview of video editing using Vegas Video, and "vegas_rondo" and "vegas_tele-vision" show many specific editing techniques I used to create those two music videos. I use Vegas, but most other professional video programs have similar features and work the same way. So the basic techniques and specific examples that follow can be applied to whatever software you prefer.

CHAPTER 16

Video Production

Video Production Basics

Most modern video editing software works similarly to an audio DAW program, with multiple tracks containing video and audio clips. Video plug-ins are used to change the appearance of video clips in the same manner as audio plug-ins modify audio. And just like a DAW, when a project is finished it's rendered by the software to a media file, ready for viewing, streaming online, or burning to a DVD or Blu-ray disk.

Modern video production software uses a paradigm called *nonlinear editing*, which means you can jump to any arbitrary place on the timeline to view and edit the clips. This is in contrast to the older style of video editing using tape that runs linearly from start to finish, where editing is performed by copying from one tape deck to another. This is not unlike the difference between using an audio DAW program and an analog tape recorder with a hardware mixing console.

Most music videos are shot as overdubs, where the musicians mime their performance while listening to existing audio. The music is first recorded and mixed to everyone's satisfaction; then the band or solo performer pretends to sing and play while the cameras roll. When overdubbing this way, the music is played loudly in the video studio or wherever the video is being shot so the performers can hear their part clearly and maintain the proper tempo. The cameras also record audio, usually through their built-in microphones, but that audio is used only to align the video track to the "real" audio later during editing.

Often, a band will mime the same song several times in a row so the camera operators can capture different performances and camera angles to choose from when editing the footage later. If the budget allows having four or more manned cameras, each camera can focus on a different player or show different angles of the entire band during a single performance. But many videos are done on a low budget with minimal equipment, and even a single camera can suffice if the band performs the same song several times. Shooting multiple takes lets a single camera operator focus on one player at a time during each performance. When editing, video of the guitar player can be featured during the guitar solo, and so forth. Live video can be cool, with the occasional blurred frame as the camera swings wildly from one player to another, but editing from well-shot clips is more efficient and usually gives more professional-looking results.

Of course, it's possible to video a group during a live performance. In that case, the audio is usually recorded separately, taken from the venue's mixing board either as a stereo mix or with each instrument and microphone recorded to a separate track to create a more polished mix later in a more controlled environment. Again, each camera's audio track is used only when editing the video later to synchronize the video tracks with the master audio. This is the method I use when recording my friend's band playing live.

Figure 16.1 shows the main editing screen of Vegas Video. The video and audio tracks are at the top, and at the bottom are a preview of built-in video effects on the left, the Trimmer window where video clips are auditioned before being added to the project, and the audio output section at right. Several other windows and views not shown here are also available, such as an Explorer to find and import files, and a list of the project's current media. The audio window at lower right can also be expanded to show the surround mixer if appropriate, and so forth. One of the many available plug-ins is in the middle of the screen.

Unlike an audio DAW where multiple tracks are mixed together in some proportion to create a final mix, video tracks usually appear only one at a time or are cross-faded briefly during a

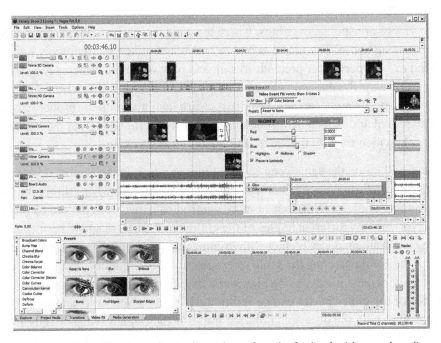

Figure 16.1: Vegas Video offers an unlimited number of tracks for both video and audio, and every track can have as many video or audio plug-in effects as needed.

transition from one camera to another. Therefore, video tracks require establishing a priority to specify which track is visible when multiple clips are present at the same time. As with an audio DAW, tracks are numbered starting at 1 from top to bottom. The convention is for upper tracks to have priority over lower tracks below. So if you want to add a text title that overlays a video clip, the track containing the text must be above the clip's track. Otherwise, the clip will block the text.

Video editing software also has a Preview window for watching the video as you work. My computer setup has two monitors, and Vegas allows the preview window to be moved to the second monitor for a full-screen view of the work in progress. This is much more convenient than trying to see enough detail on a small window within the program's main screen. Video editing and processing take a lot of computer horsepower, memory, and disk space. Unless your computer is very fast or you're working with only a few video tracks and plug-in effects, your computer may not be able to keep up with playback at the highest resolution in real time. Vegas lets you specify the preview quality so you can watch more tracks in low resolution or fewer tracks at a higher quality. Another option is to render a short section of the project to memory. Rendering a short section may take a few minutes to complete, but it lets you preview your work at full quality. You can also render all or part of the video to a file.

Live Concert Example

The project in Figure 16.1 shows a live performance of my friend's band that I shot in a medium-size theater using four cameras. This song is very funny, describing a concert the band performed at a nudist folk festival! You can see each camera's audio track below its video clip, and the height of every track can be adjusted to see or hide detail as needed. In Figure 16.1, the video tracks are opened wide enough to see their contents, and the camera audio tracks are narrower.

For this show, three cameras were at the back of the room, with a fourth camera to the left in front of the stage. The seats in this venue rise toward the rear, so the cameras had a clear shot of everything on the stage even though they were in the back. The audio in the bottom track was recorded during the show by taking a feed from the venue's mixing board into the line input of a Zoom H2 portable audio recorder. I used the H2 because it runs off internal batteries to avoid any chance of hum due to a ground loop. This is the audio heard when watching the video.

One of the three cameras on tripods in the rear of the room was unmanned, set up in the middle of the rear wall and zoomed out to take in the entire stage. This is a common technique, where a single static camera captures the full scene, serving as a backup you can switch to when editing later in case none of the other manned camera shots were useful at a given moment. A second camera was over to the right, also unmanned, set up to focus on Vin, who plays the fiddle, mandolin, and acoustic guitar. Having a single camera on the second most important player in the group ensures that this camera angle will always be available when needed to capture an instrumental solo or funny spoken comment between songs.

The third camera was also tripod-mounted near the center of the back wall. I operated this camera manually, panning and zooming on whatever seemed important at the moment. Most of the time, that camera was pointed at the lead singer, Rob, though I panned over to the piano player or drummer during their solos. The fourth camera was also manned, on a tripod down in front by the left side of the stage to get a totally different angle for more variety. Both of us running the cameras focused on whomever was the current featured player, free to pan or zoom quickly if needed to catch something important. Any blurry shots that occurred while our cameras panned were replaced during editing with the other manned camera or one of the unmanned cameras.

Speaking of blurry footage, it's best to use a tripod if possible, especially when zooming way in to get a close-up of something far away. Handheld camera shots are generally too shaky for professional results, unless you have a Steadicam-type stabilizer. These are very expensive for good models, and they're still not as stable as a good tripod resting on solid ground. Unlike still photography, where you position the tripod once and shoot, a video tripod also needs to move smoothly as you pan left and right or up and down. The best type of video tripod has a *fluid head* that can move smoothly, without jerking in small steps. Again, good ones tend to be expensive, though some models, such as those from Manfrotto, are relatively affordable. Another option is the *monopod*, which is basically a single pole you rest on the ground. This gives more stability than holding the camera in your hands unsupported, but it's still not as good as a real tripod with a fluid head.

Many bands are too poor to own enough high-quality video cameras to do a proper shoot in high definition, which was the case for this video. I own a nice high-definition (HD) Sony video camera, and my friend who ran the fourth camera is a video pro who owns three really nice Sony professional cameras. But the two other cameras were both standard definition (SD), one regular and one wide-screen. This video project is SD, not HD, so I used my HD camera for the static full-stage view. That let me zoom in a little to add motion when editing, without losing too much quality. If you zoom way in on a camera track after the fact when editing, the enlarged image becomes soft and grainy, and edges may become jagged. But you can usually zoom up to twice the normal size without too much degradation, and zooming even more is acceptable when using a camera having twice the resolution of the end product.

Color Correction

One of the problems with using disparate cameras is the quality changes from shot to shot when switching between cameras. This includes not just the overall resolution, but the color also shifts unless the camera's *white balance* is set before shooting. Modern digital still-image and video cameras can do this automatically: You place a large piece of white paper or cardboard on the set, using the same lighting that will be present during the shoot. Then you zoom the camera way in so the white object fills the frame and press a button that tells the camera "this" is what white should look like. The camera then adjusts its internal color balance to match. But even then, different cameras can respond differently, requiring you to apply color correction later during editing.

The Secondary Color Corrector plug-in shown in Figure 16.2 can shift the overall hue of a video clip, or its range can be limited to shift only a single color or range of colors. This is done using the color wheel near the upper left of the screen by dragging the dot in the center of the wheel toward the outside edge. The closer the dot is toward one edge, the more the hue is shifted toward that color.

The color corrector also has controls for overall brightness, *saturation* (color intensity), and *gamma*. If you want a clip to be black and white for special effect, reduce the saturation to zero. If you want to bring out the colors and make them more vibrant, increase the saturation. The gamma adjustment is particularly useful because it lets you increase the overall brightness of a clip, but only for those portions that are dim. Where a brightness control makes everything in the frame lighter, gamma adjusts only the dark parts, leaving brighter portions alone. This avoids bright areas becoming washed out, as can happen when

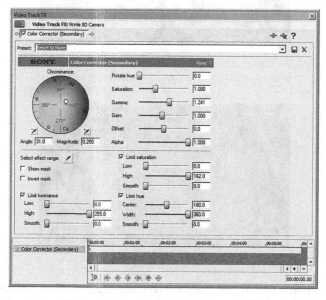

Figure 16.2: The Secondary Color Corrector plug-in lets you shift the overall color balance of a video clip or alter just one color or range of colors, leaving other colors unchanged.

increasing the overall brightness. This is not unlike audio compression that raises the volume of soft elements without distorting sounds that are already loud enough.

Figure 16.3 shows a scene from a YouTube video I made of my pinball machines before and after increasing the gamma. You can see that the lights on the play field are the same brightness in both shots, but all of the dim portions of the frame were made much brighter in the lower clip.

It's common for recording engineers to check an audio mix on several systems and in the car, and likewise you should check your videos on several monitors or burn a DVD to see it on several TVs, including the one you watch the most. Look for too dark or washed-out images or areas, and for too much or too little contrast. In particular, verify that skin colors are correct. If you plan to do a lot of video editing, you can buy a device to calibrate your video monitor. This attaches to the front face of the display during setup, and the included software tells you what monitor settings to change to achieve a standard brightness and contrast with

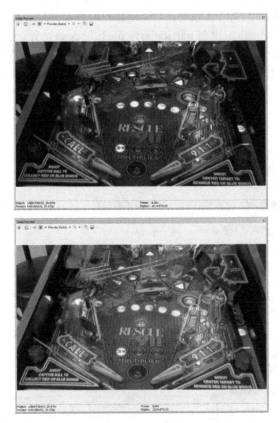

Figure 16.3: The gamma adjustment lets you make dim areas brighter, without making bright portions even brighter. The clip at the top is as shot, and the bottom is after increasing the gamma. If the overall brightness was increased, the play field lights would have become washed out.

correct colors. There are also software-only methods that use colored plastic sheets you look through while adjusting the red, green, and blue levels.

Synchronizing Video to Audio

Continuing on with the live concert example in Figure 16.1, each of the four camera's files is on its own track. When you add a video file to a project, both its video and audio are imported and placed onto adjacent tracks, so there are really two tracks associated with each camera file. By putting each camera on its own video track, you can add the color corrector or other corrective plug-ins to modify only that track. Vegas also lets you apply video plug-ins to individual clips when needed.

Importing video clips from older tape-based video cameras requires using a *video capture* utility that runs in real time as you play the tape in your camera. This program is usually included with the video editing software, and the camera connects to your computer through a FireWire port. Modern cameras save video clips as files directly to an internal memory card, and the files can be transferred via USB, FireWire, or a card reader much more quickly than real-time playback. A camera that uses memory cards not only transfers video more quickly, but it's also more reliable by avoiding the drop-outs that sometimes occur with video tape.

The first step, after importing all of the camera files, is to synchronize each camera to the master audio track by sliding the camera clips left or right on the timeline. The camera's video and audio portions move together unless you specifically ungroup them, so aligning the camera's audio to the master audio also aligns the video. Figure 16.4 shows a close-up of one

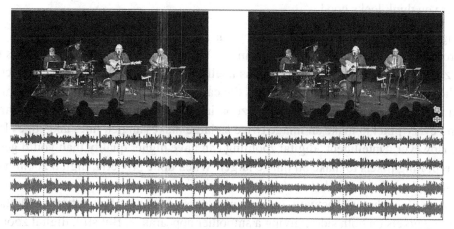

Figure 16.4: To synchronize a video track to the master audio, place the master audio track directly underneath the camera's audio track, then slide the camera clip left or right until both audio tracks are aligned.

camera's video and stereo audio tracks, with the master stereo audio track at the bottom. As mentioned, the master audio was recorded from the mixing console, and the camera's audio was recorded through its built-in microphone. Once the tracks are visually aligned—zoom way in as needed—you should listen to both audio tracks at once to verify they're in sync with no echo. Then you can mute the camera's audio or even delete that track. It's probably better to just mute the track, in case the clip is moved accidentally during editing and needs to be sync'd up again.

Once all of the camera tracks are aligned, you can watch each track one at a time all the way through, making notes of where each camera should be shown in the video. Often, the first shot in a music video will be the full-stage camera so viewers get a sense of the venue. Then you'll decide when to switch to the other cameras, depending on who should be featured at that moment. It's also common to zoom in (or out) slowly over time, which adds motion to maintain interest. Ideally, this zooming will be done by the camera operator, but it can also be done during editing as long as you don't zoom in too much.

Panning and Zooming

Generally, when running a video camera, you'll avoid zooming in or panning across quickly. Too much fast motion is distracting to viewers, and it can create video artifacts at the low bit-rates often used for online videos. It's also recommended to avoid "trombone" shots that zoom in and then zoom out again soon after. Do either one or the other, but not both over a short period of time. Of course, it's okay to zoom in fast to capture something important, but consider switching to another camera when editing to hide the zoom. Understand these are merely suggestions based on established practice and common taste. Art is art, so do whatever you think looks good.

Speaking of zooming, I'm always amused by ads for inexpensive video cameras that claim an impressive amount of digital zoom capability. What matters with video cameras is their optical zoom, which tells how much the lens itself can zoom in to magnify the subject while shooting. Digital zooming just means the camera can enlarge the video while playing back on its built-in screen, and this type of zooming always degrades quality. For example, repeating every pixel to make an image twice as large makes angled lines and edges look jagged. However, "intelligent" digital zooming can have higher quality than you'd get from simply repeating pixels to make an image larger. When done properly, digital zooming creates new pixels having in-between values, depending on the image content. A good digital zoom algorithm will actually create new colors and shades that better match the surrounding pixels on both sides, giving a smoother appearance. But still, digital zooming always compromises quality compared to optical zooming, especially for large zoom amounts.

Video Transitions

Switching between cameras during editing can be either sudden or with a cross-fade from one to the other. You create a transition by dragging the edge of one camera's clip to overlap another, and the transition occurs over the duration of the overlap. This is much like using slip-editing to cross-fade audio clips in a DAW program, and the two clips can be on the same track or separate tracks. As with audio clips, be very careful to slip-edit only the clip's edges—if you accidentally drag the clip left or right that puts it out of sync! If both clips are on separate tracks, as I usually do, you'll use fade-in and fade-out envelopes on both tracks, rather than have one clip physically overlap the other. Either way, to create a fast cross-fade, the clips will overlap for half a second or so, or the overlap can extend for several seconds to create a slow transition. The second clip on Track 6 of Figure 16.1 shows a fade-out envelope, which creates a cross-fade to the subsequent clip below on Track 8. Since Track 6 has a higher priority and hides Track 8, there's no need to apply a corresponding fade-in on Track 8; Track 6 simply fades out to gradually reveal Track 8 over a period of about one and a half seconds. You can also specify the *fade curve*, which controls how the cross-fade changes over time.

Pop music videos are usually fast-paced, often switching quickly from one camera angle to the next, or applying a transition effect between clips. Besides the standard cross-fade, most video editor software includes a number of transition effects such as Iris, Barn Door, and various other wipe and color flash effects. For example, an Iris transition opens or closes a round window to reveal the subsequent clip as in Figure 16.5. Many other transition types are available, and Vegas lets you audition all the types and their variations in a preview window. The included video "vegas_television" shows how that works.

I prefer to cut or cross-fade from one camera to another a second or two before something happens, such as the start of a guitar solo. This gives the viewer a chance to prepare for what's coming and already be focused on the player in the new perspective before the solo begins. But sometimes one solo starts before the previous one has ended, or a solo starts while the lead singer is still singing. In that case, you can use a slow cross-fade spanning two to four seconds to partially show both performers at the same time. Or you can split the screen into left and right sections, or put one camera in its own smaller window on the screen to show both cameras at the same time.

When I do these live videos for my friend, both camera operators focus on whatever seems important to us at the moment. But sometimes neither of us is pointing our camera at what's most important. Maybe we'll both think a piano solo is coming and aim there, but it was actually the guitar player's turn. So by the time we focus our cameras on the guitar player, he's already five seconds into the solo. This is where the fall-back camera that captures the entire stage is useful. When editing, I'll switch to the full-stage camera a second or

Figure 16.5: Most video programs offer a large number of transition types, including the Iris shown here that opens a round window over time to expose the subsequent track. In this example, the guitar player's track transitions to the mandolin player.

two before the guitar solo starts, then slowly pan and zoom that camera's track toward the guitarist in anticipation of the solo. As mentioned, you can usually zoom a clip up to 200 percent (double size) before the quality degrades enough to be objectionable. So I'll do a slow zoom over a few seconds that doesn't enlarge the frame too much and at the same time pan toward the soloist to imply what's coming. Then I finally switch to one of the manned cameras once it's pointing at the featured player. This type of zooming is shown in Figure 16.6 in the next section to anticipate a piano solo.

The venue where this live concert was shot is not huge, with about 300 seats. Since I recorded the audio from the house mixing board, the sound was very clean and dry—in fact, too dry to properly convey the feel of a live concert. To give more of a live sound, I added the audio recorded by the two cameras in the rear—one each panned hard left and right—mixed in very softly to add just a touch of ambience. This also increased the overall width of the sound field, because instruments and voices from the board audio were mostly panned near the center.

Key Frames

One of the most powerful features of nonlinear video editing is *key frames*. These are points along the timeline where changes occur, such as the start and end points of a zoom or pan. Figure 16.6 shows the Pan/Crop window that pans and zooms video clips. The top image shows the full frame, which is displayed initially at the start of the clip. The large hollow "F" in the middle of the frame is a reference showing the frame's size and orientation. Video software can rotate images as well as size and pan them, so the "F" lets you see all three frame properties at once.

You can see three diamond shaped markers in the clip's timeline at the bottom marked Position: one at the far left, another at the four seconds marker, and the last at eight seconds. In this case, the full frame is displayed for the first four seconds of the clip because both key frames are set the same. The lower image shows that a smaller window is displayed at the eight seconds mark. Since a smaller portion of the clip is framed, that area becomes zoomed in to fill the entire screen. When this clip plays, Vegas creates all the in-between zoom levels to transition from a full frame to the zoomed-in portion automatically. The other timeline area marked Mask is disabled here, but it can be used to show or hide selected portions of the screen. The video "vegas_basics" explains how masks are used.

As you can see, key frames are a very powerful concept because you need only define the start and end conditions, and the software does whatever is needed to get from one state to the next smoothly and automatically. If you want something to change more quickly, simply slide the destination key frame to the left along the timeline to arrive there earlier. Again, key frames can be applied to anything the software is capable of varying, including every parameter of a video plug-in.

Most video runs at 30 frames per second (FPS), which is derived from the 60 Hz AC power line frequency. In most US localities, the frequency of commercial power is very stable, so this is a convenient yet accurate timing reference for the frame rate. The timeline is divided

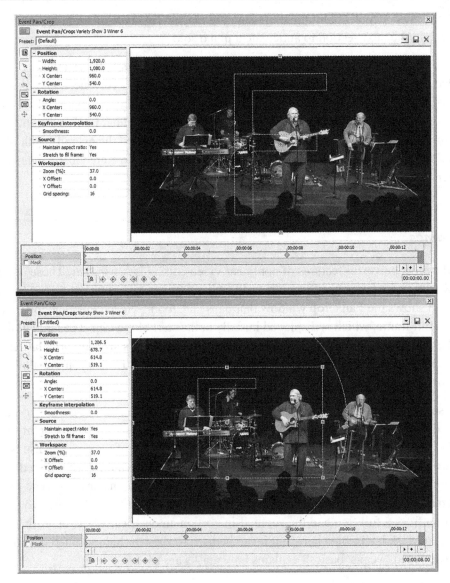

Figure 16.6: Key frames indicate points on the timeline where something is to change. This can be a pan and zoom as shown here, or changes in color, brightness, text size, or literally any other property of a video clip or plug-in effect.

into hours, minutes, seconds, and frames, with a new frame every 1/30th of a second. The PAL video format used in Europe divides each second into 25 frames, because AC power there is 50 Hz. NTSC video uses a frame rate of 29.97 Hz—the explanation is complicated, but the simple version is it allowed making TV sets cheaper.

Blu-ray disks run at 24 frames per second, so when creating those you have to shoot at 24 FPS or your software will convert the data when it burns the disk. The process is similar to audio sample rate conversion, dropping or repeating frames as needed. Many professional HD video cameras can also shoot at 60 FPS. This is not so much to capture a higher resolution but to achieve smoother slow motion when that effect is used. If you slow a 30 FPS clip to half speed, each frame is repeated once, which gives a jerky effect. If you shoot at 60 FPS, you can slow it down to half speed and still have 30 unique frames per second.

Orchestra Example

Pop music videos often contain many quick transitions from one camera to another, sometimes with various video effects applied. But for classical music, a gentler approach is usually better, especially when the music is at a slow tempo. In the orchestra example linked at the start of this chapter, you'll see that most of the cross-fades from one camera to another are relatively slow, and with slower pieces, camera cross-fades can span four seconds or even longer. You'll also notice that many of the camera shots constantly zoom in (or out) slowly on the players and soloist, as described earlier. This video was shot in high definition using four cameras, with full 5.1 surround sound, though the YouTube clip is standard resolution and plain stereo.

My friend and professional videographer Mark Weiss was at the front of the balcony at the far right, and I was also in the balcony but at the far left. This way Mark could capture the faces of players on the left side of the stage and the cello soloist's left side. My position let me do the same for players on the right of the stage and zoom in on the cellist's right side. In other words, we each mostly shot across the stage at an angle, though we could capture players on our own side too. Another camera operator was on the main floor near the front to the right of the audience. A fourth unmanned camera was placed high on a ledge at the side of the stage pointing down toward the conductor. Having a dedicated camera lets Mark switch to the conductor's face occasionally during editing, which is not possible from in front where the camera operators were stationed. The three of us have shot many videos for this orchestra, though for some videos the third camera operator was on the stage itself, off to one side, to get better close-ups of the players.

Figure 16.7, taken from the stage looking up toward the balcony at the rear of the hall, shows the surround microphone rig Mark built for these orchestra videos. This consists of a metal frame to which five microphone shock-mount holders attach. It's hung from thin steel wires 18 feet in the air, centered above the third row of the audience. The microphones are connected by long cables to a laptop computer with a multi-channel audio interface in a room at one side of the stage.

Figure 16.7: These five microphones are placed high up over the audience. Three of the mics point forward and down for the left, center, and right main channels, and two more mics point toward the rear left and right to capture a surround sound field from the back of the hall.

Cello Rondo and Tele-Vision Examples

The demo "vegas_rondo" shows many of the editing techniques I used to create that video, but it's worth mentioning a few additional points here. The opening title text "A Cello Rondo" fades in, then zooms slightly over time, using two key frames: one where the text is to start zooming and another where it stops at the final larger size. Likewise, the text "By Ethan Winer" uses key frames to slide onto the screen from left to right, then "bounces" left and right a few times as the text settles into its final position.

There are three ways to apply key frames to change the size of on-screen text. If you use the Scaling adjustment in the text object itself, or resize the frame in the Pan/Crop window, each size is generated at the highest resolution when the video is rendered. You can also size and move video clips using a track's Track Motion setting, but that resizes the text after it's been generated, which lowers the resolution and can soften the edges. So with Vegas, anyway, it's better to automate the text generator directly.

As mentioned, the order of tracks determines their priority for overlapping video clips, with tracks at the top showing in front of lower tracks. Figure 16.8 shows part of the video where

Figure 16.8: In most video editing software, lower-numbered tracks appear in front of higher-numbered tracks. Here, the player on the left is on Track 7, the player on the right is on Track 11, and the cello background is on Track 24.

Figure 16.9: Each of the nine elements in this scene is sized and positioned individually, with the halo and spotlight programmed to move via key frames.

two players are on-screen at the same time. In this case, the track for the player on the left is above the track for the player on the right, which in turn is above the track holding the textured background. The result looks natural, as if one player's bow is actually behind the other player's arm, with both players in front of the background.

Figure 16.9 shows a portion of the video with nine separate elements: five cellists, my cat Bear, a golden halo over Bear's head that's automated using key frames to follow his head movement, a white spotlight that sweeps across the screen via key frames, and a static photo of my cello used for the background. I used a green screen to create what looks like a single performance from all of these video elements, letting me keep just the players and strip out the wall behind me. A green screen lets you remove the background behind the subject and float the subject on top of a new background. This is explained further in the "vegas_rondo" demo video.

Backgrounds

Vegas includes a large number of effects plug-ins, including a very useful "noise" generator. This creates many different types of texture patterns, not just snow, as you'd see on a weak TV station, which is what video noise really looks like. The Vegas noise patterns include wood grain, clouds, flames, camouflage, lightning, and many others. I used every one of those for my Cello Rondo video, but I wanted something more sophisticated for Tele-Vision. Many companies sell animated backgrounds you can add royalty-free to your projects, and I chose a product called Production Blox from 12 Inch Design. These backgrounds are affordable and far more sophisticated than anything I could have created myself using the tools built into Vegas.

One goal of a music video is to add interest by doing more than is possible with only audio. In a live concert video you can switch cameras to change angles and show different performers and use transitions and other special effects. When creating a video such as Tele-Vision that's compiled from many different green screen clips, you can also change the backgrounds. Not only can you switch between backgrounds, but you can change the background's appearance over time using key frames. The Production Blox backgrounds are already animated, but I spent a lot of time varying the stock Vegas backgrounds, such as making a checkerboard pattern change color and rotate, and animating cloud patterns. I even created an entire animated disco dance floor from scratch, which took me more than a day to complete! A big part of audio mixing is sound design, and likewise an important part of video production is thinking up clever ways for things to change over time to maintain the viewer's interest.

Time-Lapse Video

Although not directly related to music videos, a common special effect is time-lapse video, where several minutes or even hours elapse in just a few seconds. If you need to speed up a clip by only a modest amount, Vegas lets you add a Velocity envelope to a video clip to change its playback speed. You can increase playback speed as much as 300 percent or slow it to a full stop. You can even set the Velocity to a negative value to play a clip backward, as shown in the "vegas_rondo" demo. If 300 percent isn't fast enough, you can increase the speed further by Ctrl-dragging the right edge of a clip. Ctrl-dragging its right edge to the left compresses the clip to play up to four times faster. You can also Ctrl-drag to the right to stretch it out, slowing down playback to as little as one-fourth the original speed. By setting the Velocity to 300 percent and Ctrl-dragging fully to the left, you can increase playback speed as much as 12 times.

If that's still not fast enough, Vegas lets you export frames from a video clip to a sequence of still image files. You specify the start and end points of the clip to export and how much time to skip between each frame saved to disk. For one of my other music videos I wanted to speed up a clip by 60 to 1, where each minute passes in one second. So when exporting the image sequence, I kept one frame for every 59 that were discarded. Then I imported the series of image files into Vegas, which treats them as a single unified video clip. Rather than describe all of the steps required here, I created the YouTube tutorial www.youtube.com/watch?v=ibE-tviIxUg to show the effect and describe the process in detail.

Media File Formats

Audio files can be saved as either full-quality Wave or AIFF files, or in a lossy-compressed format such as MP3 or AAC. Lossy compression discards content deemed to be inaudible, or at least less important, thus making the file smaller. Raw video files are huge, so they're almost always reduced in size using lossy compression. An uncompressed high-definition AVI file occupies nearly 250 MB of disk space per second! Clearly, compression is needed if you hope to put a video longer than 18 seconds onto a 4.7 GB recordable DVD.

Where lossy audio compression removes content that's too soft to be heard, video compression instead writes only what changes from one frame to the next, rather than saving entire frames. For example, with a newscaster in front of a static background, most of the changes occur in just a small part of the screen where the speaker's mouth changes. The rest of the screen stays the same and doesn't need to be repeated at every frame. This is a simplification, but that's the basic idea. Video in North America and many other parts of the world runs at 30 frames per second, so not having to save every frame in its entirety can reduce the file size significantly.

As with audio, lossy video compression is expressed as the resulting bit-rate for the file or data stream. Most commercial DVDs play at a bit-rate of 8 megabits per second (Mbps), though high-definition video on a Blu-ray disk can be up to 50 Mbps. One byte of data holds eight bits, so each second of an 8 Mbps compressed DVD video occupies 1 MB of disk space. Therefore, a single-layer 4.7 GB recordable DVD holds about 78 minutes at 8 Mbps. You can reduce the bit-rate when rendering a project to store a longer video or use dual-layer DVDs. Note that the specified bit-rate is for the combined video and audio content, so the actual video bit-rate is slightly less. You can also specify the compressed audio bit-rate when rendering videos to balance audio quality versus file size.

As with audio files, variable bit rate (VBR) encoding is also an option for compressed video. VBR changes the bit-rate from moment to moment, depending on the current demands of the video stream. A static photo that remains on screen for five seconds can get away with a much lower bit-rate than a fast action scene in a movie or a close-up of a basketball player rushing across the court or a shot where the camera pans quickly across a crowd. Since lossy video compression encodes what changes from frame to frame, motion is the main factor that increases the size of a file. With VBR compression, the maximum bit-rate is used only when needed for scenes that include a lot of motion, so VBR video files are usually smaller than constant bit-rate (CBR) files.

The file format for DVDs that play in a consumer DVD player is MPEG2, where MPEG stands for Moving Picture Experts Group, the standards outfit that developed the format. If you plan to put your video onto a DVD, this is the format you should export to. Vegas includes a render template for this format that is optimized for use with its companion program DVD Architect. Video that will be uploaded to a website can use other file formats, but don't use a format so new or exotic that users must update their player software before they can watch your video. Windows Media Video (WMV) and Flash (FLV) used to be popular, but the newer MP4 is now preferred and for uploading to YouTube and other places. However, the popularity of video file formats comes and goes, and new formats are always being developed. What works best for YouTube today might be different next year or even next week.

I usually render my videos as MP4 files at a high bit-rate so I can watch them on my TV in high definition, and then I make a smaller version to put on my websites. I use the excellent and affordable AVS Video Converter software to convert between formats and sizes. There are other such programs, including those that claim to be freeware, though some are "annoyware" that add their branding on top of your video until you buy the program. Also, this is one category of program that's a frequent target for malware. Video conversion and DVD extraction are popular software searches, and unscrupulous hackers take advantage of that. Beware!

Lighting

Entire books have been written about lighting, and I can cover only the basics here. The single most important advice I can offer is to get halogen lights that are as bright as possible. Newer LED lights are also available; they don't run nearly as hot as 1 kilowatt of halogen lighting, and they use less electricity. But at the time of this writing they're more expensive and are a good investment only if you do a lot of video work. Halogen lamps produce a very pure white light, so colors will be truer than when using incandescent or fluorescent bulbs. As with most things, you can spend a little or a lot. Inexpensive halogen "shop" lights work very well, though professional lights have better stands that offer a wide range of adjustment for both height and angle. Some pro lights also offer two brightness settings. Regardless of which type of halogen light you get, buy spare bulbs and bring them with you when doing remote shoots.

When lighting a video shot in a home setting, it's best to point the lights up toward the ceiling or at a nearby wall. This diffuses the light and avoids strong shadows with sharp edges. Direct light always creates shadows unless you have a lot of lights placed all around the subject, with each light filling in shadows created by all the other lights. Placing lights to avoid problems with interaction is a bit like placing microphones. Figure 20.9 from Chapter 20 shows a product photo I took in a friend's apartment. I used two professional 650-watt halogen lights on stands, with the lights several feet away from the subject, raised to about two feet below the ceiling and pointing up. With video, having an additional light behind the subject is also common, often set to point at a person's head to highlight his or her hair, which adds depth to the scene. Watch almost any TV drama or movie, and you'll notice that many of the actors have a separate light coming from one side or behind, pointed at their head.

Earlier I mentioned that modern cameras include automatic white balance, which is a huge convenience for hobbyists who don't have the time or resources to become camera experts and learn how to set everything properly manually. To get the best results, however, it's important to use the same type of lights throughout on a set rather than mix halogen, incandescent, and fluorescent lights. Each bulb type has a different *color temperature*, which affects the hue the camera captures. If the colors of some parts of a room or stage vary compared to others, the colors will shift as the camera pans to take in different subjects. Don't be afraid to turn ordinary room lights off when setting up the lighting for your video shoot.

Summary

This chapter explains the basics of video production, including cameras, editing, media file formats, and lighting. Besides the four example music videos, three demo videos let you see video editing in action in a way that's not possible to convey in words alone. Modern video software works in much the same way as audio DAW programs, using multiple tracks to organize video and audio clips. And as with audio, video plug-ins can be used to change the appearance of the clips or to add special effects. But unlike audio mixing, video tracks are usually shown one at a time, with tracks at the top of the list hiding the tracks below.

Most music videos are performed as overdubs, where the players mime their parts while listening to an existing audio mix. If you don't have enough cameras available to capture as many angles and close-ups as you'd like, you can use one or two cameras and have the band perform a song several times in a row. Then for each performance, the cameras will feature one or two different players. It's common to dedicate a single unmanned camera to take in the entire scene to serve as a backup in case all the other camera shots turn out flawed at a particular moment. However, when using disparate cameras, the quality and color can change from shot to shot, especially if the cameras vary in quality. Thankfully, many cameras can set their white balance automatically before shooting to even out color differences. Besides being able to adjust color after the fact, a gamma adjustment lets you increase the overall brightness of a clip, without washing out sections that are already bright enough.

Once you've imported all of the video files from each camera, they need to be synchronized with the final audio track. After that's done, the camera audio is no longer needed, so you can mute (or delete) those tracks. Switching between cameras during editing can be abrupt or with a cross-fade, and most video software includes a number of transition effects. When editing video, one of the most powerful and useful features is key frames that establish the start and end times over which a change occurs, and the software creates all the in-between points automatically. Key frames can vary anything the software is capable of, including video plug-in parameters.

This chapter also explained that video files are always reduced in size using lossy compression, though the resulting degradation is not usually a problem unless you compress to a low bit-rate. Proper lighting is equally important. Using lamps that are bright and white and are diffused by bouncing the light off a wall or ceiling is a good first step to achieving a professional look. But no matter how good you think your video looks, it's useful to verify the quality and color balance on more than one monitor or TV set.

PART 4

Transducers

A *transducer* is a device that converts one form of energy to another. A light bulb that converts electricity to light is a transducer, as are electric and gasoline motors that convert electricity and combustion, respectively, to mechanical motion. Even a foam or fiberglass absorber used for acoustic treatment could be considered a type of transducer, because it converts acoustic energy to heat. In that case, however, the energy is discarded rather than reused in its new form.

These days most electronic devices have sufficiently high quality to pass audio with little or no noticeable degradation. But electromechanical transducers—microphones, contact pickups, phonograph cartridges, loudspeakers, and earphones—are mechanical devices, and thus are susceptible to frequency response errors and unwanted coloration from resonance, distortion, and other mechanical causes. For example, the cone of a loudspeaker's woofer needs to be large in order to move enough air to fill a room with bass you can feel in your chest, but it's too massive to move quickly enough to reproduce high frequencies efficiently and with broad dispersion.

Therefore, most loudspeakers contain separate drivers for the different frequency ranges, which is yet another cause of response errors when sounds from multiple drivers combine in the air. At frequencies around the crossover point where two drivers produce the same sound, comb filtering peaks and nulls result from phase differences between the drivers. Where most audio gear is flat within a fraction of a dB over the entire audible range, with relatively low distortion, the frequency response of loudspeakers and microphones can vary by 5 to 10 dB or more, even when measured in an anechoic test chamber to eliminate room effects. Transducers also add much more distortion than most electronic circuits.

A microphone doesn't require separate woofers and tweeters, but its diaphragm can move only so far before bottoming out in the presence of loud sounds. Further, as a microphone's diaphragm approaches that physical limit, its distortion gradually increases. The diaphragm in a dynamic microphone is attached to a coil of wire whose mass restricts how quickly it can vibrate, which in turn limits its high-frequency response. All microphones have a resonant frequency, resulting in a peak at that frequency as well as ringing. As explained in Chapter 1, a mass attached to a spring resonates at a frequency determined by a combination of the two properties. With a dynamic microphone, the resonant frequency is determined by the mass of the coil and the springiness of the diaphragm suspension. In a sealed capsule design, the air trapped inside can also act as a spring.

The same resonance occurs with loudspeakers: They can continue to vibrate after the source stops, unless they're mechanically or electrically damped. Indeed, the design of transducers is always an engineering compromise between frequency response both on and off axis, power handling capability (for speakers), SPL handling (for microphones), overall ruggedness, and, of course, the cost to manufacture.

CHAPTER 17

Microphones and Pickups

Chapter 6 explained microphone basics such as their pickup patterns and common placements, without getting overly technical. This chapter delves more deeply into the various microphone types and how they work internally. Microphones respond to air pressure, converting pressure changes into a varying voltage. When the changing voltage is sent to a loudspeaker, the speaker moves to recreate the original pressure changes at the same rate and volume. If everything goes well, sound from the loudspeaker should resemble closely what was picked up by the microphone.

Most microphones employ a lightweight diaphragm that vibrates in response to pressure from the acoustic waves striking it. However, there are also contact microphones that physically attach to an instrument such as an acoustic guitar or stand-up bass. In this design, sound waves transfer directly from the instrument's body to the pickup, rather than first passing through the air. Different parts of an acoustic instrument's body vibrate differently than others, so depending on where the pickup is attached, the fundamental pitch or a particular harmonic might be stronger or weaker.

Another type of input transducer is the phonograph cartridge. Years ago, many phono cartridges were made from piezoelectric crystals that generate voltage when twisted or bent. A lightweight thin cantilever—similar to a tiny see-saw—transfers motion from a needle in the record's groove applying pressure to the crystal. These days most cartridges are electromagnetic, using a coil of wire and magnet, though piezo pickups are still used in inexpensive phonographs.

The more common electromagnetic phono cartridge employs a *moving magnet* design, where a tiny magnet placed near a coil moves in step with the record grooves. An equivalent design instead moves the coil, while the magnet remains stationary. While more expensive, *moving coil* cartridges usually have a better high-frequency response than moving magnet types. One reason is because a coil weighs less than a magnet and so can vibrate more quickly. A moving coil design also has fewer turns of wire to minimize its weight, so its output impedance is lower and is affected less by capacitance in the connecting wires. One important downside of moving coil designs is their very low output level due to having so few turns of wire. Therefore, a moving coil pickup requires a special high-gain, low-noise preamp.

Microphone Types

The earliest microphones were made of carbon granules packed into a small metal cup with an acoustic diaphragm on top, as shown in Figure 17.1. Carbon passes electricity, though not as well as copper wire. When the granules are packed loosely, they have a higher resistance—and pass less electricity—than when packed tightly. Voltage from a battery is sent through the metal diaphragm and cup assembly. When positive wave pressure reaches the diaphragm, the carbon compresses slightly, which lowers its resistance, letting more electricity pass through to the output. Negative wave pressure instead pulls the diaphragm away from the carbon, so it's compressed less and the output voltage decreases.

Carbon microphones were used years ago in telephones, and some may still be in service today. Technically, this is considered an active microphone because the output voltage is derived from a DC power source. In old telephones, the voltage is supplied by the phone company and comes down the same wires used for the voice audio. Many telephones today contain electronics for memory-dial and other modern features, so they can more easily include a preamp suitable for use with dynamic or condenser microphones that are higher quality than the older carbon types.

Other early microphones, called *crystal* mics, use piezoelectric elements. This is a thin wafer of crystal or ceramic material, sandwiched between two thin metal plates that carry the output voltage. Like a piezo phono cartridge, a crystal microphone generates voltage when the diaphragm flexes the crystal. One advantage of piezo microphones (and phono cartridges) is

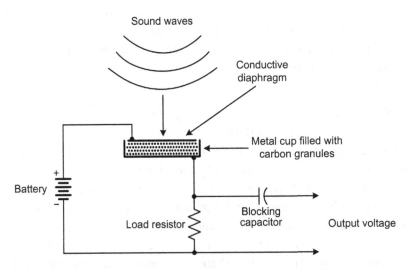

Figure 17.1: A carbon microphone generates electricity by modulating the DC voltage from a battery. A capacitor blocks the constant at-rest DC voltage from passing to the audio output, so only voltage changes get through.

their relatively high output voltage, so less gain is needed in the preamp. However, the output impedance is quite high, requiring short, low-capacitance connecting wires.

Piezo mics often have a "peaky" response that emphasizes midrange frequencies, so they're not used today for professional audio recording. However, that type of response works well for communications applications, and crystal microphones were very popular with amateur radio operators in the mid-twentieth century. I'll also mention that crystal mics were an essential part of the Chicago amplified blues harmonica sound that developed in the late 1940s and early 1950s. Home tape recorders became popular in that period, and most included a crystal mic. Because of their high impedance, high output level, and ¼-inch phone plug, these mics were an obvious match for a guitar amplifier. The combination of a convenient size and shape, a peaky response that complemented the harmonica, enough output to drive an amplifier into distortion, and the amplifier's built-in reverb formed the basis of a sound that's still with us today. I won't dwell further on older microphones types that are no longer used professionally, though it's useful to understand how these early microphones work and know their history.

The mass and weight of a microphone's diaphragm affects its high-frequency response, so dynamic microphones with an attached coil tend to roll off well below 20 KHz. The diaphragms in condenser and ribbon mics are much lighter, and without the weight of the coil, they have a better high-end response with less ringing. "Tiny" diaphragm condensers are better still, because their lighter diaphragms can respond to frequencies beyond 20 KHz, which improves their response below 20 KHz. Ribbon diaphragms are also light, but their length limits their response to less than 20 KHz.

Besides classifying microphones by how they create electricity from sound waves, they're also categorized by how they respond to changes in wave pressure. A *pressure* microphone responds to the absolute amount of air pressure reaching its diaphragm, which makes it omnidirectional. Whether a wave arrives from the front, the rear, or the side, the wave's pressure exerts the same positive or negative force on the microphone's diaphragm.

The other type is the *pressure-gradient* microphone. The diaphragm in a pressure-gradient mic is open on both sides, so it instead responds to the *difference* in pressure reaching the front and back of the diaphragm. Therefore, these microphones are inherently directional; if the same sound wave strikes both the front and rear of the diaphragm equally, the result is no physical movement and therefore no electrical output. The classic example of a pressure-gradient pickup pattern is the Figure 8, shown earlier in Figure 6.3.

Dynamic Microphones

Among modern designs, dynamic microphones are very popular because they're sturdy and have an acceptable, if not always great, frequency response. When Electro-Voice years ago introduced their 664 dynamic microphone, they showed off its ruggedness at product demos by using it as a hammer, earning this mic the endearing name "Buchanan Hammer" for Buchanan, Michigan, where Electro-Voice was located. Dynamic microphones generate electricity by placing a lightweight coil of very fine wire in a magnetic field, using a principle called *electromagnetic induction*. This is shown in Figure 17.2.

Note that the compliant diaphragm edge is shown to make the operating principle clear. In practice, many microphone diaphragms are more like a drum head; the diaphragm stretches slightly, and the center is displaced by sound waves even though the edges are secured to the housing. Also note that the round bar magnet shown is just one possible shape. The magnet for the microphone in Figure 17.3 is round, like a very thick coin, with a hole in the center to accept the coil similar to the loudspeaker in Figure 18.3.

The further the coil moves through a magnetic field, the larger the output voltage. And the faster it moves, the faster the voltage changes. Therefore, the output voltage of a dynamic microphone corresponds to the changing wave pressure striking its diaphragm, within the frequency response and other mechanical limits of the moving parts. Figures 17.3 and 17.4 show a cheap dynamic microphone I took apart to reveal the plastic diaphragm and attached tiny coil.

Dynamic microphones having a sealed enclosure as in Figure 17.2 are omnidirectional, responding to sound arriving from all angles. A pressure transducer responds to changes in atmospheric pressure, not air flow, so in theory the direction the sound arrives from doesn't matter. At any point in space, the barometric pressure is whatever it is. In practice, even the best omnidirectional microphones become slightly directional at higher frequencies. At very high frequencies, the microphone's body is larger than the acoustic wavelengths, so some

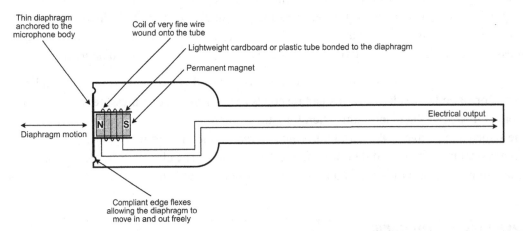

Figure 17.2: A dynamic microphone generates electricity by placing a coil of wire within a magnetic field. As the diaphragm pushes the coil back and forth through the field, a corresponding output voltage is produced.

Figures 17.3 and 17.4: These photos show a dynamic microphone's plastic diaphragm (top) and the attached coil of fine wire (bottom). This cheap mic gave its life for audio education!

sound from the rear is blocked from reaching the front of the diaphragm. The diaphragm's diameter is another factor, again mostly at very high frequencies where the wavelengths are similar to the diameter. This increased directionality at higher frequencies applies to all omni microphones, not just dynamic models.

Dynamic Directional Patterns

Achieving other directional patterns with dynamic microphones requires adding an acoustical delay for sound arriving at the rear. Figure 17.5 shows a simplification of the method Electro-Voice uses to create a cardioid response in their Variable-D series, such as the RE20 model. Sound arriving from the front strikes the diaphragm, deflecting it as usual to generate an output voltage. Sound from the rear also arrives at the front of the diaphragm, as well as entering the various port openings to pass through a labyrinth of baffles. At some frequencies, the baffles delay the sound enough to put those waves in phase with sound going around the mic's body to reach the front. Since the phase-shifted waves reaching the rear of the diaphragm now have the same polarity as at the front, the front sound pressure is canceled. Again, this is a simplified model; in practice, multiple sound paths are used to delay different frequencies by different amounts, extending the directionality over a wider range of frequencies than a single delay path.

Sound from the front also gets into the ports, but it's delayed twice—once just to reach the ports farther back along the microphone's body and again through the baffles—so some frequencies are phase shifted 180 degrees, reinforcing the front sound rather than canceling it. In other words, pressure on the diaphragm's front and rear both push the diaphragm in the same direction—one pushing and the other pulling—which increases the diaphragm's deflection and microphone's output. By using a series of spaced vents along the mic's body rather than a single port, sound arriving from the front is staggered over time. This helps minimize the *proximity effect*, an increased output at low frequencies for sources close to the microphone.

Most modern cardioid dynamic microphones create the necessary acoustical phase shift with a low-pass filter based on a mass-spring system built from a weighted fabric (the mass) and air trapped in the capsule behind the diaphragm (the spring). This is shown in Figure 17.6. As with a labyrinth, it's not possible for a single filter to create a uniform *group delay*[1] over

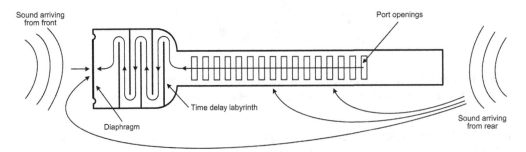

Figure 17.5: Cardioid dynamic microphones employ an internal acoustic delay, so sound from the rear arrives in phase with the same sound reaching the front of the diaphragm.

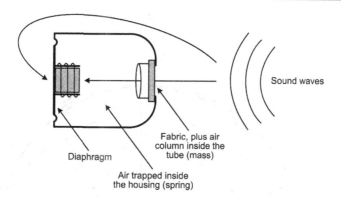

Sound waves

Fabric, plus air
column inside the
tube (mass)

Diaphragm

Air trapped inside
the housing (spring)

Figure 17.6: The common design for an acoustical phase shift network combines fabric covering a short tube with air trapped inside a sealed cavity. Together these create a mass-spring low-pass filter. This delays sound arriving through the rear port, which provides enough phase shift to cancel the same sound at the front.

the entire range of audio frequencies, so rear rejection is less effective at very low and very high frequencies. However, high frequencies arriving from the rear are blocked from reaching the front of the diaphragm by the microphone's body, so that helps maintain the rejection at higher frequencies where the acoustical delay is less effective.

Since we can't build an acoustic filter that has equal phase shift for all frequencies, directional response varies with frequency. Figure 17.7 shows the *polar plot* of a cardioid dynamic microphone's response to sound arriving from different angles. This type of graph shows how the response varies with frequency as well as angle of arrival. Polar response versus frequency is often plotted on the same graph using separate trace lines, as shown here. At 1 KHz the phase shift works as expected, rejecting sound coming from the rear almost completely. But at 100 Hz, sound from the rear is attenuated much less, partly because of the proximity effect of nearby rear-arriving sounds, which negates the cancellation.

The output from directional microphones also rises at low frequencies when the mic is closer than a few feet from the sound source. This applies to all directional microphones, not just dynamics. Chapter 1 explained the Inverse Square Law, which describes how the intensity of sound waves falls off with increasing distance. The same happens with microphones. When a singer is only a few inches from the front of a microphone, the level of the direct sound increases due to the short distance, as expected. But higher frequencies must travel farther to reach the rear ports that create the directional response, so they're effectively farther away and thus are attenuated more than low frequencies.

Another directional pattern common with dynamic microphones is the *supercardioid*, which is similar to cardioid but with a slightly tighter pickup pattern. This design also rejects sound

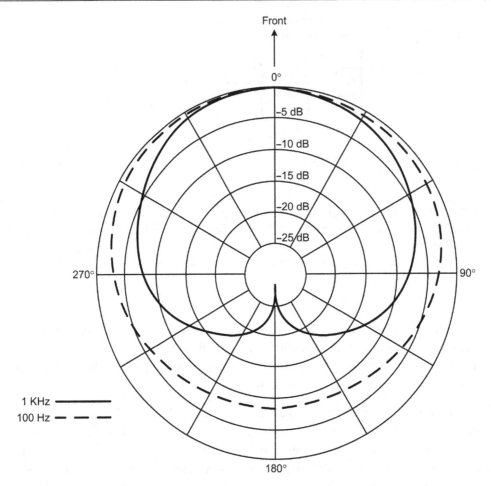

Figure 17.7: The cardioid pickup pattern is not uniform with frequency because the acoustic phase shift network attenuates high frequencies more than low frequencies.

from the rear less, shifting the maximum rejection from 180 degrees off-axis to 150 degrees, as shown in Figure 17.8.

When even more directionality is required, a *shotgun microphone* is the best choice. These are popular for TV and film use because they reject ambient noise arriving from other directions. Shotgun mics can use either dynamic or condenser capsules, and you'll often see them on the end of a long pole held by a *boom operator*.

Some vendors want to have it both ways. I often see ads for microphones claiming their sound is "warm" and "accurate" in the same sentence.

The response of most dynamic microphones falls off at the highest frequencies due to the mass of the diaphragm and attached coil. Their combined weight is simply too great to allow

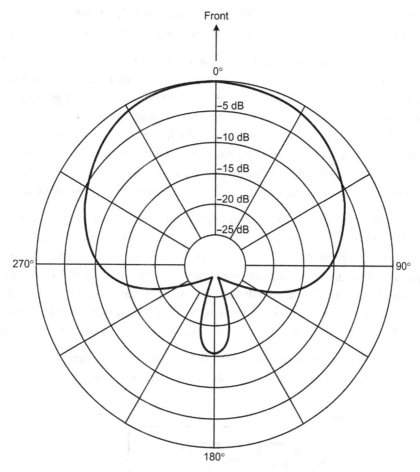

Figure 17.8: A supercardioid pickup pattern is similar to cardioid but slightly more directional.

vibrating at frequencies much higher than 10 KHz. Many dynamic microphones also have a peak in the response corresponding to a natural resonance within the capsule. Some people consider this peak to be beneficial, depending on where it falls. For example, the Shure SM57 has a peak of around 6 dB between 5 and 6 KHz, so it's popular for use with snare drums and electric guitar amps that might benefit from such a "presence" boost. Of course, adding EQ to a mic having a flat response can give the same result.

Ribbon Microphones

Ribbon microphones have been around since the 1930s. Technically, ribbon mics are classified as dynamic because they generate electricity via a metal conductor and magnet, as shown in Figure 17.9. They're constructed from a very thin strip of lightweight metal, which does double duty as both the diaphragm and coil, suspended in a strong magnetic field.

The thin ribbons on early models were fragile and easily destroyed by a blast of air from a signer's mouth or a loud kick drum in close proximity. Some early models also had a limited high-frequency response. Modern versions are sturdier and can capture higher frequencies. However, ribbon microphones have an extremely low output voltage because their electrical source is a single strip of metal rather than a coil of wire having many turns. The low output voltage is converted to a useable level by a step-up transformer, and some modern ribbon mics contain a preamp designed specifically to match the microphone. This improves the signal-to-noise ratio by raising the output level to be comparable to a dynamic mic.

A typical ribbon mic transformer has a *turns ratio* somewhere between 20 to 1 and 45 to 1, to convert 0.75 ohms to 300 ohms, or 0.1 ohm to 200 ohms. The ratio between the number of turns on the transformer's primary coil versus its secondary determines the change in voltage, but the impedance changes by the *square* of the ratio. So a ratio of 45 to 1 increases the voltage 45 times but increases the impedance by a factor of 2,025 which is 45 squared. This brings the ribbon's 0.1-ohm output impedance up to the 200 ohms expected by a mic preamp. Figure 17.10 shows the pieces that comprise a modern ribbon microphone's capsule.

Most ribbon microphones inherently have a bi-directional pickup pattern, also called Figure 8 due to the shape of the polar response when plotted. Figure 6.3 from Chapter 6 showed

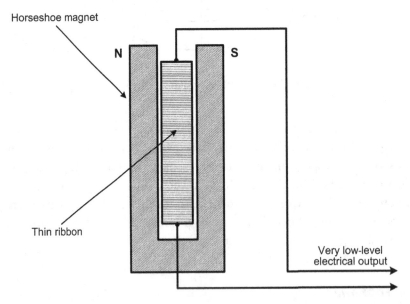

Figure 17.9: A ribbon microphone generates electricity using a metal strip inside a magnetic field. Unlike a dynamic microphone that contains many turns of wire, the single metal strip in a ribbon microphone has a very low output voltage, with a correspondingly low output impedance.

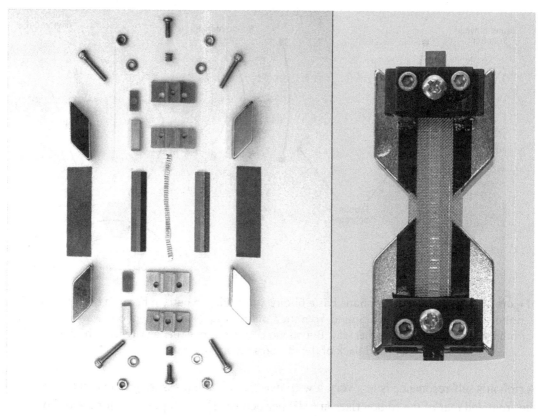

Figure 17.10: Ribbon microphones have come a long way since the 1930s! This photo shows the construction of a modern high-end ribbon mic that's more rugged than earlier models and also features an improved high-frequency response.
Photo courtesy of Royer Labs.

a Figure 8 pickup pattern, and Figure 17.11 shows why sound arriving from either side is rejected. The concept is pretty simple: Sound pressure arriving from the side impinges equally on both sides of the exposed ribbon. The waves therefore cancel, and the ribbon isn't deflected.

One important feature of ribbon mics is that their Figure 8 pattern is uniform over the entire range of frequencies, varying only slightly due to reflections inside the microphone's case and grill. Where a cardioid mic's pickup pattern varies substantially with frequency, you can place a ribbon mic to reject sound sources from both sides, confident that all of the sound will be rejected, not just the midrange, and that any off-axis sound picked up will not sound colored. Of course, reflections from a rejected source can bounce off room surfaces and find their way to the front or rear of the mic.

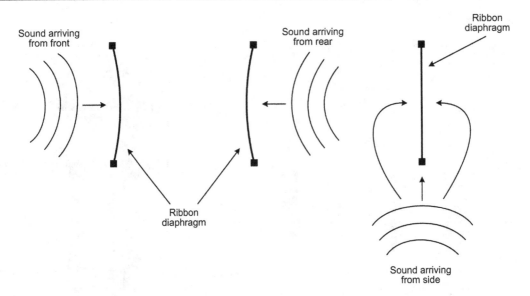

Figure 17.11: A ribbon microphone has a bi-directional pickup pattern because both sides of the ribbon are exposed to the air. Sound from the front pushes the diaphragm one way, and sound from the rear pushes it the other way. But sound coming from either side reaches both the front and back of the ribbon so it doesn't move.

A ribbon's self-resonance is at a very low frequency, and the pressure gradient between the front and rear of the ribbon rises at 6 dB per octave. This compensates for the ribbon's inherent 6 dB per octave roll-off above resonance, resulting in a net flat frequency response. Ribbon mics tend to have a uniform response that extends to high frequencies, with fewer ripples than many dynamic models. The ribbon diaphragm is very thin with a low mass, and no coil is attached as with dynamic mics, so the ribbon can move quickly. However, the frequency response of a classic "long ribbon" mic doesn't extend quite as high as a good small diaphragm condenser mic. Modern designs use shorter ribbons, often stiffened by vertical creases to vibrate more like a flat diaphragm.

Condenser Microphones

Condenser mics are the first choice of many recording professionals, favored for their extended high-frequency response. They use electrostatic properties to generate the electric signal rather than electromagnetism as with dynamics and ribbons. The word *condenser* is an obsolete name for the more modern term *capacitor*, but it's still used by recording engineers to describe this type of microphone.

The diaphragm material for a dynamic microphone can be almost anything that's light and compliant enough for the purpose because the connection to the attached coil is purely mechanical.

Figure 17.12: Modern large-diaphragm multi-pattern condenser microphone capsules employ two gold-plated Mylar diaphragms placed back to back, with a rigid metal plate between them. A switch changes the DC bias voltage on the diaphragms to vary the pickup pattern between omnidirectional, Figure 8, and cardioid. *Photo courtesy of TELEFUNKEN Elektroakustik.*

But the diaphragm in a condenser microphone must be capable of conducting electricity because it forms one plate of a capacitor. The diaphragm in modern condenser mics is typically very thin Mylar film, with an even thinner layer of gold applied to make it conductive. You can see in Figure 17.12 that some condenser microphones also feature switch-selectable pickup patterns.

Condenser microphones are often categorized by the diameter of their diaphragm—large or small. A large diaphragm is generally an inch or larger in diameter, and a small diaphragm is typically half an inch or less. Another type uses what I call a "tiny" diaphragm, often about ¼ inch in diameter, though some are even smaller. Originally designed for measurement due to their extremely flat frequency response, tiny diaphragm mics have been embraced by recording engineers since the 1980s.

Generally speaking, the smaller the diaphragm, the faster it can move and, in turn, the higher the frequency the microphone will respond to. The downside of small diaphragms

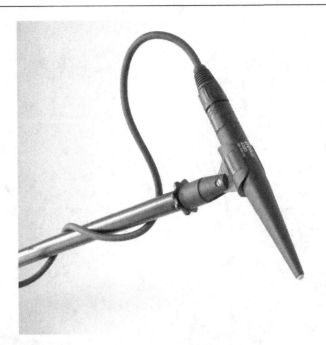

Figure 17.13: The diaphragm in this precision DPA 4090 omni condenser microphone is slightly smaller than ¼ inch in diameter.

is they capture a smaller portion of the acoustic waves, so they have less output and thus a poorer signal to noise ratio. On the other hand, receiving less of the wave means a small diaphragm mic can handle a louder acoustic volume before distorting. The DPA 4090 shown in Figure 17.13 can handle SPLs as high as 134 dB before clipping, with an equivalent input noise of 23 dB A-weighted. There are also *extremely* tiny condenser mics made by DPA and others that clip directly onto violins, trumpets, and other instruments.

Condenser microphones require DC power for two purposes: A DC bias voltage is needed to charge the internal capacitor formed by the diaphragm(s) and rigid back plate and to power the built-in preamp that all condenser microphones require. The capacitor must be charged initially, because the voltage across the capacitor changes as the capacitance varies in response to sound waves deflecting the diaphragm. It's the *change* in voltage across the capacitor that eventually appears at the microphone's output. Without an initial DC voltage to start with, changing the capacitance does nothing. However, *electret* condenser microphones are charged permanently during manufacture, so this type of mic needs voltage only to power the built-in preamp. Regardless of a condenser mic's element type, the necessary voltage can come from either batteries or an external power supply.

The basic operation of a condenser microphone is shown in Figure 17.14. The bias voltage is shown as the schematic symbol for a conventional battery, though it's labeled 48 volts,

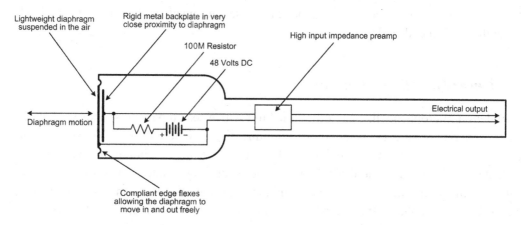

Figure 17.14: The electrical output from a condenser microphone is derived from a DC voltage that changes as the diaphragm is deflected by sound waves.

since that's the standard for phantom power. Some microphones can operate with as little as 15 volts, using an internal DC-to-DC converter to generate the higher voltage needed to polarize the capsule. Regardless of how the capsule is charged, or *polarized*, the reason the voltage changes is the physics Law of Conservation of Energy. A given polarizing voltage across a given amount of capacitance creates a stored charge whose amount is expressed in Coulombs. In this formula, the voltage (E) across a capacitor is equal to the charge in Coulombs (Q) divided by the amount of capacitance (C):

$$E = Q/C$$

So as the capacitance changes with minute movement of the diaphragm, the voltage *must* change to reflect the same amount of energy stored in the capacitor.

The output voltage of a condenser microphone is fairly large, but its extremely high output impedance provides only an infinitesimal amount of current. Therefore, the built-in preamp must have a very high input impedance to prevent loading the capsule's output. A typical value is 100M—100 million ohms—or even higher. These preamps use either an FET transistor or a vacuum tube to achieve a suitably high input impedance. Although "preamp" is the common term for the electronics in a condenser mic, it's a bit of a misnomer. It's really an active impedance converter.

Unlike the transformer in a ribbon mic that raises the output impedance along with the voltage, a condenser capsule's impedance must be reduced to a value usable in the outside world. As explained in earlier chapters, a circuit that operates at an extremely high impedance is susceptible to high-frequency losses due to wiring and other capacitance. So a condenser mic's electronics are built into the microphone close to the capsule. This also reduces the chance of picking up hum and radio signals, because the capsule and preamp are shielded inside the mic's metal enclosure.

Condenser Directional Patterns

A condenser microphone, like any other mic built with a sealed back, responds to changes in atmospheric pressure and is inherently omnidirectional. When sound is blocked from the rear, wave pressure arriving at the front of the mic deflects the diaphragm no matter which angle it comes from. A cardioid pickup is created by drilling a pattern of holes in the back plate to allow sound waves to reach the diaphragm through that path, similar to the rear vents in a dynamic microphone. If a second diaphragm is added on the other side of the back plate, other patterns can be created by combining the outputs from both diaphragms.

Figure 17.15 shows how the three most common pickup patterns are selected by applying a positive, a negative, or zero voltage to the rear diaphragm. It's also possible to obtain in-between directional patterns by varying the DC bias voltage applied to the rear diaphragm, rather than applying the full 48 volts using a switch.

Condenser microphones tend to have an excellent frequency response, thanks to the low mass of their diaphragms. Not only are the diaphragms light enough to vibrate quickly, but their self-resonance can be better controlled. So like ribbon microphones, condenser microphones

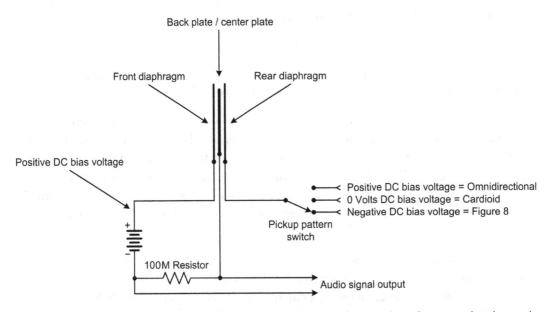

Figure 17.15: The pickup pattern of a pressure-gradient condenser microphone can be changed by switching the bias voltage applied to the rear diaphragm.

tend to have a smooth response across the audible range, with minimal rippling. Condenser microphones that have a tiny diaphragm ¼ inch or less are especially well suited to capturing very high frequencies. These are often used by acousticians to measure loudspeaker and room response.

One downside of condenser microphones is they're more fragile than dynamic mics. Another issue is the close proximity between the diaphragm and the back plate, which makes them sensitive to humidity. In very humid conditions, a condenser microphone might make a sputtering sound as the DC bias voltage arcs across the narrow gap, or the mic may not work at all.

Another type of condenser microphone uses the changing capacitance to modulate a radio frequency (RF) signal rather than generate audio directly by varying a DC bias voltage. This is exactly the same as conventional FM radio, where the frequency of a radio oscillator, or carrier, is modulated at an audio rate, then demodulated to reproduce the original sound. Sennheiser has been making RF condenser microphones this way for many years, and one advantage is less influence from humidity because no polarizing voltage is used.

Other Microphone Types

Boundary microphones are omnidirectional and are meant to be mounted directly onto a reflecting surface—generally a wall of a recording studio or other room, though they're often used on the stage floor in theaters. They can also be placed on a large surface such as a conference table or lectern. This type of mic is commonly called *PZM*, short for Pressure Zone Microphone, though that name is a trademark of Crown International, the company that licensed and produced the first commercial version in 1980. Like Kleenex, the trademarked name has become the generic name in the industry.

The main feature of a PZM is it avoids comb filtering due to reflections from nearby surfaces. Since the microphone element is aimed directly toward a large surface and is placed very close to that surface, it receives only direct sound from the surface rather than a mix of direct and reflected sound. Of course, audio sources can still sound distant and reverberant when they're far away from the microphone due to other reflections in the room.

A *parabolic microphone* combines a directional microphone capsule with a parabolic reflector to create a microphone that's extremely directional. The principle is the same as a TV satellite dish, and the dish diameters are even similar; sound waves travel much more slowly than radio waves, so acoustic wavelengths are in turn much shorter than radio wavelengths. A dish that's 18 inches in diameter is one wavelength at 750 Hz when used with a microphone, but it's also one wavelength at 650 MHz when used as a radio antenna.

Optical microphones work much like radar, but instead of radio waves, they send a laser beam toward a vibrating surface, then convert the varying reflections into an audio signal.

This type of microphone is not so useful in a recording studio, but it's great for spying on your neighbors. If you aim the mic at a closed window, sound from people talking on the other side of the window vibrates the glass ever so slightly. The changing reflections are then decoded back to the original sound.

A *USB microphone* can be any basic type, but most are cardioid condenser or dynamic. What distinguishes a USB mic is its built-in preamp, plus an A/D converter that shows as an audio sound card input in your recording software. The main advantage for home recordists and podcasters is they're simple to connect to a computer. They don't need a preamp or mixer, just a computer with a USB port. Some USB mics have a built-in earphone jack and small mixer so you can hear yourself while recording, and there are also stereo USB mics.

Phantom Power

Phantom power is a clever method of sending DC voltage to microphones that rely on external power. The most common is 48 volts; often the phantom power switch on a mixer or preamp is labeled "48 V," but phantom power as defined in the standard IEC-61938 covers 12, 24, and 48 volts. Rather than require two additional wires just for the power feed, the same two wires that send audio from the microphone to the preamp are also used to send the 48 volts to the mics. Even better, microphones that do not need power can usually be connected safely and won't be harmed or otherwise affected.

Phantom power works with balanced microphones only, which is usually the case with mics that have an XLR output connector. It can also be used to power active DI boxes. Phantom power is usually built into mixers or microphone preamps, though stand-alone units are available. Most condenser microphones that use a vacuum tube need more than 48 volts to operate the preamp, so they come with their own power supply rather than relying on phantom power. The block diagram of a phantom power system in Figure 17.16 is divided to show the preamp and microphone portions.

The key to phantom power is applying exactly the same voltage to both the plus and minus signal wires. Precision 1 percent (or better) tolerance resistors are used to avoid upsetting the balanced connection. Since transformers pass audio frequencies but not DC, the 48 volts is taken from a center tap on the output side of the mic's output transformer. It's also possible for a microphone or active DI to receive phantom power even if it doesn't have an output transformer. In that case, a corresponding pair of matched resistors inside the unit taps into both signal wires to retrieve the voltage.

You should *never* connect an unbalanced microphone of any type to a preamp that provides phantom power. You probably won't harm the preamp because the 6.8 K resistors limit the amount of current that can be drawn, but the microphone might be damaged.

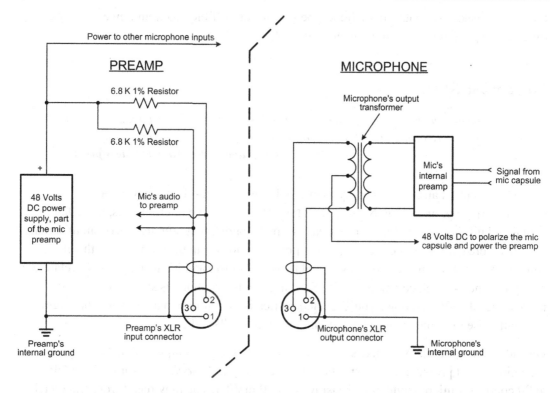

Figure 17.16: Phantom power sends 48 volts through matched resistors to both signal wires from the microphone's output transformer. That voltage is then taken from the center tap of the output transformer and used internally by the microphone. Mics that don't need power are not affected as long as they use balanced wiring and their output transformer or voice coil is not connected to ground.

There's a long-standing myth that ribbon mics should never be connected to phantom power. Like many myths, this has some basis in fact: The original RCA Model 44 microphone had a center tapped output transformer with the tap grounded to reduce hum. If this mic was connected to a phantom power source, voltage was applied across the ribbon, causing it to tear. Once users figured that out, the center taps were disconnected, solving the problem. Yet the "all ribbon mics are very sensitive and shouldn't be connected to phantom power" myth continues.

It's also possible to damage a balanced mic that doesn't need power by passing its output through a conventional Tip/Ring/Sleeve-type patch bay. When you insert a ¼-inch balanced plug into a balanced jack, it's possible for the plug's tip contact to briefly touch the jack's ring and the plug's ring to touch the jack's grounded sleeve at the same time. If that happens, the full 48 volts is sent back into the microphone's coil or ribbon. Using a ¼-inch patch panel for balanced microphones is not usually recommended, but some people do it anyway. If you use this type of patch bay with microphones, be sure to turn off the phantom power supply every time before plugging or unplugging any microphone.

Microphone Specs

I've certainly spent many hours with finicky artists trying different vocal mics, all of which sound remarkably similar, and all I have to say is that I felt it was a waste of time.
—Alan Parsons, famous recording engineer/producer

All of the standard audio specs apply to microphones as well—frequency response, distortion, ringing, and noise. Noise is a factor only with active microphones, those that contain a built-in preamp. Of course, a passive microphone that outputs a very small signal requires more gain from the preamp, so the preamp's noise can be a problem with soft sources. Microphone noise is often referred to as *self-noise*, and it's usually spec'd relative to an equivalent A-weighted ambient SPL. So if a given microphone is stated to have a self-noise of 15 dB SPL, the noise you'll get in practice is the same as placing a mic having no inherent noise in a room whose background noise is 15 dB SPL.

One additional microphone spec is *sensitivity*. This defines the output voltage for a given incoming sound pressure level, expressed as *millivolts per Pascal*. For example, the DPA 4090 condenser microphone has a sensitivity of 20 mV/Pa. You may recall from Chapter 1 that the Pascal is related to SPL because both express pressure. A volume of 0 dB SPL—the threshold of hearing—is equal to 20 microPascals (μPa), so 1 Pascal is the same as 94 dB SPL. Therefore, with its sensitivity of 20 mV/Pa, a DPA 4090 outputs about 4 mV with 80 dB SPL present at its diaphragm. In other words, 80 dB SPL is 14 dB softer than 94 dB SPL (1 Pascal), and 4 mV is 14 dB below 20 mV. If all this talk about microPascals and millivolts makes your eyes glaze over, don't worry—I promise none of this is needed to be an excellent recording engineer.

There's a common myth that condenser microphones are more sensitive to picking up room sounds and other unwanted distant sources than dynamic or ribbon microphones, but that simply isn't true. Yes, most condenser microphones output a larger voltage for a given SPL level, but that just scales the same captured signal up or down. The relation between close and far sounds remains the same. Of course, microphones have different directional patterns, and *that* will influence how loudly unintended sounds are captured. But directionality is a different parameter, and any of the common microphone types can be omnidirectional, cardioid, or supercardioid.

THD and IM distortion are also important microphone specs. Unlike electronic circuits that are usually very clean right up to the onset of gross distortion, microphone distortion often creeps up slowly at higher SPLs. A mic's diaphragm can flex only so far before it bottoms out in either direction, not unlike a loudspeaker, but tension on the diaphragm increases before hitting a hard limit of excursion. Active microphones contain electronics, so that's another potential source of distortion. As mentioned in Chapter 5, many active microphones include a built-in attenuating "pad" that can be switched on when recording drums and other loud sources. The pad connects between the mic's capsule and internal preamp to avoid overdriving the preamp. But eventually the diaphragm itself will distort. Further, even when using dynamic mics, some older preamps don't allow setting their gain low enough to avoid distortion when the input voltage is very high, so in that case you'll need an external passive pad.

Measuring Microphone Response

Measuring the frequency response of a microphone requires an anechoic chamber, as shown in Figure 17.17. Trying to measure the response of a microphone (or loudspeaker) in a regular room doesn't work because reflections from the room's surfaces skew the response. You also need a loudspeaker sound source with a response that's very flat, or at least known. Many high-end microphones come with a printed response graph measured at the factory for that specific mic. You can also send a microphone to a third-party calibration company, which they'll return along with a graph of the measured response.

It's possible to measure a mic's frequency response yourself, but it's tricky. One way is to hoist the microphone and a known-flat loudspeaker 20 or 30 feet up in the air outdoors to avoid reflections from the ground. However, finding a loudspeaker that itself is accurate enough will be a challenge, not to mention the mechanical logistics of such a test.

Figure 17.17: An anechoic chamber absorbs sound fully at all frequencies down to 100 Hz or even lower. This avoids reflections that skew the results when measuring the frequency response of microphones and speakers. Note the steel mesh floor with additional absorption below to avoid floor reflections. *Photo courtesy of Orfield Labs.*

You can also measure the response in a regular room, using a technique called *gating* to ignore the reflected waves that arrive soon after the original direct sound. Most room measuring software includes an option to specify the gate time, and this method is also used to measure loudspeakers. The problem is you need to set the input gate to turn off the audio only a few milliseconds after the direct sound arrives at the mic, unless all of the room's surfaces are far away. So unless you have a large room with a very high ceiling and you put the microphone and speaker up on a tall ladder, you'll need to set the gate time so short that low frequencies are excluded from the measurement.

I once saw a post in an audio forum by a student who used this method to measure a number of popular studio monitors. He borrowed a motorized lift to raise each speaker 20 feet above the floor in his college's auditorium and used an equally tall microphone stand to place a calibrated measuring mic directly in front of each speaker about six feet away. By setting the gate time in the measuring software to exclude reflections from surfaces farther than 20 feet away, his tests were valid to below 30 Hz.

Another method is to compare the microphone being tested against another mic having a known response and use the difference between the two measurements to determine the response of the microphone being tested. To do this you'll need to borrow or rent a calibrated microphone, then measure the microphones one after the other with both mics in the exact same place. One advantage of this approach is that the loudspeaker sound source doesn't need to be flat or even known. All you care about is the difference between the two microphones, plus the absolute response of the reference mic. However, this works best when both microphones have the same pickup pattern. Otherwise, room reflections that arrive from other angles can influence one mic's measurement more than the other. This method will be explained in more detail in Chapter 22. It's also possible to measure microphone distortion yourself, and Chapter 24 shows one way to do this with reasonable accuracy in a home setting.

> *It's a very flat microphone with a 7 dB rise between 5 and 10 KHz.*
> **—From a magazine review**

Speaking of microphone frequency response, I can't help but comment on the frequent claims by sellers that their microphones have a "pleasing" response curve. New microphone models are announced almost daily, and some include descriptions that rival empty audiophile wording such as "a smooth, full sound," yet they also claim a response flat within a dB or two across the audible range. If a microphone is flat, then by definition it has no sound of its own, nor can it be full or smooth and so forth.

Other mics show a response having an obvious "presence peak" in the treble range, with ad copy that claims such a response makes vocals and other sources cut through a dense mix.

That may be true, but you can achieve the exact same sound with a flat microphone and an equalizer. Further, using a mic that has a prominent boost at some frequency will add that same boost to everything you record with it, whether a given voice or instrument benefits from boosting that frequency or not. Chapter 2 disproved the stacking myth, but this sort of accumulated response coloration is very real. Of course, you can set an equalizer to counter the presence peak for sources that do not benefit from such a response.

Microphone Modeling

Finally, microphone "modeling" is a popular buzzword that tries to make the simple application of equalization seem more impressive than it really is. The concept seems reasonable enough: By applying an appropriate EQ curve, you can make one microphone sound like a different model. Plug-ins are available that list a large number of "source" and "target" microphones to choose from. You select the microphone model you actually recorded with from the source list, then pick the microphone you *wish* you had used from the other list. The plug-in then applies whatever EQ is necessary to make the source and destination frequency responses match.

Unfortunately, the reality is less impressive, and there are many reasons such plug-ins can't really do what's claimed. One limitation is dealing with frequencies too low or too high for the original microphone to capture. For example, a Neumann U-47 responds to frequencies higher than 20 KHz, but a Shure SM57 rolls off sharply above 12 KHz. No amount of EQ compensation can recreate content that was never captured by the SM57 in the first place. All you'll accomplish is adding a bunch of trebly hiss from the preamp. This also goes for very low frequencies that were never captured. Further, different microphones of the same model can sound different, especially if they were manufactured years apart. So then you have to wonder which specific Neumann U-47 was used as the reference.

Microphones vary in other ways that cannot be easily countered later, such as proximity effect. Unless the software knows how far a singer or acoustic guitar was from the microphone, and how both the source and target microphone's low-frequency response varies with proximity, the software can't know how much low-frequency boost to add or remove. Further, many singers "work" the microphone to maintain an even volume level, changing the distance continually as they sing louder and softer. That varying proximity response can't be determined or emulated either.

The variable off-axis response of different microphones is another factor. If you record five singers in a semicircle around a microphone, the frequency response for some of the singers will likely vary compared to the singers that were more on-axis. And some singers might be closer than others, which can also affect the response. So this, too, cannot be compensated for afterward with EQ, since all five singers were already pre-mixed together acoustically in the air. While no microphone simulator will turn your inexpensive dynamic mic into a top-shelf condenser model, it's still an interesting tool that lets you change the sound in what might be a useful way with just a few mouse clicks.

Guitar Pickups and Vibrating Strings

Like a dynamic microphone, a guitar pickup consists of a magnet and coil of wire, but with a guitar pickup, both the magnet and coil of wire remain stationary, while the steel string moves through the pickup's magnetic field. This is called a *variable reluctance* transducer, and it works on the same principle as Hammond organs described in Chapter 14. When a vibrating string passes over the face of a magnet, the amount of magnetic flux through the coil changes as the string moves nearer or farther from the magnet. So yet again, electromagnetism is the basis for an audio transducer.

Most electric guitars have two or three pickups placed at different distances from the bridge that anchors the strings at one end. Likewise, many electric basses have two pickups in

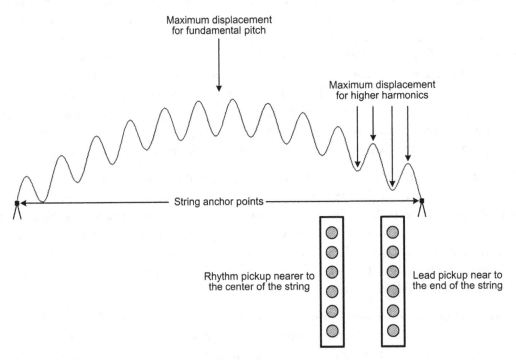

Figure 17.18: A vibrating guitar or bass string has maximum displacement for the fundamental pitch near the center of the string. But at higher harmonic frequencies, the string's motion is substantial even near the anchor points at each end.

different locations. Figure 17.18 shows the motion of a vibrating string that's anchored at both ends. This drawing is simplified to show only the fundamental pitch plus one higher harmonic. As you can see, the string's maximum displacement for the fundamental pitch is at the center of its length. But at the higher harmonic frequencies, maximum displacement occurs much closer to the end, as well as at other points along the string's length. So a pickup placed near the end is excited more by the harmonics than the fundamental, simply because there's less motion at that location for the fundamental.

The same happens when you pluck a string close to the end versus nearer to the center. Plucking near an end excites the overtones more than the fundamental, because the middle of the string doesn't move as far for the same amount of plucking energy, resulting in a thinner sound. Note that vibration near the end of the string doesn't contain more high-frequency content than elsewhere. It simply moves less distance there at the fundamental pitch, so it contains *less* energy at those lower frequencies. This is a subtle but important distinction. By the way, plucking a string harder also increases the level of harmonics relative to the fundamental because that creates a faster rise time.

Summary

This chapter explains the inner workings of microphones, including a bit of history about older carbon and piezo models rarely used today. All microphones output a voltage corresponding to changes in air pressure, and exposing or sealing the rear of the diaphragm gives different pickup patterns. Many large-diaphragm condenser mics instead employ two exposed diaphragms and vary the pickup pattern by changing how the elements are polarized.

Dynamic microphones, including ribbons, use electromagnetism to generate an audio voltage. A dynamic microphone uses a coil of wire having many turns, so it puts out a much larger voltage than a ribbon mic, which employs a single straight metal strip—in essence a one-turn coil. Most condenser microphones generate an audio signal by modulating a DC bias voltage that changes with the diaphragm capacitance, though some models use frequency modulation to vary the frequency of a radio oscillator.

Regardless of the principle used, the weight of the diaphragm is the main factor that dictates the highest frequency a microphone can capture. Dynamic mics have a coil of wire attached to the diaphragm, so they tend to have the poorest response. Besides affecting the frequency response, the excursion limits of every microphone's diaphragm determines how loud a sound it can capture without distortion.

Phantom power is an important concept that simplifies powering many microphones from a mixer or mic preamp, without having to provide an AC outlet for each or worrying about batteries running down. Better, by avoiding the need for separate AC power, hum from ground loops is avoided.

You also read that microphone performance is spec'd in the same way as most other audio gear, using the same technical parameters, though, as with loudspeakers, their frequency response is best measured in an anechoic chamber to prevent room reflections from skewing the measured response.

Finally, guitar and bass pickups capture more or less of the fundamental frequency of a vibrating string, depending on where they're placed in relation to the string's end points. A string's maximum displacement varies by frequency along its length, so a pickup near the center of the string receives more of the fundamental pitch than when close to either end.

Note

1 Group delay is the *rate of change* between phase shift versus frequency.

CHAPTER 18

Loudspeakers and Earphones

Loudspeakers are the opposite of microphones, converting electricity into equivalent acoustic energy. Even though they serve opposite purposes, many of the same technologies are used. For example, most loudspeakers use electromagnetic force to move a coil of wire back and forth inside a magnetic field. A cone made of paper or plastic is attached to the coil, and the cone increases and decreases the surrounding air pressure to create sound. This is very similar to the coil, magnet, and diaphragm of a dynamic microphone. Indeed, a dynamic microphone can function as a loudspeaker, and vice versa. Other speaker designs are based on an electrostatic principle, much like condenser mics. Piezoelectric elements are also used, as with microphones and contact pickups.

Perhaps the first question to consider is how a single speaker driver can reproduce more than one frequency at a time. The principle is pretty simple: Even though a loudspeaker's cone moves only back and forth, it can do so at more than one rate simultaneously. To help visualize this, I created the video "loudspeaker_motion," which shows the cone's movement side by side with a picture of a simulated Wave file containing two frequencies at once. As the video "plays" the audio file, you can see the cone move slowly left and right for the lower frequency, while also moving more quickly at the higher frequency. Note that the same motion occurs in reverse with a microphone, as its diaphragm vibrates in response to complex sound waves in the air.

Loudspeaker Basics

The basic design of dynamic (electromagnetic) loudspeakers has been around since before 1900, though modern versions are, of course, greatly improved and have much higher fidelity. The earliest dynamic speakers used electromagnets requiring a power supply, versus the permanent magnets used today. Clever engineers used the loudspeaker's magnet coil as the filter inductor in a radio's power supply, providing power to the electromagnet while saving the cost of an expensive component in the process. Other speaker driver designs use a flat membrane, a thin ribbon, a piezo element, and even compressed air. But the basic dynamic loudspeaker shown in Figures 18.1 through 18.3 remains the most common design.

Figures 18.1 through 18.3: Most loudspeakers are built using a coil of wire suspended within a magnetic field, which in turn is attached to a paper or plastic cone that actually moves the air to create sound. Other cone materials are fabric, metal, composite material, and even wood. Again, an innocent transducer was sacrificed for audio education.

Figure 18.4 shows the construction of a typical dynamic loudspeaker, identifying the various components. When electricity is sent to the voice coil, it pushes the cone outward or pulls it inward, depending on the polarity of the voltage at that moment. It's common knowledge that magnets repel when their polarities are the same and attract when opposite. In a dynamic loudspeaker, the permanent magnet has north and south poles that never change, while the

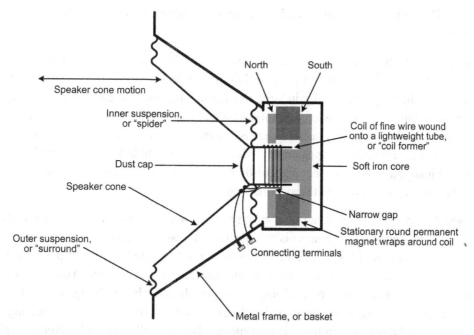

Figure 18.4: A dynamic loudspeaker driver uses electromagnetism to move the voice coil forward and back, which in turn moves the attached cone to create sound waves.

electromagnet created by current through the voice coil changes in step with the audio. Here, the magnet focuses *magnetic flux* into the narrow gap at the front of the assembly. The result is a "linear motor" based on the same principle that drives a rotating motor. Note that some speaker drivers use a round bar magnet inside the coil, where the center portion of the soft iron core is shown.

The outer surround at the front of the driver is generally a very pliable material made from soft corrugated impregnated fabric, or a plastic or rubber *roll surround*, strengthened to survive flexing back and forth literally billions of times over the expected life of the driver. The inner surround, or *spider*, is equally flexible and rugged. Together these surrounds allow the cone to move in and out, while keeping the voice coil aligned in the center of the narrow gap without scraping the sides. The surrounds also serve as springs to return the driver to its center rest position when no voltage is sent to the voice coil. If the driver will operate above a few hundred Hz, the surrounds must also absorb waves propagating in the cone itself, or resonances will color the sound.

The voice coil consists of copper wire insulated with an enamel coating (called "magnet wire"), wound around a tube called the *voice coil former*. High-power speakers use flat copper or aluminum ribbon instead of round wire because they carry a lot of current and must dissipate heat quickly to protect the driver. A flat wire's larger surface area radiates heat more efficiently. The coil former is often made of aluminum, though inexpensive drivers may use paper or plastic.

For high efficiency, the gap between the coil and surrounding magnet structure must be extremely narrow, so the pieces of a loudspeaker must fit together precisely. One design and manufacturing challenge, especially in large excursion woofers, is keeping the coil from touching the magnet and core assembly during violent motion. If the coil and magnet are not in intimate proximity, efficiency suffers because more current must be sent into the voice coil to obtain the same amount of displacement. Likewise, the strength of the magnet also affects efficiency. Finally, the center dust cap keeps dust and other particles from getting into the tiny space between the voice coil and the surrounding structure.

A bare speaker driver behaves as a mass-spring system, and one important property is its free-air resonant frequency. The mass is the speaker cone and attached voice coil, and the spring is a combination of the outer and inner surrounds that hold the cone in position. When at rest, with no voltage applied, the cone is at the center of its travel range. You can move the cone forward or back either by pushing it with your hand or by applying electricity to the voice coil. Either way, when the cone is released, it returns back to its center resting place because of the spring action of the two surrounds. In a well-designed driver, the spider suspension and the outer surround suspension will have similar resistance in both directions. But speaker drivers vary all over the map; a poorly made driver can have a stiff surround with a less stiff spider, or vice versa. Even more important than matching stiffness is the linearity of the restoring force as the diaphragm moves in and out by different amounts. The suspension linearity and the linearity of the motor's magnetic force related to cone displacement both affect the driver's distortion.

There's an inverse relationship between the frequency sent to a loudspeaker and the displacement of its cone. As the input frequency drops by half, or one octave, the cone must move twice as far to produce the same output level. The reason is simple: When driven by a voltage source, as with modern solid-state power amplifiers, a dynamic loudspeaker operates as a *constant-velocity* device. This means that for a given volume level its cone moves at the same velocity no matter what frequency is applied (within its normal operating frequency range). Since the cone moves only half as fast at the lower frequency, it must move *twice as far* to output the same total acoustic velocity. Note that this behavior is how dynamic speakers work naturally, so the driving power amplifier doesn't need EQ to increase its output at lower frequencies to maintain the constant velocity needed for a flat frequency response.

Loudspeaker Driver Types

It's impossible to design a single speaker driver that can produce the entire range of audio frequencies from 20 Hz to 20 KHz. Single-driver loudspeaker systems are available, but none actually cover the full range efficiently, and they're overly directional at high frequencies. Therefore, most loudspeaker systems use two or three different driver types, each optimized for one portion of the audible range. The incoming audio is split into separate frequency bands by a *crossover*, which will be described shortly. Occasionally you'll find a loudspeaker that uses even more frequency bands and driver types. I used to have an old pair of McIntosh XR 14 hi-fi speakers my wife bought in 1983. This is a four-way model with crossover frequencies at 700 Hz, 1.4 KHz, and 7 KHz.

Most modern loudspeakers use two types of drivers—called *woofers* and *tweeters*—for low and high frequencies, respectively. In a three-way system, the middle band is handled by a *midrange driver*. Many different frequency ranges are used. A two-way system might cross over at 3 KHz, while a three-way speaker might use 500 Hz and 4 KHz. A few speaker models include a *super-tweeter* to handle the highest frequencies, generally above 10 or 15 KHz. Some can even reproduce frequencies as high as 35 KHz, which is well past the upper limit of audibility. In my opinion, that's just a marketing gimmick, since nobody can hear such content. Loudspeaker system design requires a long list of engineering trade-offs—not just a choice of crossover frequencies and driver types. Indeed, if there was one "best" way to design a loudspeaker, they would all be made that same way.

Midrange and tweeter drivers are often sealed with a solid metal back, or they use a separate enclosure sized for their intended frequency range. Where the full-range speaker driver in Figures 18.1 through 18.3 has openings in the rear of its metal frame, cone tweeters with a sealed back have no such openings. A woofer moves much farther than a tweeter and creates much more pressure inside the enclosure. If the back of the tweeter is not isolated from the internal cabinet pressure, it's possible for the tweeter's cone to be pushed near its excursion limit by the woofer compressing and expanding the air inside the enclosure. This increases the tweeter's distortion even when its own signal is at a low volume level. Some speaker designs divide the cabinet into separate closed-off areas using interior baffles, but smaller drivers often have metal or molded plastic enclosures that cap the rear.

Most woofer drivers are conventional dynamic speakers, though models designed to produce very low frequencies often have a highly pliable outer surround and spider allowing the cone to move as much as two inches each direction. Most midrange drivers are also the standard dynamic type, and being constant-velocity devices, they don't need such a long excursion to create an acceptably loud volume. Likewise, tweeters operate with even less displacement.

Two basic diaphragm types are commonly used for dynamic tweeters: cone and dome. A cone tweeter is similar to a typical woofer or midrange driver, only much smaller. A dome tweeter's diaphragm is convex, hence the name. It has less surface area than a cone, which limits the lowest frequency it can reproduce. However, since the radiating surface is driven from the outer edge rather than from the center, the dome is less prone to flexing and deformation as it vibrates, especially at high volumes. This is similar to the structural strength of an egg or an arch holding up a bridge.

One important property of all tweeters is their *dispersion*, which describes how evenly they radiate various frequencies in different directions. Dispersion will be described in more detail later in this chapter. It might seem that a dome-shaped driver would radiate high frequencies more uniformly than a cone, but in practice dome drivers have a dispersion pattern similar to cones.

There are also full-range single-driver speakers. The Auratone 5C was hugely popular in the 1980s, and every recording studio had one as a second reference even though its frequency response was limited, to put it kindly. Small speakers are intended to get a sense of how a music mix or TV commercial will sound on the limited range built-in speakers typical of small radios and TV sets. But there's an important difference between a small speaker and a lousy speaker. A good small speaker aims to be as flat as possible over its limited range of frequencies, and with minimal resonance. A good small speaker is also able to output high enough volume levels without too much distortion in order to make informed mixing decisions. Sadly, the Auratone was not a particularly good small speaker. Perhaps the best way to judge how your mixes will sound on lesser systems is to simply listen through high- and low-cut filters on regular high-quality monitor speakers.

Planar speakers, also called *dipoles*, are often electrostatic, though some models are magnetic. In either case, instead of a cone, a large plastic membrane is suspended in a frame with both its front and rear surfaces exposed to the air. Therefore, this type of speaker is inherently bi-directional, radiating sound equally front and back. Some dipole speaker vendors claim that reflections from the front wall of the room behind the speaker, due to the bi-directional radiation, increase depth and spaciousness compared to conventional box speakers. But unless the speakers are placed far away from the wall, those reflections just create comb filtering. In my opinion, the bi-directional pattern is a *by-product* of the design, not a feature, and the rear wave should usually be absorbed or diffused.

Planar speakers also tend to be inefficient, requiring much more power than conventional box speakers for the same output SPL. Their displacement is also limited, so a large surface area is needed to produce lower frequencies. But then the large surface area increases directivity. So unless the diaphragm is divided into separate smaller areas for high frequencies, the result is a very directional speaker that's unable to deliver the same response to three people sitting on one sofa. For this reason, many commercial planar designs use a smaller surface and add a standard cone woofer that takes over below 500 Hz or so, where planar drivers are even less efficient.

Ribbon tweeters are becoming more popular, perhaps by association because ribbon microphones have made a comeback. But there are limitations: The ribbon is a single short wire that by nature has a very low impedance, so a transformer is often used that adds its own distortion. One perceived advantage of a ribbon is that its radiating surface can be larger than a small dome-type tweeter, so a ribbon can move more air for the same displacement, making it potentially more efficient than a cone or dome. However, the larger surface area increases its directivity. Figure 17.9 showing a ribbon microphone is basically the same as a ribbon tweeter; just change the "output" wires label to "input," and you'll have a pretty good picture of how ribbon tweeters are built. However, ribbon tweeters in speakers made by ADAM Audio, a popular manufacturer, work slightly differently. According to the ADAM website: The X-ART membrane consists of a pleated diaphragm in which the folds compress or expand according to the audio signal applied to them. The result is that air is drawn in and squeezed out, like the bellows of an accordion. This design was invented by Dr. Oscar Heil, who called it the "air motion transformer." One advantage over a flat ribbon diaphragm is it can be smaller in area for a similar output level, therefore improving dispersion.

There are two basic types of *horn* speakers. One is a type of driver design used to create highly efficient midrange and tweeter drivers. This is called a *compression driver*, and it's attached to a flared tube that spreads the sound outward. The other type of horn is an enclosure design that's highly efficient at bass frequencies. The latter are used as bass enclosures for traditional dynamic loudspeakers and will be described shortly. In both cases, the horn acts as a transformer, though instead of transforming voltage and current ratios keeping the same amount of power, a horn transforms *acoustic impedance* by varying the ratio between acoustic pressure and wave velocity.

A horn loudspeaker is not unlike a megaphone that acoustically amplifies the voice of a sports coach yelling at her players on the field. However, while a megaphone is an "accidental" horn, and sounds characteristically colored, a high-fidelity horn is carefully designed to fine tune the horn's flare shape over distance for a flat response and broad dispersion. In this case, the motion of the driver's small diaphragm is transformed into a larger radiating surface. The driver radiates directly into the narrow neck of the horn, which presents a high *acoustic impedance* to the driver. The wide output end of the horn therefore provides a better match to the lower impedance of free air. This creates less total wave motion, or velocity, but the motion is spread over a larger radiating surface, so the total acoustic power remains the same. Acoustic impedance will be explained in more detail in the Electronics section.

One potential problem with midrange and tweeter horn speaker drivers is increased distortion at high volumes due to nonlinearity of the air in the horn's throat. Air is mostly a linear medium, and in turn room acoustics are usually linear; whatever happens at a soft volume happens the same at louder volumes. But at the very high pressure levels that can develop inside a horn's throat, the air itself can become nonlinear and cause acoustic distortion. This usually happens only in large, high-powered PA systems. In fact, when hearing horn drivers in a home or studio setting, people often perceive a sense of increased dynamic range. Further, the directional control afforded by horns makes them a good fit for listening rooms with large reflecting areas that, for reasons of decor, cannot be covered with acoustic treatment. Indeed, some horn speaker models have surpassed even highly respected cone or dome systems in double-blind preference tests.

A *piezo* tweeter is similar to a piezo microphone, only run in reverse: Voltage applied to metal plates bonded to either side of a crystal or ceramic element causes the element to flex. The element is coupled to a larger cone, or a horn, to increase the size of its radiating surface. Piezo speaker drivers are generally lower quality than other designs, because their frequency response is not very flat and often includes a resonant peak. But piezo speakers are cheap to manufacture, and they're rugged. The impedance of piezo tweeters also rises at low frequencies, so less current is drawn, letting manufacturers avoid the cost of a crossover capacitor. However, it's still a good idea to add a series capacitor anyway, because that reduces IM distortion in the tweeter by removing low frequencies the tweeter won't be handling.

A *whizzer cone* tweeter is a second small cone that's bonded to the voice coil at the center of a full-range driver's regular cone. Since it's a separate structure, it can vibrate independently from the main portion of the cone. This design extends the frequency range of a simple one-driver speaker system, though a whizzer cone tends to ring at its self-resonant frequency, making it suitable only for low-fidelity applications. Where a normal speaker cone has an outer surround attached to a metal frame, a whizzer cone is attached only at its narrow center, leaving the entire cone to flop around and resonate.

Unlike whizzer cones, *coaxial* speakers can be excellent. These have a proper tweeter mounted in the center of a woofer, just in front of the dust cap. One big advantage of coaxial speakers is they're time-aligned, so frequencies around the crossover point emit from the same central location and thus arrive together at your ears to avoid a skewed response caused by acoustic interference. The Altec 604 coaxial loudspeaker developed in the 1940s was used in UREI 815 speakers, a staple in many recording studios through the 1960s and 1970s. Indeed, loudspeakers based on this type of driver are still used today, though modern implementations are better than the original models sold years ago.

In some speakers, a rigid aluminum dome in the center of the cone improves the response at frequencies toward the upper end of the driver's range. The metal cap is attached directly to the voice coil, so it can vibrate more or less independently of the main cone. Being metal and dome shaped, it's also less likely to deform than a paper dust cap as it vibrates. A metal cap therefore serves as a secondary radiator. But unlike a whizzer cone, it's better damped by virtue of being bonded securely all around its outer edge. The metal cap is often damped with thin open-cell foam attached to the back side to further reduce resonance. It can also serve as a heat sink, conducting heat away from the voice coil to the surrounding air better than a paper dust cap. When a voice coil heats up, its resistance increases. This causes the driver to draw less current for a given applied voltage, which reduces efficiency and in turn lowers the output SPL. This effect is known as *thermal compression*. So in this case, a metal center cap does more than simply keep dust out of the tiny gap around the voice coil.

Finally, a few oddball driver designs are worth mentioning just for laughs. I used to see inexpensive speaker drivers that mount on a wall, which create sound by vibrating the wall itself. I haven't seen these in many years, which is probably just as well. However, proper in-wall speakers can be excellent, and they're popular with hi-fi and home theater enthusiasts who prefer to avoid the high-tech look of loudspeakers on the floor in their living rooms. Mounting speakers in a wall has other advantages too, as explained in Chapter 19.

Years ago I bought a loudspeaker that used compressed air from an electrical pump as the sound source to achieve extremely high output levels. It had a dynamic driver with a leather diaphragm and was driven by a conventional power amp. The fidelity was atrocious, even when used for distorted electric guitar, which is why I bought it. But it required a relatively small amount of electrical power to operate, and it was *extremely* loud!

Loudspeaker Enclosure Types

An un-mounted dynamic loudspeaker suspended in free air sends sound more or less equally out the front and rear. Like a ribbon microphone, its radiation pattern is inherently bi-directional. However, at low frequencies whose wavelengths are much longer than the cone's diameter, sound from the rear gets around to the front, and vice versa, canceling both the front and rear radiation. So some means of isolating the front and rear radiation is needed to achieve a useful output level at lower frequencies.

One solution is a sealed speaker cabinet called the *closed box*, *infinite baffle*, or *sealed baffle*, which is the simplest cabinet design—a plain rectangular box with a hole in the front to mount the speaker driver. In theory, an infinite baffle is simply a flat panel so large in both dimensions that the lowest desired frequency cannot go around and cancel sound radiating from the other side of the driver. In practice, a large enclosure "folds" the baffle into a sealed box that contains the woofer's rear radiation. Note that the term "infinite baffle" derives from long ago, and many modern loudspeaker designers prefer to call this a *sealed baffle* or *sealed box* enclosure.

As mentioned earlier, a standard dynamic driver has a free-air resonant frequency related to the mass of its cone plus coil, combined with the springiness of the two surrounds that hold the cone in place. But air trapped inside a sealed enclosure increases the spring's resistance, raising the resonant frequency compared to when the driver is in free air. Take a rubber band and stretch it a bit, then pluck it like a guitar string and note the pitch. Now pull it even tighter and pluck it again, and you'll see that the frequency is higher. Air trapped inside a speaker cabinet resists the cone's motion, with that resistance increasing as the box is made smaller. This is why the bass response rolls off at higher frequencies in a small enclosure versus a larger cabinet. The greater air spring tension of a smaller box also increases the Q of the driver's resonance.

An *acoustic suspension* enclosure is a variation of the infinite baffle and was pioneered by Acoustic Research in their AR line of loudspeakers first introduced in the 1950s. The design uses a *high-compliance* woofer, having a massive diaphragm and very loose suspension, to obtain a very low free air resonant frequency. Since the springiness of the driver's surrounds is very low, air trapped inside the box serves as the primary spring. The resulting enclosed resonance is higher than the free air resonance, but because of the high-compliance suspension, it is still quite low. This lets a high-compliance driver in a small enclosure reproduce low frequencies well, even though the trapped air raises its resonant frequency. The name of this design comes from the fact that air inside the cabinet is the "acoustic suspension," or spring, that forms part of the speaker's mechanical system. Acoustic suspension loudspeakers work well but require more amplifier power than conventional large enclosure systems. This was a consideration in the era of tube amps, but it's no longer an issue today where high-powered solid-state amplifiers are common.

Again, for all sealed-box designs, the larger the box, the lower in frequency the speaker can play to with a given efficiency. Absorption in the form of fiberglass, mineral wool, or acoustic foam inside the box serves to increase the apparent size of the cabinet in the same way that acoustic absorbers increase the apparent size of a room. When a negative signal causes the speaker driver to move backward into the enclosure, the increased air pressure raises the temperature inside ever so slightly. The smaller the box, the more pressure that's created, and in turn, the larger the temperature increase. Adding insulation serves as a *heat sink*, reducing the amount of temperature change. This is the same as having a larger enclosure that develops less internal pressure for a given cone displacement.

Another important consideration for speaker enclosures is to be rigid. If the cabinet flexes as the woofer is drawn in and pushed out, energy is wasted moving the cabinet's walls, which reduces efficiency. Further, a wall that's able to move self-resonates at a frequency determined by its mass and also weakens the spring action of the air trapped inside the box. Finally, if the enclosure walls are allowed to flex, the walls themselves become another sound source that can add acoustic interference and skew the frequency response.

A third type of design is the *bass reflex*, or *ported* cabinet. This design can produce even lower frequencies for a given box size and with more efficiency than a sealed design. The basic idea is to add an intentional resonance slightly below the driver's own low-frequency roll-off. The added peak combines with the natural roll-off of the driver to extend the flat portion of the response as much as half an octave lower, though the additional filter poles cause the response to eventually roll off twice as fast at 24 dB per octave. This is shown in Figure 18.5.

Figure 18.6 shows two basic bass reflex enclosure designs. The drawing at the left shows an internal round tube coupled to a port opening, and at the right is a wider slot-shaped opening that serves the same purpose. Reflex designs let both the front and rear waves from the driver

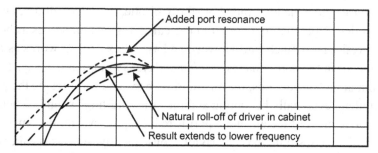

Figure 18.5: Adding a tuned port to a speaker cabinet boosts the driver's output at the tuning frequency, which counters the driver's natural roll-off, extending the overall response to a lower frequency.

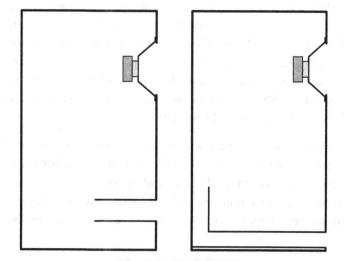

Figure 18.6: A bass reflex cabinet has a port opening that's tuned slightly below the driver's natural low frequency roll-off. The port in the enclosure at the left is a simple round tube that protrudes into the cabinet's interior. The port at the right is a wide rectangular slot with a wood shelf that wraps around inside the box to be longer than would otherwise fit. A subwoofer cabinet that uses a rectangle slot and folded shelf is shown in Figure 18.8.

contribute to the overall output. For this to work, the rear waves must be phase shifted to augment the front sound rather than cancel it. Air within the box acts as a spring in a mass-spring system, and air trapped inside the port tube is the mass. The phase shift added by this tuned system delays the waves coming from the rear of the speaker enough to be in phase with the direct waves from the front at frequencies near the port's resonance. In other words, at resonance the rear waves are shifted 180 degrees, reinforcing sound from the front. But at very low frequencies, there's no spring action from the trapped air to restrict cone movement, so even small amounts of amplifier power at those frequencies can create excess excursion possibly damaging the driver.

Most round ports use a tube made of metal or sturdy plastic, though cardboard tubes are used in inexpensive models. Note that in a ported loudspeaker, trapped air serves as both the mass *and* the spring. Physics is so much fun! However, a bass reflex cabinet works as intended only as long as the air in the port behaves as a *lumped element* of mass. This is true for low volume levels, but at higher volumes, the air shuttling back and forth inside the port becomes increasingly turbulent, eventually becoming audible as a "chuffing" sound. When that happens, the effective mass of the air in the port is reduced, the system becomes detuned, and it's difficult to predict the resulting bass behavior. Consequently, serious speaker designers have investigated the aerodynamic performance of ports. With a properly contoured tube, much higher sound levels can be reached before port turbulence dominates.

A *passive radiator* works similarly to a tuned port, except the mass of air inside a tube is replaced by a loudspeaker driver with a cone but no coil or magnet. The spring is a combination of the passive driver's outer and inner surrounds, plus the air trapped inside the cabinet just as with an open port. A big advantage of using a passive radiator is there's no air turbulence coming from a port opening, which avoids port chuff and wind noise at high volume levels. However, a passive driver must be capable of the same excursion limits as the main driver. Otherwise, distortion will occur when the passive driver's cone bottoms out, even if the main driver is still within its linear range.

A variant of the ported bass reflex is the *transmission line* enclosure, as shown in Figure 18.7. This is similar to the slotted style opening of a reflex cabinet, but a labyrinth provides a simple delay to create the needed phase shift, instead of using the mass-spring resonance of a port. A transmission line system is essentially a quarter-wavelength organ pipe tacked onto the back of the woofer. Unlike a horn that flares, its cross section remains constant

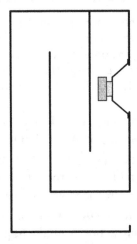

Figure 18.7: A transmission line enclosure is a folded quarter-wavelength resonator that behaves much like an organ pipe.

over distance. To be useful at low frequencies, the pipe must be long, so it's folded—usually several times—making the cabinet complex and heavy. Straight pipes resonate at multiple frequencies, so the resonance is tamed by filling the pipe with insulation. The trick is to damp standing waves in the pipe sufficiently, without reducing the useful bass assistance that radiates from the opening.

Chapter 1 introduced the term *filter pole* to describe the behavior of a single electronic filter section having a slope of 6 dB per octave. Acoustical and mechanical filters also have poles, with each rolling off at either higher or lower frequencies at the same 6 dB per octave rate. Another term for the number of poles a filter contains is its *order*. So a second-order filter has two poles, third-order means there are three poles, and so forth. High-pass and low-pass filters using a single resistor and capacitor contain one pole, but band-pass filters and simple mass-spring resonating devices have two poles with a center frequency. The mass provides one pole, and the spring provides the other, with one rolling off toward higher frequencies and the other rolling off at lower frequencies.

More complex systems have multiple poles, as is the case with a loudspeaker in an enclosure, where two components contribute one pole each to the low frequency roll-off. A two-pole filter, whether electrical, mechanical, or acoustical, can have either one pole up and one pole down at 6 dB per octave as described, or both poles rolling off for a combined 12 dB per octave in the same direction. Which type of roll-off you get depends on the arrangement of the components.

The low-frequency response of a sealed box speaker rolls off naturally at 12 dB per octave, because two filter poles are involved. One is due to the speaker cone's mass, and the other is the combined springiness of the driver's two surrounds and air trapped inside the enclosure. Ported speakers roll off even faster at 24 dB per octave, because two additional poles are added by the mass of the port's air and the air spring within the box. Further, at frequencies below the port's resonance, the driver's rear radiation begins to combine out of phase with sound from the front. The phase shift needed to reverse the driver's rear radiation and reinforce its front output is greatest at the port's resonant frequency. At lower frequencies where the wavelengths are longer, the rear radiation comes out of the port with less phase shift. So at those frequencies, it's as if the driver is not in a box at all, and the rear radiation cancels the front.

Note that in both closed-box and reflex designs, the roll-off does not suddenly begin at a single frequency. Rather, there's a gradual transition from a flat response to an *eventual* attenuation rate of either 12 or 24 dB per octave. By intelligently manipulating the mechanical parameters of the driver and the acoustical parameters of an enclosure, it's possible for loudspeaker designers to generate many different roll-off characteristics, ranging from under-damped (resonant) to over-damped (non-resonant). Bass-reflex systems have more variables and therefore offer more opportunities to experiment with the bass cutoff characteristics. However, that also offers more opportunity to get it wrong, as sometimes happens.

Some people prefer non-ported speakers because they believe ported models add a resonant peak at the low cutoff frequency. But as just explained, a reflex system can be designed as under-damped to avoid ringing. However, a bass reflex speaker can make *port chuff* and wind sounds due to air turbulence at the port opening as described earlier. Ringing can also be avoided by tuning the enclosure port to such a low frequency that it's below the audible range, or at least below the range of expected program material. For a superb example of this concept, Figure 18.8 shows a subwoofer enclosure with a port tuning of 8 Hz, designed by my friend and audio expert Mark Weiss. This speaker, shown at the far right in the photo, is flat without ringing to well below 20 Hz. Mark is also a state-of-the-art video producer, and these speakers are part of his production room, which he rightfully calls the *Bass Pig's Lair*. When I asked Mark to hear his demonstration of 8 Hz playing at 130 dB SPL, it was like standing in front of a strong fan!

The tall enclosure at the left holds a large midrange driver that operates between 124 Hz and 4 KHz, and the tweeter above handles frequencies 4 KHz and higher. Because three large speaker enclosures are used for each side of this system, the broad range handled by the midrange driver keeps the sound source focused to a single location, rather than from two locations several feet apart. Very low frequencies are not well localized by our ears, so having that content come from three cabinet locations doesn't harm imaging. This midrange/tweeter enclosure is divided internally, and the bottom section holds a separately sealed woofer

Figure 18.8: The dual 18-inch subwoofer cabinet shown at the right has a slotted port tuned to 8 Hz to avoid resonance within the audible band. The drivers have a maximum excursion of four inches, and the six power amplifiers for this entire system can put out 15,000 watts. An identical set of speakers for the left channel, not shown, is on the other side of the video monitor.

whose slot-shaped port at the bottom is tuned to 16 Hz. Two more woofers tuned to the same frequency are in the center cabinet, and together the three woofers cover the bass range from 16 to 124 Hz.

The twin BassMaxx ZR-18 18-inch subwoofer drivers in the enclosure at the right handle frequencies up to 65 Hz, and this is the cabinet whose port is tuned to 8 Hz. In most systems, the crossover would not allow different drivers to output overlapping frequency ranges. Here, the range from 16 to 65 Hz is handled by both the three woofers and two subwoofers. But these woofers are capable of outputting down to 16 Hz, so it's more efficient to let them cover that range along with the subwoofers. Equalization within the crossover restores a flat response in the overlapping range.

Horn bass cabinets are large and extremely efficient. Using a speaker enclosure that's highly efficient lowers distortion because the driver's cone doesn't have to move very far, which keeps the cone away from its hard excursion limits where distortion rises rapidly. A highly efficient speaker can also play very loudly without requiring thousands of watts to drive it. The *folded bass horn* in Figure 18.9 is similar, but with the horn constructed as a labyrinth inside the cabinet. This type of bass horn is the most common and was popularized by Paul Klipsch in his designs from the 1940s.

The problem with horn enclosures for domestic settings is that low-frequency wavelengths are very long. In order to benefit from the horn principle at very low frequencies, one would need a mouth literally the size of a whole wall. In fact, a few wealthy enthusiasts have actually done this. The Klipsch design uses the existing room walls, with the speaker placed in a corner that expands the horn, though the flare rate of a 90-degree corner is less than ideal.

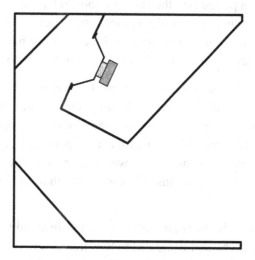

Figure 18.9: A folded bass horn is built with a labyrinth to keep the size of the enclosure reasonable for the low frequencies it must handle.

Finally, a *line array* is a series of many identical drivers placed adjacent to one another, as is often seen in large-venue PA systems. As long as all of the drivers are extremely close together, they'll act as a unified whole with minimal acoustic interference when all of the separate driver outputs combine in the air. A line array can be oriented either vertically or horizontally, though most are arranged vertically with one driver above another. This creates a dispersion pattern that can be well controlled—fairly wide horizontally and narrow vertically—to focus sound toward the audience where it's needed, but away from the ceiling and side walls, which causes unwanted reflections and wastes energy. As drivers are added to an array, its frequency response extends lower simply because the total cone surface area is larger.

Subwoofers

One of my earliest memories of hearing loud bass that sounded exciting was in high school when the marching band practiced outside, and the deep, full sound of the bass drum came through the classroom windows loud and clear. For others, attending a live fireworks display was their first introduction to deep bass that's very loud but sounds great rather than irritating. Indeed, deep bass at high volume levels is a very important part of many movies. Imagine how much less impressive *Jurassic Park* would be without the ominous sound of dinosaurs stomping around!

A *subwoofer* is a regular dynamic woofer driver but with a larger diameter or a greater cone excursion, or both. Subwoofers are usually added as a separate component, so they warrant their own category here. Some subs are monsters, like my powered SVS PB12-Ultra/2, which has two 12-inch drivers in a ported box the size of a short refrigerator. Other subwoofers use drivers 15 or even 18 inches in diameter, though most subs meant for home use are smaller. One big advantage of a subwoofer is it reduces the burden on the main woofers at low frequencies, allowing higher volume levels and with lower distortion.

The Carver Sunfire series is a good example of a small but effective subwoofer. I used to own one of their smaller models, a cube-shaped enclosure only 11 inches on each side. This sub has an active driver on one side face and a passive radiator on the opposite side. Even though the 11-inch Sunfire is tiny by subwoofer standards, it's capable of very high output to below 25 Hz thanks to its 3-inch excursion and 2,400-watt power amp. The power amp is as much a design marvel as the speaker itself, because it's small enough to fit inside the same tiny box, along with the power supply.

Subwoofers are important for home theater multi-channel surround systems because movie soundtracks include an *LFE* channel dedicated to low-frequency sound effects. Here, LFE stands for Low Frequency Effects, and it's the ".1" in a 5.1 or 7.1 system because its limited bandwidth handles only frequencies from 120 Hz and lower. For recording studios and normal stereo listening, my personal preference is a single pair of full-range speakers that can play down to at least 50 Hz or lower.

When you add a subwoofer to a stereo music system, another crossover frequency is needed that splits the bass range to come from three physical locations instead of only two—the left and right main speaker's woofers, plus the subwoofer. This can skew the response around the crossover frequency more than the regular crossover in a two-way speaker, because the sub is not usually adjacent to the main woofer. So now sound from three distant locations must combine in phase to a unified whole at the listening position. As loudspeaker expert Floyd Toole would say, this is basically a "lottery" scheme; there's no simple way to predict how these three disparate low-frequency sources will couple to excite the room modes.

Where you place a subwoofer has a huge effect on the amount and quality of bass you'll hear. The best way to know where to put a sub is to measure the low-frequency response at the listening position while experimenting. There are just too many variables to rely on a simple formula. The bass response from a sub depends on where it's placed, where the main speakers are placed, where you sit while listening, what crossover frequency is used, and how much bass trapping you're able to put in the room.

When I bought my first subwoofer, the Carver Sunfire mentioned earlier, I tried several locations in my 25 by 16-foot living room home theater. Then I noticed that the manual suggested one of the front corners. Bingo—that was clearly the best place. A year later, I replaced the Sunfire with the larger SVS also mentioned previously, and I noticed its manual also said a front corner is best. But by then I didn't even need to experiment. I put it there, and it's even more fabulous than the Sunfire. Another advantage of corner placement is the extra gain from that location lets the subwoofer work less hard, which lowers distortion. That said, a front corner is clearly the loudest location, but it won't be the flattest unless you have a fair number of bass traps. Loud works for me! But I also have many bass traps in my living room, and they reduce the low-frequency peaks and ringing that are emphasized by corner placement.

Another method often suggested is to put the subwoofer up on a chair at the listening position, then play bass-heavy music and crawl around on the floor listening for where the bass is the most even. Once you find the most even-sounding bass by ear, you put the subwoofer there. One problem is this works only with smaller subs that can fit on a chair. (My SVS sub is enormous and weighs 190 pounds!) Another problem is the key of the music affects what you hear. If the music happens to contain bass tones that align with a room's natural peaks, this method can work pretty well. But if the music is in a key that doesn't excite the room modes, other music that does excite the modes will sound unbalanced.

Really, the only way to know for sure which location is best is to *measure* the response at high resolution as you experiment. This can be time consuming because moving the sub even an inch or two can make a real difference. So you end up measuring, moving, measuring, moving, and so forth for the better part of an afternoon. Chapter 22 describes an efficient method for optimizing subwoofer placement using a real-time analyzer.

Many subwoofers include a polarity switch, or sometimes a continuously variable phase control, to optimize the combining of waves from the sub with those from the main speakers. The idea is you flip the switch, or vary the knob, while measuring the room response, looking for the setting that gives the loudest output for frequencies around the crossover point. When the output level in the room is loudest, the waves are combining in phase properly.

Some home theater systems have two subwoofers, sometimes with each near the left and right mains to reduce the distance between the main speaker's woofer and its subwoofer. Other setups use four or even more subs placed around the room, which can reduce peaks and nulls due to room acoustics. But four high-quality subs can be very expensive, especially compared to bass traps that flatten the response and reduce modal ringing, which multiple subs do not. Multiple subs are also more difficult to calibrate because the volume and crossover phase shift of each all combine to affect the low-frequency response at the listening position. However, when performance matters more than anything else, using multiple subwoofers—plus plenty of bass traps—will get you closest to the ideal response.

Enclosure Refinements

In some speaker cabinet designs, the tweeter is recessed slightly into the box to time-align its output with the woofer for frequencies around the crossover point. The low-pass portion of the crossover that drives the woofer adds phase shift, which delays the signal to the woofer. Setting the tweeter into the cabinet delays the tweeter's sound acoustically, so sounds near the crossover frequency arrive from the woofer and tweeter at the same time. A related concept is the *stepped enclosure*, where the cabinet has different depths becoming shallower toward the top. In this design, the upper portion of the cabinet containing the tweeter doesn't extend as far forward as the lower portion. In a three-way system, two steps are used, with the midrange driver farther back than the woofer and the tweeter above that even farther behind.

Diffraction describes how sound waves behave as they travel along a surface that suddenly ends, which happens with a driver mounted in a cabinet. The sudden change in acoustic impedance from a solid surface to free air causes reflections at the cabinet's edge, and those reflections become a new sound source that combines with the direct sound to skew the frequency response. Diffraction also bends sound waves around the edge of a surface, which is why you can hear someone talking from around a corner or on the other side of a tall dividing wall. If the front face of the speaker enclosure is very large, diffraction occurs only at lower frequencies whose wavelengths correspond to the distance between the driver and the enclosure's edge. But with a small cabinet, diffraction effects extend up into the midrange. Some speaker cabinets use rounded rather than square edges to provide a smoother transition from one acoustic impedance to the other. This is also one reason the large main speakers in a professional studio's control room are mounted into the walls.

Crossovers

As mentioned earlier, a loudspeaker's woofer cone must be large enough to move a lot of air to fill a room with bass. But then it's too heavy to move fast enough to reproduce high frequencies. The large diaphragm also bends and wrinkles, creating resonant modes on its surface, and dispersion also narrows. Therefore, smaller drivers are employed to produce higher frequencies, using a *crossover* to split the audio into different frequency ranges. In the frequency range where one speaker rolls off and another becomes active—the crossover region—design engineers aim to ensure that the acoustical summation of the two sound sources remains smooth. The frequency response should not exhibit peaks or dips due to acoustic interference, either on- or off-axis.

Ideally, the woofer and tweeter outputs will each be 6 dB down at the crossover frequency, to yield a net flat output for frequencies emitting from both drivers at once. Figure 18.10 shows simple 6 dB per octave two-way and three-way passive crossovers that use capacitors and inductors to block frequencies outside the range of each speaker driver type.

The two-way crossover uses a series inductor to block high frequencies from reaching the woofer and a capacitor to block low frequencies from the tweeter. The three-way type is similar, but it adds an inductor and capacitor in series to block both low and high frequencies to the midrange driver. The particular inductor and capacitor values used depend on the chosen crossover frequencies and the impedance of each driver. A well-designed crossover will also compensate for less than ideal driver properties.

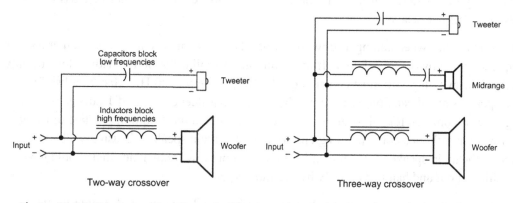

Figure 18.10: Passive crossovers use inductors and capacitors to pass only the frequencies that a given driver can handle.

Another potential problem with crossovers is distortion added by the inductors and capacitors at high power levels. One solution is an *active crossover*. This splits the signal into different frequency ranges before the power amplifiers, rather than afterward with passive components inside the speaker cabinet. An active crossover requires separate power amplifiers for each frequency range, as well as the crossover itself, which is a separate electronic device. So an active crossover is more expensive to implement, but it offers many advantages. When an active crossover is used with two separate power amps, the system is said to be *bi-amplified*, or *bi-amped* for short. A three-way system is considered *tri-amped*.

Active crossovers let you experiment with different crossover frequencies by turning a knob instead of having to solder new capacitors and inductors. The slopes can also be easily varied by flipping a switch, if that feature is available. One reason to experiment with different crossover frequencies and slopes is to optimize the response around the crossover point where the same content comes from two drivers at once. Most commercial active crossovers employ *Linkwitz-Riley* filters, which are 24 dB per octave. This design places two second-order *Butterworth* filters[1] in series, yielding 360 degrees of phase shift at the crossover frequency. Since 360 degrees is similar to 0 degrees, this minimizes response errors when both driver sound sources combine in the air. But active crossovers can use other slopes, in multiples of 6 dB per octave.

Different crossover slopes suit different applications. Steep slopes—12 dB per octave or greater—can produce an audible peak or null when set to frequencies that are inappropriate for the drivers used. Some people believe that a slope of 6 dB per octave is more agreeable to the ear because response errors around the crossover point are more gradual and occur over a broader range of frequencies. However, using a broad slope requires each driver to operate over a larger frequency span outside its intended range. So it's often best to use a steep slope to maximize the power handling of a tweeter or midrange driver, cutting it off well above the bass range. When a driver doesn't have to reproduce to as low a frequency, you can have louder playback with less distortion. For this reason, most commercial active crossovers are fixed at 24 dB per octave.

The same applies when adding a subwoofer to a stereo system. The most common crossover frequency for subwoofers is 80 Hz. If the mains start to roll off at 80 Hz with a slope of only 6 dB per octave, they still must contribute substantial energy at 40 Hz. Many bookshelf-type speakers are already pretty far down by 40 Hz, so a steeper slope of 12 dB per octave or greater is common for subwoofers. I should mention that the common practice of using a subwoofer without a corresponding crossover on the main speakers is usually a bad idea, because the main speaker's woofers still strain to output low frequencies that should have been filtered out and handled entirely by the subwoofer.

An active crossover also lets you easily adjust the balance between woofer and tweeter to better integrate drivers that have different sensitivities. Adjusting the volume with a passive crossover requires placing a resistor in series with the more efficient driver, which wastes power. With an active crossover, you simply adjust a volume control. Also, an active crossover doesn't lose power through passive components, especially inductors whose internal coil resistance is in series with the woofer. Avoiding this resistance also helps preserve an amplifier's high damping factor.

Note that some speaker designers and DIY enthusiasts reverse the polarity of the tweeter compared to the woofer. A single filter pole adds a maximum of 90 degrees phase shift for frequencies far away from its turnover frequency, but the shift is only 45 degrees at its −3 dB point. Therefore, when using a crossover that rolls off at 6 dB per octave, the woofer lags by 45 degrees at the crossover frequency, while the tweeter leads by 45 degrees. For content near the crossover frequency that's reproduced by both drivers at the same time, the driver cones are not in phase or out of phase with each other but are somewhere in between. With a slope of 12 dB per octave, the disparity between woofer and tweeter is 180 degrees at the crossover frequency, though it increases farther away. So in that case, it *probably* makes sense to reverse the tweeter's polarity. But at 18 dB per octave, the two drivers differ by 270 degrees at the crossover point, and at 24 dB per octave, they're shifted a full 360 degrees, which restores the correct polarity.

There are other factors, such as when the crossover frequency is near a driver's natural roll-off. In that case, phase shift due to the driver's own roll-off adds to that of the crossover, further confounding matters. So in practice, whether one driver should be reversed or not depends simply on which polarity gives the flattest response. I believe this is best determined by measuring rather than by listening alone, as some speaker designers seem to prefer.

Understand that the preceding explanations are intended only to describe crossover basics. In truth, modern state-of-the-art designs are much more sophisticated, and powered speakers in particular can include DSP (digital signal processing) to compensate for driver irregularities such as peaks and dips, and an impedance that changes with frequency. For example, just because a crossover network rolls off at a given slope, the total system comprising the driver and crossover may have a different curve that DSP can improve. Indeed, modern skilled loudspeaker design engineers use computer simulation and anechoic measurements to optimize a crossover for the best performance possible in ways that go far beyond the basics described here.

Active versus Passive Speakers

A relatively recent design concept for loudspeakers is to include an active crossover plus power amplifiers built into the speaker enclosure. Mackie was one of the first modern companies to offer this in its HR824 and HR624 models, though studio monitors with active crossovers were produced as long ago as 1967 by Klein+Hummel. Today all-in-one active loudspeakers are very common. There are many advantages of active monitors for the typical project studio, besides a simpler hookup with fewer pieces to carry if you ever do remote recordings: Active speakers are typically bi-amped, which yields less distortion as already explained. Bi-amping also offers more ways to optimize the crossover performance because it uses active rather than passive components, as was also explained. Further, using an active crossover increases headroom within each band by segregating the bands. That is, if the amplifier that powers the bass range clips at a very loud volume, high frequencies are still reproduced cleanly unless that amplifier is also driven into distortion.

Just as important, the power amplifiers can be well matched to the speakers, they won't have a fan, and the wires between each power amp and its speaker are shorter, which might improve damping. (Amplifier damping is explained in Chapter 23.) An active loudspeaker can also contain DSP circuitry to counter frequency response errors in the drivers themselves and to add any needed delays so frequencies near the crossover point emit from both drivers at the same time. This not only improves frequency response, but it can also reduce the radiating directivity problem known as *lobing*, which will be described shortly. But to me, the overwhelming advantage of powered monitors, as implemented in Mackie speakers anyway, is that the woofer cone's mechanical motion is included within the power amplifier's electrical feedback loop. This improves low-frequency response, reduces ringing at the port's resonant cutoff frequency, counters thermal compression in the voice coil, and reduces driver distortion.

Negative feedback will be explained more fully in Chapter 23. But briefly for now, negative feedback lets an amplifier circuit self-correct its own output signal to reduce distortion and frequency response errors. A portion of the output is fed back to the input, but with the polarity reversed. So if the amplifier's output is not an exact voltage-multiple of its input at a given point in time, its input receives more or less of the feedback signal. The amplifier in turn sends more or less signal to its output to compensate, which forces the output to better match the input. With an active speaker, the amplifier can be designed to sense the amount of current being drawn by the driver and compare that to the amount of current that *should* be drawn for a given input signal. If the two do not match, which means the driver is distorting or ringing, the amplifier varies its output signal to compensate.

Finally, pre-distortion as described in Chapter 6 for use with analog tape recorders can also be applied to active speakers. As long as the driver's characteristics are known—which it should be in a commercial design!—a circuit can apply amplitude boost to exactly counter the compression that occurs as a driver's cone approaches its excursion limits. However, this is not the same as including the driver's motion within the amplifier's feedback loop allowing correction in real time, so external effects such as thermal compression won't be countered.

Room Acoustics Considerations

Chapter 21 explains the interaction of loudspeakers and listening rooms in detail, but it's worth mentioning a few issues here. The sound of every speaker is dominated by the room you put it in. No matter how flat a response the speaker's published measurements may claim, you'll never actually get that response in a room unless that room is an anechoic chamber. The main cause of a non-flat response—peaks and nulls—is reflections from the room's surfaces as they combine in the air with the direct sound from the speakers and with one another. Indeed, *all* room acoustic problems are caused by reflections off the walls, floor, and ceiling: peaks and nulls, reverb and echoes, comb filtering, ringing—everything!

The most damaging reflections at mid and high frequencies come from surfaces located between your ears and the speakers. Figure 3.16 in Chapter 3 shows how early reflections in a small listening room reach your ears a few milliseconds after the direct sound, which creates comb filtering. Reflections off the rear wall behind you can also be early if that wall is 10 feet from your head or closer. As you move your head from side to side, or forward and back, the peak and null frequencies shift, which affects the stability of stereo imaging. Thankfully, reflections at mid and high frequencies are easily tamed using relatively thin absorption at key places on those surfaces.

Comb filtering also happens at bass frequencies, though reflections from any room surface can be the cause because sound propagates around a room differently at low frequencies. At mid and high frequencies, waves travel more or less like a flashlight beam that spreads outward. But at low frequencies where the wavelengths are longer, the waves tend to expand outward until they strike a room boundary and reflect. So the root cause of peaks and nulls— reflections—is the same at all frequencies, though the solutions are slightly different because taming bass reflections requires larger and much thicker absorbers.

Bass also builds up near the walls, floor, and ceiling, increasing to a maximum of 6 dB at each room boundary. This buildup occurs because wave pressure—what our ears respond to—is greatest at a boundary. Imagine a boxer throwing a punch at his opponent. While his fist is in the air approaching his opponent's nose, there's no pressure. But once contact is made, there's a lot of pressure! The same happens with acoustic waves as they approach a room boundary. In truth, boundary pressure buildup occurs at all frequencies, not just the bass range. But the longer wavelengths that correspond to lower frequencies spread the increase over a greater distance.

As shown in Figure 1.21 in Chapter 1, peaks and nulls occur at predictable quarter-wavelength distances from a boundary. One quarter-wavelength at 40 Hz is seven feet long, and the pressure increase is active over some of that distance from the boundary. But at

1 KHz, one-quarter of a wavelength is about 3½ inches, at which distance the pressure is minimum. The pressure then rises again toward 7 inches away, then falls at 10½ inches away, and so forth at 3½ inch intervals. So considering the big picture, bass tends to build up more consistently near a boundary. This is why many active speakers include switches labeled "half-space" and "quarter-space" for use when the speaker is placed near one wall or near two walls at a corner, respectively. These switches reduce only low frequencies to counter the boundary buildup of bass waves due to the speaker's close proximity to the wall or corner. The omnidirectional nature of most loudspeakers at low frequencies is also a factor, as you'll see in the section in this chapter about loudspeaker specs.

Loudspeaker Impedance

One important property of a loudspeaker is its *impedance*. This is similar to resistance, and it determines how much current a speaker driver will draw when a given voltage is applied. The only difference between resistance and impedance is that resistance stays the same regardless of frequency, whereas impedance changes with frequency. So a loudspeaker might have an equivalent resistance of 8 ohms at 1 KHz, but only 6 ohms at 500 Hz, and 12 ohms at 100 Hz. When coupled with the relatively high output impedance of tube-based amplifiers, the varying impedance of a loudspeaker can have a very real effect on the sound.

Most modern speaker drivers have a nominal impedance of either 4 or 8 ohms, though some older designs are 16 ohms. Years ago, my friend and expert electrical engineer John Roberts tried to convince Rudy Bozak, a prominent hi-fi speaker manufacturer at the time, to produce a series of 2-ohm speakers for use in car audio systems. A car battery puts out 12 volts, which can drive a 4-ohm speaker to less than 36 watts at most. But a 2-ohm driver would draw twice as much current from 12 volts, and thus output twice the power. (This is also why headphones used with smart phones and other battery-operated devices are low impedance.) Today, digital "switching" power supplies generate higher voltages from a 12-volt car battery. But back in the 1970s, this was an excellent product idea.

A loudspeaker's impedance is not usually a concern to end users unless it dips substantially below its rated value at some frequencies or it's connected to a tube amplifier. In that case, the power amplifier driving the speaker might suffer increased distortion as it tries in vain to deliver more output current than it was designed for. But competent solid-state power amplifiers can comfortably drive 4 ohms, and some can drive 2 ohms without straining.

Most hi-fi type speakers and professional recording studio monitors contain two or three drivers, with each handling a different frequency range. Whether active or passive, the impedance of the drivers and how they're wired are determined by the manufacturer. Choosing the appropriate impedance of each driver and the passive crossover components

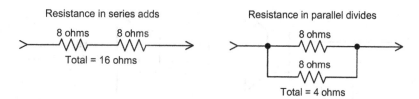

Figure 18.11: Resistance and impedance add to a larger value when connected in series, and in parallel the result is smaller. If the resistors in parallel are all the same, you can divide their value to find the result. But if the values are different, a more complicated formula is needed.

is part of the design process. But many guitar and bass instrument amplifiers use two or four full-range loudspeakers of the same size and type. These speakers do not use a crossover, but they must be wired correctly to obtain the correct total impedance and to distribute the amplifier's power equally to each speaker.

Before considering how a multi-speaker guitar amp is wired, let's take a step back and see how resistance combines when connecting multiple devices. Ignoring changes with frequency, impedance combines the same as resistance, so I'll use simple resistance for the following examples. As shown in Figure 18.11, wiring two resistors in series yields a higher resistance that's the sum of the two values. This is true whether the resistors are equal or not. So two 8-ohm resistors in series yields a total resistance of 16 ohms, which draws half as much current as 8 ohms when connected to a power amplifier. But when placed in parallel, the result is a lower resistance, more like a short circuit, which draws more current for a given voltage. If two equal-value resistors are wired in parallel, the result is simply half the value of each. Combining three equal resistors gives a value equal to one-third of each resistor, and so forth. So wiring four 8-ohm resistors in parallel yields 2 ohms.

If you have a guitar amp that can safely drive 4 ohms, you'd use two 8-ohm speakers and wire them in parallel. If the amplifier is rated for 8 ohms, you'd use two 4-ohm speakers and connect them in series to get 8 ohms. If you intend to use four speakers, and the amplifier can drive 8 ohms efficiently, all four of the speakers would be 8 ohms, wired in series *and* in parallel as shown in Figure 18.12. If the amplifier can drive 4 ohms, you could use four 16-ohm speakers in parallel.

When connecting more than one speaker driver, it's very important they be wired so all of the cones move in the same direction. Connecting speakers in series is similar to wiring batteries in series, where the positive terminal of one goes to the negative terminal of the next in line. But unlike batteries, which should never be wired in parallel, when using two speaker drivers or groups in parallel, you'll connect the like polarities together. Both arrangements are used in Figure 18.12.

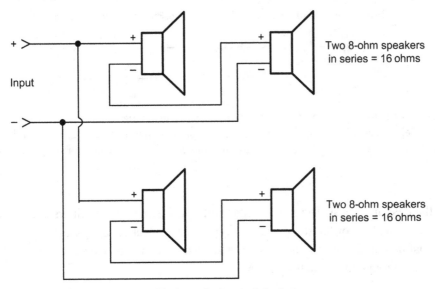

Two 8-ohm speakers
in series = 16 ohms

Input

Two 8-ohm speakers
in series = 16 ohms

Two 16-ohm pairs in parallel = 8 ohms
Note the polarity of each speaker's wiring.

Figure 18.12: To wire a four-speaker instrument amplifier cabinet, connect two speaker pairs in series, minus to plus, then connect the pairs in parallel with like polarities together.

In the interest of completeness, this simple formula shows how to calculate the combined resistance of any two resistors connected in parallel:

$$R_{total} = \frac{R1*R2}{R1+R2}$$

This is the generalized formula for any number of resistors wired in parallel:

$$R_{total} = \frac{1}{\dfrac{1}{R1} + \dfrac{1}{R2} + \dfrac{1}{R3} + \dfrac{1}{Rn}}$$

You can also calculate what value is needed for one of the two resistors to obtain a new desired total resistance:

$$R2 = \frac{R1*R_{total}}{R1-R_{total}}$$

This is useful if you need a specific nonstandard value—perhaps when designing a crossover circuit—using two standard resistors. So if you need 810 ohms, which is not a standard value, you'd start with 820 ohms, which is the next higher standard value. Then calculate the second resistor to wire in parallel with 820 ohms for the desired 810 ohms:

$$R2 = \frac{820*810}{820-810}$$

In this case R2 calculates to 66,420 ohms, which is not a standard value either. But using 68,000 ohms, which is the closest standard value, yields a parallel resistance of 810.23 ohms. At only 0.03 percent off, that's surely close enough:

$$\frac{820 * 68,000}{820 + 68,000} = 810.23$$

For those of us who are math-challenged, calculating this type of percentage is very simple:

$$Error\ percent = \frac{Actual\ Value - Desired\ Value}{Desired\ Value} * 100$$

Therefore:

810.23 − 810 = 0.23
0.23/810 = 0.0002839
0.0002839 * 100 = 0.02839 percent
After rounding = 0.03 percent

Loudspeaker Isolation

Mechanical decoupling is an important concept in recording studios and listening rooms—for example, under a drum set or turntable it avoids mechanical coupling that can transmit low frequency thumping sounds or make the record skip. But mechanical isolation is not usually needed with loudspeakers as many believe. A speaker cabinet should be massive enough, and rigid enough, to not flex or vibrate audibly at low frequencies and high volumes, or excite resonances in the surface it rests on. This is how loudspeakers are supposed to work: The box remains perfectly still, without expanding or contracting as pressure inside the box changes, and only the woofer driver's cone moves forward and back.

It's not difficult to tell if your setup will benefit from loudspeaker isolation. Invite two friends to help you, and have each lift one speaker 1/4 inch while you listen. Obviously, your friends should stand behind the speakers, or off to the side, to avoid blocking the sound. Or you could play music through only one speaker with one helper. Either way, if you can't hear a difference when the speaker is raised, it's not likely you'll benefit from isolation. But measuring loudspeakers is even better than listening because measuring tests every frequency, not just those present in a given piece of music.

The earliest loudspeaker isolation product I know of is Auralex MoPADs. The company claims these thin blocks of foam prevent the loudspeaker's vibration from transferring to the surface it rests on thereby "smearing" the sound when that surface becomes a secondary sound source. The product claims sound reasonable enough:

- Reduces coloration
- Increases clarity
- Decreases structural resonance

Before I started my own acoustics company I was active in audio forums touting the benefit of acoustic treatment to anyone who would listen. Auralex sent me a pair of MoPADs to try, hoping I'd like them and recommend them to others. The huge JBL 4430 speakers in my home studio weigh 127 pounds each (!), so I visited a friend with Mackie HR824 monitors resting on a wood riser. After putting his speakers on the MoPADs neither of us heard any difference, and that was the last I thought of it.

Over the years I started seeing ads for more and more isolation products. Not only for loudspeakers, but also expensive "isolating" equipment racks and dedicated platforms sold on the premise that CD players and power amplifiers are also harmed by vibration. This has grown into an entire industry with full-color ads in both audiophile and recording magazines. Literally dozens of products claim to improve sound quality by isolating your loudspeakers, subwoofers, and everything else you own including the connecting wires!

One fact that isolation proponents miss is that sound transmits mostly through the air. Another is that competent loudspeakers have sufficiently rigid cabinets that don't shake or vibrate very much. Yet another fact is wires and electronic components are mostly immune to vibration. So while putting a subwoofer on springs or a rubber platform might reduce coupling to the floor, the majority of sound emits from the driver's cone and almost none from cabinet vibration. If the floor shakes with loud bass notes, that's due to the cone moving the air in the room rather than cabinet vibration coupling to the floor. Sound waves in the air cause the walls and floor to vibrate, not cabinet motion. If you watch a loudspeaker playing music loudly, you won't see the cabinet expand or contract even 1/32 inch. But woofer cones can move half an inch or more, and subwoofer cones move as much as two or even four inches! The amount of sound generated by a loudspeaker depends entirely on mechanical displacement and total surface area. The cone's area might be smaller than the enclosure's surface, but it moves much farther!

An earthquake will probably make your CD player skip because it's a mechanical device, but solid-state preamps and MP3 players are mostly unaffected by vibration. As mentioned in Chapter 2, vacuum tubes can become *microphonic* over time, and when that happens vibration causes a ringing sound similar to a microphone on the verge of feedback. But again—and this is important—vibrations in the air are much stronger than vibrations passing through a table or floor. If your tubes are ringing from vibration, an isolation product won't help because the tubes are being shaken by sound waves in the air rather than through the chassis from underneath.

In all the years I've seen loudspeaker (and other) isolation products advertised, I have never seen data proving better audio quality. Magazine and online reviews contain only subjective assessments and user testimonials, as in these actual examples:

> *The Apertas minimized [bass boominess], resulting in a smoother, more natural low end. I also found that the sound-staging improved, taking on a more holographic quality. The placement of individual sections of an orchestra are rendered with greater precision.*

I immediately recognized the improvement in the sound field my monitors produced when used with the Recoil Stabilizers. Bass reproduction was tighter and exhibited none of the upper-bass blurriness that typically plagues shelf-mounted monitors. Localization in the stereo field of different elements of the mix was much more precise. Detail and depth improved significantly. Surprisingly, hard-panned electric guitars also sounded more present.

The first time I placed my Bag End M-6 Time-Align monitors on top of RX7 Recoils, I immediately heard the difference. More-and punchier-low end. Better defined mids. Clearer highs. And it wasn't some kind of immeasurable difference that self-proclaimed, golden-ear audiophiles who spend $1000 per foot on speaker cable purport to hear. The increased clarity is not at all subtle!

[Recoil Stabilizers] Tests with my Mackie HR624s produced noticeably better stereo imaging than mounting the speakers directly on the shelf surface, and the low end seemed much better controlled.

If these products really do affect the sound from a loudspeaker as claimed, it will be easy to prove using standard measurements of frequency response and ringing decay time. The IsoAcoustics testing page shows impressive photos of a "laser vibrometer" test conducted by the National Research Council in an anechoic chamber in Ottawa, Canada. But there's no audio measurement data! These photos show only that vibration doesn't travel down through the isolation device to the table the loudspeaker rests on. And the only test frequency shown is 75 Hz, so we'll never know how well other frequencies are isolated.

This is not unlike the "data" we see from power conditioner vendors. They show their product reducing a tiny amount of noise on the AC power line, but they never show what happens at the output of the connected audio equipment, which of course is what really matters. Even if these products isolate low frequency vibration completely, that's not evidence that isolation is needed or even helpful.

So I decided to investigate this for myself and for the entire audio community. I borrowed four commercial loudspeaker isolation products, plus I already had two DIY versions I made, and I also measured with no isolation for comparison. So that's a total of seven tests. These products claim to block harmful vibrations from passing from the speaker to whatever surface it rests on. So I found a rickety old table that has strong resonances at 298 and 475 Hz as shown in Figure 18.13, and used that for my tests. If loudspeaker vibration doesn't excite the obvious resonances in this flimsy table, nothing will.

For my tests I used the Room EQ Wizard (REW) program described in Chapter 22, along with my DPA 4090 precision measuring microphone and a Mackie HR624 powered loudspeaker. This speaker is typical of what you'll find in many home studios, and its published specs show that it's flat within 3 dB to below 50 Hz, with usable output as low as 25 Hz. I put the microphone one meter (39 inches) away from the speaker, which is standard for loudspeaker testing. Putting the microphone closer would have yielded a flatter response, but for these tests the absolute response isn't as important as comparing each device versus

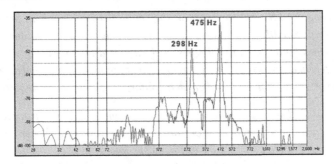

Figure 18.13: Rapping my thumb on top of the table shown in Figure 18.14 produced strong resonances at 298 Hz and 475 Hz.

no isolation. The setup is shown in Figure 18.14, and you'll see I taped the table legs to the floor to prevent the table from moving as the speaker was lifted and lowered repeatedly while placing the various isolators. You can also see small white business cards taped to the table top. These marked the speaker corners to help ensure that the speaker was in the same place for every test.

I tested the devices listed below in order of their increasing height, using the riser and plywood shims shown in Figure 18.15 to maintain the same speaker height. For the No Isolation tests I put the speaker flat on the table as well as elevated, to show how only a small difference in height affects the measured response. The DIY isolation pad is made from 2-inch thick 705 rigid fiberglass topped with 3/4-inch plywood for stability, and the empty box is typical corrugated cardboard.

- No Isolation elevated, also not elevated
- IsoNode Isolation Feet (3/4 inch high)
- Auralex MoPADs (2 inches high)
- Empty Cardboard Box (2–3/8 inches high)
- Primacoustic Recoil Stabilizer (2–5/8 inches high)
- DIY 705 & Plywood (2–7/8 inches high)
- IsoAcoustics Aperta (2–7/8 inches high)

The graphs in Figures 18.16 through 18.18 show three of the low frequency response measurements, and additional graphs are on my website[2] showing all seven measurements including waterfall and impulse graphs. Figure 18.16 shows the speaker at two heights without any isolation, so you can see how the three-inch height difference alone changes the response. In the two remaining graphs of Figures 18.17 and 18.18, the lighter line shows the response without isolation and the darker line with the named device. Again, for the No Isolation measurements the speaker was on a riser or shims to maintain the same height.

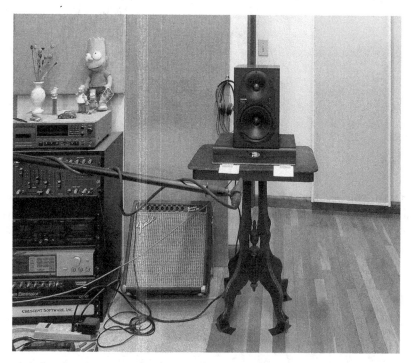

Figure 18.14: This is the setup used to test the various isolation devices. The microphone appears too low in the photo, but that's due to the camera angle. It was pointed directly at the woofer since these tests are concerned mainly with frequencies below the speaker's 3 KHz crossover.

Figure 18.15: The shims and riser were used to keep the speaker at the same height for every test, to avoid response changes due only to placement.

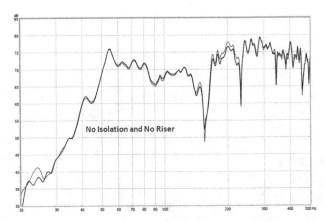

Figure 18.16: This graph compares the response with the speaker flat on the table versus raised three inches, to show how a small difference in height has a very real affect even when nothing else is changed.

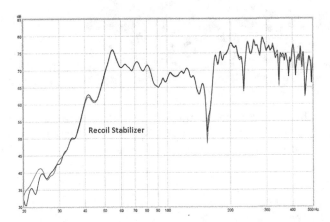

Figure 18.17: This is the response with the speaker resting on one of the commercial isolation devices, compared to the speaker without isolation at the same height.

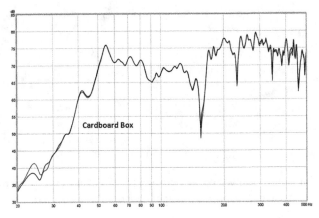

Figure 18.18: This shows the response with the speaker resting on an empty cardboard box, compared again to the speaker without isolation at the same height.

In these tests we're looking mainly for differences at the table's two resonant frequencies of 298 and 475 Hz. Differences at other frequencies are most likely due to unavoidable small changes in position. You'll notice response changes below 30 Hz, about 35 dB down, which might be air flow noise from my heating system. I doubt it's related to isolation because the change is just as pronounced with the speaker at different heights with no isolation. You'll also notice a deep null in all of the graphs at 155 Hz. This is due to the usual reflections in a room, and it's common to all the measurements based on the fixed microphone and speaker positions so it can be ignored.

It's imperative that the microphone and loudspeaker be in the exact same place for every test, which turned out to be surprisingly difficult. If either moves even 1/8 inch in any direction, the measured response changes due to both loudspeaker beaming and lobing (its off-axis response) and acoustic peak and null locations in the room. In fact, I had to do these tests twice because the first time the speaker position varied about half an inch for the different tests. So to ensure a more consistent speaker height for this second round of tests I had my friend master craftsman Phil Cramer build a small riser, plus a series of plywood shims letting me adjust the height in 1/8-inch increments as shown in Figure 18.15. To minimize lateral displacement I taped business cards to the table to mark the corners of the speaker, then sighted straight down the edges of the speaker as I repositioned it for each test.

Even with all this care I was still unable to get the loudspeaker in precisely the same spot for all seven tests. One problem was the flimsiness of the table, which shifted slightly even though its legs were taped in place. Another problem was the business cards I used to frame the speaker corners were partly covered by the larger isolation devices. I'm confident the errors were less than 1/4 inch in any direction, but even that small offset was enough to skew the responses, especially at higher frequencies. Fortunately, these tests are concerned mainly with frequencies below 500 Hz, and especially the two specific table resonances.

So why do people swear that adding isolation improved the sound of their loudspeakers? I'm convinced that the measured changes are too small to account for the "obvious improvement" so many people claim to hear after floating their speakers on isolation pads. Most of the response differences are less than 1 or 2 dB, with the biggest differences at the 155 Hz null frequency. To me, an obvious change is at least a 3 dB response difference over a wide range of frequencies, or a 30 percent change in decay times. It's possible that some people have even poorer tables than the one I used, though I doubt it! I'm pretty sure that the small changes in frequency response are due entirely to slightly different speaker positions. Further, all of the differences are 20–30 dB down, so how could this be perceived as a "more-and punchier-low end" as described by one of the magazine reviewers?

One reason the sound can change is because raising the speakers on isolators puts them higher than they had been. As shown in Figures 3.17 and 3.18, moving even a few inches changes the frequency at your ears by a very large amount. But other than the change in frequency response due to different speaker placements, which is real and easily measured, I believe the frailty of human hearing tricks people into thinking the sound improved even though it didn't. Call it placebo effect, or wishful thinking, or even a misguided allegiance to a magazine's advertisers rather than its readers. The bass didn't really become "much better controlled," nor did the speakers produce "better defined mids, clearer highs," nor did "detail and depth improve significantly" as claimed in the various quotes above. If any of those changes really occurred, they'd be clearly visible in my measurements. They are not. *Simply raising the loudspeaker three inches changed the response more than the difference between no isolation and any of the tested devices.*

Most of the product reviews and user testimonials for loudspeaker isolation products claim they make the bass sound tighter and clearer, less boomy, and better controlled. I've even seen people suggest that speaker decoupling reduces the need for bass traps. For reference, the graph in Figure 18.19 shows the bass response and ringing in a small room with and without six 2x4-foot bass traps. The lighter portion of the graph shows the response and ringing with the room empty, and the darker portion in back with shorter decays is after adding the traps. It's easy to see that the response is flatter with bass traps, and all of the modes decay much more quickly. Versus the minimal changes with any of the isolation devices.

So what can we learn from these tests? First, it's clear that moving a loudspeaker even a small amount makes a very real change in the perceived and actual frequency response. So raising your speaker on an isolating stand can change what you hear because of the height

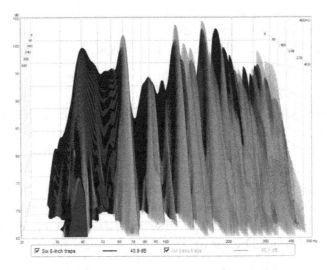

Figure 18.19: This waterfall graph shows the large improvement in response and ringing after adding six bass traps to a typical home studio room, compared to the tiny change measured with any of the isolation devices.

difference alone, even if any isolation present had no effect. (As explained in Chapter 19, the correct loudspeaker height puts the tweeter at ear level because that gives the flattest response, so please do that!) But my tests also call into question the hearing acuity of professional recording engineer endorsers, and golden-eared magazine reviewers, who claimed to hear "obvious" improvements in clarity, bass tightness, and imaging. Either they don't realize the limits of their own hearing, or they all have even worse speaker stands than the flimsy piece of crap table I used!

Loudspeaker Polarity

It should be obvious that both speakers in a stereo pair need to be wired with the same polarity so both sets of driver cones move in the same direction. If the polarity of one loudspeaker is reversed, low frequencies might be weaker, or stronger, than normal. And at high frequencies the sound will seem to come from strange places such as beyond the left-right speaker width or even from above or behind you. When using active speakers, there's less chance for one speaker to be reversed compared to the other. But with passive speakers, you have to pay attention to the plus and minus labels when connecting them.

Most power and loudspeaker "zip cord" wire includes a marker on one of the conductors so you can easily identify the plus and minus conductors at both ends. Sometimes a stripe is painted along one edge, though most wires use a raised ridge or bumpy surface. My personal preference is to consider the marked wire as the negative lead, though, of course, it doesn't really matter. As long as both speakers are wired to the power amp the same way, the left and right driver cones will move forward and backward together.

There's an easy way to test if one speaker's polarity is reversed: Play a mono source through both speakers, and then adjust the pan pot or balance control quickly from full left to full right. If the speakers are wired correctly, the sound will move smoothly from side to side. If the sound seems to come from beyond the speakers, or from other unlikely locations, the wiring for one of the speakers is reversed. If your receiver doesn't have a balance control, you can play a mono source through both speakers, then move your head quickly from side to side. After reversing the polarity of one speaker a few times, it will be obvious just by listening which wiring is correct.

A different method can be used to verify the polarity of multiple loudspeakers in a single guitar or bass amp cabinet. Connect a pair of wires to a 1.5-volt battery and touch them to the speaker terminals. Many instrument amplifiers have a connector between the power amp and speakers, so disconnect the amplifier when doing this test if possible. Then, as you touch the battery wires to the speaker terminals, watch the speaker cones and confirm they all move forward or backward together. You can also put your hand on each cone in turn and feel which way it moves. This is difficult to do with a midrange or tweeter driver because the crossover's capacitor blocks the DC voltage from the battery. So the driver's cone will lurch forward or back briefly, then quickly settle to its center resting position. But for typical guitar amps this method works well.

Speaking of batteries, here's another little tip: You can quickly test a 9-volt battery by touching both of its terminals at once to your tongue. If the battery is fresh and fully charged, you'll get a mild shock that's unpleasant but not painful. Try this once when you buy a new battery to learn how a new battery feels.

Earphones

Most earphones contain very small dynamic loudspeakers, though some models—mostly sold to the audiophile market—use electrostatic drivers. Unlike loudspeakers, earphones are capable of producing all or most of the audible range from 20 Hz to 20 KHz using a single driver. Because they're coupled directly to your ears, very little cone motion can create relatively high volume levels. Even the limited excursion of a small driver is sufficient to output very low frequencies.

Earphones are available in four basic types: sealed enclosures that fit over your ears, open-back frames that also fit over your ears, "ear buds" popular for use with portable music players, and an ear bud variant called the *in-ear monitor* (IEM) used in live concerts by musicians to avoid loudspeaker monitors on the stage. Musicians need to hear themselves clearly when singing and playing, but leakage from loudspeaker monitors can get into the microphones. Using IEMs avoids that and lets the house mixing engineer project a cleaner sound to the audience.

Sealed earphones are usually preferred in recording studios because less sound escapes to be picked up by the microphones. This works the other way as well: Sound from loud instruments in the same room is partially blocked, so the earphones mix you hear is affected less. While mixing with earphones is always a compromise, a good pair of closed back earphones is a necessary evil when recording live shows. Open-back earphones are usually lighter and more comfortable to wear for extended periods, but they won't do when you need isolation.

Loudspeaker Specs

Like microphones, loudspeaker frequency response is measured in an anechoic chamber to prevent reflections, background noise, and temperature variations from skewing the results. As mentioned in Chapter 17, microphones and speakers can be measured in a regular room by using the gating feature of most measurement software to avoid the influence from reflections. But gating limits the lowest frequency you can measure accurately, as well as limiting the frequency resolution. This in turn limits your ability to see narrow-band but audible defects at higher frequencies, unless the room is very large and all boundaries, including the floor and ceiling, are at least 15 feet or more away. So "for real" speaker tests are always done in a professional anechoic chamber.

Also as with microphones, the frequency response measured directly in front of a loudspeaker tells only half the story. The other half is how the response varies at different angles in front of the speaker, as well as how much sound at various frequencies emits toward the sides and rear. Ideally, a loudspeaker will radiate all frequencies evenly in every forward direction where someone is listening and not at all toward other directions. Sound that hits an untreated wall, floor, or ceiling can cause damaging reflections, and sending sound where it's not needed is inefficient. Radiation patterns are especially relevant with PA systems to ensure that everyone in the venue hears the same frequency balance, without wasting energy or causing echoes by sending sound elsewhere, such as toward the side walls or ceiling.

Figure 18.20 shows the *polar response* of a typical mid-sized loudspeaker found in a home or recording studio. Even though this is a conventional box-type speaker whose woofer and tweeter face forward, at low frequencies—125 Hz in this example—the response is more or less omnidirectional. You can also see that at higher frequencies the radiating pattern narrows, with a reduced output toward listeners sitting off to either side. This is one important reason to angle, or "toe-in," the left and right speakers toward the center listening position, rather than facing them straight forward as is sometimes recommended.

There's a tendency for loudspeakers to radiate very high frequencies at multiple narrow angles, an effect called *beaming* or *lobing*. This is shown in Figure 18.21. These terms also describe the radiation pattern of radio transmitting antennas, where such directionality is often intentional and necessary. As you may know, there are many more commercial radio stations than available frequencies, so most stations must share the same frequency with other distant stations. Radio stations are required by law to "protect" other stations operating on the same frequency by sending their signals in specific directions to avoid conflict.

Lobing has several different physical causes: When sound near the crossover frequency comes from two drivers at once, each driver has a slightly different phase response. So when the two sound sources combine in the air, these phase differences interact with each other and affect directionality. Using the radio analogy again, when a station must control the direction of its radiation, two or more antennas are used. All-pass filters in series with the antennas shift their relative phase to create the desired transmitting pattern. The same happens with loudspeakers, because the woofer and tweeter have different amounts of phase shift at frequencies near the crossover where both drivers are active.

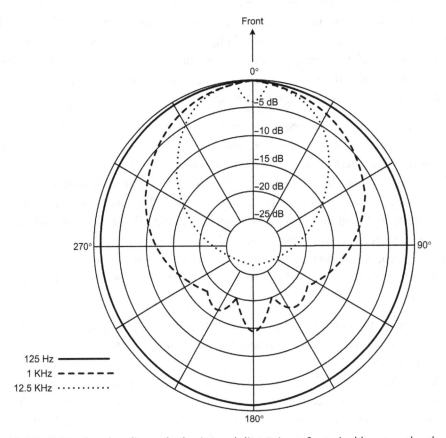

Figure 18.20: This polar plot shows the horizontal dispersion of a typical box-type loudspeaker. The vertical dispersion (not shown) is often intentionally narrower, having less output at low and high angles toward the floor and ceiling.

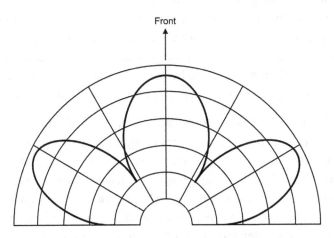

Figure 18.21: This polar plot shows the effect of lobing, where sound is sent outward in narrow zones of peaks (the lobes) and nulls, rather than radiating uniformly. The patterns become even more complex and more directional at higher frequencies.

Loudspeaker beaming also occurs because radiating a small wavelength from a relatively large surface creates acoustic interference, as shown in Figure 18.22. Even if a tweeter cone is only one inch in diameter, that's larger than the wavelength for 15 KHz. This is another reason tweeters need to be small: to behave more like a single point source, which minimizes lobing effects. Another cause of lobing is a speaker cone that flexes or deforms as it moves back and forth, rather than behaving like a rigid piston, which is the ideal.

Note that this type of lobing occurs with all drivers, not just tweeters. A four-inch midrange driver will also beam and lobe unless it's cut off by the crossover at a low enough frequency. When the off-axis response of a loudspeaker is measured in an anechoic chamber, the speaker is often placed on a rotating turntable that spins while the response is measured using a single stationary microphone. Another approach is to use an array of microphones at various locations around the chamber, though that requires a lot of mics to include enough in-between angles. So spinning a platform is better because it's more like sweeping a sine wave to measure frequency response—the resolution is much higher than spot-checking a few frequencies at third-octave intervals.

Another important loudspeaker spec is its *sensitivity*, which tells how loud it sounds when a given amount of power is applied. As with frequency response, sensitivity is also measured in an anechoic chamber. The original method for measuring speaker sensitivity was to apply one watt of power, then measure the SPL from one meter away. For an 8-ohm speaker, one watt is drawn when 2.83 volts is applied, but no speaker has the same impedance at all frequencies. The more modern method is called "voltage sensitivity," which drives the speaker with a constant 2.83 volts, rather than whatever voltage results in one watt. Then the resulting SPL is measured in the *far field* of an anechoic chamber—two meters or more for consumer

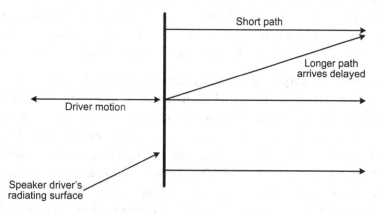

Figure 18.22: A vibrating surface that's large compared to the wavelength it's producing creates interference patterns as the different source points combine in front of the surface. Only one delayed path is shown here with the angled line, but there are an infinite number of point sources that all combine to skew the response and beam the sound toward different directions.

loudspeakers. From that the acoustic lab calculates what the output would have been at the standard distance of 1 meter.

Using an anechoic chamber avoids the influence of room reflections, which would increase the measured SPL as the reflections add to the direct sound from the speaker. Speaker sensitivity varies wildly for different loudspeaker types. Horn speakers are extremely efficient, which is why they're often used in large venues, though consumer speakers that employ horn tweeters are made intentionally less sensitive because the horns must match the conventional woofers they're coupled with. Acoustic suspension speakers are typically inefficient, though the trade-off is an extended low-frequency response. My large old-school JBL 4430 monitors are spec'd at 93 dB, and very large horn-type PA speakers can be as efficient as 110 dB. Smaller speakers meant for home use often have a sensitivity closer to 85 or 90 dB.

Loudspeaker distortion is typically ten to a hundred times higher than that of electronic circuits. Figure 2.14 from Chapter 2 showed the distortion at 50 Hz for a typical woofer playing at 86 dB SPL. Even at this moderate volume, the distortion is greater than 6 percent. If someone tried to sell a power amp like that to the pro audio market, they'd be laughed out of business. At lower frequencies and higher volume levels, distortion can easily exceed 10 percent, even for "good" speakers. As explained earlier, a driver's excursion doubles with each halving of frequency, so lower frequencies force the cone that much farther from its center rest location and closer to the hard limit.

Midrange and tweeter drivers often have much less distortion than woofers, partly because their cones don't have to move as far, but also because most music doesn't contain nearly as much energy in the midrange as at low frequencies. However, 1 percent distortion or higher is not uncommon at higher volumes. By the way, the total linear excursion of a speaker driver is its X_{max} spec, expressed in millimeters. Here, X_{max} is an abbreviation for maximum (max) and excursion (X). The absolute maximum excursion—the hard limit at which the cone bottoms out in both directions—is called $X_{lim,}$ for excursion limit, and is also stated in millimeters.

Loudspeaker distortion is very tricky to measure, especially in a home setting. Besides needing an anechoic environment to avoid the influence of room reflections, whatever microphone you use will adds its own distortion to the measured result. So at the minimum, a microphone with a known frequency response and low inherent distortion is needed, and these tend to be expensive. Note that the microphone's frequency response also matters when measuring distortion because THD and IMD components from a speaker occur at different frequencies. So if the microphone is down, say, 6 dB at some frequency, a distortion artifact from the speaker at that same frequency will appear that much lower than it really is.

No discussion of loudspeaker specs would be complete without mentioning the *Thiele-Small parameters* and their importance to modern loudspeaker designers. These parameters are named for A. Neville Thiele and Richard H. Small, who in 1976 published a method of loudspeaker analysis that's still used today. This detailed collection of parameters defines the performance of speaker drivers, including the X_{max} and X_{lim} specs mentioned above. A complete discussion of Thiele-Small parameters is beyond the scope of this book, but interested readers are encouraged to search the web or local library for further information.

Another important loudspeaker property is its self-resonance, both for speaker drivers and for the enclosures they're mounted in. As you probably know, inertia is a two-way street— a body at rest wants to stay at rest, and a body in motion tends to stay in motion. When a loudspeaker driver is moved by a voltage that then stops, it will continue to resonate for a while unless the motion is damped. Loudspeaker damping can be applied either mechanically or electrically. Chapter 23 explains electrical damping, so I'll focus here on mechanical damping.

Enclosure resonance is pretty simple. The main resonances are modal, at frequencies determined by the enclosure's internal length, width, and height dimensions—just as with rooms. Additional resonance also occurs if the enclosure's walls are not stiff enough to resist flexing as the speaker drivers modulate the air pressure inside the cabinet. Good speaker cabinets are made from wood or MDF (a composite wood board) thick enough to not bulge out or draw in as the driver motion changes the air pressure inside the box. Many speaker cabinets include internal bracing to further reduce mechanical flexing and vibration. Driver damping also requires careful selection of the material used for the surrounds. Adding absorbing material such as fiberglass or mineral wool inside the cabinet further increases damping of the driver and the cabinet, in the same way that absorption reduces acoustical ringing in a room.

In tweeters, *ferrofluid* can be inserted into the narrow gap between the voice coil and magnet. This is a liquid containing very fine suspended iron particles to prevent the fluid from draining out of the voice coil gap. The iron is attracted to the driver's magnet, which holds the fluid in place. Besides cooling the voice coil to avoid overheating by thermally coupling the voice coil to the magnet and surrounding structure, ferrofluid also helps couple the magnet and voice coil magnetically by reducing the apparent gap size. This makes the magnet and coil behave as if they were closer together, which increases efficiency. But ferrofluid also helps to mechanically damp the driver's physical motion, reducing self-resonance. However, this damping tends to reduce tweeter efficiency because extra energy is needed to overcome friction added by the fluid's viscosity.

Finally, maximum power handling is an equally important spec, since exceeding that can destroy a loudspeaker. Sustained high power is the main enemy that kills a driver by overheating its voice coil. This either melts the pieces together or melts the insulation, effectively creating a short circuit within the coil, or it blows the wire like a fuse. Another failure mode is where the glue holding the coil windings to its former softens from excess heat; either the whole coil comes unglued from the former or it bunches up and scrapes the magnet assembly. If a speaker mostly works but sounds "tizzy," this is a likely cause.

Understand that you can safely connect a 200-watt power amp to a 50-watt speaker, as long as you're sensible with volume levels. In fact, it's better to use an amplifier rated for more power than the speaker can handle, to allow brief transients through undistorted, rather than clip them. Brief transients have a high crest factor, and they come and go so quickly that they don't contribute much to the total power and resulting increase in voice coil temperature. However, power rating specifications for loudspeakers are not always trustworthy because there's no standard measuring method. Many ratings are based on anecdotal user reports, while others seem to have been invented by the marketing department. So take published power ratings with a grain of salt, and use common sense when deciding how loud to listen. If the sound is distorted, you're probably pushing the speaker too hard. Fortunately, many active loudspeakers include built-in protection, shutting down their internal amplifiers when overdriven for too long.

Many power amps can put out more than their rated continuous power for brief transients anyway, so you may not need an amplifier whose rated power is greater than what the speaker can handle just to preserve transients. For example, many power amps can put out half again more than their stated power for a short length of time—say, a tenth of a second. In fact, the disparity between continuous and instantaneous peak power output has been a contentious consumer issue. Many years ago, unscrupulous amplifier makers would state the maximum power available for short peaks, rather than the amp's continuous power capability, which is more realistic and what matters most. For example, continuous power is needed to play sustained bass notes. In 1974, Amplifier Rule CFR 16 Part 432[8] was introduced by the United State Federal Trade Commission (FTC). This law requires stating power and distortion specs for consumer audio equipment in a specific and consistent manner to avoid confusing customers with impossibly inflated claims.

Accurate or Pleasing?

A common question I see in audio forums asks about the difference between professional studio monitors and regular hi-fi speakers. Conventional wisdom says that monitors must be as flat as possible to mix accurately, while hi-fi speakers should aim for a pleasing sound. It seems to me that a hi-fi speaker should aim to recreate the same listening experience that the mixing and mastering engineers heard in their rooms while equalizing and otherwise adjusting the sound of the recording. Otherwise, the listener will not fully appreciate the artist's intent. So by that logic, a hi-fi speaker should also be as accurate as possible. If someone prefers an intentionally skewed frequency response, it probably makes sense to just buy an equalizer. Then you can also dial in different response curves to tailor the sound on a per-recording basis.

This brings up the related issue of whether speakers used for professional mixing benefit from a pleasing frequency response curve. In my opinion that's not a good idea, and I find the current trend toward "smooth sounding" non-harsh loudspeakers aimed at the professional market disturbing. If you mix on speakers that have an intentional dip in the harshness range between 2 to 4 KHz, you'll tend to add too much energy in that range to compensate, making your mixes sound brittle on good speakers that are more accurate. Indeed, loudspeaker selection is one of the most difficult and personal decisions anyone can make, whether you're a professional recording engineer or serious listener. There are many audio devices I'd buy mail-order based on specs alone, but a loudspeaker is not one of them! You might find the following anecdote interesting:

In 2009 I brought one of my Mackie 624s to the studio of a friend, a well-known mastering engineer in New York City. I wanted to measure my Mackie, which I know is very flat, and compare it to a measurement of one of his Revel Ultima Studio 2 speakers whose specs are also very flat. I had heard his Revel speakers a month before and was blown away. I just had to learn how two supposedly "flat" speakers could sound so different. Figure 18.23 shows the response I measured for both speakers. (A better comparison would be to Mackie HR824s, which are larger and play to lower frequencies than 624s.) Well, they didn't sound

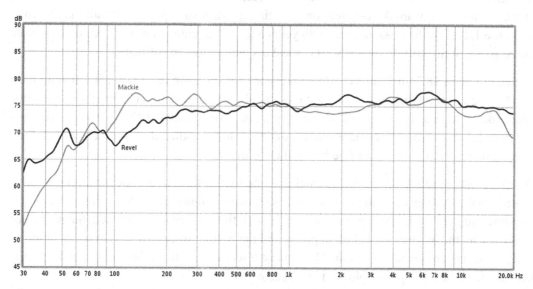

Figure 18.23: These measurements were taken with a precision DPA 4090 microphone a foot in front of both speakers. The traces are aligned to coincide at 1 KHz, and 1/6 octave averaging was applied to better see the overall response trends. Note that these home-made measurements are not as accurate as proper tests done from farther away in an anechoic chamber.

very different at all! It was just the situation, the music we played, the different room, expectation, and so forth. When both speakers were placed next to each other, with my small Mackie upside down on top of his larger Revel and the tweeters very close together, there were differences in the frequency response as measured and heard. But the differences were small. At one point I thought I was hearing the Revel, but it was actually my Mackie. The test was not blind, and it wasn't even a test. My friend did the switching, and at one point I lost track of which speaker was playing. When he asked which speaker I thought I was hearing, I guessed wrong. So I no longer lust after the Revels, though they are excellent and a little flatter at the highest frequencies compared to my Mackies. But at more than $16,000 per pair versus under $1,400 for two Mackie HR824s, I no longer have speaker envy.

Speaking of loudspeaker flatness, in the early 1980s the Yamaha NS-10 loudspeaker became popular as an "alternate" monitor in recording studio control rooms. Although this speaker was designed for home use, many recording engineers embraced it because of its prominent midrange. They reasoned, if a mix can be made to sound good on this speaker that reveals midrange detail well, the mix will sound good on other speakers too. But some engineers felt the NS-10 sounded a bit too bright, so they covered the tweeter with tissue paper. There were even heated arguments about which brand of tissue sounds best!

I recall from back then some people saying that tissue paper not only reduces high frequencies but also adds comb filtering, which of course should be avoided. At very high frequencies even soft paper can be somewhat reflective, so sound could bounce from the rear of the tissue back toward the tweeter cone creating quarter-wave peaks and nulls as shown in Figure 1.21. But that was more than 30 years ago, so I decided to re-visit this and measure the response of my still-pristine NS-10M speakers using modern software.

Notice in Figure 18.24 that my NS-10 speakers are upside down on the stands. I bought these stands long ago for a different setup, and their height isn't adjustable. Having these speakers upside down worked out perfectly because it puts the tweeters at exactly seated ear height. I often recommend this to people when, for one reason or another, their tweeters are too high. There's no reason a tweeter can't be below the woofer!

For these tests I put my DPA 4090 microphone one foot in front of the speaker, pointed directly at the tweeter, and ran the four sweeps shown in Figure 18.25. Looking at the response around 7 KHz, the top darkest trace is with the tweeter fully exposed, the next trace down is with the original grill cloth in place, and the bottom trace is with Kleenex® brand 2-ply tissue paper and no grill cloth. The lightest trace, second from the bottom, is the tweeter bare but with the *microphone* blocked by tissue paper.

I'm not sure I'd call the bottom Tissue Paper trace a "comb filtered response," but it certainly doesn't look like a response I'd want from my monitor speakers! Then again, the null around 16 KHz that starts to rise again could be a sign of comb filtering, given the half-inch distance

Figure 18.24: You can still see remnants of the masking tape that held tissue paper over the NS-10 tweeter all those years ago.

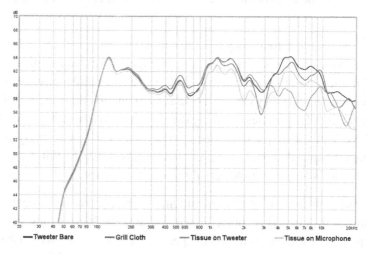

Figure 18.25: In the four responses above, 1/12th octave smoothing was applied to better see the trends.

between the tweeter cone and the tissue. This is further suggested by the more uniform reduction when the tissue was in front of the mic. That more closely follows the response curves of the tweeter bare and with grill cloth. Regardless, the tissue reduces high frequencies above 2 KHz about 6 dB compared to the tweeter fully exposed. But if a reduction of high frequencies is the goal, I think an equalizer is a more sensible way to achieve that because the reduction would be more uniform and easier to control. Or just choose a better loudspeaker!

House Curves

The concept of a *house curve* (sometimes called a *room curve*) arose in the late 1930s as a way to get a handle on loudspeaker sound quality and frequency response variations in movie theaters. The problem is that when the response measures flat two-thirds of the way back in the room—the standard location to measure when using one microphone—the sound is subjectively too bright near the front of the room. So audio engineers of the day decided it's reasonable to allow a reduced high frequency response farther back in the room as a compromise.

Part of the response disparity is due to non-uniform loudspeaker dispersion, especially the speakers available in the 1930s when this issue was first addressed. To a lesser extent, humidity in the air absorbs high frequencies over distance more than low frequencies, as explained in Chapter 1. So that can account for another few dB loss above 5 KHz in a large theater. Just as important, locations farther back in the room receive proportionally more reverb. In a large venue reverb travels hundreds of feet in multiple bounces around the room, and so loses more high frequency energy due to the accumulated distance. The common practice of putting loudspeakers behind the projection screen can reduce higher frequencies even more.

Yet another issue is that basic frequency response measurements are steady state, and can't distinguish between the direct response from the loudspeaker and accumulated reverb energy hanging around in the air. Further, all rooms—both large and small—have shorter decay times at high frequencies, which gives less accumulated energy. So if the RT60 decay time in a room is shorter at high frequencies, viewing the response on a Real Time Analyzer (RTA) will show less level at those frequencies. Seeing a non-flat response wrongly suggests that EQ is needed even though what emits from the loudspeakers may be flat. So in that case it's the *measurement* that's skewed, not the actual response in the room.

In 1976 Dolby created their "X-Curve" as an attempt to standardize the expected response in commercial movie theaters. This curve, shown in Figure 18.26, specifies that for proper sound, high frequencies should roll-off at 3 dB per octave starting at 2 KHz. (A modified curve that falls at only 1.5 dB per octave was implemented later for the smaller theaters that have since become more common.) This curve also allows sound tracks played in a small room

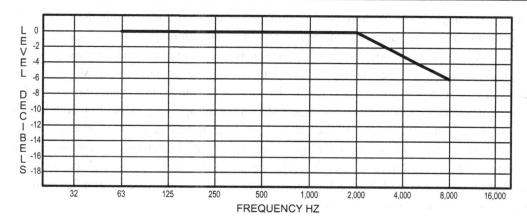

Figure 18.26: The Dolby "X-Curve" specifies a high frequency roll-off starting at 2 KHz, and falling at a rate of 3 dB per octave. This is the measured response expected when calibrating movie theaters.

having no high frequency losses to sound the same as in a larger room that has an X-Curve response. However, the X-Curve is relevant only for movie theaters and other venues that are *much* larger than anyone's home setup. It's just what happens naturally in larger spaces. For example, only theaters and other large spaces have true reverberation at bass frequencies, as opposed to the ringing at select low frequencies that occurs in home-size rooms.

To be clear, an X-Curve is the response professional movie theater calibrators aim for on their RTA when measuring farther back in the room, in order for the sound to be subjectively "correct." It's *not* a curve that's meant to be applied with an equalizer when using flat loudspeakers. I'll also mention that in large theaters the loudspeakers are typically aimed slightly upward toward the last row of seats. So seats that are farther back are on-axis and receive the flattest response, while seats close to the front receive less high frequency energy since they're off-axis. This helps balance the high frequency response between the front and rear rows because the front seats don't lose treble due to distance.

Most films are mixed in *dubbing stages*—essentially small movie theaters with mixing desks and other related equipment. It's tempting to think that EQ'ing your own home theater system to have the same response as a dubbing stage or movie theater is desirable, but it's not because home theaters don't have the same problems. Further, some films mastered for home video have a countering EQ applied, but others just duplicate the theatrical mix with no correction. Unfortunately, this information isn't printed on the DVD or Blu-ray package, so all you can do is listen and decide if the sound seems good or too bright.

Related, some people boost the bass EQ, or raise the volume level of their sub beyond flat, and call that a "house curve." However, that's just boosting the bass because you like bass.

There's nothing wrong with more bass if you like the sound! But it's not a house curve because a house curve addresses a measured reduction at high frequencies. Personally, I don't think it's a good idea to boost the bass beyond flat anyway. In my experience most movies, including musicals and "action" movies, have an appropriate amount of bass. Most have the right amount of midrange and treble too. If the balance in your system seems wrong, the first thing to consider is probably the room's acoustics. Further, applying global EQ—other than to counter known deficiencies—is like second-guessing the mixing and mastering engineers. It also defies research by acoustician Floyd Toole. He found that listeners universally prefer loudspeakers that are flat, rather than intentionally colored to sound "pleasing" as some speaker models are voiced. Now, a flat loudspeaker is not exactly the same as a flat room or a flat EQ setting, but it's closely related.

I'll mention that some audio engineers distinguish between a house curve and a room curve. By that logic, a "room curve" is an acceptable use of EQ to obtain a flat response to counter non-ideal loudspeakers, or other flaws that skew the response such as having loudspeakers behind the screen. Versus a "house curve" that represents the personal preference of whomever owns the "house" and thus controls the audio system. Personally, I think it's a mistake to impose a global EQ just because you like the sound. As mentioned in Chapter 3, boosting the treble and bass can sometimes make music sound "better," especially to unsophisticated listeners. But that's not more faithful to the original source material, or to the artist's intent. It's more like wearing too much lipstick. With experience, many listeners learn to appreciate accurate sound, and to reject unnatural response coloration.

That said, it's not wrong to apply EQ to individual movie and music sources if you think it improves their sound. My own home theater system aims for a flat response, and most things sound fabulous. Obviously, some music and movie sources sound more well balanced than others, but there's no single EQ curve you could apply that would compensate for the myriad differences in recordings. So I believe that flat is best for an overall system response, and you can apply EQ as desired on selected titles if you believe it helps.

Summary

This chapter explains how loudspeaker drivers work and are built and how they perform when mounted in various types of enclosures. The most common driver type is dynamic, using a permanent magnet and wire coil to move the cone forward and back to create sound. Other driver designs such as planar, ribbon, and piezo are used, though dynamic drivers are by far the most common. Regardless of a driver's type, its size must be optimized for the frequency range it's expected to reproduce. Larger sizes are needed to move enough air at low frequencies, and smaller radiating surfaces are needed to avoid beaming and lobing at higher frequencies.

Since each frequency range has different driver requirements, a crossover divides the incoming audio sent to drivers optimized for each range. Different crossover slopes may be used, in multiples of 6 dB per octave, with advantages and disadvantages for gentle versus steep slopes. Passive crossovers are the most common and cost-effective, but for the highest performance, active crossovers offer many advantages, including a flatter response and lower distortion.

There are many different types of speaker enclosures, including sealed baffle, acoustic suspension, bass reflex using an open port or a passive radiator, transmission line, folded horn, and line array. Each type has its own pros and cons—for example, a bass reflex extends to a lower frequency than a sealed box the same size, but it rolls off faster and so doesn't produce the very lowest frequencies as loudly. For each enclosure design, its size directly affects the lowest frequency that can be produced, which introduces yet another trade-off between size and efficiency.

This chapter also explains loudspeaker impedance and shows simple formulas to calculate the combined impedance when speakers are wired in series and/or in parallel. Equally important is loudspeaker polarity, which must be correct for every driver in a musical instrument amplifier, as well as for all of the loudspeakers in a stereo or multi-channel system. Earphones are also mentioned briefly, with descriptions of the four major types.

The same performance parameters that apply to electronic gear also apply to loudspeakers, but speakers are much more difficult to measure, requiring an anechoic chamber for dependable results. Further, as with microphones, the off-axis response of a loudspeaker is as important as its on-axis response. Indeed, loudspeakers and their enclosures are surprisingly complex, given the seemingly simple task expected of them. With loudspeakers, everything matters—stiffness of the inner and outer surrounds, the natural self-resonance and roll-off frequencies of the drivers and enclosures, the crossover frequencies and their slopes, how precisely the pieces fit together—and all of these interact and combine in complex ways.

Finally, I address the perennial debate over whether loudspeakers should be accurate or pleasing, coming down firmly on the side of accuracy. Not only is this needed for professional mixing but also for home listening to preserve the artist's intent. If someone prefers a response curve that's intentionally skewed, adding an equalizer is surely a better choice.

Notes

1 Butterworth is one of several basic filter topologies that include Chebyshev, Elliptic, Bessel, and others. Some filter types ring less at their turnover frequency, while others are flatter with less ripple in the pass-band, and so forth. Which filter type is most appropriate depends on the specific application.

2 "Testing Loudspeaker Isolation Products" by Ethan Winer. http://ethanwiner.com/speaker_isolation.htm

PART 5

Room Acoustics, Treatment, and Monitoring

After all the preceding chapters, you might think that high-quality audio is all about the gear. It is not. I once read a terrific interview with Leslie Ann Jones, chief engineer at Skywalker Sound, where she said she can get a great sound at Skywalker using *any* competent microphone. The rooms professional studios record in are designed to sound good inherently, in part by being large enough to avoid reflections from nearby surfaces that add an off-mic, hollow sound caused by comb filtering. Likewise, professional control rooms and mastering suites are usually much larger than most home studios.

Even budget microphones and loudspeakers have a better frequency response than the bedroom-size spaces often used for project studios and home listening rooms. If the goal for excellent sound is to improve the weakest links first, then your room is clearly the weakest link that needs addressing most. In my experience, problems due to poor room acoustics are the biggest cause of dissatisfaction among home recordists. Sadly, many people consider everything *but* acoustics when they're unable to make their recordings sound the way they want. Besides being able to hear what's actually "on tape" more easily, having a well-treated environment makes recording and mixing much more *fun*. It's a real eye-opener the first time you hear every note clearly articulated by an electric bass and are able to hear very small changes in EQ, panning, and reverb. Acoustic treatment will improve the quality of everything you record and mix far more than your choice of microphone, preamp, and sound card. In that one moment, when you first work in a well-treated room, it's immediately clear what you've been missing all along.

When you think about acoustics, you should consider these four issues:

- Preventing standing waves and acoustic interference from affecting the frequency response of recording studios and listening rooms
- Reducing modal ringing in small rooms and shortening the reverb time in larger studios, churches, and auditoriums
- Absorbing or diffusing sound in the room to avoid ringing and flutter echoes at mid and high frequencies, and improving stereo imaging

- Keeping sound from leaking into or out of a room—in other words, preventing your music from disturbing the neighbors and keeping the sound of barking dogs and passing trucks from getting into your microphones

Please understand that acoustic treatment is intended to control the sound quality within a room. It is *not* intended to reduce sound propagation between rooms. Sound transmission and leakage are addressed via construction—using thick, massive walls, isolating the building structures by decoupling the walls and floating the floors, and hanging the ceilings with shock mounts. Even though both fall under the umbrella of "acoustics," acoustic treatment and soundproofing are totally different. If you come across a salesperson or forum poster who confuses these basic terms, I suggest you run the other way.

CHAPTER 19

Acoustic Basics

Most budget electronic gear these days is very high quality, letting you create first-rate music that sounds as good as anything on the radio. Yet, many recording enthusiasts, unhappy with the quality of their productions, wrongly blame their gear. Of course, talent and experience matter more than anything else, but so does working in a good acoustic environment. While it may be possible to create a great mix in an untreated room, it's difficult and often frustrating. After all, if you can't hear accurately, it's impossible to know what mix elements need adjusting. This chapter explains the basics of acoustics and room setup using plain English with minimal math. Although much of the focus is on recording studios, especially smaller studios, most of the information applies equally to audiophile and home theater listening rooms, as well as large professional studios.

If you've read this far, it should be clear that I live to bust audio myths, and some of the most prevalent myths are those surrounding acoustics. Perhaps the biggest myth is that room acoustics is an impossibly complex, arcane science that only a specialist with a physics degree can understand. Nothing is further from the truth. *All* room acoustic problems—peaks, nulls, ringing, flutter echo, excess reverb and ambience—are caused by reflections off the walls, ceiling, and floor. Reflections can be absorbed or diffused, and some can be ignored. It's really that simple. Of course, the treatment solution for a 30 by 40-foot professional control room is different from that for a 10 by 12-foot bedroom because reflections from nearby walls are stronger and thus more damaging than reflections from more distant surfaces. So the types of treatment used depend on the situation. But the basics are indeed universal, so we'll begin there. Along the way I'll debunk several common acoustics myths.

> **Myth:** *A listening room should never be dead sounding, and absorption should be applied sparingly when possible.*

There are two philosophies about this. One says that a listening room should contribute to the total ambience you hear while playing music, and the other says the room should add as little of its own sound as possible. I am firmly in the second camp. Otherwise, everything tends to sound the same as the room imposes its own color onto everything you play, masking the hard work the recording and mixing engineers put into creating the environment they want you to hear. Once you remove your room's own ambience—especially early reflections—you can then hear the recording as intended because the small-room echoes no longer drown out the larger-sounding reverb present in the recordings.

Admittedly, a well-treated room can be an acquired taste for some people. But in my opinion, once you get used to the sound and learn to appreciate the improvement in clarity, there's no going back. The goal is for a *neutral* sound; when you listen in a room that adds no character of its own, you'll be closest to hearing the true tonality of the music. Indeed, in my well-treated living room, I hear *more* spaciousness and ambience in the recordings I play, not less. The sound is richer, fuller, and much larger overall.

Good acoustics is just as important in rooms where instruments and singers are recorded with microphones. The same reflections that reduce clarity when listening through loudspeakers also make live instruments sound boxy, hollow, and off-mic. When performers and microphones are close to bare reflecting walls, reflections that get into the microphones are stronger. Small-room ambience always sounds small, and that's always bad, except perhaps as a special effect such as recording in a tiled bathroom to achieve a certain vibe. When a recording studio's "live" room is properly treated, you can place microphones farther away to better capture the full sound of physically large instruments with less undesired room tone, and less EQ will be needed as well. You can easily add reverb and ambience later when mixing, and those effects will surely sound better than the boxy ambience of a bedroom-size space.

Room Orientation and Speaker Placement

Placing your loudspeakers and listening position correctly is the first step toward getting good sound, especially at low frequencies. While positioning alone won't eliminate the need for bass traps and other acoustic treatment, it's an easy and free way to help reduce low-frequency response errors and improve imaging. The first step is to identify the ideal listening position within the room, and from there you can determine the best speaker placement.

> **Myth:** *Speakers should fire the short way across the room, because the added width gives better imaging.*

Having the loudspeakers fire the short way across a room, where the room is wider left to right and shorter front to back, is an audiophile concept that dates back many years. The idea is to reduce the strength of side-wall reflections and also have them arrive later at your ears. But the damage caused by side-wall reflections affects mainly mid and high frequencies, and is easily solved using relatively thin absorber panels. However, bass peaks and nulls caused by strong reflections from a wall directly behind you are typically much worse than when the wall is farther away and the reflections are weaker due to distance. Figure 19.1 proves the point using a pair of measurements taken in a small room with the speakers facing both the long and short ways. It should be obvious that the longer orientation gives a much better bass response.

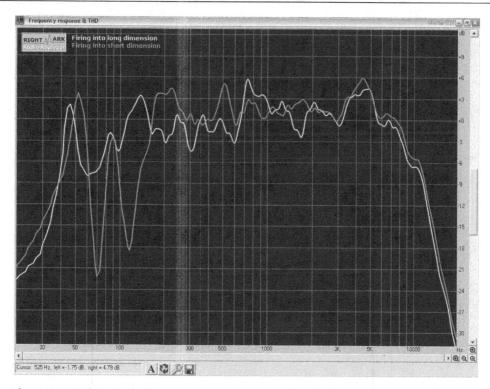

Figure 19.1: This graph shows two measurements taken in the same room. For one test, the speakers fired the longer way down the room. The other measured the response with the speakers firing across the shorter dimension. *Image courtesy of Martin Walker,* Sound On Sound *magazine columnist.*

Figure 19.2 shows the layout and dimensions for the lab room at my company's factory where I test acoustic treatment, and it illustrates the basic principles. Note that the placement method used here is based on the "38 Percent Rule," a theory popularized by acoustician and studio designer Wes Lachot. In his writings, Wes has shown that the theoretically best listening position is 38 percent into the length of the room when measured from either the front or rear wall. This offers the best compromise of peaks versus nulls for any given room size. For a two-channel listening room, you'll get the flattest low-frequency response sitting 38 percent of the way back from the front wall. However, this is not always practical in a home theater, especially one with a large screen, because that might put you too close to the screen for good visuals. Fortunately, you'll get a similar benefit sitting 38 percent of the length when measured from the rear wall.

Understand that 38 percent is a theoretical best location. It's a good starting point, but in practice it may not be best due to other factors—wall properties, speaker location, speaker

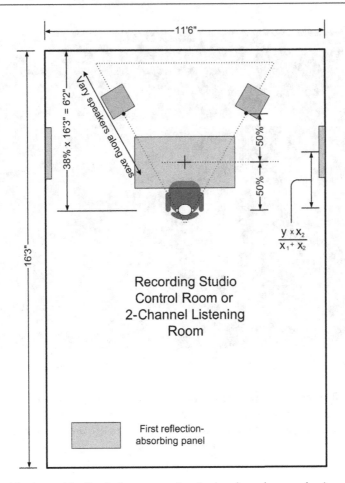

Figure 19.2: This shows idealized placements for the loudspeakers and prime listening seat in a room used mainly for stereo playback.

type, room furnishings—plus other factors that affect low-frequency response. In the end, the best way to find the optimum speaker placements and listening position is by measuring the low-frequency response at high resolution using room testing software. This is described in Chapter 22.

Once you've decided where to put your seat, the next step is placing the loudspeakers. The speakers and listening position should define an equilateral triangle, as shown in Figure 19.2. This means that the distance between the left and right speakers is the same as the distance from your head to each speaker. Note that the point of the imaginary stereo triangle is just behind your head, with the axis lines grazing your ears. Tweeters should be at ear level for the flattest response, since most speakers have less high-frequency output off-axis.

Figure 19.3 shows a similar layout but optimized for a home theater or multi-channel surround audio system. In this case, the surround speakers should also be the same distance from your ears as the front three speakers. When this is not possible, you can use the *speaker distance* setting available in most hi-fi receivers to compensate. When the same sound comes from more than one speaker, each speaker's output should reach your ears at the same time. This prevents the precedence effect from harming localization, as explained in Chapter 3. Note that in a surround setup, the center and two main speakers should form an arc, rather than all be the same distance from the front wall. This puts the center speaker slightly farther forward in the room compared to the mains, again keeping all three the same distance from you.

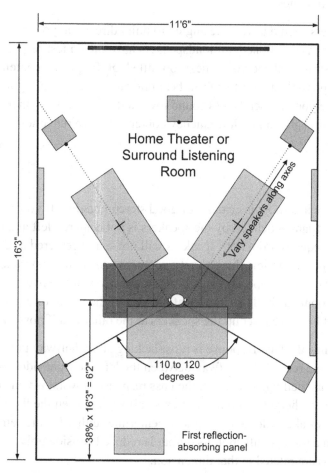

Figure 19.3: This shows the idealized placements for a home theater or multi-channel audio listening room.

Note that the tweeters in the rear surround speakers should also be at ear height. Some home theater enthusiasts place the surround speakers high up on the side or rear walls, mimicking the setup of some commercial movie theaters. But that's a throwback to years past, when a single rear channel contained the surround information rather than separate channels as in today's 5.1 soundtracks. Back then, some movie theaters placed one or more speakers high up on the rear wall to increase ambience by including the room's natural reverb. But that was long ago, and in theaters much larger than anyone's living room.

Since the front speakers define an equilateral triangle, all three angles will be 60 degrees. But the rear speaker angles should be between 110 and 120 degrees as shown. This is the standard setup as defined by the Producers & Engineers Wing of the Recording Academy, a division of the Grammys. These are the same standards used by movie studios and professional surround music mixing engineers, so it only makes sense for your setup to follow the same guidelines.

Finally, loudspeakers should always be angled to point directly at you, sometimes called *toed-in* positioning. As mentioned, all loudspeakers have the flattest response on-axis, so if they're pointed straight ahead you'll hear too little high frequency content as shown in Figure 18.20, and possibly also suffer from beaming and lobing as in Figure 18.21. Angling the speakers to face you also sends less sound toward the side walls, which has the benefit of reducing both the strength and high frequency content of those reflections.

Symmetry

Left-right symmetry in a room is critical for good stereo imaging. If you sit more to one side of the room, or the triangle defined by the speakers is not centered left and right, instruments and voices coming equally from both speakers will not sound centered as they should. When perfect symmetry is not possible throughout a room, at least aim for symmetry in the front part of the room. The most important area is along the side walls between your head and the speakers. If a room has built-in shelves at one end of the room, or other obstructions or furniture that can't be moved, set up the speakers at the other end if possible.

In rectangular rooms, the bass response is most lacking at the halfway points—halfway between the front and rear walls, halfway between the left and right side walls, and halfway between the floor and ceiling. Therefore, the bass response is worst if you sit in the exact center of the room at a height that puts your ears halfway between the floor and ceiling. You shouldn't put speakers along any of those centerlines either for the same reason; when a loudspeaker is in a room's null spot, its output is reduced considerably at low frequencies whose wavelengths are related to that dimension.

This raises a dilemma because sitting with your head centered left and right in the room gives the best stereo imaging, but it also puts you in a bass null for the room's width dimension.

Some acoustics experts consider imaging to be more important than bass response, while others disagree. There is no one right answer for this, which shows that acoustics is as much an art as a science. In my opinion, it's better to offset your listening position a few inches to either side to avoid being exactly halfway between the left and right side walls. A few inches either way won't harm imaging as much as it will reduce the depth of the left-right bass null. Further, the halfway width null is not usually as damaging as the halfway length null. You can also place absorption symmetrically—with one panel spaced slightly off the wall to be the same distance from its adjacent speaker as the panel and speaker on the other side— which maintains acoustical symmetry at mid and high frequencies. On the other hand, if you sit *precisely* centered left and right, each of your ears will be a few inches to either side of the center line anyway.

Again, the best way to know if small positional changes help or hurt is with room measuring software. This lets you experiment with different speaker distances by sliding both speakers along their axes while you measure the response at different proposed listening spots. Otherwise, simply put the speakers at a distance that's convenient and sensible for the size and layout of your room, while maintaining an equilateral triangle. Sometimes people obsess over minute details that matter only a little, while ignoring ergonomic concerns that matter just as much or even more.

Reflection Points

Now let's consider the *first reflection points*. The concept of first reflections—also called *early reflections*—was introduced in Chapter 3. Reflections from specific locations along the side walls, ceiling, and floor arrive at your ears slightly after the direct sound from the loudspeakers. Early reflections can also come from the top surface of a mixing desk or coffee table. Reflections from all of these locations can cause comb filtering that skews the frequency response and harms clarity and imaging. Depending on the difference in arrival times, some frequencies are boosted and others are reduced. In professional control rooms where music and movie sound tracks are mixed, it's a *de facto* standard to treat all nearby room surfaces with absorption to eliminate early reflections. The graph in Figure 19.4 shows the comb filter frequency response measured with and without absorption at the reflection points in a typical small room.

The general goal is to eliminate all such reflections by placing absorbing panels at key locations. Another approach is to angle the side walls and ceiling to send reflections away from your ears. It's not always practical to put an absorber on top of a console while mixing, though some people do that using a thin sheet of acoustic foam. A better solution is to put the speakers on stands that are positioned some distance behind the console. With the proper stand height and distance, reflections won't bounce off the top of the console toward your ears, as shown in Figure 19.5.

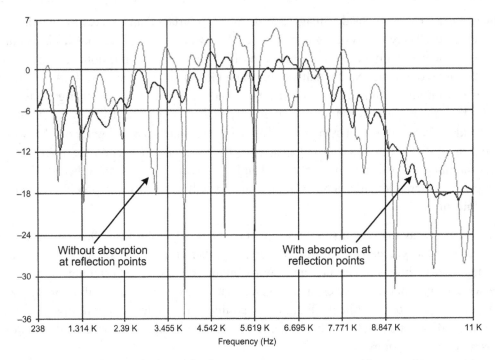

Figure 19.4: This graph shows the frequency response measured in a small room with and without absorption at the side-wall reflection points. The falling high-frequency response is due to the inexpensive SPL meter used.

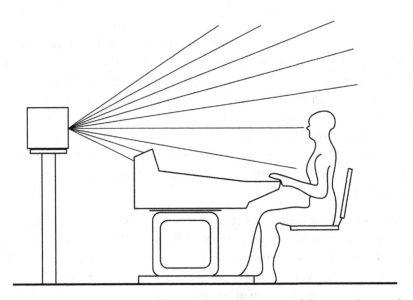

Figure 19.5: Placing loudspeakers on stands, rather than on top of the console, avoids reflections from the top surface of the console. *Drawing concept by John H. Brandt.*

When a room has acoustic treatment to absorb these reflections or if the walls are angled sufficiently, the protected listening area is said to be a *reflection-free zone*. Another related term you may have heard is *non-environment room*, which is literally that: damaging reflections are absorbed to remove the effects of the room's "environment," more closely approximating the sound of outdoors where there are no echoes.

Reflections not only skew the frequency response, but they also harm clarity and imaging— an effect known colloquially as "time smearing." When a direct sound is followed by a reflection that arrives within about 20 milliseconds, the ear is unable to distinguish the reflection as a separate sound source. So instead of hearing an echo, the two sounds are perceived as one that obscures clarity and imaging. If these reflections are present, you can still tell when an instrument is panned hard left or right, but the in-between positions are not as well defined. Listening to music free of early reflections is not unlike listening with earphones: Instruments and voices sound clearer, and their placement in the stereo field is more stable. Indeed, headphones are fully anechoic having no reflections, yet music generally sounds larger on headphones than through loudspeakers. Once all early reflections are removed, clarity and soundstage magically come to life. All of a sudden you can easily hear EQ changes of half a dB and very small changes in left-right panning.

As mentioned, some people believe that early reflections are desirable. But the logic is simple and, in my opinion, irrefutable: Ambience that's already present in many recordings is often that of a large space—a concert hall, movie scoring sound stage, large recording studio, or created using artificial reverb. But when played back in a small untreated room, strong, small-room reflections drown out the larger-sounding ambience that's present in the recording. This makes the sound field seem smaller, not larger.

The value of absorbers at reflection points is standard for professional listeners, and should likewise be the goal for an audiophile or home theater enthusiast who wants a listening environment as excellent as a million-dollar control room. Anything less and you won't experience the same clarity and quality the mix engineers heard when creating the music or movie soundtrack. Many of my friends are professional musicians, recording engineers, and composers. They all appreciate the quality of my two audio systems, and several of them bring their mixes here for a final reality check. In hi-fi and home theater forums it's common for someone to ask if they'll benefit from absorbing early reflections. I always suggest they hang folded-over bath towels on the side walls with masking tape to see if they hear an improvement. Two towel layers aren't as good as real acoustic panels that absorb more and over a wider range of frequencies. But it's an easy test to try, and you can get proper panels if you like what you hear.

More to the point, rooms vary wildly, so a recording engineer has no idea how strong the reflections will be, or their timing, when consumers hear their work. The only way to ensure consistency throughout the playback chain is for the recording engineer and end listener to both avoid excess ambience and early reflections. The eventual playback room then becomes much less important. Acoustical differences will still exist, of course, but at least they'll be minimized as much as possible. Note that the reflections that damage imaging and clarity are mainly at mid and high frequencies, above approximately 300 Hz. However, using absorption that's effective to even lower frequencies can only help, even if it's not strictly needed.

Understand that a reflection "point" is really the center point of a larger area. In a nearfield setup where the speakers are two to four feet in front of you and two to four feet apart, a two by two-foot absorbing panel centered at each reflection point along the side walls is sufficient. But as you get farther from the speakers, and in turn the speakers are farther apart, you need to cover more surface area to ensure a reflection-free zone where you listen. Treating a larger area also lets you move around without leaving the protected zone. In a professional control room the mix engineer may roll forward and back while working, and others in the room may be seated behind the mix position. Likewise, if people might be standing up while listening, it's useful to extend the absorption higher as well as wider. So while the general term is "reflection point," which I'll continue to use, what's really meant is an area whose center is at ear height and approximately halfway between the listener's ears and the front of the loudspeakers.

The two classes of reflections are early or late, and first or second. Early reflections are those that arrive within about 20 to 25 milliseconds of the direct sound from the speakers, and first or second refers to whether the sound reached you after one bounce or two. While not technically correct, I consider reflections from a speaker on the opposite side of the room to be secondary as well—that is, sound from the left speaker that bounces off the right wall and reaches your right ear. It's only one bounce, but it comes from the opposite side. Many people confuse the difference between early and first reflections, referring to all such reflections using one term or the other. First reflections are often early, but not always; it depends on the size of the room and, in turn, the distances involved.

Note that the delay time is determined by the *round-trip distance* of sound from a loudspeaker as it travels toward a reflecting surface, then bounces toward your ears. Sound travels at about one foot per millisecond, and the delay time is the *difference* in path lengths for the direct and reflected sounds. So if the speaker is five feet from your ears, and the total path length for a reflection to reach you is eight feet, the reflection will arrive about three milliseconds after the direct sound.

Figure 19.6, a repeat of Figure 3.16 in Chapter 3, shows the three main paths by which sound from a loudspeaker arrives at your ears. The direct sound is shown as solid lines; the early first reflections—a single bounce off a nearby surface—are dashed lines; and secondary reflections from the opposite side walls are shown as dotted/dashed lines. Whether the secondary reflections are early or not depends on the total distance they travel before reaching your ears, compared to the direct sound from that same speaker. Echoes and

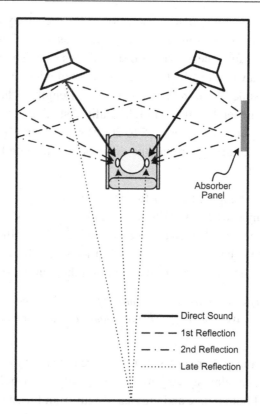

Direct Sound
1st Reflection
2nd Reflection
Late Reflection

Figure 19.6: In a typical small room, direct sound from the speakers arrives first, followed by the first reflections, then the second, and finally late reflections from the rear. Other reflections, not shown here, arrive from the ceiling and floor.

ambience arrive from the rear of the room even later, as shown by the dotted lines. These, too, can be early or late, depending on the distance. In truth, reflections from the rear of a room are more complex and dense than the single dotted line paths shown here coming from only one speaker. But this is sufficient to explain the concept.

The easiest way to tell where to place side-wall absorption to stop first reflections is with a mirror. While you sit in the listening position, have a friend place a mirror flat against each side wall and move it around while you look at the mirror. Any surface location where you can see either loudspeaker in the mirror should be covered with absorption. Once the side wall locations are identified, do the same on the ceiling. It's more difficult to slide a mirror around on the ceiling, though I've done it using a hand mirror attached to a broom or garden rake handle with rubber bands. You can optionally calculate the locations instead of using a mirror, and that will be described shortly. There's also a simple shortcut cheat: Absorbers on the ceiling are exactly halfway between you and the speakers, and absorbers on the sides are slightly forward of halfway.

In narrow rooms it's often useful to place absorption at the secondary reflection points as well as the first reflection points, where sound from the left loudspeaker bounces off the right wall and arrives at the right ear, and vice versa. Floor reflections can also harm clarity if there's no carpet. Those parts of a wood or tile floor can be covered with throw rugs, preferably as thick and plush as possible. What matters most is the strength of the reflections after applying absorption. The general goal for a high-quality listening environment is to have all early reflections arrive at your ears at least 15 to 20 dB softer than the direct sound from the speakers. If a reflection is 20 dB softer than the direct sound, the comb filtering peaks and nulls are both less than 1 dB.

I'll also mention that some people prefer diffusion rather than absorption at reflection points. But when I tried that in my own 25 by 16-foot listening room, my wife and I both agreed it sounded no better than a bare wall. Perhaps in a very wide room diffusion could be useful. Then again, personal taste is just that. So while I won't say that diffusion is always bad at reflection points, I can report that I preferred absorption there in my room.

> *Myth: Rooms designed for stereo listening should be treated differently from home theaters. A hi-fi room should be live sounding, whereas home theaters sound better with more absorption.*

As explained previously, many domestic-size rooms are too small to benefit from natural ambience. More to the point, almost every movie has plenty of music! So for that reason alone, it makes sense that both music and movies benefit from the same basic layout and acoustic treatment types. However, a home theater has more speakers to handle the additional channels, so absorption is needed in more places. But by and large, I believe the basic goals should be the same. The only change I make when listening to stereo music versus movies is to switch the playback mode in my receiver.

Figure 19.7 shows the front portion of my living room home theater, which I use both to play stereo music and watch surround sound movies and concerts on DVD or Blu-ray. This room has five acoustic panels to absorb early reflections. The panels in the foreground on the left and right sides are on stands, and three more panels are on the ceiling to prevent vertical sound radiation from the front three speakers from reaching the listening position.

Just as early reflections from the side walls can harm imaging and create comb filtering, so can the rear wall if it's less than ten feet behind you. Figure 19.8 shows the rear wall of my living room, which is about ten feet behind the couch.

Many people have no choice but to place the listening position directly in front of the rear wall. Those reflections are especially damaging, not only because they're so early, but they're also very strong due to the close proximity. Figure 19.9 shows four absorbers on the wall directly behind the couch to reduce the strength of those reflections.

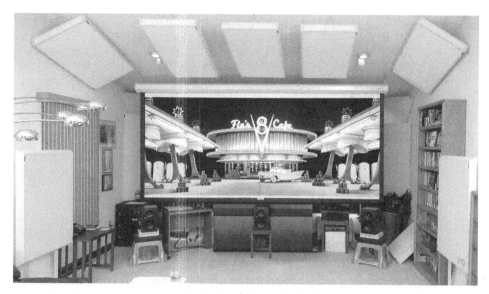

Figure 19.7: This photo shows the front of Ethan's living room home theater. Stand-mounted absorbers at the left and right handle the side reflections, and three more panels hang under the ceiling. Bass traps are in the left and right front wall-wall corners, and more bass traps are placed at the top of the side walls in those wall-ceiling corners.

Figure 19.8: The rear of Ethan's living room has four diffusers on the center portion of the wall, plus bass traps in the rear and side wall-ceiling corners. The center diffusers rest on more bass traps that handle the rear wall-floor corner. Thick terry cloth covers the leather couch seat backs to reduce those reflections, too.

Figure 19.9: These panels directly behind the couch absorb early reflections from the wall.
Photo courtesy of Perry Perone.

Even reflections from a couch or chair back can cause comb filtering if it rises above your head. As a test, I measured the response directly in front of the leather seat back of my couch—first with the leather exposed, then again with an absorber I made from ultra-soft "baby fabric" stuffed with cotton batting to make a sandwich one inch thick. The audible difference was substantial, though when measured, the response change might not seem as significant as it sounds. Figure 19.10 shows the response four inches in front of the leather couch back when bare, and again with the thick cover in place. I was surprised to see the response become a little worse at a few frequencies after adding the cover, so I measured again and got the exact same results. Such is the nature of reflections and comb filtering. But the sound is much better with the cover, and it remains on my couch to this day, replacing the terry cloth bath towels in the photo. Ideally, when listening to surround movies and music, the seat backs will not be above head level anyway, since that blocks sound from the rear speakers. But sometimes it's nice to slouch down and relax.

Won't All That Absorption Make My Room Sound Dead?

I believe the notion "I don't want a dead sounding room" is mostly misguided. There are 44 RealTraps acoustic panels in my living room shown in Figures 19.7 and 19.8, and it's not at all dead sounding. Many of these panels are bass traps with a semi-reflective surface to reduce absorption at higher frequencies. But there's still a fair amount of broadband

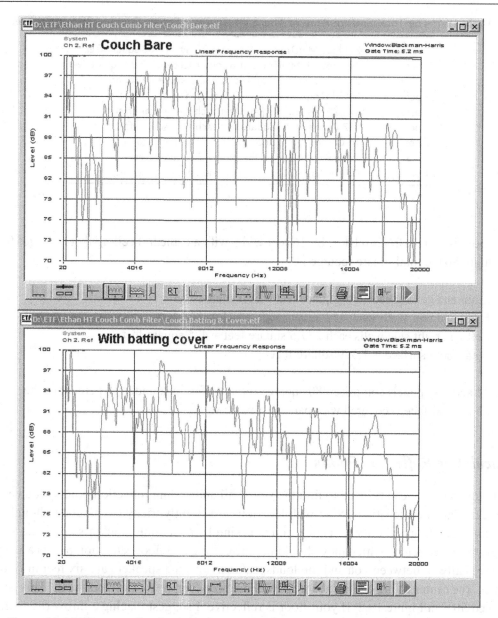

Figure 19.10 These graphs show the frequency response measured four inches in front of a reflective couch back, with and without a thick absorber pad in place.

absorption in this room, and the sound is exceptionally full, clear, and coherent. The room is not at all dead sounding, and conversation sounds absolutely normal. When this came up in a hi-fi forum I created the list of panels in Table 19.1. As you can see, all those panels account for less than 18 percent of the total room surface.

Table 19.1: The Total Surface Area of These 44 Acoustic Panels Equals 280 Square Feet

11 pcs.	2x4 foot MiniTraps at 8 square feet each = 88 square feet
3 pcs.	2x4 foot HF MiniTraps at 8 square feet each = 24 square feet
11 pcs.	2x2 foot MiniTraps at 4 square feet each = 44 square feet
2 pcs.	MondoTraps at 9.5 square feet each = 19 square feet
8 pcs.	MegaTraps at 5.6 square feet each = 45 square feet
2 pcs.	RFZ Panels at 9.3 square feet each = 19 square feet
2 pcs.	Tri-Corner Panels at 3 square feet each = 6 square feet
4 pcs.	Diffusers at 8 square feet each = 32 square feet
1 pc.	Planter Bass Trap at about 3 square feet = 3 square feet

The room is 25 feet front to back, by 16 feet wide, with a cathedral ceiling that averages 9.5 feet high. So the total square footage is:

Floor and ceiling each 400 square feet = 800 square feet
Front and rear walls each 152 square feet = 304 square feet
Left and right side walls each 237.5 square feet = 475 square feet
Total room surface is 1,579 square feet

The 44 acoustic panels comprise 280 square feet, therefore just under 18 percent of the room's total surface is covered with panels.

Calculating Reflection Points

Although a mirror can be used to find the side-wall and ceiling reflection points, you can also calculate their locations. Side-wall locations are slightly tricky to calculate, though finding the ceiling and floor reflection points is simple: Assuming your ears are at the same height as the tweeters, which they should be, the ceiling and floor reflection points are exactly halfway between you and the loudspeakers. So if the speakers are six feet in front of you, the center of the ceiling absorber will be three feet in front of your head. This is marked 50% in Figure 19.2. If your floor is reflective, I suggest placing a throw rug there to handle the floor reflections the same distance away, which is directly under the ceiling absorbers. For larger rooms, or when the speakers are farther away or farther apart, you'll need two overhead panels (and two throw rugs), with one for each speaker as shown in Figure 19.3.

You can calculate the placement of side-wall absorbers using the formula shown in Figure 19.11, which refers back to the earlier drawing in Figure 19.2 showing the listening position 38 percent away from the front wall. Figure 19.11 shows how to determine the reflection point based on the distance between the side wall and listener, and the side wall and tweeter. Only the right speaker

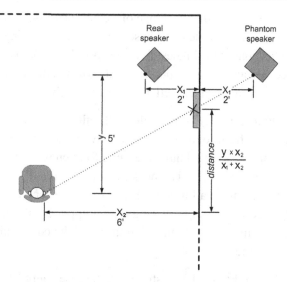

Figure 19.11: This formula shows how to calculate the placement of RFZ panels on a side wall.

is shown, but the calculations are identical for the left side and for a center speaker, too, if present. The panels are centered vertically at tweeter height, which should be ear height as well.

Let's say the speaker's tweeter is two feet from the side wall ($X1$), and you're six feet from that same wall ($X2$), and the loudspeaker's tweeter is five feet in front of you (Y). The distance along the wall from the point directly to the right of your head to the center of the panel marked "X" is solved as follows:

$$\text{Distance} = \frac{Y * X_2}{X_1 + X_2} = \frac{5*6}{2+6} = \frac{30}{8} = 3.75' = 3'9''$$

Another way to look at this is to pretend you have a mirror image of the room on the other side of the wall. Imagine a "phantom" loudspeaker placed on the other side of the wall at the same height and distance from the listening position. This is the extra loudspeaker shown in the figure. If you draw a line from the phantom speaker's tweeter to the listening position, that line passes through the wall at the reflection point. As you can see, the side-wall panels are placed slightly forward of the halfway point, with the exact distance determined by the formula. Note that the tweeter locations are used because reflection control concerns mainly mid and high frequencies.

This brings up a related issue, because the notion of a reflection point, or area, is valid for only a single listening position. If you have more than one listening seat, you should determine all of the reflection points and make sure there are enough absorber panels to handle all of the seats. In smaller rooms where one person listens near field, a single two by four-foot panel on each side wall and the ceiling is probably adequate. But larger rooms require more panels, or larger panels. If you have a second row of seats, even more coverage is needed to keep everyone in a reflection-free zone.

Finally, it's important to understand that the range of frequencies absorbed affects the quality of the reflection-free zone. Some people use thin sculpted foam or moving blankets, but those thin materials absorb reflections at the highest frequencies only. Using panels two inches thick is therefore a better choice, because the absorption extends down to the lower midrange frequencies. As mentioned, it's best if absorption at reflection points is effective down to 300 Hz or lower.

The last reflections myth to consider is that side-wall reflections are acceptable when they're "neutral" sounding, coming from loudspeakers that have a flat off-axis response. In fact, if reflections emanating from a side-wall have the same frequency content as the direct sound, the comb filtering will be most severe. The peaks will be up to 6 dB and the nulls will be very deep. But if the loudspeakers have a limited off-axis response such that the wall reflections contain less high frequencies, the comb filtering in that upper clarity range will be less severe. In other words, the more closely the original and delayed sounds match in spectrum and level, the more severe the comb filtering.

The three response graphs in Figure 19.12 show a pink noise source, the same noise mixed with a simulated reflection delayed 15 milliseconds to emulate a loudspeaker with a flat off-axis response, then again with the reflection's high frequencies rolled off at 6 dB per

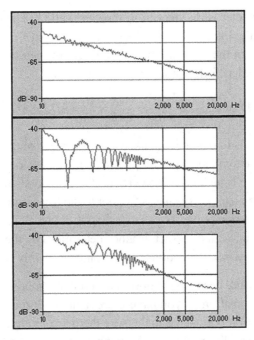

Figure 19.12: On top is the pink noise source, then the same source with a simulated reflection delayed 15 ms, and at bottom is the same source again but with the reflection filtered to remove high frequencies.

octave. You can clearly see that the comb filtering is less severe when the reflections have less high frequency content. This is basic math, and a rolled-off response at high frequencies is the same as having high frequency absorption on the walls. So while a flat off-axis response is an important goal of all loudspeakers, that actually harms audio quality when early reflections are allowed. In other words, if you have good loudspeakers, it's even more important to absorb reflections, not less important as some people claim.

Angling the Walls and Ceiling

When the budget allows for dedicated construction, early reflections can be avoided by angling the side walls and sloping the ceiling upward. A great example of this is shown in Figure 19.13. Given a large enough angle—at least 35 degrees—reflections are directed behind the listening position. This retains more room ambience by avoiding absorption on the walls and ceiling. But most people do not have the luxury of building new walls, so the only option is to use absorption. Allowing room ambience makes sense only in larger rooms like this anyway.

Understand that professional control rooms, with a live front end like this one, still need to control ambience using absorption and diffusion elsewhere in the room. For example, you can see the large absorbing cloud above the console, even though the steep angle would have been sufficient to direct the ceiling reflections over the mix position toward the rear of the room. Most professional control rooms are also larger than a typical bedroom, so this doesn't contradict the goal of avoiding small-room ambience. Further, a control room like this needs

Figure 19.13: This control room uses angled side walls to direct early reflections away from the listening position to the rear of the room. However, the ceiling above the console is absorptive rather than reflective to provide additional bass trapping. *Photo courtesy of Wes Lachot Design Group.*

the side walls to be clear for windows into the live rooms on both sides, making absorption on the side walls impractical.

One common layout mistake I often see is facing a corner while listening. Although the wide angles do avoid side-wall reflections, in most rectangular rooms this puts another corner behind you. Corners tend to focus sound, much like a satellite dish or cupping a hand behind your ear. So having a corner behind you reflects sound coming from anywhere in the room directly at your head. Focusing is the opposite of diffusion, and it's always best avoided. If there's no choice but to set up facing a corner, the entire corner area behind you should be treated with thick broadband absorption. How large an area depends on the size of the room. But for a bedroom-size space, covering both walls from floor to ceiling at least four to six feet out from each corner will help to minimize the focusing effect.

Another common mistake is placing racks or other studio furniture directly behind the mix position, or at reflection points, or other locations that prevent sound from the speakers reaching your ears unobstructed. Figure 5.5 from Chapter 5 shows the equipment rack in the professional recording studio I built in the late 1970s. That rack was off to the side rather than directly behind the mix position. This let the mix engineer wheel back in the chair to work the patch panel and tweak the knobs, while remaining centered left and right in the room. Likewise, having a video screen on a desk between the near field monitors is okay, as long as it doesn't block the speakers.

Low-Frequency Problems

Bass frequencies are the most difficult to tame in a small room because the wavelengths are long, which requires thick absorbers called *bass traps*. One of the most common problems is mixes that sound great in your room sound too bassy when played elsewhere. The culprit is one or more nulls in the response; nulls as deep as 20 or 30 dB are not only common, but typical. Most small rooms have many nulls in the range below 300 Hz, but some people fear that adding bass traps will reduce the amount of bass even further.

> **Myth:** *Bass traps absorb low frequencies, so putting them in a room reduces the amount of bass you'll hear.*

In truth, bass traps lower peaks and also raise nulls, so they make the response *flatter*, rather than add or remove bass. In most small rooms, the main problem is deep nulls caused by reflections from the wall behind you combining out of phase with the direct sound from the loudspeakers. Adding bass traps therefore increases the perceived (and measured) level of bass. But in some rooms, especially those that are square or cube shaped, peaks can dominate the response. In that case, adding bass traps reduces the peaks, again making the response closer to flat. Whether peaks or nulls are the bigger problem also depends on where in the room you listen.

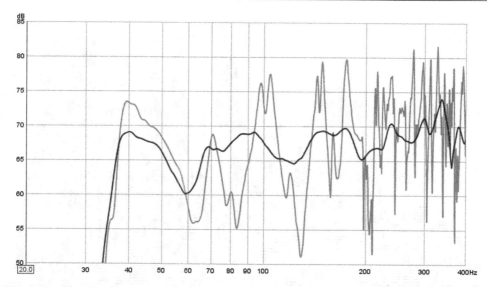

Figure 19.14: The low-frequency responses shown here are typical for most small rooms, before and after adding bass traps. It should be obvious which graph line is before and which is after.

The Before/After frequency response shown in Figure 19.14 was measured in the same small room as Figure 19.2, which is 16 by 11.5 feet with an 8-foot ceiling. The graph shows the room's low-frequency response with and without bass traps, and you can see three severe nulls around 64, 84, and 140 Hz in the lighter Before trace, as well as additional nulls at higher bass frequencies.

Since these deep nulls cause you to hear less bass than is really in the mix, you'll tend to add too much bass level and EQ to compensate. As you can see, the finest loudspeakers in the world are of little value if your room skews everything you hear this badly. When bass traps are added to a room, the low-frequency response becomes flatter and also changes less around the room. The most effective place for bass traps is in corners where bass waves tend to gather, though other locations are also viable. Note that rectangular rooms have 12 corners: four where each wall meets another wall, four where each wall meets the ceiling, and four more where each wall meets the floor. After treating as many corners as possible, the front and rear walls are good candidates for even more bass traps.

After bass traps are added, the response becomes not only flatter and more full-sounding but also tighter and clearer because the decay times are reduced. Figures 19.15 and 19.16 show the reduction in *modal ringing* after adding bass traps, using the same measurement data as above. This type of graph is called a *waterfall plot,* where the "mountains" come forward over time to show how long certain frequencies continue to ring after the initial sound stops. Ringing peaks is the main cause of the problem commonly known as "one-note bass," where every bass note sounds like the same pitch, no matter what note was actually played.

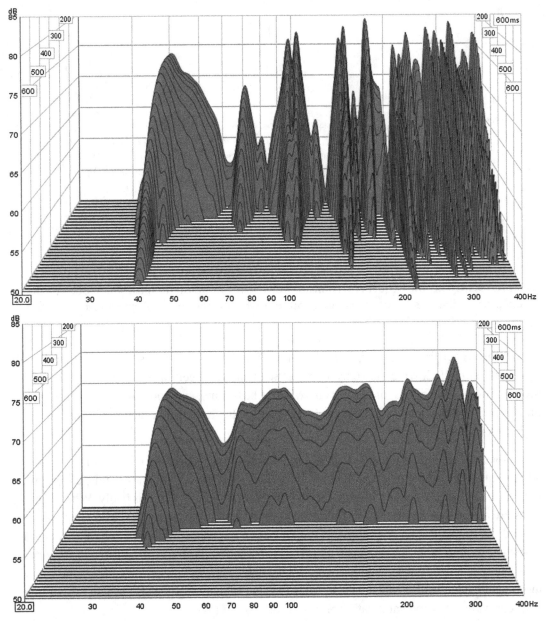

Figures 19.15 and 19.16: These graphs are derived from the same measurement data as Figure 19.14, but they show the peak decay times as well as their amplitude. The upper graph shows the room when empty, and the lower graph is after adding bass traps.

It's impossible to make any small room perfectly flat, especially small rooms like this, but the more bass traps you add, the closer you'll get. It's really that simple. The only trade-off is how good you want versus how much effort and expense you're willing to endure. The response and ringing shown in these graphs is about as good as can be expected in a room this size, short of lining every single inch of room surface with extremely thick absorption.

Reverb Decay Time

Another important goal of room treatment is to achieve a uniform decay time versus frequency. If some frequencies take longer to decay than others, the effect is similar to having a skewed response because in both cases there's more energy in the room at those frequencies. Most small rooms don't really have true reverberation unless they're totally empty. Rather, they foster a series of individual reflections that decay quickly, usually after only one or two bounces. Real reverb tends to swell and build over time, as happens in a gymnasium or auditorium. The distinction between discrete reflections and true reverb is subtle, but important.

The graph in Figure 19.17 shows the reduction in reverb decay time versus frequency after adding a large number of bass traps, broadband absorbers, and diffusers. This is the same 16 by 11.5-foot room as the previous graphs showing low-frequency response and ringing. Not only are the decay times much shorter, which improves clarity, they're also much more uniform.

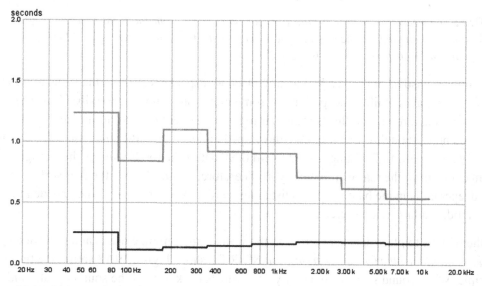

Figure 19.17: This graph shows reverb decay time versus frequency in octave bands and was derived from the same measurement data as in Figures 19.14 through 19.16.

Indeed, having uniform decay times versus frequency is almost as important as controlling them at all. Even though this room is too small to have true reverb, we use the same reverb metrics as for larger rooms.

Reverb decay time is called *RT60* because it's defined as how long it takes for the reverberation to decay by 60 dB after the original sound stops. In many situations it's not possible to measure 60 dB below the normal listening level because of background noise from air handlers or passing traffic outdoors. So it's common to instead measure the time it takes the reverb to decay by 30 dB, then extrapolate how long it would have taken to fall by the full 60 dB. Decay time is linear, so you simply double the time it took the sound to decay by 30 dB to get RT60.

Stereo Monitoring

Figure 18.20 in Chapter 18 showed the off-axis response of a typical loudspeaker, with reduced output at high frequencies for listeners off to either side. At 45 degrees off-axis, the response at 12.5 KHz is down a full 10 dB. Many speakers are even more directional vertically, losing highs as they radiate toward areas above and below the tweeter. Therefore, loudspeakers should be placed so their tweeters are at ear level and aimed directly at the listening position. It's also a mistake to place speakers higher up, angled down to compensate, as you sometimes see. While angling the speakers downward does put the listener on-axis, the response you hear changes continually if you move forward or back even slightly because the angle keeps changing.

> **Myth:** *Loudspeakers should be placed as far as possible from the front and side walls. Putting them one-third of the way into the room from the front, and listening one-third of the way into the room from the rear, is ideal.*

Conventional wisdom says it's a bad idea to place loudspeakers close to the walls of a room. This is another audiophile concept that has some merit, but is not necessarily ideal. The idea of dividing a room into thirds is similar to the 38 percent rule, to minimize the effect of peaks and nulls, though it was probably developed by listening rather than via calculations. As shown in Figure 19.2, sitting more toward the front of the room reduces the strength of reflections from the rear wall behind you. Therefore, in most rooms the rear suffers from deeper nulls in the bass range, so listening there makes sense only when needed to accommodate a large video screen.

Another reason to place speakers closer to the walls is to minimize peaks and nulls due to *speaker boundary interference response*, or SBIR for short. Figure 19.18 shows the basic principle. When sound strikes a boundary and reflects back toward the source, peaks and nulls occur at predictable quarter-wavelength distances. At one-quarter wavelength, a null is

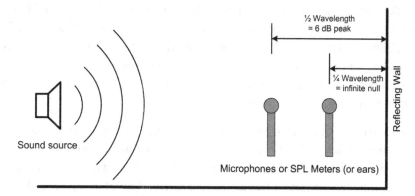

Figure 19.18: Peaks and nulls occur at predictable quarter-wavelength distances from a reflecting boundary. Measuring half a wavelength away from a perfectly reflecting boundary yields a 6 dB increase at the corresponding frequency. At a distance of one-quarter wavelength, the result is a null that's infinitely deep.

created because the round-trip distance is one-half wavelength. This puts the reflection 180 degrees out of phase with the original, canceling the sound at that location.

Figure 18.20 also shows that the radiation from many loudspeakers becomes omnidirectional below about 300 Hz. So sound from the rear of a speaker strikes the wall behind, creating a null at whatever frequency is one-quarter wavelength for that distance. For frequencies where that distance is one-half wavelength, the delay from a round trip adds 360 degrees of phase shift, which reinforces the sound, creating a peak in the response. As you can see, SBIR is just another way to consider comb filtering.

When a speaker is placed farther from a reflecting boundary, the inevitable peaks and nulls occur at lower frequencies where they're more difficult to tame with absorption. The closer a speaker is to the wall, the higher the frequency of the lowest null. At only six inches away, the first null is at 565 Hz, so absorption on the surfaces behind and to the side of the speaker is highly effective. Indeed, the ideal place for a speaker is mounted directly *in* the wall, as shown in Figure 19.13. This eliminates SBIR peaks and nulls entirely, and also avoids diffraction effects from the speaker cabinet's edges, because a wall usually extends much farther from the speaker driver.

Note that SBIR occurs both for sound sources near a boundary and for listeners near a boundary. When referring to the peaks and nulls created by boundaries near a loudspeaker, it's called SBIR. If the peaks and nulls are considered as related to the listener's proximity, it's called LBIR, for *listener boundary interference response*. This term was coined partly in jest by Canadian acoustician Andre Vare, but it's totally appropriate.

Placing a speaker near to a boundary, but not in the wall, gives an overall bass boost as explained in Chapter 18. This is why many powered monitors include "half-space" and "quarter-space" switches to reduce bass output. Not only is the bass boost easy to counter with these switches, or outboard EQ when using passive speakers, the "free" bass boost means the speakers don't have to work as hard to produce the same output level. This in turn lets the speakers play louder and with less distortion.

The *Frequency-Distance Calculator* shown in Figure 19.19 can be downloaded from the website for this book. I wrote this Windows program for the work I do with my acoustics company. You enter a distance, and it calculates a series of related quarter-wavelength frequencies. It optionally converts a frequency to a series of quarter-wavelength distances. The program is intended to help analyze a room's low-frequency response to determine if a particular peak or null is modal (related to a room dimensions) or non-modal (related to SBIR). The sine wave image on the screen shows the level of the wave as it leaves the wall at the right. Since these reflections are out of phase with the source, the maximum levels produce the deepest nulls.

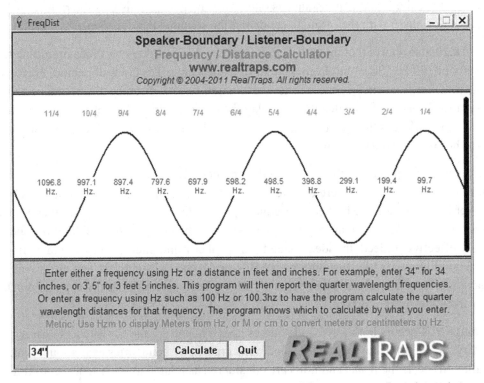

Figure 19.19: The Frequency-Distance Calculator converts distances to wavelengths and vice versa automatically, depending on what you enter.

For example, if you measure a deep null at 100 Hz 34 inches in front of the rear wall behind the listening position, the calculator will confirm that 34 inches is indeed 1/4 wavelength at 100 Hz. So in this case the null is not necessarily caused by a room mode, but rather by the listener's proximity to the rear wall. Nulls occur at other odd multiples from the rear wall, such as 3/4 wavelength, 5/4, 7/4, and so forth. This program displays all quarter-wavelength multiples from 1/4 through 9/4 as either frequencies or distances, and it's very clever because you can enter either a distance or a frequency. The software figures out what you entered and displays the opposite series of parameters. Note that this program doesn't need to be installed; just unzip all of the files into any folder, then run freq-dist.exe.

> **Myth:** *When mixing in a small room that's not treated with bass traps, it's better to avoid the influence of peaks and nulls by using small speakers that can't reproduce the lowest frequencies.*

It's true that using speakers that are unable to reproduce the lowest frequencies prevents you from hearing peaks and nulls and ringing in the room at those frequencies. But in my opinion this is like an ostrich sticking its head in the sand. If you can't hear those frequencies when mixing, it's impossible to know if the bass levels and EQ sound good or not. Further, damaging peaks and nulls occur at all frequencies, not just the lowest few octaves where small speakers roll off. As you can see in Figures 19.14 and 19.15, peaks, nulls, and ringing extend all the way up to 400 Hz and even higher. Even stuffing some pillows in the corners to serve as marginal bass traps gives better results than hiding the room's problems by using small speakers!

Another common monitoring mistake is putting near field monitors sideways on top of a mixing console's meter bridge. Most speakers have a tweeter above the woofer, and should not be placed on their sides because that delays the sound from the farther driver at frequencies around the crossover point. Sound combines in the air from both drivers at those frequencies, and the staggered arrival times skew the response. However, this is less of a problem with active speakers, or passive speakers that are bi-amped. Passive crossovers typically transition at 6 or 12 dB per octave, so half an octave or more emits from both speakers around the crossover point. But active crossover slopes are usually 24 dB per octave, so a smaller range of frequencies is negatively affected.

It's also a good idea to tighten the mounting screws for your woofers once a year or so to prevent rattles and buzzing from developing over time. Slowly sweep a loud sine wave from 20 Hz to 300 Hz while listening for buzzing not only from your speakers but also from windows, shelves, equipment racks, and so forth. I promise you it will be a real eye-opener! In fact, this is further proof of the power of the masking effect, since many people don't even notice the buzzing in their room until they isolate it this way.

Speaking of rattles and buzzes, it's worth noting that our ears perceive low frequencies as being omnidirectional. With content below about 100 Hz, it's difficult, if not impossible, to tell where the sound is coming from. Some people believe that years of acoustical research are wrong and that people can perceive bass direction at very low frequencies. But I'm convinced the real issue is buzzing and rattling from a woofer or subwoofer, or port noise from a reflex enclosure, or maybe vibration from a nearby window. If you can hear where a subwoofer is placed, and the crossover is 100 Hz or lower, then something else is going on. However, some all-in-one speaker systems use too-small "satellite" speakers with a subwoofer that crosses over at 200 Hz or even higher. In that case, you can definitely hear that the bass is coming from the subwoofer.

Surround Monitoring

Most home studios are set up for stereo playback only, but home theaters and many hi-fi listening rooms have additional speakers to play 5.1 surround music and movie soundtracks. Some systems have additional speakers to accommodate six or seven main channels, or even more. Modern consumer receivers support at least 5.1 channels, where the "5" refers to the number of main channels, and the ".1" is the LFE effects channel for frequencies 120 Hz and below that are handled by a subwoofer. Other newer formats offer more channels and potential speaker placements, though not all have been fully accepted as standards. Even budget receivers include a surprising number of advanced features:

- Discrete 5.1 analog inputs via RCA connectors
- One or more digital inputs using HDMI, optical, and/or RCA connectors
- Circuitry to decode multi-channel Dolby and DTS format signals from the digital inputs
- Multiple stereo inputs for CD and DVD players, and cable or satellite TV
- AM and FM receivers
- A loudness switch that boosts low frequencies automatically at low volumes to compensate for the Fletcher-Munson effect
- A "midnight mode" that compresses the volume of loud action scenes in movies to not disturb others late at night
- Bass management that routes low frequencies from any combination of main channels to a subwoofer by specifying the speakers as "small" rather than "large"
- Separate power amplifiers for all of the main channels
- Variable delay for each channel and the subwoofer to compensate for different speaker distances
- Surround special effects that add artificial spaciousness to stereo and surround music
- EQ that adjusts itself automatically for "room correction" at low frequencies
- Video switching, so selecting TV or DVD switches both the audio and video
- Dolby ProLogic to create a *faux* center channel from normal stereo

This last feature is actually an unintended side effect of what was originally meant to decode an older form of Dolby surround sound used in the 1980s. But it's useful today with TVs because it routes mono dialog and other sounds to the center channel instead of coming equally from the left and right speakers. This way, no matter where someone is sitting, the voices are anchored to the center rather than seeming to come from whichever speaker you're closer to.

As a bonus, Dolby ProLogic is also useful for separating a singer from a backing track, rather than use the vocal removal method described in Chapter 13. Content panned to the middle is routed to the center channel and also removed from the left and right channels. If your receiver has separate left, right, and center analog outputs, as many do, you can use this process to create karaoke backing tracks from regular stereo recordings by using only the left and right channels.

SPL Meters

Originally SPL meters were developed to measure absolute volume levels, but some models can do much more. As explained in Chapter 1, sound waves are rapid changes in air pressure to which our ears respond. Therefore, SPL stands for *Sound Pressure Level*, though you'll sometimes see them called Sound Level Meters (SLM). Note that SPL meters are "reference" devices, so they must be calibrated at the factory (and periodically thereafter) using a known absolute volume level. The Galaxy meter shown in Figure 19.20 is accurate to within +/− 2 dB, which is reasonable for an inexpensive meter. SPL meters can be calibrated by an independent lab for a fee, but that's not needed for casual use such as matching loudspeaker levels in a stereo system or home studio. What matters there is consistency from one reading to another more than absolute accuracy.

In truth, SPL meters are quite simple—a built-in microphone and preamplifier connect to a readout that displays the sound level in decibels. This Galaxy meter can report levels between 40 and 130 dB SPL, which is adequate for most situations. Many SPL meters have a switch to accommodate a range of sound levels, though some can cover a large span without switching. When used to measure absolute volume levels, some common applications are:

- To be sure your rock band's outdoor gig doesn't exceed the maximum volume allowed by local ordinance
- By law enforcement for the same reason
- By highway engineers to assess pavement surface noise, or barriers that shield nearby homes
- In a recording studio to ensure consistent monitoring levels from one mixing session to another
- As a reality check for how loud you're listening to avoid hearing damage
- To brag to your friends how loud your hi-fi system can play!

An SPL meter is also useful for assessing relative levels, for example, to calibrate your stereo or surround system to ensure that all loudspeakers play at the same volume level.

Figure 19.20: This popular Galaxy SPL meter is affordable, and works well for non-critical applications.

Calibration products, such as the *Digital Video Essentials* test DVD[1] I use, include signals that play through each speaker one at a time. This lets you set identical levels from each speaker, and also adjust your subwoofer's volume relative to the main speakers. You simply place the SPL meter where your head would be while listening, then play that section of the DVD. As each speaker sounds the test signal, you adjust your receiver so the volume from each speaker is at the same level. Many recording studios use active speakers, so in that case you'd use the speaker's own volume controls.

Unlike the directional microphones commonly used by singers and radio announcers, SPL meters contain an omnidirectional microphone that responds more or less equally to sound arriving from all directions. In theory it shouldn't matter which way the meter is pointed, though all omni microphones have a flatter response for sound arriving from the front. When balancing speaker levels in a stereo system, pointing the microphone straight ahead favors the left and right speakers equally. Neither speaker is on-axis to the microphone, but the angle and resulting response errors are the same. However, when balancing speaker levels in a surround system, the convention is to point the microphone straight up, with the microphone's element at the same height as the tweeters (which should all be the same). Again, this ensures the same response for all speakers, even if that response is not perfect. However, when verifying the frequency response of loudspeakers it's better to test them individually with the microphone pointed straight at their tweeters, and from about one foot away to minimize room reflections that could skew the measurements.

When using an SPL meter to calibrate loudspeaker levels, you can use either static sine waves or pink noise as the test signal. A sine wave used for level adjustment is usually 1 KHz, which is the center of the midrange. The meter displays a number corresponding to the volume at its microphone. However, sine waves are not a good choice for matching speaker levels because of *standing waves* in the room. Standing waves create peaks and deep nulls in the response that are highly positional. If you play a 1 KHz sine wave and note the level on the meter, then move the meter only an inch or two in any direction, the level will likely be very different.

Room acoustic treatment—especially absorption at reflection points—reduces the level difference at nearby locations, but doesn't avoid it completely. Therefore, a much better signal source for matching loudspeaker levels is pink noise which contains all frequencies. If the SPL meter's microphone happens to be in a null location for 1 KHz, a nearby frequency, such as 1.1 KHz or 912 Hz, will not be in the same physical null. Therefore, the main advantage of pink noise is its inherent averaging of volume level versus frequency versus location. Taken as a whole, the measured volume will be fairly accurate. The *Digital Video Essentials* DVD that I mentioned earlier takes this one step further and filters the pink noise to contain only midrange frequencies. The bass response in most domestic-size rooms varies wildly with position—even more than in the midrange—so filtering the noise to remove those frequencies ensures more consistent and reliable readings when used for matching speaker levels. For your convenience I created a similar file containing filtered pink noise, which is described in the section Calibrating Loudspeakers in Chapter 22.

While pink noise is better for speaker level matching than sine waves, there's one drawback: Because noise is by definition random, the volume constantly fluctuates. So when viewed on a conventional level meter, such as the VU meter in a cassette deck, the needle dances around making it difficult to read. The variance is typically several dB, though it can be 5 dB or even more at very low frequencies. So the meter might display 80 dB for a moment, then 77 dB, then 84 dB, and so forth. To get the true picture, you need to watch the meter carefully for ten seconds or longer and mentally average all the numbers. This is another advantage of filtered pink noise that contains only midrange frequencies. It's also why I prefer modern digital SPL meters over older analog models. A digital meter can average the variations over time for you.

It's worth mentioning that inexpensive SPL meters use inexpensive microphones that are not as accurate as professional microphones. Fortunately, most budget meters are fairly accurate at bass frequencies, though above 1 KHz they're typically much worse. I use a precision DPA 4090 microphone for room testing; but it costs a lot more than the Galaxy meter shown earlier! However, even if the microphone in an inexpensive meter is not highly accurate, it's still useful to assess *relative* changes, for example, to see how the response and ringing in your room improve as bass traps and other acoustic treatment are added.

The Weighting switch on an SPL meter selects one of two or three response curves that bias the display. As explained in Chapter 2, our ears hear midrange frequencies more readily than bass or treble, so when the goal is to assess *perceived* volume—how loud something actually sounds—the A-weighting curve shown in Figure 19.21 is preferred. With A-weighting, sound containing mainly midrange frequencies displays a higher value than very low or very high frequencies at the same SPL. But when measuring the frequency response of a loudspeaker or room you'll use C-weighting (or Z-weighting which is perfectly flat) because you want to learn the actual response. Most SPL meters also let you select a Fast or Slow display response time. A slow response is usually better because it keeps the displayed level more consistent.

When absolute accuracy is needed, professional acousticians use a calibrated SPL meter with a known-flat frequency response. The NTI Audio XL2 meter shown in Figure 19.22 is that

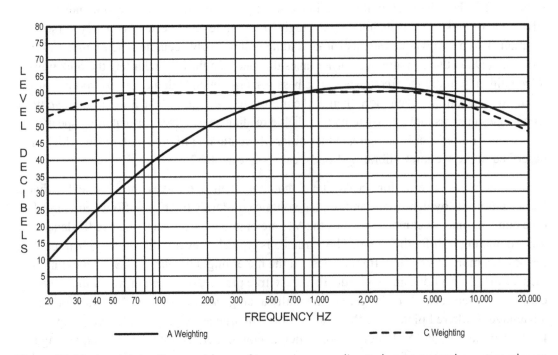

Figure 19.21: A-weighting favors midrange frequencies according to how our ears hear at moderate volume levels, and C-weighting is closer to a flat response.

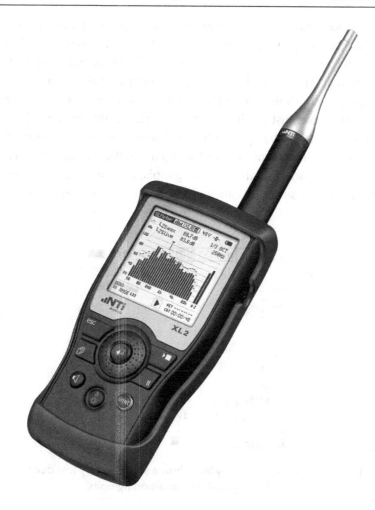

Figure 19.22: The NTI Audio XL2 provides a comprehensive set of measurement types required by professional acousticians.

and much more. The XL2 boasts a frequency response (−3 dB points) from 4.4 Hz to 23.6 KHz, over a volume range of 17 to 153 dB SPL depending on the microphone option chosen. In addition to serving as an SPL meter with an accuracy of 0.5 percent, the XL2 also displays frequency response, reverb times in third-octave bands, harmonic distortion, and absolute polarity. It also offers RTA and FFT spectrum analysis, audio recording for later analysis, and can even serve as an oscilloscope. As the British say, this is a serious piece of kit. Of course, a professional meter like this comes at a professional price. Other companies that make professional quality SPL meters include B&K, 3M SoundPro, ScanTek, and Pulsar Instruments.

The main advantage of a professional meter like the XL2, besides accuracy, is that it's small and totally self-contained. For a professional acoustician who measures sound every day in disparate locations both indoors and out, carrying around a laptop, microphone, and external sound card with a phantom powered preamp and calibrated gain gets old quickly. Some external sound cards can be powered by a computer's USB port, but many laptops do not provide enough current. The USB 1.0 and 2.0 specs require USB ports to provide up to 500 milliamps, but many laptops fall short of that and thus cannot power an external sound card. So then you need AC power just for the sound card. (Thankfully, the USB 3.0 spec increases the available current to 900 milliamps, but that still doesn't mean every computer can provide that much.) The XL2 will run for more than four hours on the included rechargeable battery. And while its built-in display is smaller than a laptop's, it shows everything needed very clearly. It can also save measurement data and screen images for later transfer to a computer.

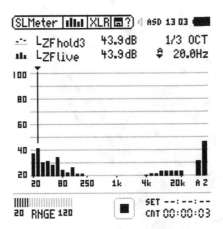

Figure 19.23: The residual noise level in my large home studio is very low due to its distance from the heating and air conditioning units.

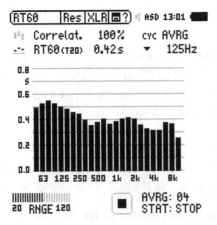

Figure 19.24: This graph displays the third-octave reverb times measured in my home studio.

Figures 19.23 and 19.24 show the residual background noise and RT60 reverb time measurements in my large home studio (about 6,000 cubic feet). This room is on the second floor of my home, and the furnace and air blowers are in the basement two floors below. So even when the heat or air conditioning is on, the ambient noise is very soft at most frequencies. Figure 19.23 shows the noise in third-octave bands, but note the A and Z level bars at the bottom right. The A level is the sum of noise at all frequencies with A-weighting applied, while the Z level is unweighted reflecting a greater contribution from low and high frequencies. Although the RT60 times might seem a little long for a home studio, they're actually just right for a room this size.

Summary

This chapter explains why the room is usually the weakest link in any audio system, varying far more than even budget microphones and other gear. If you can't hear what's really in a track, it's very difficult to get a good mix. The same applies to audiophiles who listen for enjoyment. Most untreated rooms have half a dozen peaks and deep nulls in the bass range below 300 Hz, comb filtering at higher frequencies, as well as excess ambience that masks low-level detail in the music.

The first step toward good sound is determining where to place the loudspeakers and listening position, which can be fine tuned later with room measuring software. The 38 percent "rule" identifies the best place to listen, and from there you position the speakers. Loudspeakers should fire the long way down a room to reduce the strength of peaks and nulls by putting the rear wall farther away from your ears. Once you've optimized the bass response, any remaining problems due to reflections at mid and high frequencies can be corrected with relatively thin absorbers.

Symmetry is also key to good stereo imaging to ensure that voices and instruments played by both speakers equally are perceived as coming from a phantom center point midway between the speakers. This requires placing the speakers and listening position at the points of an equilateral triangle that's centered left-right in the room. For a surround system, all loudspeakers should be the same distance from your head when possible, though a receiver's speaker distance settings can compensate if needed.

A room should add as little of its own sound as possible, though a well-treated room can be an acquired taste for some people. In my opinion, a neutral-sounding room lets you hear what's in the recording and better appreciate the artist's intent. One of the keys to achieving neutrality is absorbing early reflections that not only skew the frequency response but also harm clarity and imaging. Reflections from the rear are equally damaging if the wall is closer than about ten feet behind you, and even a reflecting seat back creates early reflections. You can find where to put absorption either with a mirror or by calculating the locations, though angling the walls and ceiling works well when such construction is feasible.

Finally, bass frequencies are the most difficult to tame, requiring many large, thick absorbers called bass traps. Bass traps not only improve the frequency response, but they also reduce modal ringing, which is just as damaging. Reducing decay times at higher frequencies is equally important, as is having the decay times be uniform with frequency.

Note

1 Digital Video Essentials: www.videoessentials.com

CHAPTER 20

Room Shapes, Modes, and Isolation

Two of the most important properties of a room are its size and shape, which directly affects its low-frequency behavior. Generally speaking, large rooms are better acoustically than small rooms, because not only are the walls and ceiling farther away from listeners and microphones, which reduces the strength of reflections, but in large rooms the *mode* frequencies are closer together. Most acoustics experts recommend a minimum volume of at least 2,500 cubic feet for rooms where music will be produced with high quality.

A room mode—short for *mode of vibration*—is acoustics-speak for a natural resonance whose frequency depends on the distance between two opposing surfaces. For example, a room that's 14 feet long front to back has Length modes at 40 Hz, 80 Hz, 120 Hz, and subsequent multiples of 40 Hz. Acoustic waves at these frequencies fit exactly between the front and rear walls, so they bounce back and forth repeatedly, reinforcing their energy at each cycle. This is unlike other frequencies where reflections don't combine in phase at each cycle. Therefore, modes create response peaks inside the room, and they also foster *modal ringing*, which sustains those frequencies unnaturally by extending their decay times.

In many ways, a room can be viewed as a set of resonant band-pass filters or, more precisely, three such filter sets in parallel, with one for each dimension. As with resonant electronic filters, room modes not only extend decay times, but they also build energy in the room over time. If you play a sine wave through a loudspeaker at a mode frequency and then shut it off, the wave inside the room doesn't start and stop suddenly. Rather, it swells over time, sustains until the source sound stops, and then decays over time. The swell and decay times depend on the strength and Q of the mode, which is determined by the mass and stiffness of the room boundaries for that dimension. Unlike a simple band-pass filter with a single center frequency, a room has multiple harmonically related modes for each dimension, which is why it's considered a set of filters.

There are two classes of room modes: *Axial* modes occur between two opposing surfaces, and *non-axial* modes take a more circuitous route, traveling like a cue ball around a pool table in a diamond pattern as it bounces off the room surfaces. The two non-axial mode types are called *tangential*, where the reflected wave touches four surfaces, and *oblique*, where it hits all six surfaces. A rectangular room has three axial modes, with one each for the length, width, and height. Axial modes are more important than non-axial modes simply because they're stronger and contribute more to peaks, nulls, and modal ringing. Tangential modes are 3 dB weaker than axial modes, and oblique modes are 3 dB weaker than tangentials. Therefore, axial modes are the most important type to consider when planning the dimensions for a new room. Note that axial modes develop between all opposing surfaces, even if they're not parallel. However, it's more difficult to predict the mode frequencies of nonparallel boundaries.

> **Myth:** *It's not possible to reproduce deep bass in a small room because long waves need sufficient distance to develop.*

A related myth is that you must be some distance from the loudspeakers or other sound source before the low end can be heard properly. But if that were true, headphones wouldn't work, nor would you be able to hear deep bass from your car stereo. As mentioned in Chapter 6, many large musical instruments send different frequencies off in different directions, including toward the rear. So it's often best to place microphones farther away to capture those parts of the sound after they bounce off room boundaries and "develop" in the room. But that's not the same as needing some minimum physical distance to accommodate a given wavelength. Sound is created entirely by changes in air pressure, which occurs even within the small sealed cavity of a closed-back earphone. When a loudspeaker or other sound source produces a frequency whose wavelength is too long to fit within a room's longest dimension, the room is said to be operating in *pressure mode*. But the wave still exists in the room, and it's still heard at its normal amplitude.

Modal Distribution

Figure 20.1 shows the modes for one dimension—let's say the length—of two different rooms. The larger room at the top has a length of 28 feet, so its fundamental (lowest) mode frequency is 20 Hz. Subsequent modes occur at 20 Hz intervals. Even though this creates many resonant peaks in the response, the peaks are close together, so the average response is fairly flat. As one peak is falling, the adjacent peak is rising, which helps fill the void.

Now consider the length modes for the smaller room, shown at the bottom of Figure 20.1. Here the first peak is at 60 Hz, which corresponds to a half-wavelength of about 9½ feet. Therefore, subsequent modes occur at 60 Hz intervals, making the overall response less uniform because there are wider gaps between the boosted frequencies. However, modal peaks are not always bad. Some people consider peaks at very low frequencies to be useful because they provide a "free" bass boost called *room gain*. Modal peaks are also referred to as providing *modal support* for the same reason.

Room modes that lie near, but not exactly at, standard musical note pitches can sometimes make music sound out of tune because a nearby bass note can excite the mode to ring at its

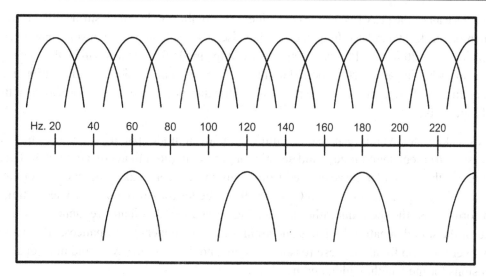

Figure 20.1: In a large room (top), the resonant peaks caused by modes are closer together than in a small room (bottom). The closer spacing yields an overall flatter response.

natural frequency. This is most likely to occur when a mode has a high Q, as happens with walls made from cement or other rigid materials that reflect strongly at low frequencies.

Recall that Chapter 3 included an audio example of a high-Q equalizer boosting a frequency near the natural pitch of a pair of wood claves. In that example, the EQ's resonance made the claves seem to sound higher in pitch.

When the Q of a mode is high, its resonance is more pronounced and sustains longer. A short bass note at 110 Hz might excite a mode at 112 Hz to ring for half a second or more, making the note sound sharp even though it's not. A strong room resonance can also shift the perceived pitch at higher frequencies. Everyone who works at my company is a musician, and one of the guys keeps a drum set at the far end of our large warehouse space for after-work jam sessions. I recall the time I whacked his cowbell and noticed that the flutter echo between the floor and ceiling sustained at a pitch slightly lower than the cowbell's own frequency!

Room Ratios

Another important factor is the ratio among a room's length, width, and height. The worst room shape is a cube, where all three dimensions are the same. A cube has the fewest number of peaks, because all three dimensions resonate at the same series of frequencies. This yields a greater distance between peaks because there are no other dimensions to contribute in-between peaks.

Worse, since all of the peak frequencies coincide, they combine to be much stronger than a room having more peaks at unrelated frequencies. Likewise, it's best to avoid dimensions having even multiples such as 20 feet by 10 feet because many of the mode frequencies will be the same. There's no such thing as a perfect room, but if there were, each dimension would contribute peaks at different frequencies, evenly spaced with little distance between them, as shown at the top of Figure 20.2.

Besides making the overall response less uniform, uneven mode spacing can make one note on a bass instrument sound louder and sustain longer in the room than adjacent notes. This is much worse than a gentler curve created from many in-between peaks that, even if not flat, affect a wider range. As explained in Chapter 19, it's better for a room's reverb decay time to be uniform across the spectrum rather than be higher at a few dominant frequencies, which can color the sound unnaturally. The same applies to low-frequency resonances, which also decay over time. So for all of these reasons, a room should have different, and non-related, dimensions for the length, width, and height.

Professional studio designers agree that certain ratios of room height, width, and length are most desirable. Three of these ratios, developed by L. W. Sepmeyer, are shown in Table 20.1. Good-sounding rooms often have a height, width, and length similar to one of these Sepmeyer ratios. The dimensions in the table are between solid, reflective walls. There are other good ratios, but these are probably referenced most often.

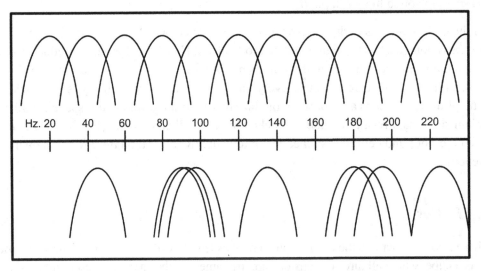

Figure 20.2: The modes for a theoretical room having ideal ratios (top) give a more even response than a room with poor ratios (bottom). When a room's ratios are poor, some of the natural resonances are spaced far apart, while others are clustered close together.

Table 20.1: Preferred Room Ratios

Height	Width	Length
1.00	1.14	1.39
1.00	1.28	1.54
1.00	1.60	2.33

Note that when a room has a suspended tile ceiling, the real height for the modes is the solid surface above the tiles. Bass waves go right through ceiling tiles, so the distance that defines the modes is to the upper rigid boundary. Likewise, in a basement with exposed joists, the modal height is to the solid floor above, even if the joists are filled with fiberglass insulation. When a room has an irregular shape or angled walls, modes still exist, but they're more difficult to calculate. However, you can average the dimensions to get a rough idea of the frequencies. For example, if a side wall is angled so the room is 10 feet wide at one end and 12 feet wide at the other, you could use 11 as the average for the width. Rooms with irregular shapes, such as with a large alcove, have more than three sets of modes and so are more difficult to calculate. This is why measuring an existing room is generally more useful than trying to predict its response with a mode calculator.

All of that said, I believe the importance of room ratios is often overstated, sometimes ignoring ergonomic concerns that are important, too. Even if a room has a "perfect" ratio, which doesn't exist, bass traps are still needed to minimize the inevitable peaks, nulls, and ringing. Regardless of a room's size and shape, standing waves, peaks, and deep nulls occur at a broad range of frequencies, not just those related to the room's dimensions. So you still need bass traps that handle the entire bass range rather than absorb only mode frequencies. Indeed, with enough bass traps, even very small rooms can sound excellent, so don't despair if that's your situation. More to the point, calculating room modes is needed only when designing a new space. Whether a ratio is good or not is irrelevant when treating an existing room, because you can't do anything to change the modes unless you're willing to build new walls.

Modes, Nodes, and Standing Waves

A common source of confusion is the difference between modes, nodes, and standing waves. A *mode* is a frequency defined by a room dimension that resonates when the same frequency is played in the room through a loudspeaker or other sound source. In other words, a mode is a propensity to vibrate. If you have a mode at 100 Hz but never play that frequency in the room, the mode will never be excited, and thus it will never ring.

A *node* is a location in the room where two waves collide out of phase creating a null. For example, nodes occur at predictable quarter-wave distances from the rear wall behind a listener, as explained in Chapter 19. An *antinode* is similar, but it occurs at locations where two waves combine in phase to create a peak in the response. When two waves combine in phase, the result is an increase in pressure, hence the peak.

A *standing wave* also occurs when two opposing waves combine in phase, but it defines the state of the wave motion at the location—not the location itself, or the frequency, or any ringing that may result. Most acoustic waves are considered *traveling waves* because they travel through the air from one place to another. But at the point of collision, where they butt heads so to speak, a peak is created by the increased pressure at that location. Many people, including some acousticians, confuse modes and standing waves and consider them the same. But they're not the same; one is a natural inclination to vibrate, but only when excited by sound waves, and the other is a static condition that occurs when opposing wave fronts collide in the air. Indeed, this confusion is so rampant that in his book *Recording Studio Design* (Focal Press, 2003), acoustics expert Philip Newell addressed it thus:

> *It should be stressed that standing waves always exist when like waves interfere, whether a resonance situation exists or not, and that the common usage of the term "standing wave" to describe only resonant conditions is both false and misleading.*

If you've ever used an ultrasonic cleaner to clean jewelry or small electronic components, you've probably seen standing waves in action. When you drop a pebble into a pond, a series of waves is created that extends outward from the point of impact. Since a pond is large, the waves dissipate before they can reach the shore and be reflected back to the place of origin. But in a small contained area like the tub of an ultrasonic cleaner, waves bounce off the nearby surrounding walls and create a pressure front that makes them literally "stand still" within the cleaning solution. The same thing happens in a room when your loudspeakers play a sustained sine wave such as from an electric bass. Standing waves cause static nodes and antinodes to develop at various places around the room, depending on the loudspeaker's position, the room's dimensions, and the frequency of the tone.

ModeCalc Program

Several mode calculators are available for free on the Internet, but most give only a long list of numbers, making it difficult to see the big picture. Since I needed a good mode calculator for my acoustics work, I created the ModeCalc program shown in Figure 20.3, which is available on the website for this book. ModeCalc runs on all Windows computers and calculates the first 16 axial modes up to 500 Hz for any rectangular room using dimensions you enter as either feet and inches or meters. Its main purpose is to determine good dimensions for a proposed new room or to estimate the low-frequency behavior of an existing room. It will not help you treat an existing room, nor will it predict the frequency response at the listening position. For that you need to measure. Further, reality can differ substantially

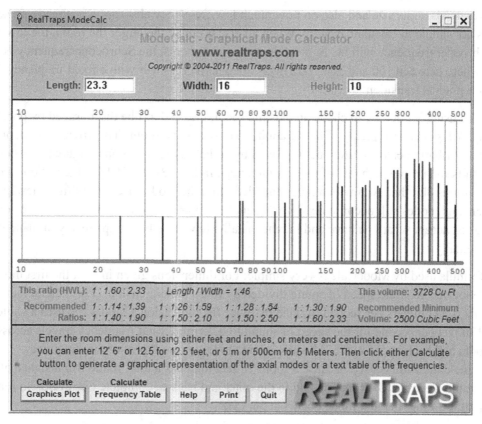

Figure 20.3: ModeCalc runs on all Windows computers and displays the axial modes for any rectangular room based on its dimensions.

from mode predictions, depending on the room's construction. Predictions can be skewed when very low frequencies pass through thin walls to a more rigid boundary beyond, or by the presence of bass traps and other objects in the room.

As with the Frequency-Distance calculator, there's nothing to install. Just unzip all of the files into any folder, then run ModeCalc.exe. I don't have a version of ModeCalc for Macs, but you can still get some benefit by making your room similar to one of the eight recommended ratios shown on the program's screen.

Note that modes naturally become closer at higher frequencies. Acousticians consider low and high frequencies in a room to split at a crossover point known as the *Schroeder frequency*, after German physicist Manfred R. Schroeder (1926–2009). This is the frequency above which individual modes no longer dominate the response and traditional reverberation takes over. In other words, above the Schroeder frequency, the modes are so close together that they're no longer considered individual resonant peaks. The Schroeder frequency for a room depends on its size and also on how much low-frequency absorption is already present, so it's not simple to calculate exactly. When all else is equal, the larger the room, the lower its Schroeder frequency will be. In most domestic-size rooms, the Schroeder frequency is somewhere between 100 and 300 Hz. This is not a hard crossover with a fixed frequency, but rather a gradual transition.

If you're using ModeCalc to check an existing room, please don't be discouraged by poor results. All rooms need plenty of bass trapping anyway, and a room that suffers from poor modal distribution can definitely be improved by adding more bass traps. Figure 20.3 shows the results for a room with one of the recommended ratios, 23.3 by 16 by 10 feet. You can see from this plot that even with a recommended ratio, the mode spacing is still somewhat uneven, and modes still occur at nearby frequencies. So unless you're willing to move your walls, just accept what you have and maybe install a few more bass traps than you planned for originally.

The formula used by ModeCalc is very simple: For dimensions given in feet, the first mode frequency occurs at the speed of sound divided by twice the dimension. (For meters, the formula is 344 meters per second divided by twice the dimension.) ModeCalc uses a value of 1,130 feet per second for the speed of sound to determine what frequency travels across the room and back in one cycle. All subsequent modes are multiples of that frequency. Twice the dimension is used because a room 10 feet long has a total round trip distance of 20 feet; the wave travels from one end to the other and back to complete one cycle. So for a room 10 feet long, the first mode occurs at 56.5 Hz:

$$\frac{1130}{10*2} = 56.6$$

The second mode for that dimension is two times 56.5 or 113 Hz, the third is three times 56.5 or 169.5 Hz, and so forth. Instructions at the bottom of the screen explain how to use the program, and a complete tutorial is available in the program's Help.

Room Anomalies

For most smaller project studios, a rectangular room is fine, especially if that's all you have. It's pointless to build angled walls in a small room. Unless you can use the space behind the new angled wall—for example, as storage—you'll only make the room even smaller. The same goes for an angled ceiling. If you have the luxury of building a new room, and a budget to make that room large enough for professional results, then adding angles can make sense. In that case, the room should become wider and/or higher toward the rear rather than narrower or lower. Concave surfaces are best avoided because they focus sound.

The same type of focusing occurs with a peaked ceiling, where it's higher in the middle rather than at one end. This focuses sound to an area under the peak, though absorption can reduce

Figure 20.4: The ceiling in Ethan's home studio is peaked, rising from 8 feet high at the front and rear walls to 12 feet in the center. Absorbers hanging a few inches under the peak reduce the focusing effect that occurs with concave surfaces.

the focusing effect. Figure 20.4 shows the peaked ceiling in my home studio, with absorbers hanging directly under the peak.

A peaked ceiling isn't all bad, though, because it avoids the flutter echoes and ringing that would occur between opposing reflective surfaces when the ceiling is parallel to the floor. When my home was being built in the 1990s, I had the contractor create this high angled ceiling instead of a traditional flat ceiling with an attic above. At 33 feet long by 18 feet wide, with a ceiling peak 12 feet high, my one-room home studio is large enough to have "good" ambience, which I've used to great advantage. A standard 8-foot ceiling would have been too low for a room this long and wide.

My ceiling rises from the front, then goes back down in the rear. This is better than having the ceiling rise from the left and go back down on the right because that puts the listening position directly under the peak. If the peak is centered front to back as in my room, the listening position will be forward of the peak's focus. If your peak rises from each side rather than front to back, just treat a wider area under the peak, especially directly above your head. Figure 20.5 shows the floor plan for my studio, which is on the second floor of my home. Chapter 6 showed several photos of my home studio in the section about common microphone placement. The video "studio_tour" explains how my studio is set up and treated, and why, and brings together many of the room design concepts and theory from this chapter.

Even though rooms should become wider and taller at the rear, that wasn't practical in my studio. If the bathroom was in the rear, the rear would have been even narrower. But I wanted

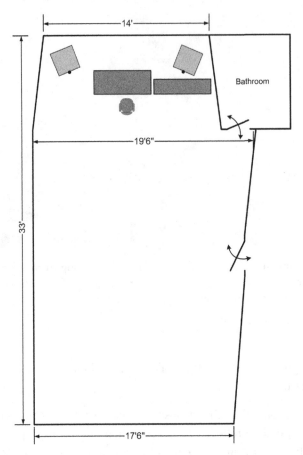

Figure 20.5: Ethan's one-room home studio is large enough to record more than a dozen musicians at once.

to angle the long side walls slightly to avoid flutter echo without requiring additional absorption. Further, the only type of angled high ceiling possible in a normal house puts the peak in the center. Again, this room is large enough that neither of these deviations from ideal is a real problem.

Myth: Walling off corners at an angle avoids the need for bass traps.

The belief that putting drywall across the room corners at an angle avoids bass problems is surprisingly common. It's true that bass waves generally collect in room corners, so by extension bass traps work best when placed there, too, giving the traps something to absorb. But cutting off the corners of a room with drywall doesn't avoid bass buildup; all it does is remove the single best places to put bass traps! Just as angled walls don't eliminate room modes or standing waves, neither does walling off a corner with a solid reflective material.

Likewise, solid soffits around the ceiling perimeter are best avoided because that's also where bass traps work best. When you build out a corner with a drywall soffit, two corners are created, requiring even more bass traps. If you like the look of soffits, or need to hide air ducts along one wall and want a symmetrical look, build the remaining soffits as bass traps. The best way to do this is with a wood or steel stud frame that outlines the soffit, and then stuff the cavity fully with fluffy fiberglass insulation. You can stretch fabric around the exposed side and bottom or cover those openings with cardboard painted to match the walls and ceiling. Even a soffit meant to hide ducts will reflect less bass if it's made of cardboard. If the ducts are made of fiberglass, or thin metal lined with fiberglass, they'll absorb some at low frequencies rather than reflect like drywall.

Odd Room Layouts

People sometimes fret unnecessarily about their odd-shaped rooms. One common shape is the L shown in Figure 20.6. As explained in Chapter 19, it's best to have the speakers fire the longer way down the room. So if this room were wider left to right than top to bottom, the better listening position would have been in the lower right of the drawing with the speakers facing to the left.

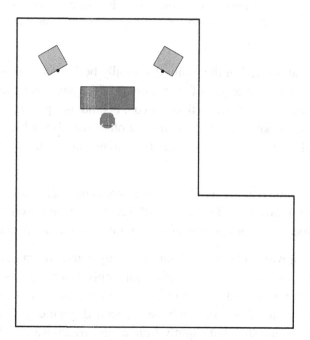

Figure 20.6: This is the best way to set up an L-shaped room, where the speakers fire down the longest dimension, and the left and right sides are symmetrical near the loudspeakers and mix position.

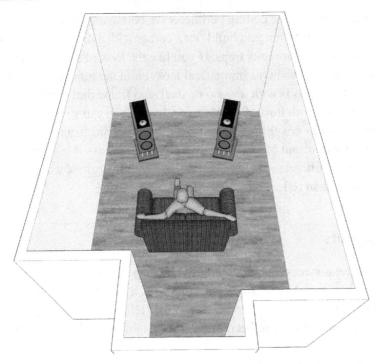

Figure 20.7: A stepped wall in the rear of a room helps break up reflections at low frequencies. A closet, or even a vocal booth, can serve a similar purpose. *Drawing by James Lindenschmidt.*

If you have a window at one end of the room, it's usually best to face the window while mixing, unless that puts the sun on your face for much of the day. Even then, adjustable blinds work well. A reflective surface in front of you, behind the speakers, is not usually a problem because most speakers send their sound the other way. But when a window is behind you, it's difficult to place absorbers or diffusers there to handle reflections from the rear of the room.

A stepped rear wall as shown in Figure 20.7 helps to break up reflections at bass frequencies, staggering their arrival time much like a QRD diffuser. Most people won't build a wall like this, but a closet or vocal booth in the rear of a room gives a similar advantage.

As you can see, dealing with odd-shaped rooms is mostly common sense, and basics such as firing the speakers down the longer dimension still apply. If a room has an odd angle at one end, or part of a corner is blocked by a built-in bookcase that can't be moved, set up the speakers at the opposite end where you can have the needed symmetry. Further, small things like radiators and ledges that stick out slightly from a wall are invisible to bass waves, which go right around them. The same applies for ceiling beams, unless they hang down so far from a low ceiling that they reflect sound from the speakers toward your ears. What matters most with acoustics is the big stuff. A small portion of a wall that juts out a foot or so into the room might form a corner, but it doesn't collect much bass.

One Room versus Two Rooms

If you record mostly yourself or small groups of friends for fun, a studio where you record and mix in the same room is more convenient than constantly running back and forth between rooms. This is especially true in smaller rooms; it's much better to have one room large enough to get a decent low-frequency response, rather than two rooms, each too small to sound good. Dividing a small room is always a bad idea. Even if you record others occasionally, it's not necessarily a problem to use one room, and I do that all the time. Everyone wears headphones while tracking, including me; then later the performers hear the same playback I hear through the loudspeakers. Indeed, many professional engineers record vocals in the control room because it's convenient and easy to communicate with the artist. When recording others in a one-room studio, you'll put them behind you, away from loudspeaker reflection points and the front of the room where symmetry matters. You could even put a small mirror on the wall in front of you so you can see the performers without having to keep turning around.

Many people believe two rooms are needed for sound isolation when recording, though this is not as useful as it might seem. In the old days when most studios were limited to 16 or 24 tracks, it was common to sub-mix drum microphones while recording to preserve tracks for subsequent overdubs. So maybe ten drum mics would be recorded to a total of four tracks, with the kick and snare microphones on separate tracks, and all the others pre-mixed to two tracks in stereo. In that case, it's useful to be in a separate room to hear the sub-mix accurately because you can't change the drum balance later when mixing. But these days most people use DAW software having an unlimited number of tracks. Once you know your room and where to place the microphones, you can record each mic onto its own track and sort out the balances later. Of course, you can make a short trial recording to confirm the mic'd sources sound as intended before recording for real.

Further, many people with home studios use MIDI samplers instead of real drums, so isolation is even less necessary. But even with separate rooms, you still won't have true isolation in most homes. Effective isolation at very low frequencies is difficult to obtain, so the kick drum and electric bass are likely to bleed through the walls and common floor anyway. In that case, you can't really trust the balance you think you're hearing through the speakers. All of that said, as much as I prefer a one-room layout for smaller project studios, it can be tiring to record long sessions with a live drummer or guitar player playing loudly in the same room.

Vocal Booths

> **Myth:** *For best results, you should record vocals in an enclosed booth because that's how professionals do it.*

To the uninitiated, a studio isn't a "real" studio unless it has a vocal booth. But in my opinion, a vocal booth isn't always necessary, or even desirable, for typical project studios. For large, commercial studios a vocal booth makes sense; even their "small" booths are the size of an entire bedroom studio, and they're acoustically treated to sound good. Having a large vocal booth lets the engineer isolate the vocalist during full-band tracking to prevent leakage from the rest of the band getting into the singer's microphone.

But vocal booths in home studios are usually the size of a closet. Even if you treat all of the booth's surfaces with absorption, a room that small will have severe problems at bass and low midrange frequencies. So you need not only absorption on the walls and ceiling to avoid the small-room boxy sound, but also extensive bass trapping to eliminate pronounced resonances that add an unnatural, "chesty" quality. The same applies to small booths used for bass or guitar amps.

Even in "real" studios, engineers and producers often track vocals in the control room. My friend Richard Hilton, chief engineer for Nile Rodgers, has recorded Michael Bolton, Simon Le Bon, Tina Arena, and many other famous singers in Nile's control room using headphones with excellent results. At about 20 by 25 feet, this room, shown in Figure 20.8, is a typical size for a home studio. So even for fully professional serious studios, recording in the control room is a good solution. Using a booth makes sense only if it's large enough to sound good

Figure 20.8: Richard Hilton, chief engineer for producer Nile Rodgers, regularly records vocals in Nile's home studio shown here.

Figure 20.9: The RealTraps Portable Vocal Booth absorbs excess ambience in a room, and it is large enough to effectively block computer fans and other noise coming from behind it.

and you truly need isolation to record vocals or soft instruments while a band plays live at the same time, or to avoid disturbing your family.

One potential problem with one-room recording is ambient noise, often caused by fans in a computer or power amplifier. If you have this problem, the Portable Vocal Booth shown in Figure 20.9 or another commercial or homemade equivalent, placed between the musician and the source can reduce the amount of noise getting into the microphone. Placing the singer behind the mix position, facing the front of the room, puts the computer in the null of a cardioid microphone, and the barrier further blocks sound from coming from behind the mic.

If you must have an enclosed booth in a small room, place it behind the mix position, centered left and right in the room against the rear wall. This mimics a QRD diffusor having three sections similar to the stepped wall in Figure 20.7, and it helps to break up low-frequency reflections from that wall. Reflections still occur, but the waves come back at different times due to the varying distances, weakening bass peaks and nulls in the room.

Surface Reflectivity

Myth: *I have a small recording studio in my basement, and I plan to add a wood floor because wood sounds much nicer and "warmer" than cement.*

The conventional wisdom is that wood surfaces impart a warm sound to a room, compared to concrete, linoleum, or drywall. But in the context of wall and floor surfaces, what matters most is the surface's *reflectivity*. That is, how much of the sound that strikes the surface is reflected, and at what frequencies. I've seen people argue that wood surfaces add a pleasing quality to a room in the same way that wood affects the tone of a fine violin. But that's a false analogy because the thin wood in a violin is meant to vibrate and add pleasing resonances, while wood on a floor or wall is much thicker and is anchored solidly to the backing. Indeed, resonances in a musical instrument are desirable and even necessary, but good recording and listening rooms must aim to avoid resonance in order to sound neutral.

I've also seen people claim that the sound of wood versus cement is so obviously different that you need only listen to hear it. But this, too, is false logic, unless you can audition (and hopefully measure) two identical rooms where the *only* difference is the surface materials. Many professional recording studios these days have stained cement floors instead of wood for various reasons, including cost. Linoleum reflects sound about the same as cement or wood and is common, too. To illustrate this, Figure 20.10 shows the strength of reflections in standard octave frequency bands for several common floor and wall materials. This table is included in the "Surface Reflectivity.xls" spreadsheet available on the website for this book.

Each surface reflectivity number in Figure 20.10 is the reciprocal of its absorption, so if a material absorbs 1 percent of the sound, then 99 percent is reflected. This assumes the surface

	A	B	C	D	E	F	G
1	Frequency	125	250	500	1000	2000	4000
2							
3	Marble or glazed tile	0.01	0.01	0.01	0.01	0.02	0.02
4	Reflections dB down	**0.04**	**0.04**	**0.04**	**0.04**	**0.09**	**0.09**
5							
6	Concrete sealed or painted	0.01	0.01	0.02	0.02	0.02	0.02
7	Reflections dB down	**0.04**	**0.04**	**0.09**	**0.09**	**0.09**	**0.09**
8							
9	Vinyl tile or linoleum on concrete	0.02	0.03	0.03	0.03	0.03	0.02
10	Reflections dB down	**0.09**	**0.13**	**0.13**	**0.13**	**0.13**	**0.09**
11							
12	Wood parquet on concrete	0.04	0.04	0.07	0.06	0.06	0.07
13	Reflections dB down	**0.18**	**0.18**	**0.32**	**0.27**	**0.27**	**0.32**
14							
15	Wood floor on joists	0.15	0.11	0.10	0.07	0.06	0.07
16	Reflections dB down	**0.71**	**0.51**	**0.46**	**0.32**	**0.27**	**0.32**
17							
18	Glass small pane	0.18	0.06	0.04	0.03	0.02	0.02
19	Reflections dB down	**0.86**	**0.27**	**0.18**	**0.13**	**0.09**	**0.09**
20							
21	Gypsum board on masonry	0.01	0.02	0.02	0.03	0.04	0.05
22	Reflections dB down	**0.04**	**0.09**	**0.09**	**0.13**	**0.18**	**0.22**

Figure 20.10: Surface reflectivity of several common building materials. The upper row across for each material shows its absorption coefficient, and the boldface numbers below are computed to show how much of the sound is reflected in each frequency range.

is anchored to a solid backing, not suspended in air. When a material is free to vibrate, some of the sound—mostly lower frequencies—passes through the material rather than gets reflected back toward the source. From this it's easy to convert the percent reflectivity to reflection strength in decibels. Here's the formula, where AbsCo is the material's absorption coefficient:

$$\text{Reflection Strength in dB} = 10 * \text{Log}(1 - \text{AbsCo})$$

This brings up an important point. When a glass mirror is attached to a rigid sheet rock wall, the glass simply reflects the highest frequencies a little more than the drywall alone. Likewise, drywall glued to a solid cement or smooth brick will reflect high frequencies a little less due to the paper facing that absorbs very high frequencies slightly. But glass and drywall are often mounted in a way that allows flexing. When a large sheet of window glass is free to flex in a frame, or drywall is attached to wood or metal studs, sympathetic vibration can increase their absorption at some lower frequencies. This is evident in rows 15 through 19 in Figure 20.10, which show more absorption and less reflection at 125 Hz for glass and a wood floor on joists.

You can see that all of these surfaces reflect the same to within less than 1 dB, and in most cases well under half a dB. In my opinion, surfaces that reflect the same to within a dB or less will sound pretty much identical. Some people believe that wood and cement and linoleum all sound different, but any difference heard is most likely due to other factors such as the size and shape of the rooms, other surfaces and objects present, or even expectation bias and placebo effect.

One failing of the data in Figure 20.10 is it goes no higher than 4 KHz. I suspected that any differences between materials will be greater at the highest frequencies, so I measured the reverb time inside a small box measuring about 38 by 23 by 15 inches made of MDF. Then I lined the entire box with window glass and measured again. At 5 KHz and below, the RT60 decay times were very similar. But at 8 KHz and above, decay times with the glass present were about one-third longer. So this confirms that glass and similar surfaces such as ceramic tile are slightly more reflective than wood in the upper treble range. But even at those high frequencies, sealed concrete is likely similar to a polished or laminated wood parquet floor.

When I first started writing about the myth of surface reflectivity in 2007, it caused virtual fist fights in several audio forums. Dozens of people argued about it for weeks, with many insisting that the difference between wood and cement floors is huge. But a legitimate apples-to-apples comparison is nearly impossible because it requires two identical rooms with only the floor surfaces changed. Hence this spreadsheet. I'm pleased to report that in the years since those posts, my measurements have become accepted in all of the acoustic forums I visit. Nobody disputes anymore that all of these surfaces sound more or less the same, and people now realize that this is yet another example of bias and flawed perception being refuted by hard data.

Calculating Reflections

Figure 19.18 from Chapter 19 shows that when a surface reflects 100 percent of the sound at some frequency, a null of infinite depth is created 1/4 wavelength away from that surface. But if you reduce the reflection strength by even 1 dB, the null becomes only 19 dB deep. Reduce it by another dB, and the null is only 14 dB deep. If you reduce the reflection strength to −6 dB, the null is now only 6 dB deep. You can look up the formulas used in the spreadsheet to see how to calculate the extent of peaks and nulls based on absorption amount. But this can be calculated directly using the Decibels.xls spreadsheet first mentioned in Chapter 1.

The voltage dB calculations were already explained, but this spreadsheet also calculates acoustic decibel relationships. When you enter a material's absorption coefficient, the spreadsheet calculates the reflection strength, as well as the resulting null depth and peak increase at 1/4 and 1/2 wavelength boundary distances, respectively. The last section accepts a reflection reduction you measured using room testing software with and without an absorber in place and tells you the absorption coefficient of that material and the peak and null sizes. It also calculates peak and null amounts from a dB difference between two identical signals having the same or opposite polarity.

However, calculating the absorption and reflectivity of materials is not as simple as it might seem, even with a spreadsheet to do the math. In-room testing can skew the results measured with a microphone or SPL meter due to reflections from other room surfaces. The distances between the loudspeaker, the absorbing surface, and the measuring microphone also influence the results. Not only will multiple reflections skew the results, but absorption also varies with the *angle of incidence*, depending on the direction from which sound arrives. So if you intend to use this spreadsheet for homemade absorption testing, the microphone should be near the reflecting boundary and far from the sound source to obtain peak and null amounts that represent what's actually absorbed. Further, the absorption calculations assume a boundary that reflects 100 percent, which is not usually the case at all frequencies.

Isolation and Noise Control

As explained in Chapter 19, acoustic isolation to reduce leakage between rooms is very different from acoustic treatment used to improve the sound quality within a room. Sound isolation requires specialized construction in the form of thick, rigid walls that are decoupled from the building using various techniques and materials. Understand that sound isolation is a two-way street. You don't want your Marshall amp stack disturbing your neighbors, nor do you want your neighbor's lawnmower or barking dog picked up by microphones that record your delicate acoustic guitar parts.

Myth: Foam is a good soundproofing material. If you cover the entire wall of your apartment, your neighbors won't complain about hearing your stereo.

The main ingredients for effective sound isolation are mass, rigidity, and air spaces filled with insulation. Acoustic treatment such as foam and rigid fiberglass improves the sound quality within a room but does almost nothing to block sound from traveling through the walls or a ceiling. You might reduce sound leakage by a small amount at mid and high frequencies, but only because absorption in a room reduces the room's natural reverb, which removes energy from the room. The most objectionable sound leakage occurs at bass frequencies, and true soundproofing requires walls that are too stiff to flex when sound waves strike them.

A wall that's stiff and massive flexes less than a thin, lightweight wall. Avoiding flexing is key because a wall that flexes on one side also flexes on the other side, creating a new sound source on the other side. Indeed, sound doesn't actually pass through objects, but rather, it's regenerated as the wall vibration creates a new source. However, walls that don't flex are also more reflective at low frequencies. When a wall is limp enough to allow flexing, some of the bass energy passes through, reducing the extent of peaks and nulls inside the room. Therefore, construction methods that improve sound isolation often make the acoustics inside a room worse, and vice versa. So when isolation is not needed in a project studio or home theater, it's best to avoid thick, rigid walls.

The specification that defines sound isolation is called *sound transmission class*, or STC for short. This is a single number that specifies how much sound is blocked in the range between 125 Hz and 4 KHz. For example, an STC rating of 30 means sound on the other side of a dividing wall is 30 dB softer than on the source side for frequencies within that range. However, STC values can specify multiple frequencies, which are needed for professional applications. As you know, low frequencies propagate through walls more readily than mid and high frequencies. So you may not hear a next-door neighbor yelling loudly inside his house, but you'll surely notice his stereo playing loudly, because low bass passes more easily through the walls of both your homes. Therefore, a single STC spec is not nearly as useful as one that lists the amount of isolation in octave bands or, even better, third-octave bands.

When a wall dividing two rooms is the only source of sound transmission, doubling the mass of that wall increases the STC by 6 dB. But in practice, sound travels not only thorough walls via their flexing but also through what are known as *flanking paths*. This is a fancy term for mechanical coupling between two objects that touch each other. Musicians know that you can amplify the sound of a tuning fork by touching its base to a tabletop. This couples the fork's vibration to a much larger radiating surface, increasing its volume substantially. A violin works the same way. A vibrating string displaces very little air, but when coupled mechanically through the bridge and sound post to the violin's large wooden top and bottom plates, the entire plates vibrate. Sound travels through buildings in a similar manner via coupling between a wall and the floor it rests on, as well as through air ducts, water pipes, and electrical conduits that pass through and touch the wall or ceiling separating two rooms.

Like many homes heated with hot air, mine is divided into separate zones using three metal ducts that leave the furnace. Figure 20.11 shows one of the rubber decouplers attached to a duct that sends air from my basement furnace throughout the house. It's made from flexible rubber and replaces a short section of the main duct. Without these isolators, furnace vibrations would transfer to the ducts and vibrate the floor above they're attached to, which in turn would create audible rumble in the rooms upstairs.

When I built a large professional recording studio in the 1970s, our acoustic consultants had our contractor saw a thin vertical gap from floor to ceiling in the common corridor walls joining the control room and live room to prevent sound from traveling along those walls. It's also common to saw a gap in a cement floor or, with a new building, pour each room's concrete floor separately, leaving a small gap between them. In 1992, when I started playing the cello, I lived in a house that had a finished basement with a cement floor and carpet. As you may know, cellos have a pointed end pin that digs into the floor for stability while you play. I remember playing my cello for a friend, and he could feel the vibrations from my cello as they passed through the cement floor to where he stood several feet away. You might think that a solid cement floor is too massive to be vibrated by an acoustic instrument, but clearly that's not true.

Mechanical decoupling is an important part of sound isolation, but having an air gap between the dividing boundaries is equally necessary. Filling the gap with insulation increases

Figure 20.11: Rubber decouplers are attached inline to metal air ducts to prevent furnace vibrations from being transmitted throughout a building.

isolation further and helps to damp the walls so they won't resonate when excited by low-frequency sounds within the room. Even if you build two walls that don't touch each other, they're still connected through the floor at the bottom and the ceiling at the top, unless you decouple those junctions by resting both walls on rubber pads or similar.

It's also important to avoid what's called *triple-leaf* construction, where three wall sections are used instead of only two. Many people wrongly assume that building a wall with more internal dividing sections increases isolation. Figure 20.12 shows six different types of wall assemblies and lists the STC rating for each. The drawings and captions speak for themselves, and you can see that the difference between walls C and F is a substantial 23 dB, even though the exact same materials are used and occupy the same amount of space.

Another way to increase isolation is by decoupling the two sides of a single wall so sound doesn't pass to the other side through their common studs. Figure 20.13 shows a construction method known as *staggered studs*, where separate studs are used for each side of the wall. However, coupling still occurs where the studs are joined at the floor and ceiling base plates.

Edge view

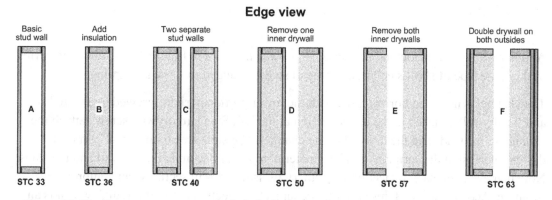

Basic stud wall — A — STC 33

Add insulation — B — STC 36

Two separate stud walls — C — STC 40

Remove one inner drywall — D — STC 50

Remove both inner drywalls — E — STC 57

Double drywall on both outsides — F — STC 63

Figure 20.12: Contrary to common belief, the isolation from two separate walls (C) is nowhere near as good as when only two barriers are present (E and F).

Top View

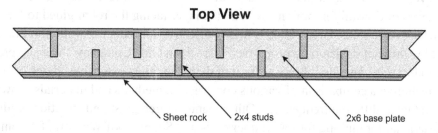

Sheet rock   2x4 studs   2x6 base plate

Figure 20.13: With staggered studs, each side of the wall has its own set of studs to avoid coupling through the studs.

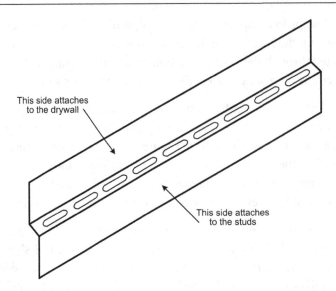

This side attaches
to the drywall

This side attaches
to the studs

Figure 20.14: Resilient channel serves as a spring to decouple drywall from the studs it's attached to.

Another wall isolation method uses *resilient channel*, shown in Figure 20.14. These are strips of the same type of sheet steel used for metal studs, arranged to serve as a spring.

The channel is mounted horizontally, with the lower portion at right screwed to the studs and the drywall attached to the upper section at the left. Standard drywall screws attach the channel to the studs and the drywall to the channel. The slots are an important part of the design, increasing the springiness of the material to better decouple the drywall and the studs. Note that even one nail connecting two otherwise isolated wall sections can compromise isolation. Acousticians call this type of mechanical connection a "short circuit," which is an appropriate use of the electrical term.

Another decoupling product is *Green Glue*, which is a special type of caulk that bonds well yet never hardens. Rather than hang drywall on springy resilient channel, Green Glue uses two layers of drywall for more mass, with the layer facing the room glued to the layer that's attached to the studs. Green Glue is highly effective for reducing wall vibration and resonance by virtue of its damping properties. The Green Glue Company's website provides detailed product data, and the Library section of their site offers extensive advice on sound isolation, including a comparison of various construction methods and materials showing their STC rating at different frequencies. Other manufacturers of sound isolation products are Zero International for high-performance door seals and Soundproof Windows for complete door and window systems.

Air Leaks

Another common cause of poor sound isolation is air leaks through a wall or door. If you do the math, you'll find that a 1/8-inch crack under a door three feet wide has the same surface area as a hole in the door 4.5 square inches. Even very small holes in a wall or ceiling for light fixtures and outlets can compromise isolation. The same is true for microphone and earphone wires that pass through a wall dividing the control room and live room. Never pass wires straight through a wall, even if you plan to caulk around the openings. Rather, stagger the openings so the wires come out the other side several feet away, passing sideways through holes in the studs to get from one wall opening to the other.

Two other "air" concerns are sound and vibration from an air conditioner blower motor passing through the ducts, as well as air noise itself. Vibration from a blower motor can be reduced by using flexible duct decouplers as shown in Figure 20.11, and the blower motor and its fan can be attached to the floor or hung from the ceiling using rubber or spring shock mounts to minimize vibrating the building structure itself. But vibration from the blower can also pass through the ducts as sound waves. Lining the ducts with insulation helps, as does extending the ducts through a longer, more circuitous path, rather than directly from the blower motor to the room outlets. This is also needed to prevent sound in one room from passing to another room through the ducts. If a ceiling air outlet in one room connects to a ceiling outlet in another room only a few feet away, any isolation provided by the walls is totally lost because sound goes up and over through the ducts. Likewise, a duct that passes through a double wall must not touch the walls, or it will short circuit the walls as described. In professional studios, separate ducts for each room are often run all the way back to the main blower.

Although not strictly a sound isolation issue, wind noise at duct intake and outlet openings can get into microphones and be noticeable when recording soft sources. This is solved by using oversized ducts that pass the needed amount of air at a lower velocity. In other words, the same amount of air flow, specified in cubic feet per minute, can be sent through small ducts at high velocity or through larger ducts at a lower velocity. Air also tends to "whistle" when passing through a register grate. In a home studio, consider replacing register outlet grates with an open material such as half-inch screen wire.

When I built the pro studio mentioned previously, we used all of these techniques: oversized air ducts lined with insulation that looped around, with separate runs to the blower rather than going directly from one room to the next. We also had a slow-speed, low-velocity air blower that hung from the ceiling with serious shock mounts. Yet, after all that, believe it or not, the level of vibration and air noise is lower in my home studio! Even though my ducts aren't oversized, they're a newer type made of fiberglass insulation wrapped in flexible plastic. So these ducts don't pass vibration the way metal ducts do. Also, the air outlets are high up on the ceiling, far away from microphones that point in other directions. Further, my studio is on the second floor, with the blower motor in the basement two floors below at the opposite end of the house.

Room within a Room

With a fully isolated *room within a room*, as shown in Figure 20.15, none of the inner room's surfaces touch the outer room. Think of a cardboard box inside another cardboard box, with rubber bumpers supporting every corner to isolate the inner box from shock as you see with packaging for fragile items. Indeed, this is exactly the same principle, because in both cases the goal is to prevent shock and vibration from passing from the outside to the inside. The same happens in an automobile: The car's cabin where you sit is isolated from the lower frame and wheels by resting on springs with shock absorbers. Air in the tires absorbs some of the impact from hitting a pothole, but the springs and shock absorbers do the most to prevent bumps from being transmitted to the car's interior.

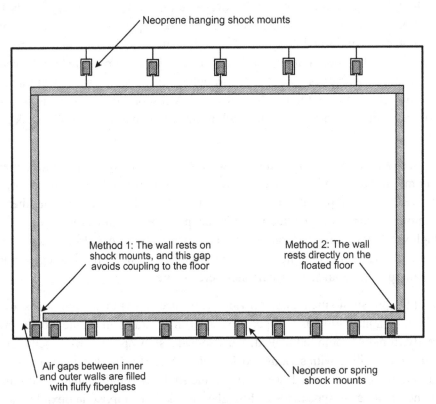

Figure 20.15: Achieving maximum isolation requires building a room within a room, with the inner structures fully isolated from the outer walls, floor, and ceiling.

A room such as this is typically built using two or possibly three layers of 5/8-inch drywall for the walls and ceiling, and the floor is usually a cement slab. The ceiling is suspended from neoprene or metal spring hangers attached to the upper ceiling, with a hole through the rubber or spring for the wire that supports the ceiling below. Figure 20.16 shows how such a ceiling hanger is constructed. Neoprene has the advantage of built-in damping versus springs that tend to continue bouncing. It's the same for the shock mounts under a cement slab floor.

A floor floating on neoprene shock mounts acts as a low-pass filter whose turnover frequency depends on the weight of the floor and the springiness of the shock mounts. Frequencies at or below the system's resonant frequency pass through the shock mounts, but above resonance vibrations are blocked to the real floor below and vice versa. Imagine lying on a mattress in an elevator that goes up and down slowly. Of course, the mattress and you will go up and

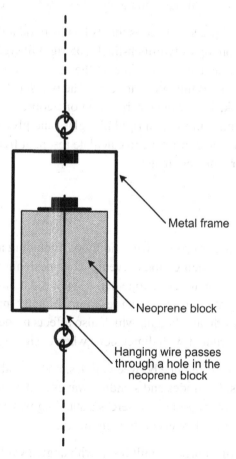

Metal frame

Neoprene block

Hanging wire passes through a hole in the neoprene block

Figure 20.16: Many ceiling hangers are made from neoprene, though some use a metal spring. Either way, vibration is blocked from passing through the hanging wires to the ceiling above, and vice versa.

down slowly, too. But if someone else jumps up and down on the elevator floor, much of that vibration will not pass through the mattress to you.

This is also like hanging a weight from a spring you hold in your hand. If you raise and lower your hand slowly, the weight will rise and fall, too. But if you move your hand quickly, inertia keeps the weight in place as the spring stretches and contracts. The standard method for tuning a floor on shock mounts is to set the turnover frequency an octave below the lowest frequency you want to block. So if you need to prevent sounds at frequencies 20 Hz and above from coming up through the floor, you'll specify a shock mount tension that tunes the system to 10 Hz based on the weight of the floor and furnishings and the number and type of shock mounts used. Note that the optimum compression for Neoprene rubber blocks and shock mounts is 15–20 percent. If the neoprene compresses less, then it's too dense for the weight and it behaves more like a solid block of wood or metal without isolating. But if it compresses too much it will bottom out, and again act as a solid block.

When I built my pro studio, our acoustic consultants had us build a floated floor using two layers of 3/4-inch plywood on top of two-inch-thick 705 rigid fiberglass. The two plywood layers were screwed together and also glued in case the screws loosened over time, which could cause the floor to squeak. Being glued together, the two sheets of plywood behaved as one solid sheet twice as thick. The cement slab resting on neoprene blocks in Figure 20.15 is better, though that costs much more than rigid fiberglass and plywood. A plywood floor is also easier to build, because you don't have to calculate resonant frequencies based on the weight of the floor and the room's contents.

Summary

This chapter explains the basics of room sizes and ratios, including axial and non-axial modes, and the importance of spacing modes evenly when designing a new room. Where a large room has many modes that are close together, a small room has fewer modes spaced farther apart, so the room's natural response varies more. Equally important is the ratio between a room's length, width, and height, which also affects mode spacing. The ModeCalc program lets you easily experiment with dimensions when designing a new room.

This chapter also clarifies the difference between modes, nodes, and standing waves. Although some people consider modes and standing waves as the same thing, they really aren't. A mode is a propensity to resonate, whereas a standing wave is a static condition that occurs when two opposing wave fronts combine in the air.

When you have the luxury of a purpose-built room with angled walls and ceiling, it's better to be wider and higher in the rear. Concave surfaces are best avoided because they focus sound, as does a peaked ceiling, though absorption under the peak reduces focusing.

Further, angled walls and ceilings are not useful in an existing small room, since that makes the room even smaller.

This chapter also busts a few acoustics myths, such as putting drywall across room corners to reduce the need for bass traps. Decorative soffits, if you like that look, are better if they're built as bass traps. The notion that bass waves need space to develop inside a room is also debunked, as is the belief that wood surfaces sound "warm" compared to drywall and cement. What really affects the sound of a material is its surface reflectivity.

I explain my preference for one-room studios, especially if you don't have space for two rooms each large enough to have a good bass response. A single room is also more convenient, because it avoids having to go back and forth between rooms to move a microphone or to press Record. Modern DAW software lets you record as many tracks as needed, so you can put each mic onto its own track and sort out the balances later. Further, in most home settings, having two rooms won't provide enough isolation to make valid balance decisions anyway. Vocal booths aren't as useful as many people think either, mostly because home studio booths are usually too small to sound good.

Finally, the basic principles of sound isolation are described. Sound isolation is defined by its STC rating, which is a single number that specifies how much sound is blocked in the range of 125 Hz through 4 KHz. But STC specs can and should include the dB reduction at different frequencies. Unlike acoustic treatment, isolation is achieved using specialized construction, including rigid walls that are decoupled from the rest of the building. Effective isolation requires identifying and isolating flanking paths and using methods and products such as air duct decouplers, staggered studs, resilient channel, floor and ceiling shock mounts, saw cuts, and Green Glue. The ultimate soundproof construction is to build a room within a room, but this is very expensive and thus impractical for many home studios.

CHAPTER 21

Acoustic Treatment

Acoustic Treatment Overview

Many studio owners and audiophiles install thin acoustic foam all over their walls, mistakenly believing that is sufficient. After all, if you clap your hands in a room treated entirely with foam (or fiberglass or moving blankets), you won't hear any reverb or echoes. But thin treatments do nothing to control low-frequency peaks, nulls, and ringing, and hand claps won't reveal that. So you might hear an absence of echoes and wrongly conclude the acoustics in the room are good but miss that bass frequencies still bounce around untamed. Basement studios and rooms with brick or concrete walls are especially prone to this problem; the more rigid the walls, the more reflective they are at low frequencies.

If you're a recording engineer you may ask why you need acoustic treatment at all, since few people listening to the music you create will be in a room that's acoustically treated. The reason is simple: All rooms sound differently, both in their amount of liveness and frequency response. If you create a mix that sounds good in your room, which has its own particular frequency response, it's likely to sound very different in other rooms. For example, if your room lacks deep bass, your mixes will probably contain too much bass as you incorrectly compensate for what you hear. And if someone plays your music in a room that has too much deep bass, the mixing error will be exaggerated and they'll hear *way* too much deep bass. Therefore, the only practical solution is to make your room as accurate and neutral as possible so any variation others experience is due solely to the response of their own room. Most people are probably used to the non-flat response of their room anyway.

Even if you don't produce music and only listen, having a room that sounds good lets you better enjoy the work of others, since you'll hear more closely their artistic intent. Mixing engineers spend hours, days, and sometimes even weeks on a single song, honing its tonal balance. While making those mixing decisions, they worked in a well-treated room that sounds neutral, and the mastering engineer's room was probably even more neutral. If you play that music in an untreated room whose response varies by 10 or 20 dB, which is typical, what you hear will be very different from what they heard and approved.

One fly in the ointment for many is the appearance of acoustic treatment. Beauty is in the eye of the beholder, and I've heard many audiophiles say they refuse to make their living room "look like a recording studio." Likewise, people with project studios often set up shop in a spare room of their house, and they don't always have the final say on such matters. Whether you—or your spouse or parents—will accept the look of acoustic treatment is another matter entirely. This truism definitely applies:

With acoustic treatment you can have (1) effective, (2) attractive, or (3) affordable. Pick any two, because having all three is not possible!

You've probably seen magazine photos of high-end home theaters or million-dollar recording studios and control rooms that look absolutely gorgeous, without a single acoustic panel or bass trap in sight. But behind those beautiful fabric walls are many bass traps and other panels, at least if the designers know what they're doing. *That* is the way to have effective acoustic treatment that also looks good. But you have to pay more for the extra work and material. Then again, some professionally built rooms look great but sound terrible. That's another story.

Myth: I rent an apartment, and I plan to move eventually, so I can't treat my room now.

In truth, bass traps work perfectly well resting on the floor, tipped back to lean into a corner. Absorber panels for mid and high frequencies can be propped up on cardboard boxes or milk crates to put them at ear height, and thinner absorbers can be hung like a picture using tiny nails that leave only a small hole that's easily patched or ignored when you move. More to the point: when you move, you can take acoustic treatment with you, just as you would your speakers and other gear.

Even hanging doubled-up bath towels at the side-wall reflection points using masking tape will improve clarity and stereo imaging. Lay a few stuffed laundry bags on the floor in the front left and right corners as minimal bass traps, and you'll enjoy even more improvement. For sure, that won't be as good as proper acoustic treatment, but it's a start and it's cheap. For still better results, a bale of raw fiberglass insulation, left in its plastic wrapping, can serve as a bass trap. Just put one bale on the floor in each of the room's four wall-wall corners. In fact, this is exactly what I suggest to people who are hesitant to invest in acoustic treatment, unsure if the difference they'll hear will really be worthwhile. I promise you it will be! And then you can get proper treatment.

Buy or Build?

Whether you should make your own acoustic treatment—DIY, for Do It Yourself—or buy commercial products, depends on how much money and effort you're willing to invest. People who can afford only DIY should do that, no question about it. Even if you can afford to buy acoustic products, if you enjoy hands-on projects and are good at them, DIY also makes sense. But DIY requires doing all your own research and being at the mercy of advice from anonymous forum posters and blog writers whose knowledge may be questionable. You'll need to learn what types of acoustic materials work well and where to put them for best results. When you buy from a knowledgeable treatment vendor, all of that goes away. Aside from fraud, which exists for acoustic products just as anywhere else, you know you're getting bass traps, diffusers, and other products that work well. Plus you get free advice from acoustic experts as part of the sale. But if you have more ambition than cash, DIY is a good option, and indeed that's how I learned many years ago.

This chapter therefore includes enough detail to understand how acoustic treatments are designed and built and what materials are most effective. I'll start by describing some common acoustic problems not yet covered in preceding chapters, then explain what acoustic materials and products do and how they work, and finally show typical placements for various types of treatment using many photo examples.

Flutter Echo

Chapters 19 and 20 explained modal ringing at bass frequencies, using waterfall graphs to show how bass traps reduce decay times. *Flutter echo* is a similar resonance that occurs at mid and high frequencies. Flutter echo colors the sound in the room by emphasizing frequencies whose wavelengths correspond to the distance between parallel walls and between the floor and a level ceiling. Flutter echoes are identified as a "boing" sound that has a specific pitch. If you clap your hands in a small, empty room or a narrow stairwell or tunnel, you can easily hear this tone. If the room is large, you'll probably notice a more rapid rat-a-tat-tat type echo, which is the flutter.

Flutter echo and modal ringing are intimately related, and in both cases the delay time and pitch depend on the distance between opposing surfaces. With a narrow spacing—for example, in a hallway or stairwell—a flutter echo's pitch is directly related to the spacing between walls. I have a long stairwell in my home that's 36 inches wide. When I clap my hands loudly, I can hear a distinct tone at an F# note. This F# has a pitch of 185 Hz, and the half-wavelength for 185 Hz is 36 inches. But for wider wall spacing, the frequency you hear may seem higher than the spacing would indicate, depending on what sound source excites the echoes. For example, when you clap your hands or otherwise excite a room with midrange frequencies, the only resonances that can respond are also at mid and higher frequencies. So if the distance between parallel walls fosters a fundamental resonance at, say, 50 Hz, you'll hear a series of clicks at that rate rather than 50 Hz when you clap your hands.

Depending on the room, sometimes it's difficult to identify which pair of surfaces is the main source of flutter echo, especially when the floor is reflective too. The graphic in Figure 21.1 shows the relation between musical notes and distances to help identify which surfaces in a room are causing flutter echo. Simply clap your hands, note the pitch on a piano or other

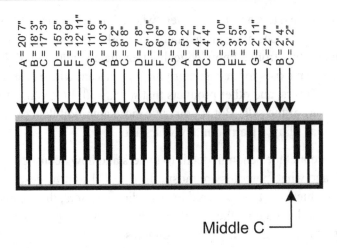

Figure 21.1: This graphic relates room dimensions to musical note frequencies, to help identify which surfaces are causing flutter echo.

musical instrument, and from that you can tell which opposing surfaces will benefit from applying absorbers or diffusers. Flutter echo is less common across very large distances, but you can still use this chart for distances greater than 20 feet by doubling a listed spacing. For example, a distance of 23 feet yields the G note an octave below the low G shown as 11'6" in the figure.

Note that flutter echo in a control room or listening room isn't always a problem; what really matters is if the loudspeaker sound sources actually excite those echoes. Even if you hear flutter echo in the back of a control room when you clap your hands there, that doesn't mean it's a problem for sound from the speakers. You can test this by playing a recording of hand claps or a similar impulse-type sound through your speakers and see if *that* creates audible flutter echo where you sit while listening. The file "reverb_test.wav" from Chapter 11 is perfect for this test. However, if you record vocals, acoustic guitars, or other instruments elsewhere in the room, absorption on nearby opposing walls and the ceiling will be needed.

Absorb or Diffuse?

Both absorbers and diffusers reduce flutter echo and comb filtering caused by reflections at mid and high frequencies. Diffusion reduces flutter echo without reducing a room's reverb time as much and also helps to make a room sound larger than it really is. But absorption is almost always needed in addition, to increase clarity. A listening room that's full of gorgeous-sounding reverb is still full of reverb. Further, if you make mixing decisions in a room that's too reverberant, you'll probably add too little reverb electronically, because what you hear includes the room's own reverb. Likewise, if the ambience is overly bright-sounding due to insufficient absorption, your mixes will tend to sound muffled when played on other systems because the tonal adjustments you make are based on what you hear. Indeed, diffusion is very useful, but absorption is also needed in most rooms, especially at bass frequencies.

Unfortunately, good diffusers are much more complex than good absorbers, so they cost more whether you buy commercial products or make your own. Further, diffusion is difficult to achieve at low frequencies, and you don't really want that anyway because bass reverb sounds muddy. So it's better to absorb bass frequencies. The stepped rear wall shown in Figure 20.7 from Chapter 20 is a type of diffusion that helps break up reflections at bass frequencies. But adding enough diffusion to create true bass reverberation below the Schroeder frequency is not recommended in rooms the size you'll find in most homes.

> **Myth:** *You don't need to worry about room acoustics when using near-field monitors, and playing music at low volumes further reduces the need for treatment.*

Acoustic behavior is mostly linear, so whatever happens at low levels happens the same at loud volumes. It's true that listening very softly reduces the strength of individual reflections, simply because they're then below the threshold of audibility. But the response peaks and nulls caused by reflections remain the same proportionally. Listening close to the speakers does improve the ratio of direct versus reflected sound, but even at only two or three feet away—which is typical for near field listening—the low-frequency response in an untreated small room is still poor.

Rigid Fiberglass

Forget packing blankets, egg cartons, thin acoustic foam, and carpet on the walls; those materials are too thin to absorb low frequencies, which every music room needs. They don't even handle the midrange very well. Likewise, non-acoustic packing foam is not suitable either. Real acoustic foam is *open celled*, which means the small crevices at its surface continue deep into the material. But packing foam is closed cell, so it absorbs only a little. Stiff, rigid Styrofoam panels meant for thermal insulation have no acoustic value either, because they have no pores at all. There's a popular DIY type website with plans to build a bass trap by filling a cardboard concrete form with sand and—surprise!—that doesn't work either. Not even a little.

The best bass traps and absorber panels are made from rigid fiberglass. This is a *porous absorber*, because sound gets into the pores inside the material where it's absorbed. Rigid fiberglass is the product of choice for most professional studio designers and builders because it absorbs more, and to a lower frequency, than most other porous absorber materials. Mineral wool is similar and less expensive, but it's more difficult to work with because it doesn't hold

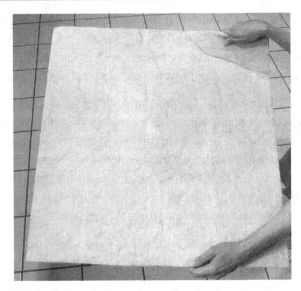

Figure 21.2: Owens-Corning 703 "rigid" fiberglass isn't rigid like wood or plastic, but it holds its shape when unsupported and is easily cut with a sharp knife.

its shape well, and it tends to crumble over time if disturbed. High-quality acoustic foam of an equal size and shape performs similarly to rigid fiberglass, as does acoustic cotton that's made from recycled denim. However, only rigid fiberglass and mineral wool are 100 percent fireproof, and this is an important concern for commercial installations open to the public such as professional recording studios. Even with a home studio, using materials that aren't fire-retardant may affect your insurance coverage in the event of a fire.

There's some confusion about the term "rigid fiberglass," because it's not really rigid like a piece of wood or hard plastic. The term *rigid* is used to differentiate products such as Owens-Corning (OC) 703 from the fluffy fiberglass commonly used in homes as wall and attic insulation. Rigid fiberglass is made from the same material as soft fiberglass, but the fibers are compressed and bonded to reduce its thickness and increase its density. Rigid fiberglass one inch thick contains the same amount of glass fiber material as three to six inches of regular fiberglass. Figure 21.2 shows a piece of 703 one inch thick. As you can see, it's rigid enough that it doesn't flop over when not supported (left side of photo), but not so rigid that it can't be bent or squeezed (top right corner).

Absorption Specs

Absorber products and materials are rated by their *absorption coefficient*, which is basically the percentage of sound they absorb at a given frequency. A material whose absorption coefficient is 0.5 absorbs half of the sound that strikes it. There's a technical difference between an absorption coefficient and percent absorption, but the two are close enough to ignore for most purposes. Understand that all materials pass, absorb, and reflect some amount—all at the same time. This is true not only for acoustic materials but for every physical object. Fiberglass absorbers work by converting acoustic energy to heat, due to friction as the sound waves pass into and through the porous material. Some absorbers meant only for bass frequencies use a vibrating wood panel or vinyl membrane, rather than porous material like foam or fiberglass. But acoustic energy is still converted to heat as part of the process.

The absolute measure of absorption is the *sabin*, after Wallace Clement Sabine (1868–1918), an early American acoustics pioneer who developed the concept of reverb decay times. His famous formula predicts the RT60 time in a room based on its volume, plus how much of its surface area is covered with absorption:

$$RT = 0.161 * (V/A)$$

Here, V is the volume of the room in cubic meters, and A is the area covered with absorption in square meters. This formula assumes that the absorption is effective at the frequencies of interest. Expressed as cubic and square feet instead of meters, the formula becomes:

$$RT = 0.049 * (V/A)$$

One sabin of absorption is equal to a one square foot opening to the outdoors. Therefore, an open window that's two by four feet has an absorption of eight sabins because the open area is eight square feet. All of the sound reaching the opening goes outside, and none is reflected back. The value of using sabins to specify absorption is that it's an absolute amount, versus a percentage as is commonly used to spec acoustic materials. One limitation of absorption coefficients is they're dimensionless, so a very small product that appeals to audiophiles because it doesn't intrude visually in a living room can claim the same performance as a large product that performs much better.

Note that absorption is often affected by the angle at which sound waves arrive at the material. At severe angles, sound waves that strike an absorbing panel may be absorbed less than when they strike it straight on. This is due to the *grazing* effect that causes the waves to bounce off the surface, much like skipping stones across a pond. This is one reason most acoustic foam has a sculpted face: It presents a perpendicular surface for sound waves that strike it at an angle, as happens at side-wall reflection points. Likewise, fluffy fiberglass absorbs a little better than denser rigid types at very high frequencies simply because its surface is softer.

Sculpted foam comes in several varieties, with most having either a wedge- or pyramid-shaped surface. The pyramid type absorbs well for sound arriving from almost any angle.

But wedge foam works better at reflection points if the pattern lines run vertically from top to bottom rather than horizontally left to right. When placed on a side wall with the wedges vertical, sound from the loudspeaker strikes the foam more or less perpendicular to its edges and is better absorbed. When mounted at reflection points on a ceiling, you'd reverse the orientation to go side to side, again presenting a perpendicular surface to the sound waves.

Material Thickness and Density

For a given thickness, 703 absorbs about twice as much as sculpted foam at lower frequencies, mostly because sculpted foam has half as much material than a solid slab. In fact, you can increase absorption at lower frequencies by placing sculpted foam with its flat side facing the room, because that puts more of the material's mass away from the wall or ceiling. Since the angled surface of sculpted foam improves its absorption for grazing angles as was described, a piece of thin sculpted acoustic foam can be placed in front of thicker rigid fiberglass to obtain the best properties of both materials.

Even better at low frequencies is 705-FRK, which absorbs much more than 703 at 125 Hz and below. FRK stands for Foil Reinforced Kraft paper, and it serves as a type of membrane that increases absorption at low frequencies compared to plain rigid fiberglass. FRK is similar to the brown paper that grocery bags are made from, but with a thin layer of metal foil bonded to one side. The FRK facing is not intended for acoustic purposes, but rather to serve as a vapor barrier in homes to keep moisture out. It just happens to be useful for acoustics, too. Be aware that the paper reflects mid and high frequencies when installed with that side facing the room, so you shouldn't use FRK rigid fiberglass at reflection points.

For appearance and to prevent glass fibers from escaping into the air, 703 and 705 rigid fiberglass panels are usually covered with fabric. This adds to the expense and difficulty of building and installing acoustic panels made from rigid fiberglass, though in practice fiberglass particles are not likely to escape unless the material is disturbed. A comparison of 703, 705-FRK with the reflective side toward the room, and typical acoustic foam is shown in Table 21.1. Each material sample is two inches thick and was measured in an acoustics lab. The data was obtained from the respective manufacturer's published literature. Again, most foam sold as acoustic treatment is sculpted for appearance and to better absorb sound arriving at an angle. But removing half of the material reduces its effectiveness at lower frequencies. When rigid fiberglass is compared to solid foam panels the same thickness, the disparity in low-frequency performance is less.

In Table 21.1, the column labeled NRC—for *noise reduction coefficient*—shows the absorption averaged over the range 125 Hz through 4 KHz. This is not unlike a single STC rating stating the isolation of walls and sound blocking materials, as described in Chapter 20. And like a single STC value, it's not nearly as useful as knowing the absorption amounts in

Table 21.1: Absorption Coefficients at Different Frequencies for 703, 705-FRK, and Acoustic Foam 2 Inches Thick

Material	125 Hz	250 Hz	500 Hz	1 KHz	2 KHz	4 KHz	NRC
Owens-Corning 703	0.17	0.86	1.14	1.07	1.02	0.98	1.00
Owens-Corning 705-FRK	0.60	0.50	0.63	0.82	0.45	0.34	0.60
Typical sculpted foam	0.11	0.30	0.91	1.05	0.99	1.00	0.80

Table 21.2: Absorption at Different Frequencies versus Density

Material	125 Hz	250 Hz	500 Hz	1 KHz	2 KHz	4 KHz	NRC
Johns Manville 814, 3 PCF	0.24	1.0	1.11	1.08	1.06	1.05	1.05
Johns Manville 817, 6 PCF	0.38	0.93	1.10	1.07	1.07	1.07	1.05

each frequency band. Note that absorption coefficients can exceed 1.0 due to the side edges being exposed during testing.

It's not difficult to understand why 705 fiberglass is so much more absorbent than typical sculpted foam at low frequencies. Sculpted foam loses half its mass when material is removed to create a sculpted surface. Another consideration is the material's *density*, expressed in *pounds per cubic feet*, or PCF. Owens-Corning 705 and Johns-Manville 817 fiberglass have a density of 6 PCF, compared to acoustic foam whose density is less than 2 PCF. Test data published by several manufacturers of rigid fiberglass and rock wool show greater density absorbing more at low frequencies, and the data in Table 21.2 is typical. The data is for panels two inches thick at two different densities.

My own tests in a certified acoustics lab confirmed this, showing that dense, rigid fiberglass absorbs as much as 40 percent more than less dense types at 125 Hz and below. However, it's important to understand that the true arbiter of a porous absorber's effectiveness is its *gas flow resistance*. When one material absorbs better than another less dense material, it's not due to the density itself. In other words, there's a correlation between density and absorption, but not necessarily a causation. I'll also mention that the advantage of higher density is reduced with thicker material. Once you get to a foot thick or so, fluffy fiberglass is just as good as rigid fiberglass, and it costs a lot less.

You probably won't find rigid fiberglass at your local hardware store or lumber yard, but many insulation suppliers stock it or can order it for you. Start by looking online or in your telephone directory under "Insulation" or "Heating/Air Conditioning Suppliers." You can find the name of an Owens-Corning dealer near you at the Owens-Corning website. Other companies, such as Knauf and Armstrong, make similar products, and they often cost less than rigid fiberglass from Owens-Corning. Other manufacturers include Roxul, Johns-Manville, CertainTeed, Ottawa Fibre, Delta, and Fibrex. You can contact them directly to find a distributor near you. Some companies call their products mineral wool, mineral fiber, or rock wool, but acoustically their absorption is comparable to rigid fiberglass, even if they don't look and handle exactly the same. For simplicity, I refer to all such products generically as rigid fiberglass.

When comparing rigid fiberglass for price and performance, it's important to assess equivalent materials. Not surprisingly, Owens-Corning 705 costs twice as much as 703 because it contains twice as much glass fiber material. Products from other companies that have a similar density should have similar absorption characteristics at the same frequencies, though this is not always the case. It's best to go by the published absorption data. If you come across a material that claims to be a useful absorber, but no official lab data is offered, it's probably not worth your consideration.

Rigid fiberglass is commonly available in panels two by four feet, in thickness ranging from one to four inches. Larger four by eight-foot sheets are also available, but two by four feet is more convenient for most studio applications, and it can be shipped more economically. As with all absorbent materials, the thicker it is, the lower in frequency it will absorb to. Rigid fiberglass one inch thick absorbs well down to about 500 Hz. The same material two inches thick is equally effective to an octave lower at 250 Hz. Doubling the thickness again to four inches extends the useful absorption yet another octave lower, without affecting absorption at higher frequencies.

Rigid fiberglass is great stuff, and you can cut it fairly easily with a razor knife or table saw, but it's not very pleasant to work with because the fibers can make your skin itch. When handling it, you should wear work gloves, and you won't be too cautious if you also wear a face mask. The usual way to mount rigid fiberglass to a wall is with sheet rock screws and large-diameter washers having a small hole, called *fender washers*. These washers are available at major hardware stores, and they're needed to prevent the screw heads from pulling through the soft front surface of the fiberglass. If your walls are cement or brick, you can use construction glue like Liquid Nails to attach narrow strips of wood to the wall, then attach the fiberglass to the strips using wood screws and washers. Rigid fiberglass works better when spaced away from a wall or ceiling, so using wood strips makes sense even when you are able to screw directly into the wall.

Acoustic Fabric

Once the fiberglass is attached to the wall, you can build a wooden frame covered with fabric and place the frame over the fiberglass for appearance. If that's too much work, you can cut pieces of fabric and staple them to the edges of the wood strips. How you do it is up to you, and figuring this out is half the fun of DIY. Nearly any soft, open-weave fabric is appropriate, and one popular brand is Guilford type FR701. Unfortunately, it's very expensive. One key feature of FR701 is that it's made of polyester, so it won't shrink or loosen with changes in humidity when stretched on a frame. But polyester is a common material available in many styles and patterns at most local fabric stores for less than the cost of FR701.

Another feature of Guilford FR701 is that it's one of the few commercial fabrics rated to be acoustically transparent. But since you're not using it as speaker grill cloth in front of a tweeter, that feature is not needed. If a soft material absorbs a little more on its own, all the better. However, hard, shiny fabrics having a tight weave, such as silk and synthetic equivalents, should be avoided because they reflect higher frequencies. The standard test for acoustic fabric is to hold it to your mouth and try to blow air through it. If you can blow through it easily, it will pass sound into the fiberglass. Burlap and muslin are two inexpensive options, but nearly any soft fabric works well to keep the glass fibers safely in place. I like burlap because it's inexpensive and can be bought already dyed in a variety of colors.

Wave Velocity, Pressure, and Air Gaps

One easy and free way to improve the low-frequency performance of any porous absorbing material, besides making it thicker, is to space it away from the wall or ceiling. For a given material thickness, adding an air gap behind lets it absorb to a lower frequency. For example, 703 rigid fiberglass two inches thick attached directly to a wall has an absorption coefficient of 0.17 at 125 Hz. However, spacing the same material 16 inches away from the wall increases that to 0.40—more than a twofold improvement! Few people are willing to give up that much space in their rooms, though smaller gaps are useful, too.

Porous absorbers are also called *velocity absorbers*, because they act on a wave's velocity as it transits from low pressure to high, and vice versa. The wood panel bass traps mentioned earlier are *pressure absorbers*, because they operate like shock absorbers for acoustic wave pressure. Figure 21.3 shows a wave striking a wall that reflects back toward the source. At the wave's positive and negative peaks, the pressure is greatest. This is also what our ears respond to. But for the brief moment where the wave crests, its velocity is minimum because the *rate of change* is lowest. Conversely, when a wave is near a zero crossing, there's very little pressure, and it changes the room's quiescent (at-rest) pressure the least. In other words, the sound is very soft when the wave is near zero. But the rate of change is greatest there, and that's what porous absorbers act on as the wave tunnels into and through the material.[1]

For a given thickness of absorbent material, the theoretical ideal air gap is equal to that same thickness because it avoids a hole in the range of frequencies absorbed. For example, if you install fiberglass that's four inches thick with a four-inch gap, some high-velocity portion of a high frequency falls within the four-inch material thickness, and will be absorbed regardless

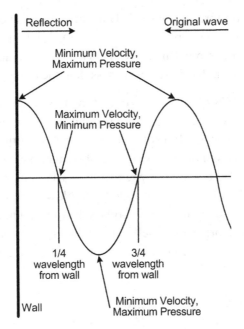

Figure 21.3: When a sound wave strikes a reflecting boundary, the pressure is greatest at the boundary, but its velocity is greatest 1/4 wavelength away. Half a wavelength away, the velocity is minimum, and then it rises again at 3/4 wavelength, and so forth.

of the gap size. This is shown in Figure 21.4 at the top in A. For lower frequencies whose quarter wavelength is between four and eight inches, the fiberglass is also at the proper distance from the boundary as in B. But at C the wavelength is so long that, at that distance, the velocity is too low for the material to be able to absorb much. This is one reason thicker fiberglass is needed to absorb low frequencies.

The Frequency-Distance calculator introduced in Chapter 19 will tell you the quarter-wavelength distances for any frequency, but in practice, you don't really need to measure wavelengths and calculate air gaps; the first few inches yield the most benefit. Most people aren't willing to give up two or more feet all around the room anyway, so just make the gap as large as you can manage. Even though the velocity is indeed highest at 1/4 wavelength from a boundary, there's still plenty at 1/8 wavelength away and even closer. These explanations are somewhat simplified and apply only for a 90-degree *angle of incidence* where sound strikes the material straight on, which is not always the case.

Bass Traps

The main purpose of bass traps is to improve the low-frequency response in a room by reducing peaks, raising nulls, and reducing modal decay times. Peaks and nulls always

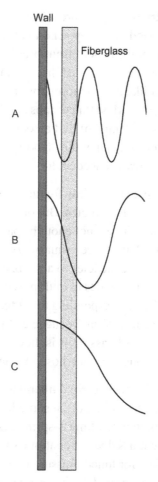

Figure 21.4: When a porous absorber is mounted with an air gap, higher frequencies are absorbed well because the wave's maximum velocity, where it passes through zero, falls within the material's thickness. Waves A and B both have a high velocity as they pass through the fiberglass. But the lower-frequency C at the bottom hasn't yet achieved much velocity at that short distance, so it's absorbed less.

vary in level around a room, so adding bass traps also makes a room more consistent for listeners at different locations. Bass traps can be used to reduce the low-frequency reverb time in large spaces, but they're more commonly used in recording studios and listening rooms to reduce modal ringing and flatten the low-frequency response. Small rooms don't really have true reverb below their Schroeder frequency anyway, because ringing at the individual mode frequencies dominates. But in very large recording studios, auditoriums, and churches where music is performed, reducing low-frequency reverb is an important use of bass traps.

The two ways to categorize bass traps are pressure versus velocity, and *broadband* versus *tuned*. Porous materials are broadband, and they act on velocity, while pressure absorbers are tuned to a center frequency. Most rooms benefit from broadband bass trapping because peaks and nulls can occur at all low frequencies, not just those related to the room's dimensions. A broadband bass trap might seem like an oxymoron, but this just means that it absorbs over a wide range of bass frequencies. Most tuned traps are useful over a range of about one octave and highly effective over a range of half an octave or even less. Tuned traps are useful to target a single problematic frequency that's too low for porous absorbers to handle efficiently, but broadband traps are always needed, too.

Although tuned traps can be effective at very low frequencies with less depth, audio rooms need substantial absorption at *all* low frequencies. But once you have enough tuned traps to sufficiently tame the room modes, there's not enough space for the broadband bass traps that are also needed. Further, rooms have three primary (axial) modes: one each for length, width, and height. So there are three frequencies to absorb, and that's for just the fundamental modes. Then there are the higher-order modes at multiples of each fundamental frequency. So if the main length mode is 30 Hz (corresponding to 19 feet), there are also modes at 60 Hz, 90 Hz, and 120 Hz. Now we have four frequencies to deal with just for the length dimension! Another problem with tuned bass traps is they're more difficult to design and build, especially for a room under construction whose response can't yet be measured.

Bass traps built from porous material such as rigid fiberglass are considered broadband because they work over a broad range, falling off at some lower frequency, depending on their thickness. Tuned traps use some type of mass-spring resonance to establish their center frequency and add damping so the mass doesn't continue to vibrate after the source sound in the room ceases. Common materials for tuned pressure bass traps are wood and vinyl sheets for the mass, with air trapped inside a sealed cavity serving as the spring. Rigid fiberglass is often used inside pressure traps to add damping.

One type of broadband bass trap sold by many companies is a foam wedge meant to be installed in room corners. Most foam wedges have a front face 15 inches across, which is wide enough to absorb well down to about 200 Hz. High-quality acoustic foam is a decent absorber, but most commercial products are simply too small to absorb well at lower bass frequencies.

DIY Bass Traps

There are several ways to create a bass trap from raw materials. The simplest and least expensive is to place thick, rigid fiberglass panels straddling the room corners or flat on the walls spaced away with an air gap. Rigid fiberglass four inches thick and spaced well away from a wall is very effective to frequencies below 125 Hz. But many rooms have problems far below 125 Hz, and losing a foot or more all around the room is unacceptable to most studio owners and audiophiles. Since bass builds up most in the corners of a room, this is an ideal location for any bass trap. Mounting two-foot-wide rigid fiberglass panels straddling corners as shown in Figure 21.5 loses only a small amount of space.

Figure 21.5 shows the corner viewed from above, looking down toward the floor. When rigid fiberglass is mounted straddling a corner, the large air gap behind the panel helps it absorb to fairly low frequencies. For this application 705-FRK is better than 703, and panels four or even six inches thick are better than thinner panels. You can either absorb or reflect higher frequencies by facing the FRK paper backing toward or away from the corner to control the amount of liveness in the room. Note that stacking two adjacent two-inch panels absorbs the same as one piece four inches thick, so you can double them up if you're unable to find material that's four inches thick. You don't even need to glue them together. However, if you're using FRK fiberglass, you should peel off the paper from one of the pieces so only the one outside surface has a paper facing. Also note that rigid fiberglass bass traps straddling a corner must not have a solid backing made of plywood or similar. The fiberglass must have both sides exposed to take advantage of the natural air gap behind the panel.

One nice feature of the bass trap in Figure 21.5 is that the air gap behind the fiberglass varies continuously, so at least some amount of fiberglass is spaced appropriately to cover a wide

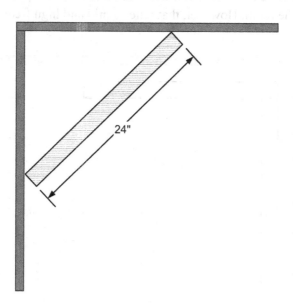

Figure 21.5: The most efficient place for a bass trap is straddling a corner at an angle. Every corner should have at least one two by four-foot trap, though covering the entire corner from floor to ceiling is even better.

range of frequencies. If you cover a wall-wall corner from floor to ceiling, you'll also have absorption at the tri-corners where three boundaries meet, which is especially effective. Understand that rectangular rooms have 12 corners, not just four where two walls meet each other. So besides wall-wall corners, bass traps are equally effective in the corners at the tops of walls where they join the ceiling and at the bottoms of walls where they meet the floor.

Filling a corner fully with rigid fiberglass is only a little better than using a four-inch-thick panel straddling the corner. If you can afford only a limited amount of material, it's better to have more panels straddling additional corners, rather than fewer corners filled solid. Since the material in the deepest part of a corner is near to the wall boundaries, there's less wave velocity for the material to act on. But when performance matters more than cost, filling a corner fully does maximize absorption. A good compromise is to place rigid fiberglass panels four inches thick straddling each corner, with the cavity behind each panel filled with less expensive fluffy fiberglass.

Using two traps adjacent in a corner as in Figure 21.6 works well, too, and takes even less space away from the room. This method is great when a room has a door in a corner, because one of the traps can be mounted directly onto the door. Spacing the panels two to six inches off the walls as shown improves their absorption further. As with bass traps that straddle corners, it's best to cover the entire corner from floor to ceiling if possible.

When using 705-FRK rigid fiberglass, you'll achieve more low-frequency absorption if the paper side faces into the room. However, that reflects mid and high frequencies somewhat.

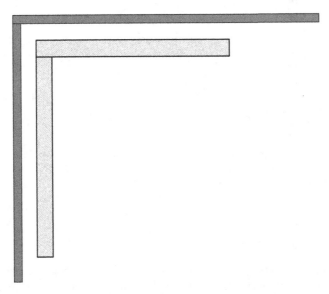

Figure 21.6: Using two traps on adjacent walls works very well and impinges less into a room than straddling a corner.

Corners are not usually at reflection points, so it makes sense to use FRK panels with the paper side toward the room. This lets you put a lot of bass traps into the room, without making it too dead-sounding at mid and high frequencies. For traps mounted flat on the walls, away from reflection points, you can alternate the panels so every other panel has the paper facing toward the room, again to avoid making the room too dead.

For a typical unfinished basement ceiling, you can take advantage of the gap between the support beams and the floor above by placing rigid fiberglass between the beams. Short nails or screws can support the fiberglass, making it easy to slide each piece into place. Then cover the fiberglass by stapling fabric to the joist bottoms as shown in Figure 21.7.

Another method is to pack the entire cavity with fluffy fiberglass one foot thick. Any part of the ceiling that's at a reflection point should not use FRK-type fiberglass or should have the FRK paper facing up toward the floorboards above. The same goes for parts of a live room ceiling above where you record drums and other instruments: It's best to absorb mid and high frequencies rather than reflect them. But for the perimeter of the room, near where the ceiling meets the walls, FRK fiberglass gives the most bass trapping.

If you have a drop ceiling with standard office-type tiles, you should replace the tiles at reflection points with rigid fiberglass or attach rigid fiberglass or acoustic foam under those tiles. Most office tiles absorb speech frequencies only and are too reflective at high frequencies to use at reflection points. Do the same for any parts of the ceiling above where instruments and microphones will be placed, especially if the ceiling is low. Then lay batts of fluffy fiberglass insulation as thick as will fit (up to 12 inches) above the tiles for additional

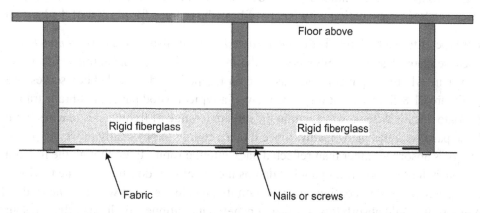

Figure 21.7: Rigid fiberglass between the joists in a basement ceiling absorbs very well and is not difficult to install. You can optionally fill the entire cavity with fluffy fiberglass. Either way, stapling fabric to the joist bottoms gives an attractive finished appearance, and you can glue or nail thin wood strips to hide the staples.

bass trapping. If you don't want to cover the entire ceiling above the tiles with fluffy insulation, at least do that around the perimeter, where bass traps are most effective.

Another popular type of bass trap is the *tube trap*, invented and sold by ASC. There are DIY plans on various websites, though most plans wrongly claim that the top and bottom end caps of the tube must be sealed airtight. Tube traps are made from rigid fiberglass, which is porous, so sealing the tube ends is pointless. Tube traps work very well if they're large enough—at least 16 to 20 inches in diameter. But like typical foam corners, small versions a foot or less in diameter are simply not large enough to absorb the lowest frequencies very well. The larger sizes work well in corners, in part because the tube's diameter serves to space much of the fiberglass away from the corner boundaries.

Yet another type of bass trap is the *Helmholtz resonator*. Unlike acoustic foam and rigid fiberglass, a Helmholtz resonator can absorb very low frequencies. This type of trap works on the principle of a tuned cavity, and it can be very efficient if designed properly. Think of a glass soda bottle that resonates when you blow across its opening, and you have the general idea. Although a Helmholtz bass trap can absorb well, it works over a narrow range of frequencies, and like all bass traps, it must be large to be effective. The frequency range can be widened by filling the cavity with fiberglass or by creating several openings having different sizes.

One common Helmholtz variation is the *slat resonator*. This comprises a sealed box filled with fiberglass, with a large front opening partially covered by a series of thin, separated wood slats. Another design also uses a sealed box filled with fiberglass, but with a cover made of pegboard containing many small holes. These traps are tuned by adjusting the number of holes and their sizes or the spacing between the slats. Since all rooms need broadband absorption, Helmholtz traps are best used to target a single problematic low-frequency mode, in conjunction with plenty of broadband traps for the rest of the bass range.

Another type of tuned bass trap is the *membrane absorber*, also called a *wood panel bass trap* because many designs use plywood for the front panel. Wood panel traps are a mass-spring system, where the panel is the mass and air trapped inside a sealed box serves as a spring. Figure 21.8 shows a cutaway side view of a typical wood panel membrane trap, built directly onto a wall. When a wave within the effective range of frequencies reaches the front panel, the panel vibrates in sympathy. Since it takes energy to physically move the panel, that energy is absorbed rather than reflected back into the room. Even though the fiberglass doesn't touch the plywood front panel, it damps the panel so it doesn't continue to vibrate. Were the panel allowed to continue vibrating on its own, less energy would be needed to keep it moving, so it would absorb less. Further, a panel that continues to vibrate after the source sound stops *adds* resonance into the room rather than removes it, which obviously is not desirable.

Side View

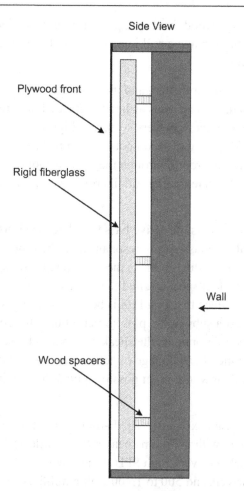

Plywood front

Rigid fiberglass

Wall

Wood spacers

Figure 21.8: Sound striking the front of a wood panel bass trap causes the plywood panel to vibrate. The fiberglass then damps that vibration, which increases resistance to the sound waves and also prevents the panel from adding new resonance.

One advantage of wood panel membrane traps is they don't need to be thick to absorb very low frequencies. Like all pressure traps, they work best mounted directly onto a wall or ceiling, rather than spaced away as benefits porous velocity absorbers. The center frequency absorbed by a wood panel bass trap is a function of the panel's mass and the depth of the air gap, which serves as a spring. A trap four inches deep with a 1/4-inch-thick plywood front absorbs 100 percent at around 90 Hz, which is more than the same thickness of rigid fiberglass at that frequency. The audible bass range spans several octaves, and panel traps absorb only part of the bass range. Therefore, a mix of traps is required, with some tuned to absorb the lower bass frequencies and others for the higher bass range. Besides absorbing low frequencies very well, the wood front on a panel trap is reflective at higher frequencies. So installing enough of them to make a real improvement at low frequencies will not make the room too dead-sounding.

Wood panel bass traps are an older design, and more modern thinking prefers porous absorbers as thick as needed, with air gaps as large as needed, to target the lowest frequencies. However, panel traps work very well, and Figure 5.5 from Chapter 5 shows the pro studio I built in the 1970s using these on both the control room and live room walls. The vertical stripes visible in the live room, through the control room glass behind the console, are wood panel bass traps alternated with rigid fiberglass behind fabric to absorb mid and high frequencies.

The last type of trap I'll mention is the active bass trap, because it uses active electronics rather than porous materials or mass-spring resonance. At the time of this writing I'm aware of only two such commercial products. One is the E-Trap sold by Bag End, a company well known for its low-frequency loudspeakers, and the other is a newer model called AVAA sold by PSI Audio. It's no coincidence that Bag End is best known for its loudspeakers because an active bass trap is at heart a subwoofer, plus a microphone that senses sound in the room. When bass waves reach the microphone, the speaker responds by sending a countering wave back into the room. One big advantage of active bass traps is they can target very low frequencies—much lower than would seem possible given the relatively small size of its woofer driver's cone.

As Jim Wischmeyer at Bag End explains: "The E-trap's output is very low in terms of SPL and does not have to keep up with the room speakers. The little 100 watt amplifier and the 10-inch speaker is not capable of very high SPL at the low frequencies. In our subwoofers we use more and bigger drivers and 500 to 1,300 watt amplifiers to reach high SPLs. What is actually happening in the E-trap is damping, so all it has to do is knock a few dB off the top of the resonance mode you're addressing. It's also working at only one or two frequencies, so it's not a very taxing job for the amp and speaker. Damping down a few dB (even 10 dB as is sometimes realized) and the associated ringing as the mode decays does not actually take much power or speaker excursion. It doesn't have to match the room speaker's level so it doesn't take a big amplifier or push the speaker driver very hard. That's part of the elegance of the invention."

Free Bass Traps!

Well, okay, not *free* bass traps, but there are ways to get bass trapping in a room without adding physical panels. First, most windows serve partly as bass traps, depending on how much low-frequency energy passes through them to the outside. Thin glass passes bass more readily than very thick glass. Windows on the front and rear walls are especially useful because the length mode is often the strongest, creating the worst peaks and deepest nulls at the listening position. A window that passes bass energy rather than reflects it also reduces modal ringing.

Likewise, a standard sheet rock wall with fluffy fiberglass insulation inside can give some diaphragmatic absorption, similar to a wood panel bass trap. The interior walls of most homes are not filled with insulation, so having a contractor add blow-in insulation to interior walls around a listening room is a good idea when possible. This is not the expanding foam type that hardens inside the wall, but soft, fluffy insulation. The insulation also damps wall resonance, helping to reduce buzzing and rattling from the walls themselves. The same applies for adding insulation above a drywall ceiling.

People sometimes obsess over peaks at very low frequencies—say, below 30 or 40 Hz—where many bass traps are ineffective. But little music contains much energy down there, so those peaks are more visually upsetting when seen in room measuring software than audibly damaging. Indeed, if a peak at 25 Hz increases the power of low-frequency effects in an action movie, that could be seen as a benefit rather than a problem! I'm not usually keen on using equalizers to reduce bass peaks, but at very low modal frequencies, EQ can make sense. As mentioned in Chapter 19, an acoustic peak that increases bass output lets you reduce the volume to the speakers at those frequencies, which in turn lets them play louder and with less distortion.

Note that adding bass traps to a room lowers the mode frequencies slightly, and this also occurs when sheet rock walls absorb as described. This is one reason mode predictions can vary from what actually occurs and is measured. Figures 19.15 and 19.16 from Chapter 19 show waterfall plots of a small room before and after adding bass traps. If you look carefully at the peaks near the 40 Hz and 70 Hz marker lines, you can see that the Before peak frequencies are slightly higher, to the right of the lines, than the After peaks to the left of the lines. Adding bass traps lowered the room's modal frequencies in much the same way that adding insulation inside an acoustic suspension speaker enclosure increases the apparent size of the box.

    Myth: *Using two or more subwoofers reduces or even eliminates the need for bass traps.*

There's no question that using two or more subwoofers can improve the low-frequency response in a room. When placed properly they not only improve both peaks and nulls, they also reduce response variations around the room. So I never argue that using multiple subs is a bad idea. But multiple subs can't replace bass traps for several reasons:

First, subs generally operate below 80 Hz, which is the standard crossover frequency as shown in Figure 21.9. But that's only half the bass range! The other half between 80 and 320 Hz is arguably more important because it's part of the "speaking" range for bass instruments where clarity and articulation are needed most. (Some people set even lower crossover

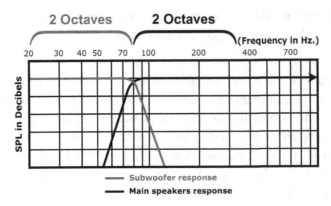

Figure 21.9: Most subwoofers are active only below 80 Hz.

frequencies, and those subs handle an even smaller portion of the bass range.) So while using more than one sub can indeed help to flatten the response below 80 Hz, it does nothing for bass instrument clarity, only fullness. Even frequencies as high as 150–200 Hz are perceived mainly as fullness, and nothing you do with subwoofers can improve problems there either.

It's also a myth that multiple subs can reduce modal ringing as much as bass traps. As explained earlier, modal ringing is similar to reverb in that certain notes continue after the source sound stops. Modal ringing is often shown using a waterfall plot as in Figures 19.15 and 19.16. In this type of graph the response peak "mountains" come forward as they decay in volume over time. Multi-sub proponents claim that placing subs away from peak locations in a room avoids "energizing the modes" and thus avoids ringing. But unless a sub is placed in a null location for a mode frequency—which then gives so little output it's a useless location—frequencies that align with that mode will still ring. So that could reduce the *level* of a peak, which is good, but not its time component.

While we're talking about room modes and extended decay times at select frequencies, here's an interesting fact that may not be obvious: The same resonance that extends decay times also extends rise times the same amount. The top trace in Figure 21.10 shows a 100 Hz sine wave sound source, and the lower wave shows the result you'll hear in a room having a mode at 100 Hz. The source starts suddenly, but in the room it grows over time, typically some fraction of a second. Then when the source stops, that frequency continues fading out over the same amount of time.

> **Myth:** *Bass traps are inefficient because they remove important energy from the room requiring much more amplifier power to compensate.*

Some people believe that adding bass traps to a room removes energy making low frequencies softer and less impactful, which in turn requires more power from the amplifiers

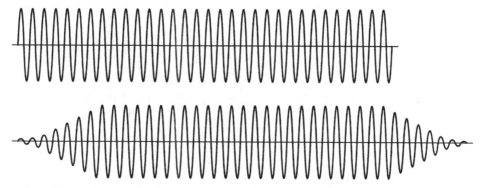

Figure 21.10: Room mode frequencies not only decay over time after they stop, but also grow slowly when they start.

and loudspeakers. This is totally the wrong way to consider what bass traps do! It's true that all absorbers remove acoustic energy because they work by converting sound waves into heat. But they absorb only the reflected energy, not the main sound coming out of the loudspeakers. When you place an absorber on a wall or ceiling, or mount a bass trap straddling a corner, all of the direct sound from the speakers still reaches your ears unchanged.

In most home-size rooms the low-frequency response is riddled with deep nulls as shown earlier in Figure 19.14. One of the most common problems in home recording studios is mixes that seem great in your room sound boomy and bassy elsewhere. When too little bass reaches your ears due to deep nulls, the tendency is to crank the bass in the mix to compensate for what you hear. But then there's too much bass and you don't realize it.

Bass traps reduce the strength of reflections that create these nulls, raising the volume at those frequencies. So adding bass traps to a room usually gives the perception of *more* bass, not less. Now, in some rooms the peaks are more prominent than the nulls, especially rooms that are square or "virtually square" such as 10 by 20 feet. In rooms like these, multiple resonances at the same frequency combine to create coincident peaks larger than 6 dB. The room measured in Figure 21.11 is 16 feet long and 8 feet high. So those dimensions create double resonances at 70 Hz and its multiples (140, 270, etc). But square rooms like these have deep nulls too, so bass traps still restore at least as much energy as they remove.

Figure 21.11 shows the response of the same room before and after adding bass traps. After adding bass traps the peaks were reduced and the nulls were raised. If you consider the shaded areas above and below the flatter "with bass traps" line near the center, more energy was restored to nulls than was removed from peaks, especially below 200 Hz. Most peaks are around 6 to 10 dB, but nulls can be very deep. So raising a null that's 20 dB

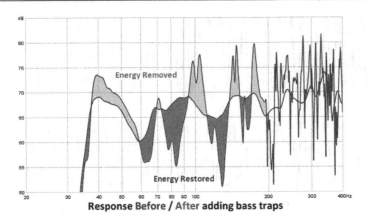

Response Before / After adding bass traps

Figure 21.11: Adding bass traps restores more energy lost due to nulls than the amount of energy removed from peaks.

down gives a huge increase in the amount of bass that's heard. Further, peaks, nulls, and ringing are all caused by reflections, so using bass traps to reduce the strength of reflections is a Good Thing. Any energy that's removed is energy you want removed! This is just like the improvement from absorption that reduces unwanted echoes and ambience at higher frequencies, which also "removes energy" from the room.

So clearly, adding bass traps usually increases the perceived amount of bass, and this is not some psychoacoustic trick. The actual measured amount of bass also increases. The room measured for Figure 21.11 was loaded with acoustic treatment. Yet even with nearly complete coverage of every surface, you can see that more energy was restored than removed because the nulls are deeper than the peaks are high. The only time adding bass traps makes the overall bass level softer is with square and cube rooms that have severe coincident peaks as described above. But these rooms are excessively boomy so, again, the energy that's removed is energy that should be removed.

Diffusers

Myth: Bookshelves and other random objects in a room are effective diffusers and help to break up reflections in a room.

Diffusion avoids the damaging echoes and comb filtering caused by reflections off nearby walls, but without reducing ambience as absorption does. However, a real diffuser is not simply a random surface. It might be possible to get some amount of diffusion from a shelf full of books, but the books would need to be arranged in a way that actually diffuses sound. One key aspect of diffusion is having an irregular surface to stagger the time between reflections. Most books are a similar depth, so a shelf full of books, unless artfully arranged, will reflect more or less the same as a plain, flat wall.

The simplest type of diffuser is one or more sheets of plywood attached to a wall at an angle to prevent sound from bouncing repeatedly between the same two walls. Alternatively, the plywood can be bent into a curved shape. A curved convex surface is called a *poly-cylindrical diffuser*, or just "poly" for short, and Figure 21.12 shows a set of three polys I built for my *All About Diffusion* video on YouTube.[2] A convex surface spreads sound outward, which is the opposite of the focusing you get from a concave surface or peaked ceiling. In truth, polys are really deflectors, not diffusers, because they deflect sound rather than truly diffuse it. However, the deflection you get from an angled or convex surface is better than a bare wall, and it can be sufficient to eliminate flutter echoes between parallel surfaces.

A true diffuser sends different frequencies in different directions, rather than deflecting all frequencies in the same directions like a poly or slanted wall. This is an important distinction, because flat surfaces give the boxy-sounding response peaks and dips caused by comb filtering, even if they're angled or curved. A real diffuser reduces the coherence of reflections, giving a more open, transparent, and natural sound.

Besides sounding less colored than an angled or curved wall in a control room, diffusers serve another useful purpose in a recording live room: reducing leakage between instruments recorded together at the same time. Whereas an angled wall deflects all of the sound reaching it in one direction—possibly toward a microphone meant to pick up another instrument—a real diffuser scatters the sound over a wide area. So whatever arrives at the wrong microphone

Figure 21.12: These poly deflectors are being built from cardboard concrete forms 24 inches in diameter.

is reduced in level simply because the original sound is distributed in many directions, with less sound going where you don't want it.

One of the most popular diffuser designs is the one-dimensional *Quadratic Residue Diffuser*, or QRD, shown in Figure 21.13. This type of diffuser is also called a *Reflection Phase Grating*. It's one-dimensional because it scatters sound in one dimension only—perpendicular to the long strips. The concept was invented by the same Manfred Schroeder who determined the transition frequency from room modes to reverb described in Chapter 20. The varying well depths stagger the reflection times, but the actual scattering is due to phase shift that creates beaming and lobing at the well edges—similar to loudspeakers and radio antennas as described in Chapter 18. A QRD comprises a series of narrow "wells" of varying depth, whose dimensions are based on a mathematical formula.

Figure 21.13: One-dimensional QRD diffusers have narrow vertical wells of varying depths to scatter different frequencies horizontally in different directions and to stagger the reflection times. Most QRD diffusers are made of wood, but this RealTraps model is made from hard cardboard, with rigid fiberglass behind to serve as a bass trap for frequencies below its effective diffusing range.

QRD diffusers are effective down to frequencies where the well depth is 1/4 wavelength and up to frequencies where the well width is 1/4 wavelength. So a shallow diffuser with wide wells operates over a more limited frequency range than one with deep and narrow wells. The most important operating range for a diffuser is around 500 Hz up to maybe 3 or 4 KHz. Note that QRD diffusers also absorb slightly because of energy losses due to diffraction at the well edges.

Another type of QRD diffuser is two-dimensional, made from square towers of varying heights instead of wells. These are called *skyline diffusers*, because the varying heights resemble a city skyline, as shown in Figure 21.14.

Yet another design is the *Binary Amplitude Grating Diffuser*, or BAD. These are built by placing a reflecting material having many small holes arranged in a specific mathematical pattern in front of absorbing material such as rigid fiberglass. The principle is similar to wood slats over insulation as shown in Figure 21.15, but with holes instead of long slotted openings. Any mix of absorbing and reflecting surfaces will scatter sound, though some combinations and patterns diffuse more effectively than others. Simply installing absorber panels on a wall in a striped or checkerboard pattern diffuses rather than simply absorbs some and reflects some.

The walls in Figure 21.15 are absorbent but covered partially with wood strips to form a hybrid design called a "diffsorber." The size and spacing of the strips is determined mathematically. The BAD diffuser made with holes over absorption mentioned previously

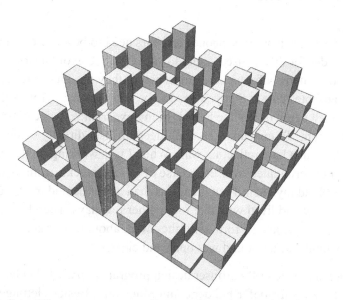

Figure 21.14: Skyline diffusers scatter sound in two dimensions rather than only left and right as with one-dimensional well diffusers.

Figure 21.15: The striped pattern of hard and soft surfaces on the walls of this room not only looks really cool, but it also serves as a diffuser. *Photo courtesy of Wes Lachot Designs.*

is also a diffsorber. The amount of absorption from a diffsorber is similar to an equivalent thickness of rigid fiberglass.

Although placing slats and holes over an absorbing surface doesn't diffuse as well as purpose-built QRD diffusers, it works over a wider frequency range than the thin profile would suggest. When the goal is pure diffusion, however, standard QRD-type diffusers are a better choice.

Conventional wisdom says that microphones and ears need to be some distance away from QRD diffusers in order for the diffused sound to "develop," to avoid hearing individual groups of reflections at narrow frequency ranges before they have a chance to combine in the air. The generally accepted rule is that you need to be one foot away from a diffuser for each inch of maximum well depth or skyline height. In my experience, you can be much closer than that and still benefit. To my ears, a six-inch-deep diffuser directly behind your head sounds vastly better than a bare reflecting wall. When I was working on the music for my *Tele-Vision* video, I experimented over the course of several weeks, using two large diffusers on stands directly behind my chair. I put them less than a foot behind me for a few days, then moved them back a foot and lived with that for another few days. Then I moved them back another foot for a few days, and so forth until they were about five or six feet away. Even when very close behind me, the sound was open and clear.

Finally, I'll mention a terrific QRD diffuser design program called QRDude, available as a free download at www.subwoofer-builder.com/qrdude.htm. Besides letting you easily experiment with QRD well dimensions before building them, the website contains a wealth of practical information and detailed theory about diffusers that goes far beyond what I can include here.

Treating Listening Rooms and Home Theaters

In this context, "listening room" applies to both recording studio control rooms and audiophile listening rooms, as opposed to venues or studio live rooms where music is recorded. Those will be covered separately. Room treatment is not as complicated as many people believe, and the photos that follow show the basics. All rooms need the following:

- Bass traps straddling as many corners as you can manage, including wall-ceiling corners and wall-floor corners when possible. Additional bass traps on the rear wall behind the listening position helps further, as does even more traps on the front wall. You simply cannot have too much bass trapping in a home-sized environment. Where you stop adding traps depends on how flat and clear you want the low end to be, versus your available space and budget.
- Mid- and high-frequency absorption at the side-wall and ceiling reflection points, plus thick throw rugs on the floor at those reflection points if possible.
- Additional mid- and high-frequency absorption and/or diffusion on any large areas of bare parallel surfaces, especially opposing walls, or the ceiling if the floor is reflective. Diffusion on the rear wall behind where you listen is also useful.

That's the basics in a nutshell! When treating a larger room or venue to reduce reverb decay times, the general goal is to spread the absorption more or less evenly around the room, rather than put it all on one wall or the entire ceiling. Further, when treating large surfaces, it's not necessary to cover every inch. Many acousticians recommend covering between 20 and 40 percent of a room's surface, depending on the size of the room and furnishings already present. Small rooms usually need a larger percentage of coverage than large rooms.

Figure 21.16 shows a two-channel listening room that's well treated with corner bass traps, as well as broadband absorption spaced uniformly around the room. Note the bass traps on the floor, straddling the wall-floor corners. Not everyone is willing to accept the appearance of bass traps on the floor, and in small rooms they might be underfoot. But this is a great location for traps when possible, and it avoids mounting holes in the wall; just rest the traps on the floor and tip them back against the wall. Wall-ceiling corners are also a great place for bass traps, as exemplified by the professional mastering room shown in Figure 21.17.

Acoustic panels don't have to be boring and unattractive either. Figure 21.18 shows absorbers in a piano performance room that an artist painted using watercolors directly onto commercial panels. Panels can be covered with custom art rather than painted directly, though it's important to avoid reflective fabrics and hard paints. But watercolor or dye

Figure 21.16: This room is extensively treated with bass traps in both the wall-wall and wall-floor corners, plus thick stand-mounted panels on the front and side walls to absorb well to very low frequencies. *Photo courtesy of Claudio Cavicchioli.*

Figure 21.17: Bass traps work well in wall-ceiling and wall-floor corners, as well as corners where two walls meet. Note the triangle-shaped bass traps near the ceiling above the flat traps that straddle the wall-wall corners to treat the important tri-corners. *Photo courtesy of Kevin McNoldy and Cphonic Mastering.*

applied onto white felt or some other soft fabric is fine, which can then be placed over the panels.

Again, small rooms need more treatment proportionally than large rooms, and mastering rooms need to be even flatter than control rooms because that's where the final decision about tonal balance is made. Figure 21.19 shows a modest size (21 by 12 by 8.5 feet) mastering room that's been treated fully with absorption at all reflection points, bass trapping in all 12 corners, and diffusion everywhere else to make the room sound larger than it really is.

Figure 21.18: The RealTraps panels in this room were custom painted using ink rather than normal paint to avoid reducing their absorption at mid and high frequencies. *Photo courtesy of Elizabeth Kelly.*

Figure 21.19: This mastering room has produced a number of top-selling CDs, including Brothers by The Black Keys, which won two Grammys. *Photo courtesy of Brian Lucey and Magic Garden Mastering.*

Figure 21.20: This is the rear of the control room in Figure 19.13 from Chapter 19. *Photo courtesy of Wes Lachot Design.*

Reflections from the rear wall are more damaging than from other locations, because sound from the speakers is reflected directly back toward the listening position. Most professional control rooms have diffusion across the entire rear wall, as shown in Figure 21.20.

Bass in the Place

When it comes to bass traps, you can never have too many. I suppose this is the audio equivalent of "You can never be too rich or too thin." It's impossible to make any room perfectly flat at low frequencies, so all you can do is get as many bass traps as you can manage and accept the results. If you're able to line every corner and tri-corner with bass traps, you're ensured of enjoying an excellent low-frequency response with minimal ringing. But most people can't stuff every single corner of their room with bass traps, so it's useful to know which corners will actually benefit the most.

The best place to put bass traps is wherever bass builds up in a room. If you put a trap where there's little low-frequency energy, it won't have anything to act on. Low frequencies propagate around a room in various patterns, depending on the specific frequency and where the speakers are located, so it's difficult to predict where bass traps will do the most good. The best approach is to experiment with placement while measuring the resulting low-frequency response and ringing, but there's another, simpler method.

Some people recommend playing music, listening in each corner for where the sound is most bassy. To find the best ceiling corners for traps, you'll climb up on a ladder and listen there.

The problem with using music is that the bass notes are constantly changing, and only some low frequencies are present in any given song. If the modes in your room don't happen to match the frequencies present in the music, you're just wasting your time. A better sound source is band-limited pink noise. Noise contains all frequencies, not just a few tones as with music, so it's more suitable than music for identifying effective trap locations. With the higher frequencies filtered out, you can more easily hear the bass portion, and it's a lot less irritating to listen to for the five or ten minutes it takes to walk around and listen. Even better than listening is using an SPL meter to measure the relative bass levels at various corners. As explained in Chapter 19 you don't need a laboratory-grade instrument for this, and there are several affordable smart phone apps that work fine.

The "bassy_noise.mp3" available on the website for this book plays for about 80 seconds. If you have a computer connected to your system, play the file in looped mode so it runs continuously while you walk around measuring or listening. Otherwise, you can burn the file to a CD and play that through your system. If you make a CD, add the file ten times in a row so you don't have to keep pressing Play.

It's worth mentioning an interesting situation that sometimes occurs where adding a bass trap makes the measured response worse rather than better. As explained in Chapter 3, the sound at any given place in a room is the sum of the direct sound from the loudspeakers, plus many reflections arriving from different locations. All of the reflections are at different volume levels, and they combine with varying amounts of phase difference. Where two waves are more or less in phase, the result is a peak, or a null if they're out of phase. The bass response at any single point in the room is affected by all of the sources, and often more than two waves are present. This is why the response changes drastically over short distances, even at very low frequencies.

It's possible, and even common, for a reflection coming from one location to interfere with a reflection coming from somewhere else, such that a reflection that *would* have caused a deep null is partially cancelled by the other reflection. If a bass trap absorbs the countering reflection, the null caused by the original reflection will be worse at the location you put the measuring microphone. This is less likely in rooms that have many bass traps, but I've seen it happen in rooms with only a few traps. This is yet another reason to measure the room's low-frequency response while experimenting with trap placement, rather than just going by ear alone.

Front Wall Absorption

> **Myth:** *The best way to treat a home theater is to cover the entire front wall with rigid fiberglass one inch thick, then cover the remaining three walls entirely from the floor up to ear height.*

This myth is surprisingly common, which is a shame because so many people follow that advice. First, side-wall reflection panels should be centered vertically at ear height, extending above and below by at least one foot, preferably two feet or more. Sound radiates outward from loudspeakers at an angle both horizontally and vertically, so by the time it reaches the side walls, a lot of the sound is well above ear height. If the side walls are treated only up to ear height, there's still plenty of higher wall surface to reflect sound back toward listeners. Further, thin absorption down by the floor gives little benefit because it's far from the path reflected sound travels to reach your ears. Worse, treating an entire room using mostly thin absorption skews the reverb decay times versus frequency, as already explained.

When people recommend treating the front wall, their thinking is that loudspeakers radiate sound not only forward but also from the sides and rear of the enclosure. Since loudspeakers are often close to the front wall, reflections from that wall would arrive soon after the direct sound, causing interference that skews the response and harms imaging. When you need to reduce ambience generally in a room, the front wall is as good a location to treat as any other. But it's not a reflection point, except when using speakers that radiate sound from both the front and rear. So for normal "box"-type loudspeakers, the main way sound reaches the front wall is by bouncing off the rear wall behind you, then back toward the front, and finally back to your ears. Unless the rear wall is bare and fully reflective, which is not recommended, little sound should reach the front wall via that path.

As you read in Chapter 18, box-type loudspeakers radiate omnidirectionally at low frequencies. Depending on their cabinet size and construction, driver type, and other factors, most speakers start to become directional around the upper bass range and are even more directional at higher frequencies. Therefore, the real issue is at what frequency a loudspeaker radiates from the rear versus the front. Unfortunately, it's difficult to find polar plots for consumer loudspeakers—even very expensive models. Most speaker manufacturers don't want you to know how directional their speakers really are at higher frequencies, preferring you to believe you'll get a lush, full sound with great imaging no matter where you are in the room. However, loudspeakers sold to the professional audio market often do include polar data, so that's what is presented below.

Figure 18.20, which shows a typical loudspeaker polar plot, uses 5 dB per division, as do most polar plots. So for that example speaker, sound leaving the rear at 125 Hz is only 2 dB softer than what leaves the front. Table 21.3 lists a number of loudspeakers, showing at what frequencies their rear output is 5 dB less than sound coming from the front. The eight speakers in this table are all commercial models, most meant for public address use, but their radiation patterns are similar to hi-fi and studio loudspeakers. The data is derived from the manufacturer's published polar plots. As you can see, speakers do radiate sound out the rear, but only at frequencies much lower than thin rigid fiberglass absorbs.

Table 21.3: The −5 dB Rear Radiation Points for Eight
Commercial Loudspeakers

Speaker Model	Rear −5 dB Crossover Frequency
Apogee AE-8	250 Hz
Apogee FH-4	125 Hz
Bogen A2	500 Hz
Bose Panaray LT 9402-III	200 Hz
EAW MK5164	500 Hz
Electro-Voice Sx100+	250 Hz
Electro-Voice Xi 1152	250 Hz
TOA F-500WP	500 Hz

When I first questioned the importance of treating the front wall, several audio industry friends suggested I add absorption in front of the large (65-inch) glass TV screen in my living room as a test, expecting me to notice improved imaging. Even though room measuring software showed no significant reflections at the listening position, I tried that anyway. A friend and I took turns, with one of us listening while the other first held and then removed a large absorbing panel in front of the TV. The only change either of us noticed was a very slight reduction in overall room ambience, which makes sense given that the room is already free of major reflections.

All of that said, one situation where absorption on the front wall could be useful is in a small room with surround speakers that face toward that wall. If the total round trip from a surround speaker to the front wall and back to the listener is less than 20 feet, as happens in a room only 10 to 12 feet long from front to back, the reflections could be strong and early. So in that case absorption at the front of the room might really help. But the absorption should be thicker than one inch.

Treating Live Recording Rooms

Although there are different philosophies about how much natural reverberation a recording studio live room should have, all professional acousticians and designers agree that periodic echoes between parallel walls are best avoided. Therefore, diffusion is often used in addition to absorption to tame those reflections. Such treatment is universally accepted as being better than making the room completely dead by covering all of the walls with absorbers. For me, the ideal recording room has a mix of absorptive and diffusive surfaces, with no one large area all live or all dead sounding. Understand that "live" and "dead" as described here concern only mid and high frequencies. Low-frequency treatment is another matter entirely, and at those frequencies less reverb and ringing is almost always the goal.

Figure 21.21: This recording room uses extensive diffusion to capture a large, lively sound, while avoiding coherent reflections that create comb filtering. *Photo courtesy of Paul Andrews.*

Important places for absorption or diffusion in a recording room are surfaces within 10 feet of instruments and microphones, including the ceiling area above a drum set. Important places for bass traps are all 12 corners, then the four walls, then the ceiling. Figure 21.21 shows a studio live room treated extensively with diffusers, plus absorption in the lower portion at the wall-floor corners.

As you can see in the preceding photos, one-dimensional QRD diffusers are typically placed on the walls with their slots vertical. There's little point in sending sound up toward an absorbent ceiling or down toward the floor as occurs with 2D skyline-type diffusers. For a control room or listening room, the diffusers would be centered vertically at ear height, but for recording room walls, they'll be at whatever height is appropriate for the sound sources and microphones. For a ceiling, two-dimensional skyline diffusers are more appropriate to spread sound reaching the ceiling outward around the room in all directions. Figure 21.22 shows the large live room at Criteria Recording's Studio A mentioned in Chapter 6 that described micing instruments from a distance. The entire ceiling is covered with skyline diffusers to avoid the flutter echo that would otherwise occur between the ceiling and reflective floor.

Hard Floor, Soft Ceiling

Speaking of reflective floors, this is another question that comes up frequently in audio forums. Many people recommend studio floors that are reflective rather than carpeted, but

Figure 21.22: The entire ceiling in Studio A at the Hit Factory/Criteria studio in Miami, Florida is covered with RPG skyline diffusers. *Photo courtesy of RPG, Inc.*

with no further explanation. There are a number of reasons to favor a reflective floor. First, carpet absorbs only high frequencies but does little for the bass and midrange. As you know, one of the goals for all rooms is for absorption to be as broadband as possible to keep the RT60 decay times uniform with frequency. Small patches of carpet won't have a major effect on a room's overall decay time, but when the entire floor is carpeted, that can represent a significant percentage of the room's total surface area. The result is a sound that's dull yet boomy at the same time, much like covering all the walls with too-thin foam or rigid fiberglass.

There are other, practical reasons for a hard floor: Equipment can be rolled easily, spills can be cleaned up easily, and you can have a bright sound when you want that, or use throw rugs when you'd rather avoid floor reflections. Further, floor reflections aren't always damaging, even if they're early. When you record an acoustic guitar, drum set, or maybe a woodwind quartet, reflections off the floor add to the illusion of "being right there in the room" on the recording. This is why many reverb plug-ins have an option to add early reflections as part of the effect.

Reflections from nearby surfaces are most damaging when there are *many* such surfaces, all causing different comb filter responses at the microphones. A single reflection from the floor can give a desirable ambience, and every stage in every auditorium in the world has a reflective floor for this reason. With full absorption or diffusion on the ceiling, you can put microphones above the instrument, even close to the ceiling, without risking additional comb filtering. Note that absorption or diffusion on the ceiling is absolutely needed when the floor is reflective because one of those opposing surfaces must be treated to avoid flutter echo.

Variable Acoustics

Figure 21.23 shows the live room of the large pro studio I built in the 1970s. You can see the heavy curtains on two adjacent walls in the rear of the photo, open on the left and closed on the right, to allow changing the room sound from live to dead or anywhere between. Most curtains aren't thick enough to absorb below the treble range, but these are very thick "stage" curtains. Further, both curtains are three times longer than the wall area they cover, so when closed, they form pleats a foot deep to absorb well down to the low midrange. The curtains are also a foot away from the walls, which extends their absorption even lower.

Besides the curtains that allow varying each wall from absorbing to fully reflecting, this photo also shows a good example of a ceiling treatment that's appropriate for a large room. The entire ceiling is covered in a checkerboard pattern using two by four-foot absorbers three inches thick, hanging four inches below the ceiling. The alternating hard and soft surfaces could be considered a primitive type of diffsorber, though it doesn't follow a mathematical formula.

Another way to vary the acoustics of a live room is with large wood panels mounted on hinges along the walls. If one side of each panel is absorptive and the other side reflective, the panels can be flipped to expose one side or the other, or even left half open at an angle so the hard side deflects sound around the room to avoid flutter echo.

Figure 21.23: This large live room has very thick pleated "stage" curtains covering two entire walls, allowing a large variation in reverb time between both curtains open fully or drawn fully.

Treating Odd Shaped Rooms

One of the most common "odd"-shaped spaces to treat is an A-frame room with a low ceiling, such as an attic or a bonus room above a garage. Both have angled side walls, though all the usual principles still apply. Figure 21.24 shows such a space that's been treated fully. Broadband absorbers hang below the ceiling to avoid the focusing that would otherwise occur, and the side-wall reflection points are likewise fully treated. Although the side walls are angled severely, the corners formed by each side wall and the front wall are still 90 degrees, and so benefit from bass traps straddling those corners. Those two front corner bass traps have reflective membranes, as does the partially hidden trap under the peak directly above the window. The remaining panels are all broadband, without a membrane, to absorb mid and high frequencies.

Alcoves are another common cause for concern, but they needn't be. Figure 21.25 shows a small alcove in a larger room, where sound tends to focus back toward the player and microphones on the piano. Although the etagere in the middle of the wall is not much of a diffuser, having objects on the shelves helps a little. But the real work is done by the broadband absorbers in each corner.

As you can see from these example photos, even when a room has severe angles or an odd-shaped alcove, it's still arranged and treated using the same basic principles as a plain rectangle room: Set up the speakers to fire the longer way down the room and where the left and right sides are most balanced in the front where you listen. Loudspeaker reflection points always need absorption, whether the walls or ceiling are angled vertically or not. Further, all corners are viable for bass trapping, including where walls meet the ceiling and floor. Even if

Figure 21.24: Now *this* is the way to treat an attic studio! *Photo courtesy of John Heeley.*

Figure 21.25: Bass traps in the alcove behind the piano improve the sound at the piano, as well as elsewhere in the room. *Photo courtesy of Ed Dzubak and Sharon Epstein.*

a corner isn't 90 degrees, bass traps are still useful there. And if a door or window is near to a corner, you can tilt the panel so it's more parallel to one wall or the other. Bass traps don't have to straddle corners at exactly 45 degrees.

Treating Large Venues

Most of this book has focused on treating small rooms for project studios and home theaters, and much of the advice also applies to larger studio live rooms. But in the interest of completeness, the following section explains how to determine the amount of absorption needed to obtain a specific RT60 time in a large venue such as an auditorium or church.

The general goal for applying absorption is to spread it more or less evenly around the room, though some specific locations may require extra attention. For example, the rear wall of an auditorium is a prime source of slap-back echo that, if loud enough, can be distracting for the audience as well as the performers on stage. A college near me has a wonderful auditorium where I've played in orchestras many times. The entire rear wall is covered with thick rigid fiberglass behind a metal screen, both on the main floor and the balcony above.

The following formulas and steps show how to determine the amount of absorption needed to obtain a target RT60 decay time based on the current RT60 time that was measured. You'll replace the example numbers to suit your own needs. Since many absorber products and raw materials are available as two by four-foot panels, we'll calculate how many panels that size are needed. I'll use a room that's 100 by 60 by 16 feet for an example, and we'll assume that each panel has an absorption coefficient of 1.5 when spaced four inches away from a room boundary. Although an RT60 of 2.5 seconds is way too long for a small room, it's not inappropriate for a venue. Indeed, many concert halls considered to sound good have an average RT60 between 1.5 and 2.5 seconds.

The example room has a cement floor, and all four walls plus the ceiling are standard sheet rock. Measurements show the RT60 is currently 7.5 seconds, and the goal is to get it down to 2.5 seconds. As always, the math will be easy using only basic operations: addition, multiplication, and division. Parentheses indicate the calculation order for multiple operations, with the interior calculations done first. The first step is to determine the room's total surface area and volume based on its dimensions.

A 100 by 60 by 16-foot example room has a total surface area of 17,120 square feet and a volume of 96,000 cubic feet. The surface area for both the floor and ceiling is 100 by 60 feet, two of the walls are 100 by 16 feet, and the other two walls are 60 by 16 feet:

$$\text{Surface} = ((100 * 60) * 2) + ((100 * 16) * 2) + ((60 * 16) * 2) = 17,120 \text{ square feet}$$
$$\text{Volume} = 100 * 60 * 16 = 96,000 \text{ cubic feet}$$

Next, determine the room's absorption coefficient as it's now furnished, based on the current RT60 time that was measured. First, compute the sabins of absorption already present in the room using the sabins constant 0.049 (for feet, not meters):

$$\text{sabins} = \frac{0.049 * \text{Volume}}{\text{current\_RT60}} = \frac{0.049 * 96000}{7.5} = 627$$

From that, derive the absorption coefficient (AbsCo) of the untreated room:

$$\text{AbsCo} = \frac{\text{sabins}}{\text{Surface}} = \frac{627}{17120} = 0.036$$

Next, determine the AbsCo the room requires to achieve an RT60 of 2.5 seconds. The following assumes the absorption will be spread around the room evenly, which may or may not be the case:

$$\text{Desired\_RT60} = \frac{0.049 * \text{Volume}}{\text{Surface} * \text{AbsCo}} = \frac{0.049 * 96000}{17120 * \text{AbsCo}} = \frac{4704}{17120 * \text{AbsCo}}$$

$$2.5 = \frac{4704}{17120 * \text{AbsCo}}$$

Now force the left side of the equation to 1 by dividing both the 2.5 on the left and the 4,704 on the top right by 2.5:

$$1 = \frac{1881.6}{17120 * \text{AbsCo}}$$

Finally, it's easy to solve for the total surface amount that must be covered with absorption to achieve an RT60 of 2.5 seconds:

$$AbsCo = \frac{1881.6}{17120} = 0.1099$$

Since we're starting with a room having an AbsCo of 0.036 rather than zero, instead of needing to cover 0.1099 of the room's surface area with absorption, we really need only an *additional* AbsCo of 0.1099–0.036 = 0.0739. We can therefore achieve a 2.5-second RT60 by covering 7.39 (let's call it 7.4) percent of the total surface with panels having an AbsCo of 1.0 over the frequency range of interest. Rigid fiberglass four inches thick does a good job over most of the audible range, so that's sufficient.

The total surface area is 17,120, so 7.4 percent of that is 17,120 * 0.0739 = 1,265 square feet of absorption needed. Since each panel is 8 square feet, 1,265/8 = 158 panels are needed. But we can squeeze an absorption coefficient of about 1.5 from our panels by spacing them a few inches off the wall and leaving space between them, as shown in Figure 21.26. So assuming that type of mounting, even fewer panels are actually needed:

$$\frac{1}{1.5} = 0.67$$

Therefore, 158 * 0.67 = 106 two by four-foot panels are sufficient when mounted this way. As mentioned earlier, absorption should be spread more or less evenly around a room, and that's the best approach for large spaces, too.

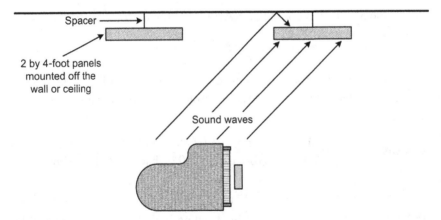

Figure 21.26: When an absorber panel is spaced away from a wall or ceiling, its effective surface area is increased because sound striking the surface near the panel gets into the rear side, which also absorbs. Sound waves also strike the panel edges, further increasing absorption.

Room Equalization

Myth: Electronic room correction is an effective substitute for bass traps and other acoustic treatment.

For more than 30 years, studio owners have tried, and ultimately rejected, using equalization to improve room acoustics. Although EQ can compensate for response deviations inherent in loudspeakers, it's not very helpful for correcting room problems caused by acoustic reflections. You simply can't EQ away an echo, nor can EQ do much for the bass response either, which is the main problem in small rooms. Further, every location in a room has a different response, so no single EQ curve can help everywhere, not even a few inches away from where you measure while adjusting the equalizer. If you adjust a precision parametric equalizer for a perfectly flat response at your left ear, the response at your right ear will be different. When using an equalizer to improve a room's low-frequency response, the more correction you apply, the narrower the physical area that's improved becomes. This is just the way it works, as governed by the laws of physics.

Modal ringing is just as damaging as a skewed low-frequency response, and only bass traps can reduce ringing. Some EQ proponents claim that EQ can *in theory* reduce ringing, but I've never seen this proven in practice over a usably large area, such as two adjacent seats in a home theater. Even if your goal is to improve the response only where you sit, it's impossible to counter nulls. If there's a 15 dB dip at 60 Hz, adding that much boost with EQ will likely blow up your speakers, or at least increase low-frequency distortion significantly and reduce headroom. And at other locations where 60 Hz is too loud, an EQ boost will make the problem worse.

Several "room correction" products claim to do more than equalization by using sophisticated DSP (digital signal processing). They promise to not only flatten the frequency response but also reduce ringing and early reflections, and to do so successfully over the full range of audio frequencies for multiple seats in a room. One such product is the Audyssey MultEQ system, though there are others. Unfortunately, the popular audio press gushes uncritically over product claims like these, printing press releases as fact and never actually testing the validity of the claims. The appeal of replacing large, visually imposing acoustic panels with a small electronic device is undeniable. But wishful thinking does not make it so.

Even the vendors themselves offer no real proof that their products work as claimed. I emailed Audyssey in 2006 regarding the technical descriptions and graphs on their website. I asked if they had any measurement data to support their claims of reducing ringing, and I also asked for clarification about how the tests on their site had been performed—for example, the size of the room and how far away the measuring microphone was from the walls. They never replied.

As it happens, a friend of mine is a columnist for a well-known audiophile magazine, and he had an Audyssey MultEQ system for review. I was thrilled when he invited me to help test it, and a few days later I arrived at his home with my Dell laptop, room measuring software, and DPA 4090 precision microphone. My friend's room is especially problematic, being nearly square at 15 by 16 feet. His ceiling is 8 feet high, which makes the room behave modally as if it were cube shaped. In addition to a few well-placed high-frequency absorbers, my friend has a modest amount of commercial and DIY bass trapping. I wanted very much to measure the room without any bass traps to see what the MultEQ could do all by itself. But it would have been too much effort to remove the bass traps built onto the walls, so we removed only two traps that weren't attached. Had we been able to remove all of the traps, the "without" graphs that follow would surely have been even worse.

Audyssey claims to flatten the response and reduce ringing over an area large enough to encompass multiple seats, so I measured at three adjacent locations on my friend's couch. It turns out this was not necessary because the MultEQ was unable to reduce ringing even at the same place it was calibrated for. Figure 21.27 shows the test setup, with the DPA microphone I used to measure the response and ringing placed nose to nose with the microphone Audyssey provides for calibrating the system.

As you can see in Figure 21.28, the main improvement is a 6 dB reduction of the lowest response peak around 35 Hz. Figure 21.29 shows a separate measurement made with only the subwoofer engaged, and you can see that two of the nulls were made worse.

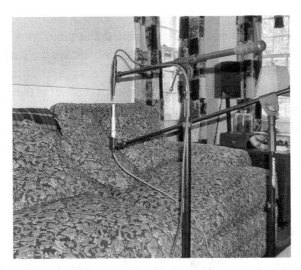

Figure 21.27: Both microphones were placed at the listening position, face to face. This allowed measuring the room at the exact same location the Audyssey system used to create its best response.

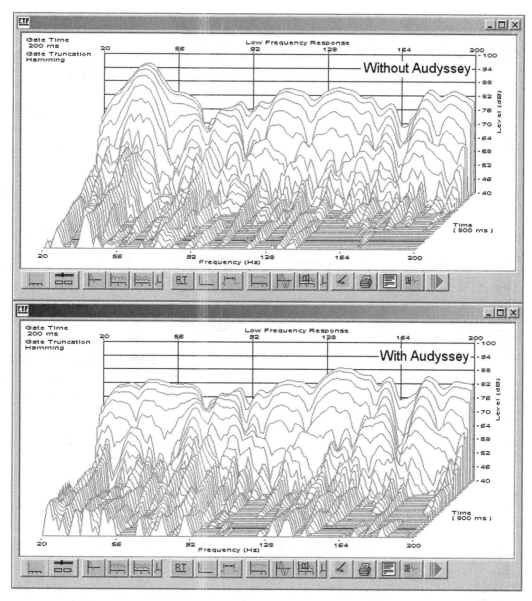

Figure 21.28: The top graph shows the response and ringing without the Audyssey MultEQ, and the lower graph is with the Audyssey engaged. The Audyssey helped flatten the two lowest response peaks, but did nothing to reduce modal ringing. At some frequencies the ringing appears even worse, lingering longer than the Without measurement.

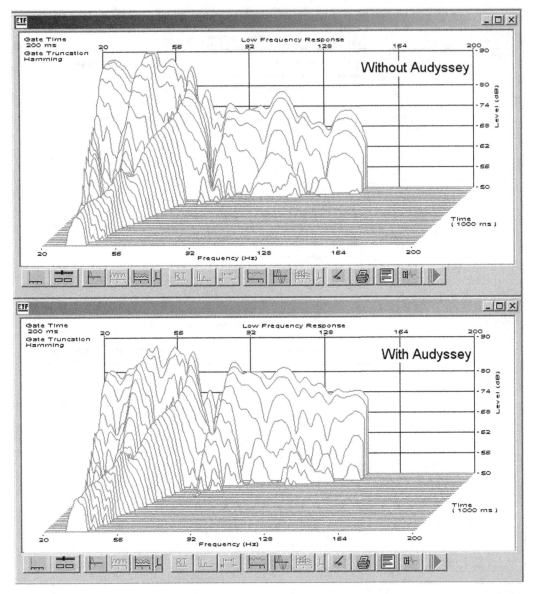

Figure 21.29: Measuring just the subwoofer, only the frequency response was improved, but not the ringing. Although the lowest peaks were flattened satisfactorily, the nulls between the 56 and 92 Hz markers became slightly worse with the Audyssey.

Listening to a variety of music, there was no question that the sound improved with the Audyssey engaged. In a nearly cube-shaped room like this, reducing the large modal peak at 35 Hz via EQ removed the boomy quality that was apparent in all of the music we auditioned. Flattening the bass also increased clarity in the low midrange by contrast, since the low midrange was no longer masked by excess bass.

The Audyssey MultEQ is certainly an effective equalizer, and since it adjusts itself automatically, it has the potential to be easy for end users to set up. However, Audyssey does not sell this device to consumers but instead requires they hire a licensed installer. It's also very expensive for an equalizer. In contrast, for about $150 you can buy a Behringer parametric equalizer and use the freeware Room EQ Wizard software to optimize the EQ automatically from your own computer. Even if you buy a separate computer just to control the equalizer, the total cost is still lower than the Audyssey system, and you own the hardware and can recalibrate your system whenever you want without paying a professional. Room EQ Wizard also performs a very thorough room analysis, showing much more information than the software bundled with this Audyssey system.

Figure 21.30 shows the bass response I measured on a second visit to my friend's home. Since he had already tweaked the Audyssey in preparation for his magazine review, its microphone was no longer set up, and I had to guess where to place my microphone. I placed it at ear height above the center of the couch, and we both agreed it was "in the ballpark" compared to where we had placed both mics at my first visit.

As you can see in the graph, some of the corrections applied by the MultEQ are inappropriate for this location. The peak at label A was reduced about 8 dB more than it should have been, and the 3 dB peak at B was boosted by 5 dB rather than reduced 3 dB. A few other frequencies in the range between 164 Hz and 200 Hz were also made worse compared to the Without graph. If you plan to apply EQ manually to reduce low-frequency peaks, I suggest you split the difference and apply only half as much reduction as your measurements indicate is needed.

To be clear, I am not opposed to using EQ to reduce the one or two lowest modal peaks in a room, especially below around 50 Hz, where conventional bass traps are less effective. Even if an equalizer or DSP cannot reduce ringing, just lowering a peak's level and amount of ringing—if not its actual decay time—improves the sound in a very real way. Indeed, I have many bass traps in my living room home theater, but I also use the one-band cut-only EQ built into my subwoofer to tame the worst modal peak around 40 Hz by 3 dB.

Some EQ proponents acknowledge that room EQ can't substitute for bass traps and agree that many of the vendor claims are unfounded. They further agree that room EQ is most useful in rooms that already have as many bass traps as is practical, and I don't disagree with that. I also can't disagree with people who claim their mixes sound better, and translate better to

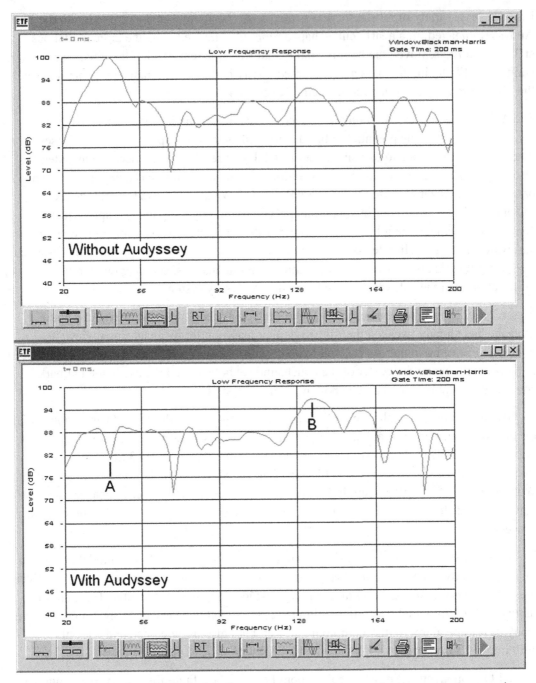

Figure 21.30: Considering only the low-frequency response but not ringing, corrections made at one location are inappropriate only a few inches away.

other rooms, after adding room EQ. In most small rooms the dominant problem is deep nulls, which EQ cannot improve. But peaks can be the larger problem in some rooms, such as my friend's room that's nearly cube shaped.

I'll add that since my test of the Audyssey system I've tested two other such systems in my own home studio, with basically the same results. These "automatic equalizers" can find and reduce peaks a little, but they do nothing for nulls or ringing. One of the systems also boosted the overall low end too much making music sound slightly tubby.

Summary

This chapter describes acoustic treatment strategies for many different types of rooms and explains how various materials and commercial products work. Most listeners don't have any acoustic treatment, and the sound of their rooms varies wildly. Therefore, the most practical solution for mixing engineers is to make their own room as accurate as possible to avoid compounding frequency balance errors heard by their audience.

One common problem in rooms used both for listening and recording is flutter echo. Like modal ringing, flutter echo is a repetitive resonance fostered by opposing parallel surfaces, except it affects mid and high frequencies. Treating one or both surfaces with either absorption or diffusion reduces the strength of these echoes, but absorption is always needed at bass frequencies where diffusion is not practical or desirable. Further, the same basic treatment advice for rectangle rooms applies to odd-shaped spaces, too, such as attics.

The best bass traps and absorber panels are made from rigid fiberglass or equivalent materials. Not only do they absorb well, unlike most foam products they're also fireproof. However, high-quality acoustic foam can be useful if it's thick enough. The effectiveness of all absorber materials is measured in sabins, which is an absolute value, though most acoustic products are instead rated by their absorption coefficient, which is a percentage. Although rigid fiberglass works very well, it absorbs even more at low frequencies when a paper FRK facing is bonded to the fiberglass. Rigid fiberglass comes in different densities, and, generally speaking, higher densities absorb better at low frequencies. However, what really affects a material's absorption is its gas flow resistance.

Porous absorbers such as acoustic foam and fiberglass are velocity absorbers that absorb a broad range, falling off at lower frequencies. The thicker the material, the lower in frequency their absorption extends. Foam and fiberglass benefit further by being spaced off the wall or ceiling, which doubles their effective thickness for free but takes away even more space in a room. Pressure absorbers work on an opposite principle, but they must be tuned, and they're effective over a range of only one octave or less. However, pressure absorbers can be relatively thin yet absorb well at low frequencies, versus foam and fiberglass that must be thick to target the bass range.

Most listening rooms benefit from having as many bass traps as possible. Thankfully, windows can absorb bass by letting some of the sound pass through to the outdoors, and sheet rock walls stuffed with fluffy insulation give some additional free bass trapping. Although most photos of treated rooms show bass traps mounted in corners where two side walls meet, they're equally effective in corners where walls meet the ceiling and floor. Playing bass-heavy pink noise lets you listen or use an SPL meter to find which corners will benefit the most from having traps.

Diffusion is a great way to avoid flutter echo and make a room sound larger than it is, but good diffusers are more expensive than good absorbers, whether you build them yourself or buy commercial products. Curved and angled surfaces can reduce flutter echo, but the best type of diffuser is the QRD because it scatters different frequencies in different directions. Other "real" diffusers can be made from specific patterns of alternating absorption and reflection, either with slats or round holes in front of rigid fiberglass.

We also saw treatment strategies for recording studio "live" rooms. Important places for absorption or diffusion are surfaces within ten feet of instruments and microphones, including the ceiling area above a drum set. It's also useful to be able to vary the liveness of recording spaces, using very thick pleated curtains that can be opened or closed as needed. Besides showing how to treat a large recording space, steps were given to determine how much absorption is needed to achieve a target RT60 time in larger venues.

Finally, this chapter debunks a number of common acoustic myths, including the notion that listening at low volume near loudspeakers avoids the need for treatment, and the value of applying thin absorption all over the front wall of a listening room. Another important myth is that an equalizer can correct room acoustics. In truth, only bass traps improve nulls and reduce modal ringing. Further, bass traps improve the sound at all locations in a room, where EQ helps some places more than others, and often makes the response at other locations even worse.

Notes

1 For the math-inclined, another way to think of this is that the phase of a wave's velocity differs from its pressure by 90 degrees. Figure 8.2 from Chapter 8 shows how a capacitor charges over time when fed a DC voltage through a resistor. When the voltage is first applied, the current into the capacitor is maximum, and the voltage across the capacitor is zero because it hasn't yet begun to charge up. As the capacitor charges over time, its voltage rises, but the current lessens and reaches zero once the capacitor is fully charged. So the voltage lags behind the current with a simple DC voltage, and it lags by 90 degrees when considering audio, which is AC. Acoustic wave pressure is similar to voltage, and wave velocity is like current that flows, so the two are offset by 90 degrees.

2 www.youtube.com/watch?v=vb30CICG68c

CHAPTER 22

Room Measuring

Myth: You must measure your room's acoustics in order to know how best to treat it.

Why We Measure

I measure rooms as part of my acoustics business, so it may seem strange for me to say that measuring isn't always needed. But there are some valid reasons to measure a room: First, many people have no idea how skewed their rooms are, so measuring can provide a badly needed wake-up call. Figure 22.1 shows the low-frequency response in a typical small, untreated room. It makes no sense to fret over the response of a sound card that's half a dB down at 20 KHz, while ignoring peaks and nulls as large as 35 dB due to room acoustics.

Another reason to measure is to confirm the improvement after adding bass traps and other acoustic treatment. Not that you won't easily hear the change. The graph in Figure 22.1 shows only the low-frequency response up to 350 Hz, but room measuring software shows other properties including comb filtering at higher frequencies, reverb time in octave or third-octave bands, the strength of individual reflections and when they arrive, as well as ringing at specific mode frequencies. So by measuring before and after adding treatment, you can assess the improvement in all of these areas and decide whether further treatment is warranted.

Measuring also helps you find the best speaker and listener placements. But you don't usually need to measure to know what or where to treat. Simply adding bass traps in the corners, and broadband absorbers at the side-wall and ceiling reflection points, will get you 90 percent of the way there.

How We Measure

In the old days, the common way to measure a room was to play pink noise through the loudspeakers, then measure the frequency response at the listening position in third-octave bands using a real-time analyzer and high-quality calibrated microphone. Adding a third-octave equalizer to the playback chain to flatten the measured response was the usual fix. This practice was common in recording studio control rooms from the 1970s through the early 1990s, but it has since been abandoned by most professional acousticians simply because it doesn't work very well.

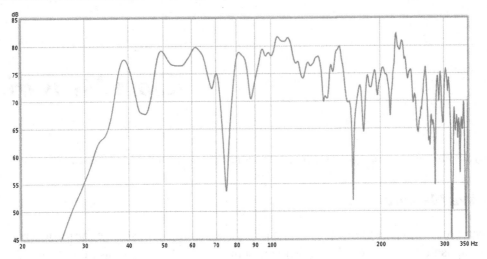

Figure 22.1: This roller-coaster low-frequency response measured in a typical bedroom size space is not only common, but typical. The span from the deepest null to the highest peak over this limited range is more than 35 dB!

There are two basic problems with measuring the response with a resolution of only one-third octave. First, the low-frequency response in a room can vary wildly, with several peaks and deep nulls within a range narrower than one-third octave. Figure 22.1 shows the response I measured in my company's test lab when empty, with no bass traps or other treatment. That room is 16 by 11.5 by 8 feet, and you can see the horribly skewed low-frequency response typical of all untreated rooms. Look at the range between 70 and 90 Hz, where there's a peak, a null, another peak, and another null, all very close together. It's impossible to see this response with third-octave analysis, and likewise impossible to correct using a third-octave equalizer whose only available boost and cut frequencies are 63, 80, and 100 Hz. Figure 22.2 shows the same graph overlaid with a version that's been third-octave averaged. As you can see, averaging the response completely hides the true extent of the peaks and nulls.

The other big limitation of third-octave analysis is that frequency response is only half the story. The other half is timed-based problems—discrete reflections, reverb, and modal ringing. If a room's reverb time is longer in the range between 500 and 1,000 Hz than at other frequencies, that range will sound louder simply because more energy is present over time in the room. Likewise, modal ringing extends the duration of some bass frequencies, but not others, which also makes the ringing frequencies more prominent. Therefore, another important metric for room measurement is decay time versus frequency.

Figure 22.3 shows the same low-frequency measurement again, but this time as a *waterfall plot*. With this type of graph, each mode's decay comes forward over time. Here, the decay times are shown over a span of 800 milliseconds.

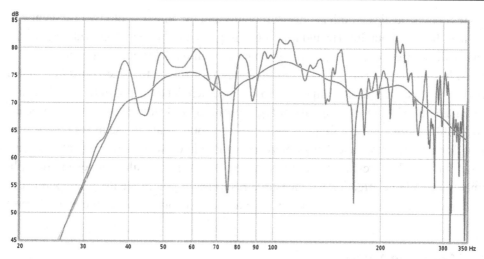

Figure 22.2: Both lines are derived from the same measurement, but one is at high resolution and the other is averaged to third-octave resolution.

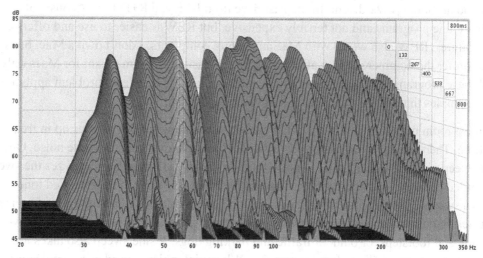

Figure 22.3: This waterfall plot is derived from the same measurement as Figure 22.1, but it also shows each peak's decay time.

Another inadequate method for measuring rooms is the typical "test tone" CD. These are often sold as a product, but some are available as files you can download for free and burn to a CD. Most of these CDs play single sine waves at the standard third-octave frequencies, which again completely misses what happens at frequencies in between. (Table 1.3 in Chapter 1 lists all of the standard octave and third-octave frequencies.) It's also tedious to play a bunch of tones, note the SPL meter reading for each, and then plot the data manually on graph paper to see the response. By contrast, good room measuring software lets you measure once, which takes about ten seconds, then display the results in many different ways.

I separate room acoustics—both for measurements and treatment—into two frequency ranges: bass below about 300 Hz and mid/high frequencies above that. In truth, the correct way to divide the frequency range for a given room is at its Schroeder frequency, as explained in Chapter 20. But 300 Hz is close enough for rooms the size you'll find in most homes. For low frequencies it's important to see as much detail as possible, which means measuring and displaying the response at high resolution to learn the true extent of peaks and nulls. But at mid and high frequencies, it's more appropriate to use averaging. A graph of the high frequency response that's not averaged is riddled with so many peaks and nulls it's difficult to see the forest for the trees, so to speak. Further, small changes in microphone placement have a huge effect on the measured response; if you move the microphone even one inch, the response changes drastically. Averaging lets you better see the overall trend.

Room Measuring Software

My room measuring software of choice is Room EQ Wizard (REW), available as a free download at roomeqwizard.com. In the past, I've used ETF and R+D from Acoustisoft, which are fine programs and not terribly expensive, but REW is easier to use and offers even more features. REW also works with Mac and Linux computers. I don't own a Mac, but some people prefer the commercial (but affordable) FuzzMeasure program meant for Macs only. All of these programs offer the same basic feature set, so you'll have no problem applying the REW examples that follow to whatever software you choose.

Like most room measuring software, REW generates a sine wave that sweeps up in frequency over time. There are many advantages to a swept sine wave over pink or white noise. One is that a sweep yields a higher signal to noise ratio. When the software later analyzes the sweep as recorded through your microphone, it can apply a *tracking filter* to the recorded tones. This is a sweepable band-pass filter that passes only the frequency of interest at that moment, thus filtering out unwanted sounds such as loudspeaker distortion, preamp hiss, your own breathing and footsteps, or outdoor traffic and barking dogs. A sine sweep also takes less time to measure, especially at low frequencies. When pink noise is used as a signal source, the noise has to play for longer and be averaged. Pink noise varies continually in level, so the meter dances around and is difficult to read. Indeed, one huge advantage of dedicated room testing software—versus an old-fashioned RTA with pink noise—is you can measure the room once, then display many different types of graphs later.

Configuring Room EQ Wizard

The first step, after installing REW, is configuring it to work with your computer's sound card. Figure 22.4 shows the configuration dialog, accessed from the Preferences menu at the

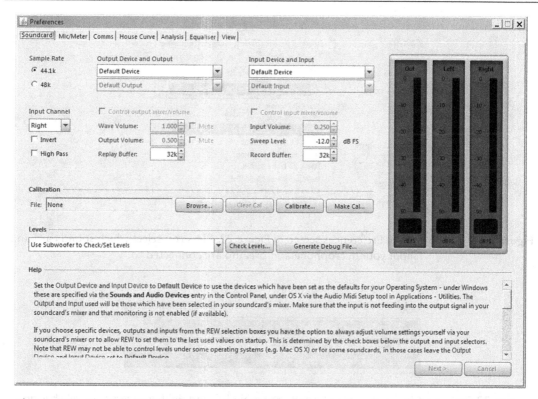

Figure 22.4: The REW Preferences section lets you select which sound card to use for input and output, the sample rate to record at, and other parameters.

top of the main screen. I'll address only the choices needed to prepare REW for normal use. Anything not described here can be left at the default setting, and REW's extensive Help explains all the other options and features in detail.

I generally use a sample rate of 44.1 KHz, simply because that's what I use to record music, though 48 KHz is fine, too. Most computers have only one sound card, but you may have an external audio interface with multiple inputs and outputs. You'll record a microphone with the software, so be sure to choose an input device that accommodates a mic. Figure 22.5 shows a close-up of the drop-down menu to select which input to use.

As you can see, my studio computer has two physical sound cards: a SoundBlaster X-Fi, and a multi-port Delta 66 made by M-Audio. I use channels 1 and 2 on the Delta 66 for REW, but I could just as easily use channels 3 and 4. A similar drop-down selector lets you pick the input sound card, which can be different from the output. Input recording can be from either the left or right channels. In this case I use the right channel. The output sweep tone is sent to both channels at once, so a selection for that isn't needed.

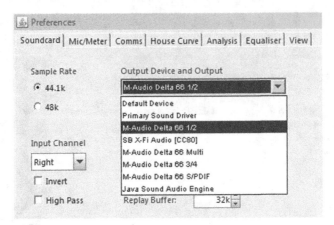

Figure 22.5: This tab of the Preferences screen in Figure 22.4 lets you select which sound card and input your microphone and preamp are connected to.

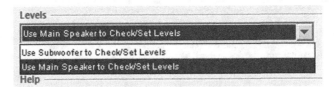

Figure 22.6: Here's where you tell REW to use the main left and right speakers for level checking, rather than the subwoofer.

You also need to tell REW which speakers you plan to use for setting the playback and record levels—the main speakers or subwoofer. This selection is shown in Figure 22.6, though you'll read below why I don't use this to set levels.

Using Room EQ Wizard

When measuring the low-frequency response in a room, it's important to measure using both the left and right speakers sounding at once, plus the subwoofer. If you measure using only the subwoofer, data within an octave or so of the crossover frequency isn't accurate because the main speakers are not contributing to the measurement, even though they will be when playing music. I'll have more to say about measuring with one versus all speakers playing in a moment.

You can now close the Setup screen and do a test sweep. Click the Measure button in the upper left of the main REW screen, and you'll see the Measurement screen shown in Figure 22.7. In most cases, you'll set the Start and End sweep range limits to 20 Hz and 20

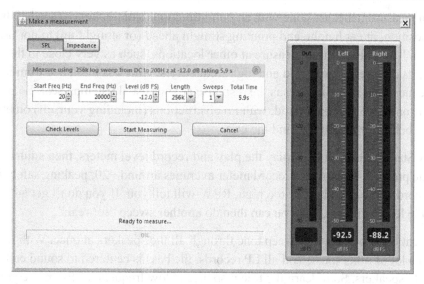

Figure 22.7: The sine wave sweeps used for measuring are configured and run from this screen.

KHz, respectively. If you have a subwoofer that goes below 20 Hz, you can specify a lower Start frequency. If for some reason you feel the need to measure the response to higher than the 22,050 Hz limit of a 44.1 KHz sample rate, you can go back to the setup screen and select 48 KHz to measure up to 24 KHz. The only time I limit the sweep range is when doing many low-frequency measurements in a row and I don't want to wait for a full-range sweep to complete every time. If you're using REW to find the best location for a subwoofer, you could set the upper limit of the sweep to 200 Hz, though you still should measure with both the subwoofer(s) and main speakers active.

The Length setting determines the speed of the sine wave sweep, with larger values giving a better signal to noise ratio but taking longer to measure. You can also tell REW to sweep more than once, which averages the results and improves the signal to noise ratio further. For most room tests the default settings are fine.

The Check Level button plays pink noise instead of a sine sweep, and the volume is different than for the sweep tone. So I just run a sweep to set levels, then cancel the measurement. The playback meter should read around −12, and you'll adjust the volume on your mixer or receiver so the sweep sounds fairly loud in the room. But don't play it so loud it hurts your ears or risks damaging your speakers. Common sense applies here. The tones must be fairly loud to drown out the ambient room noise by at least 30 or 40 dB. Nulls 30 dB deep or even more are common, especially in rooms having little or no acoustic treatment. So if the ambient room noise is only 20 dB softer than the sweep at a null frequency, the true null depth will be hidden and filled in by the background noise. The tracking filter built into REW mentioned earlier helps reduce stray noise, but it's better to deal with this at the source.

Most room measurements are done with an omnidirectional microphone placed at the listening position, at ear height, and pointing straight ahead (or straight up) to not favor either loudspeaker. It can be useful to measure at other locations, such as very close to the speaker to measure its own response at mid and high frequencies, with less influence from the room. But for the most part, what matters is the response at your usual listening seat. So place your microphone or SPL meter on a stand, with no obstructions (including yourself) on either side, or above or below, or directly behind the mic.

Next, click Start Measuring to display the play and record level meters, then adjust your microphone preamp's gain so the record meter averages around −20, peaking safely below 0. If the record level is too low, or too high, REW will tell you. If you don't get an error message, the levels are fine, and you can then do another sweep "for real."

Earlier I mentioned playing the sweep tone through all the speakers at once. With most pop music, and a lot of other music and all LP records, the bass is centered to sound equally through both speakers. So to learn the true response at low frequencies for music you listen to, you should do the same and play the sweep through both the left and right speakers at the same time. If you use a subwoofer, that should also be engaged for the same reason. However, it's useful to also test each speaker separately. This will quickly reveal unusual problems such as a blown midrange driver, or a severe peak or null due to positioning that affects only one speaker.

Interpreting the Data

We're all familiar with frequency response graphs, as shown previously in Figure 22.1. In that graph the display is limited to the range below 350 Hz, since that's where speaker placement and bass traps have the most effect. Graphs like this help you assess the improvement after adding bass traps and changing their position, and changing the placement of loudspeakers and subwoofers, and even moving the listening seat. The speakers used for that test are flat to just below 40 Hz.

Note that the graph in Figure 22.1 is shown at a high resolution with no averaging, also called *smoothing*. REW also lets you apply averaging at various resolutions, such as one-third octave. But that's not appropriate at low frequencies, because it hides the true extent of peaks and nulls. However, averaging is useful at mid and high frequencies, as mentioned earlier. The response measured at higher frequencies is often riddled with numerous peaks and nulls due to comb filtering caused by reflections from the side walls, floor, and ceiling. But comb filtering exists even in well-treated rooms, due to slight differences in arrival time from the left and right speakers. If the measuring microphone is not precisely centered, equidistant to both speakers, that alone can cause a series of peaks and deep nulls. This is shown in Figure 22.8.

After applying third-octave averaging, we get the graph in Figure 22.9, which is much easier to read and lets you see the overall response clearly without distraction. The low-frequency portion lacks too much detail to be useful, but the overall response trend is easier to interpret.

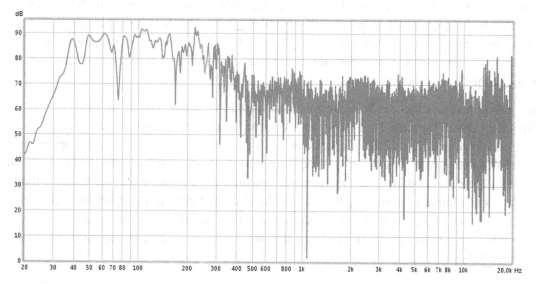

Figure 22.8: This graph shows the same measurement in Figure 22.1, but for the entire audible range up to 20 KHz.

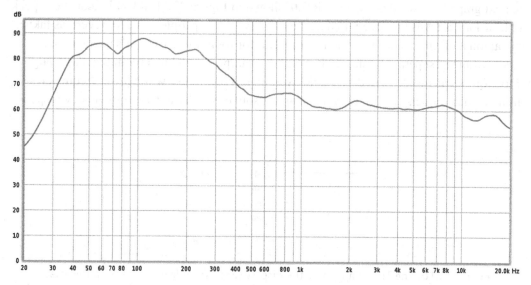

Figure 22.9: This graph shows the same data as Figure 22.8, but with third-octave averaging applied to more easily interpret the higher frequencies.

Waterfall Plots

A *waterfall plot*, shown previously in Figure 22.3, is much more useful for assessing low-frequency problems than seeing just the raw response, because it also shows how some frequencies linger in the room, taking longer to decay. As explained in Chapter 20, modal ringing is caused by resonances within the room itself at frequencies related to the room's dimensions. The longer the dimension, the lower the frequency. In this type of graph, the peak "mountains" come forward over time. I generally set waterfall plots to show a 30 to 40 dB span vertically, with the Time Range set long enough for most of the peaks to decay through the floor of the graph.

It's also important to set the Window time to at least 200 milliseconds to see enough frequency detail. Both the Time Range and Window parameters are set in the Controls section shown in Figure 22.10, displayed by clicking the Controls button near the upper right of REW's main screen. The Window time for waterfall plots is similar to the fractional octave (1/3, 1/12, etc.) smoothing used for frequency response graphs. Extending the Window time increases resolution by narrowing the analysis bandwidth. I generally use 300 milliseconds for a resolution of 3.3 Hz, though this example is set to one full second for even higher resolution. Using one second gives a resolution bandwidth of 1.0 Hz, as displayed to the right of the Window setting.

RT60 Reverb Time

The next graph type we'll consider is RT60, shown in Figure 22.11. RT60 is acoustics-speak for "how long it takes the reverb to decay by 60 dB." Most people consider reverb as taking some amount of time to decay, which is true. But for room measuring, it's useful to see the decay time in separate frequency bands. For example, sound in a room that has a large number of thin absorbers will decay quickly at high frequencies, while lower frequencies continue much longer. This gives an unbalanced sound, similar to a high frequency response

Figure 22.10: The Time Range sets the display range from front to back, in this case 800 milliseconds. The Window setting establishes the resolution bandwidth, with larger values showing more detail.

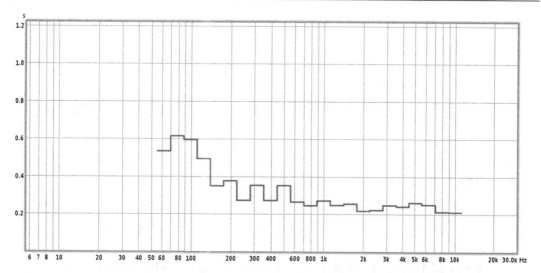

Figure 22.11: This graph shows the RT60 decay times in third-octave bands I measured in a large, professional control room. The general goal for RT60 is to be uniform over most of the audio range, but having longer decay times at low frequencies is common, and not necessarily a problem in larger rooms.

roll-off. That is, if high frequencies don't sustain for as long as lower frequencies, the overall energy in the room is lower even though the absolute volume levels coming from the loudspeakers are the same. So by displaying RT60 in third-octave bands we can see problems such as this.

Energy Time Curve

The last display type we'll consider is the *energy time curve*, or ETC, shown in Figure 22.12. Note that REW calls this type of display an *impulse response*. An ETC graph shows individual reflections, how strong they are, and how much later they arrived after the direct sound from the speakers. Unlike a frequency response plot, the horizontal axis of an ETC represents time rather than frequency. This measurement was made in the same large control room as the RT60 graph in Figure 22.11. The volume of the direct sound is shown at Time Zero near the left edge of the graph, and the reflections are shown to the right as they arrive over a period of ten milliseconds. In this graph you can see one strong reflection arrive about 4.5 milliseconds after the direct sound, and then another softer reflection arrives just before 8 milliseconds. A third reflection arrived at the right edge of the display 10 milliseconds after the direct sound. Of course, the graph can be altered to display longer or shorter time periods as needed.

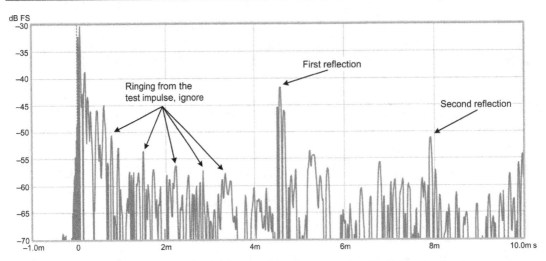

Figure 22.12: The ETC display lets you see the strength of individual reflections and when they arrive.

What matters most with early reflections is how loud they are relative to the direct sound from the speakers. This graph shows 5 dB per vertical division, so the first early reflection at around 4.5 milliseconds is 12 dB softer than the direct sound, and the second reflection, just short of 8 milliseconds, is about 21 dB softer. The goal for early reflection levels is to be at least 15 to 20 dB softer than the direct sound, with lower levels giving less severe comb filtering. Recall the Decibels.xls spreadsheet from Chapter 1, which determines the extent of peaks and nulls based on the strength of a reflection. In this case, the first reflection is 12 dB softer than the direct sound, so the resulting nulls are 2.5 dB deep and the peaks +2 dB. The second reflection is 21 dB down, so the peaks and nulls are each less than 1 dB.

The beauty of an ETC display is it shows when each reflection arrived, which helps determine where they came from. Remember that the first peak on the graph at time Zero is not when the sound left the speaker, but rather, the time the direct sound reached the microphone. Sound travels at approximately one foot per millisecond, and the first reflection arrived after 4.5 milliseconds. This means the first reflection traveled about 4.5 feet *farther* than the direct sound from the speaker. A reflection travels to and from a boundary before it reaches the microphone. Therefore, 4.5 milliseconds is the *difference* between how long it took the direct and reflected sounds to arrive at the measuring microphone, not the distance from the speaker to the reflecting surface then on to the microphone. If any reflections are significant, look around your room to see which boundaries or objects might be the cause, such as the floor, a coffee table, or the top of your mixing desk. Try placing an absorber at likely locations, run the test again, and see if the reflection is affected.

Using the Real Time Analyzer

Finally, I'll share a clever trick that can help you to place speakers, subwoofers, and even bass traps more efficiently. It takes REW only ten seconds to perform a full-range sweep and display the results, though you can limit the sweep range to 300 or 400 Hz, which takes even less time. Still, it's tedious to move a speaker or bass trap a few inches, measure, display, and repeat. Going back and forth constantly between handling the speaker and working your computer gets old quickly, and you need to constantly delete all the intermediate measurements to avoid cluttering up the display.

To speed up the process, I created a Wave file that sweeps a sine wave from 20 to 400 Hz. Set the file "sweep-20–400 Hz.wav" to run continuously in Windows Media Player, or any other program that cooperates with Windows sound drivers and shares your sound card with other programs. Or repeat it a bunch of times in a CD burning program, and play it through your CD player. Then click the RTA button at the top of the REW main screen to view the Real Time Analyzer (RTA). Figure 22.13 shows just the top right portion of the RTA screen.

Set all of the RTA parameters as shown in Figure 22.13 to display at high resolution, and then click the square red Record button to enable the RTA display. Next, start the sweep playing in your media player program with the volume fairly loud in the room. Adjust the microphone preamp level until you can see the frequency response graph line on the screen. (You may also need to adjust the Graph Limits dB range to see the entire response line on the screen.) Now you can experiment with speaker placement, or microphone placement, or bass trap locations, and see the result of your changes immediately. Pretty cool!

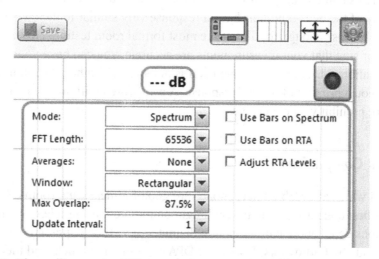

Figure 22.13: After calling up the RTA screen, the square Record button at the upper right puts REW into its RTA mode, and updates the screen continuously in real time.

Measuring Microphones

The standard microphone type for measuring rooms is an omnidirectional small diaphragm condenser. Small diaphragm condenser mics tend to have the flattest high frequency response, and their omni pattern lets you collect sound from two or more speakers at once without favoring one speaker over the others. You don't need to pay a lot of money for a competent measuring microphone, and even a budget SPL meter will do in a pinch. Often, what matters most is seeing a relative change after adding bass traps and other acoustic treatment. For this purpose, the mic's absolute frequency response is less important. Of course, a known-flat microphone is needed to assess the actual frequency response of your speakers and room.

Many high-quality measuring microphones come with a printed frequency response chart that was measured for that specific mic. In that case, you can adjust your readings to take into account any deviation from a flat response in the microphone. But REW makes this even easier for you, letting you load a *calibration file* if one is available for your microphone, or you can create your own as a standard text file. These files contain a list of frequencies and their plus or minus dB values that the software applies to the graphs it displays. REW's Help shows the format of these files, though most microphones are flat enough that you don't need to bother. Rooms vary by 30 dB, and even modest microphones vary much less. If you have a budget mic and want to know its true response, you can send it to a lab to be measured. Then you can create a custom calibration file and load it into REW to counter the mic's own response errors.

In my experience, microphones don't need to be calibrated unless you're a professional acoustician who's being paid to provide accurate readings. The following section shows the test results for ten small diaphragm microphones commonly used for room measuring. As you'll see, many of the budget models have a response very similar to the most expensive brands—certainly close enough for all but the most formal room testing. However, if you want to know for sure that your measurements are accurate, you can have your microphone professionally calibrated. The microphone itself is not altered! Rather, the lab measures the mic and sends you a printout of its true response. A web search will find companies that do this, and it's not prohibitively expensive.

Microphones Comparison

I'm often asked which affordable microphones are good for measuring rooms. As mentioned, room testing is best done using a small diaphragm omnidirectional condenser mic. A small diaphragm is considered about half an inch in diameter, but "tiny diaphragm" microphones such as those made by Earthworks, B&K, and DPA are around 1/4 inch, and their response can extend to well past 20 KHz. This type of microphone is ideal, but many are very expensive.

With the help of some friends, I tested ten popular microphones ranging from an inexpensive DIY model to a Josephson with Microtech Gefell capsule that cost $1,800. Note that these tests did not attempt to evaluate anything other than raw frequency response. We didn't measure distortion, off-axis response, maximum SPL capability, or residual noise, nor did we assess their build quality. We didn't record any musical instruments either; our only goal was to determine how suitable these mics are for measuring loudspeaker response in a room.

I created a sort of "anechoic chamber" shown in Figure 22.14 using six RealTraps MiniGobos in a semicircle around a Mackie HR824 active monitor, plus two more MiniGobos (not visible in the photo) to avoid potential reflections from the front of the room. One more broadband bass trap was placed on the floor in front of the loudspeaker to minimize reflections from the floor.

For these tests, each microphone was suspended from a lighting rig 25 inches in front of the loudspeaker, whose published response (–3 dB points) extends from 37 Hz to 21 KHz (see Figure 22.15). The microphone capsule tips were pointing downward, placed on-axis directly opposite the front of the tweeter. To ensure that every microphone was in the exact same place, within 1/8 inch in all directions, a piece of string was stretched to set the height and distance from the loudspeaker. An ink mark on the string identified the horizontal location. Each microphone was lowered so its tip was exactly one inch below the string to avoid potential minor reflections from the string itself.

Since this is not a true anechoic chamber, you'll see an influence from the room, especially at low frequencies. So even if a microphone is perfectly flat, its frequency response will not

Figure 22.14: This setup ensured identical placement for each microphone when comparing their frequency responses.

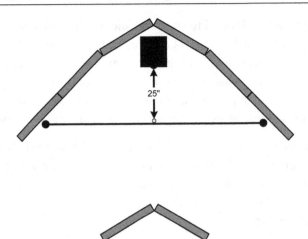

Figure 22.15: For these microphone tests, the loudspeaker was surrounded with absorbing panels, and another pair of panels was placed farther away to avoid reflections coming back from the front of the room toward the microphones.

appear flat in the graphs due to the room. The loudspeaker's own response also influenced the results. However, the *relative* responses are valid, letting you compare how closely the inexpensive microphones match the expensive calibrated models that are known to be highly accurate. Figure 22.16 shows all of the microphones tested:

- Mitey-Mike (DIY)
- Nady CM 100
- Behringer ECM8000
- Radio Shack Analog SPL Meter
- Radio Shack Digital SPL Meter
- dbx RTA-M
- Neutrik 3382
- DPA 4090
- AKG C 451 EB with CK 22 capsule
- Earthworks QTC-1
- Josephson C617 Set

We actually tested 11 microphones because two different Radio Shack SPL meters were included: an analog model about 20 years old and a more recent digital version. Both SPL meters were set for C rather than A-weighting. For all of the tests, the microphones were hung by their audio cable to point straight down, though we also tested one of the SPL meters pointing toward the speaker for comparison.

Although omnidirectional microphones supposedly receive sound equally from all directions, most become slightly directional at higher frequencies. So when measuring rooms and

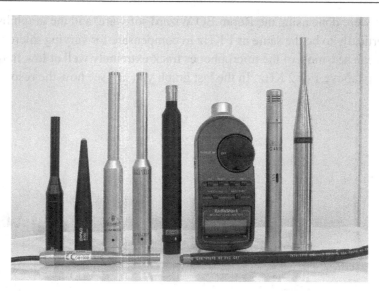

Figure 22.16: The mics tested are from left to right (rear): dbx, DPA, Behringer, Nady, Josephson, Radio Shack, AKG, and Earthworks. In the foreground are the Neutrik (left) and Mitey-Mike.

loudspeakers, the convention is to point the microphone upward. No omni microphone has exactly the same frequency response from all directions, though microphones with tiny diaphragms and slim bodies are often omnidirectional to higher frequencies than larger models. Regardless, when balancing loudspeaker volume levels on a surround system, or measuring the response of two speakers playing together in stereo, pointing the microphone up toward the ceiling or down toward the floor avoids favoring any one loudspeaker.

The Results

As you can see in Figures 22.17 through 22.20, even inexpensive omnidirectional condenser microphones are quite flat at bass frequencies. What separates the men from the boys is how high their response extends, how flat they are generally, and how well their omnidirectional pattern is maintained at higher frequencies. We measured the response up to only 20 KHz, though some of these microphones can capture higher frequencies.

It's difficult to show 11 microphones all on one graph, especially in a black-and-white reproduction, so they're divided into smaller groups. The first two graphs in Figures 22.17 and 22.18 show the four "high-end" microphones—both the raw response and then averaged at third-octave resolution. The raw data is more accurate at low frequencies but shows too much detail above 1 KHz. Therefore, the remaining graphs are all smoothed, repeating the Earthworks as an arbitrary reference in each graph, plus the four other microphones in the group.

All of the tests were done using the Room EQ Wizard software, and the graph line heights were offset vertically to be the same at 1 KHz to compensate for varying microphone output levels. As you can see, most of the microphones track extremely well at low frequencies, deviating mainly above 1 or 2 KHz. In the last graph you can see how the response of the

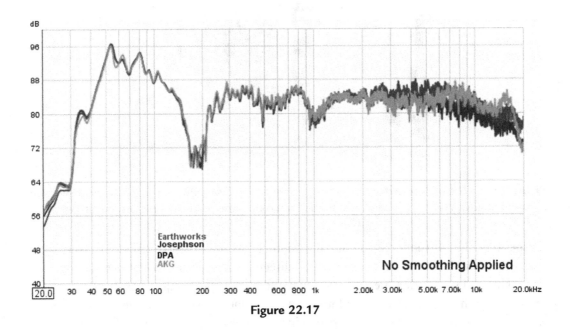

Figure 22.17

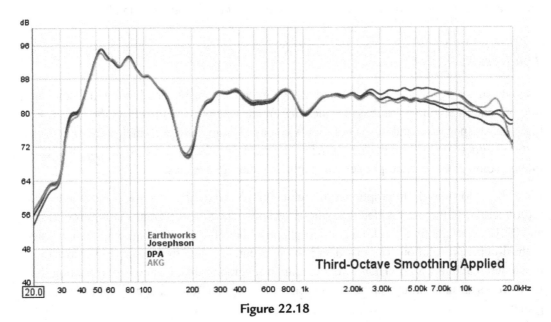

Figure 22.18

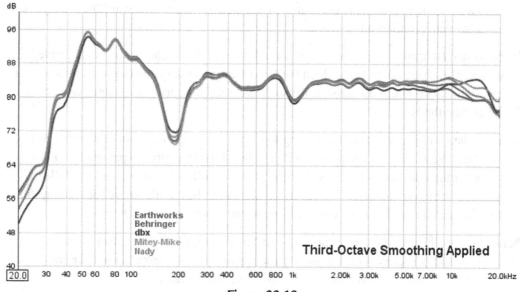

Figure 22.19

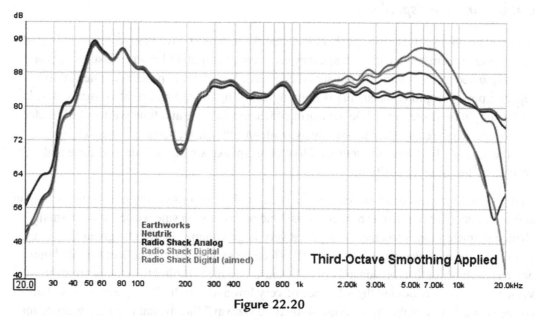

Figure 22.20

Figures 22.17 through 22.20: These graphs show the frequency response measured for 11 microphones split into smaller groups to avoid having 11 traces all on top of one another.

Radio Shack digital SPL meter is much more extended when aimed toward the loudspeaker source rather than pointing up or down.

The Nady and Behringer microphones are visually identical except for the name printed on the side. However, I didn't take them apart to examine the capsule or electronics, so it's possible only the cases are the same. I also noticed that the Behringer's output was about 3 dB higher than the Nady.

From this data it appears that for measuring loudspeaker setup and room treatments, the Nady or Behringer are quite adequate—within a few dB of the most expensive models—and far better than the Radio Shack SPL meter, which is often recommended but falls off substantially at high frequencies. However, if you plan to buy a mic to use for room testing and also for recording, you might consider one of the more expensive models.

Thanks to Doug Ferrara, Grekim Jennings, and Mike Barney and Pete Basel of the Connecticut Audio Society for their participation. And special thanks to Fletcher and Mercenary Audio for supplying the Josephson/Gefell microphone.

Calibrating Loudspeakers

It should be obvious that both speakers in a stereo setup must have their volumes balanced, so sending the same signal to each speaker gives equal output. Otherwise, instruments and voices panned to the center will not sound centered. For passive speakers connected to a separate power amplifier, balancing is done by adjusting the left and right volume controls on the amplifier, if present, while measuring each speaker one at a time with an SPL meter. Active speakers are adjusted the same way, with an SPL meter, using the speaker input volume settings. In a surround setup, all five main speakers must be balanced for equal volume at the listening position.

REW is great for measuring the response of loudspeakers in a room, but pink noise is simpler and even better for balancing speaker levels. One problem with sine waves is that a small change in microphone placement can give a very large volume change at a given frequency, so moving the mic even an inch or two will affect the reading. However, band-limited pink noise is even better than broadband noise, because it reduces further the change in level versus position by removing the influence of very low and very high frequencies. For your convenience, I created the "pink_noise-400–2000 Hz.wav" file, available on the website for this book. This file contains pink noise, but with frequencies below 400 Hz and above 2 KHz filtered out at 6 dB per octave. For completeness, the file "pink_noise.wav" containing normal pink noise is also available, and both of these files can be looped to play seamlessly for as long as needed. The difference between pink noise and white noise was explained in Chapter 1.

In an ideal surround setup, the listening position will be on-axis for all speakers. In other words, all of the speakers should point directly at your head where you sit. But some people can't do that for whatever reason, and some people have different speaker models for the front and surround speakers. The largest response variations that occur when off-axis or when using different model speakers are at high frequencies. Low frequencies are also highly positional, and moving the measuring microphone even a few inches can change the measured volume a lot when using static tones. So by using noise and omitting high and low frequencies in the noise, leaving only the midrange, SPL meter readings are affected much less by differences in each speaker's low- and high-end response. That is, you're balancing the volume level as it will be heard, with less influence from response differences of the speakers and their placement.

To calibrate loudspeaker volume in a stereo or surround system, you'll send the same signal to each speaker one at a time and adjust each channel's volume control for the same reading on an SPL meter placed where you listen. Most consumer receivers have a Setup mode to adjust the volume for each speaker, as well as compensate for its distance from your ears, as mentioned in the Surround Monitoring section of Chapter 19. If you don't have an SPL meter, an omnidirectional microphone plugged into any device that has a record level VU meter works just as well. Again, the SPL meter or microphone should be placed where you sit while listening, pointed straight up to avoid favoring any one speaker.

Summary

This chapter explains that the main reasons to measure a room are to learn how bad it really is and to assess the improvement after adding acoustic treatment. Measuring also helps find the best places for loudspeakers and listeners. In the past, rooms were usually measured in third-octave bands using pink noise. But more modern thinking uses dedicated room analysis software that can measure the frequency response at much higher resolution and also display time-based information—waterfall plots to see modal ringing, RT60 to assess reverb in multiple bands, and ETC graphs to identify individual reflections and when they arrived.

Although I used the freeware Room EQ Wizard software for my examples, other programs work similarly and offer similar features. Modern room measuring software uses a swept sine wave rather than pink noise to more quickly achieve a higher frequency resolution and better signal to noise ratio. Dedicated software also lets you measure only once, then see many different views of the same measurement data. Further, using a custom swept sine wave with the Real Time Analyzer feature of REW lets you quickly optimize placements of speakers and bass traps.

This chapter also explains that small diaphragm omni condenser microphones are the best type for measuring rooms, and the comparison of 11 microphones ranging in price from very affordable to expensive shows that budget microphones are adequate for measuring rooms.

Finally, I explained how to calibrate loudspeaker levels, which may seem obvious, but in fact is best done using band-limited pink noise to minimize response variations between different front and rear speaker models in a surround system or having some speakers off-axis, and minimize variations due to microphone placement.

PART 6

Electronics and Computers

In the old days, most professional recording studios had their own technical department staffed by electronics technicians who knew how to align tape recorders and could repair a compressor, console, or anything else that broke. Today, many studios are small, one-person operations. If you don't know how to do your own minor repairs, you'll have to pay someone else every time something breaks. The chapters that follow explain basic electronics, with an emphasis on practical solutions to common problems. You'll learn not only how to repair a busted guitar cord, but also the underlying principles of electricity and electronics. All audio devices are based on electronic circuits, and my intent is for readers to gain a deeper understanding of their gear.

I'm not a degreed electronics engineer, but I've done lots of tinkering and amateur circuit design, and I was a professional computer programmer for many years. I've also been fortunate to know some highly skilled electronics engineers who were willing to answer my endless questions. Years ago I noticed that the more educated someone is about basic electronics, the less likely he or she is to believe in the superiority of expensive speaker wire and other audiophile myths. I've always felt that my knowledge of electronics also made me a better recording engineer. It certainly helped me save money by building and repairing my own gear! These chapters use common mechanical analogies to explain electronic components, circuits, test procedures, and even computers, from the perspective of recording engineers and audiophiles.

Years ago, back in the 1950s and 1960s, audiophiles cared not only about high fidelity sound, but they also sought to understand how their equipment works. Many enthusiasts built their own phono preamps, power amplifiers, and loudspeakers from kits or plans in magazines and books, and some even designed their own gear. Back then, audio circuits were based on vacuum tubes, which are more complicated and difficult to design than today's devices that use mainly op-amps and transistors. With its very high voltages, tube gear is also dangerous to work on!

That was the age of Heathkit and Dynaco, and many audiophiles were also ham radio operators or short wave radio enthusiasts. The first Heathkit products were oscilloscopes and signal generators, and other test gear needed for troubleshooting audio circuits. I recall fondly building several Heathkits including a 75 watt per channel solid state power amp that worked

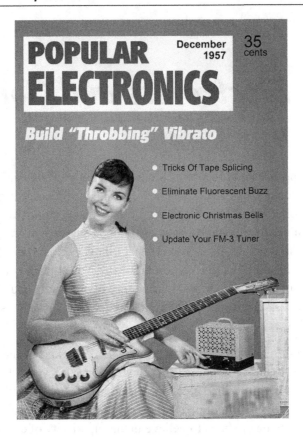

perfectly, and a 2-meter ham radio transceiver that didn't work at all. Fortunately, my older sister's friend Cliff Mills was an engineer at a local radio station, and he had it working in one afternoon. Watching Cliff figure out what was wrong, I learned a lot! And that was the beginning of my life-long quest to understand how audio and other electronics devices work.

Somewhere along the way the hi-fi hobby became less about knowing how audio gear works, and more about buying expensive stuff. Of course, an appreciation for music and high quality reproduction have always been important! But over the years interest in the technical side of audio and sound has waned. Today, few hi-fi magazines and websites offer factual information, instead presenting mainly opinions and non-technical subjective equipment and music reviews. Some of the most revered products we read about cost as much as a car. We rarely see tutorial or how-to articles anymore, and when product specs are shown they're often irrelevant, or too dumbed-down to be useful.

Although I'm a musician, and my professional experience is mainly as an acoustician, recording engineer, technical writer, and computer programmer, I also consider myself very much an audiophile. Like most audiophiles, I spent many years perfecting and fine-tuning my two audio systems to be exactly as I like. In my case I aimed for accuracy of reproduction rather than a pleasing coloration. I sought out gear and loudspeakers that are as flat as possible, and I added extensive acoustic treatment to minimize the contribution from the room. This is my preference, and whether others have the same goal is irrelevant—I'm totally pleased, and that's all that matters. You like what you like, and nobody can say you're wrong.

It seems to me that the more deeply one understands their hobby, the more satisfying the experience. Many auto enthusiasts can do a full engine tune-up or even rebuild a carburetor, and most photo buffs understand all the complex details and settings of their equipment. But these days few audiophiles or home recording engineers seem to care about the science of audio, or even the basics of music theory. Just as knowing how audio works increases our pleasure of owning the gear, knowing how music works increases the enjoyment of music. After all, enjoying music is the whole point of hi-fi! Even some audio dealers, who you'd expect to understand the products they sell, seem surprisingly clueless.

> *Q: What's the difference between an audio dealer and a used car salesman?*
> *A: The used car salesman knows he's lying to you.*

I contribute to many online audio forums, including those meant for audiophiles and home theater enthusiasts, as well as forums frequented by amateur and professional recording engineers. I'm often surprised by not only a lack of interest in how audio gear works, but also hostility expressed by some people towards science-based explanations. I'm curious to know what changed since the old days that caused audio and music enthusiasts to lose interest in the technical aspects of their hobby.

CHAPTER 23

Basic Electronics in 60 Minutes

All electrical circuits require two wires to operate. Where water flows through a pipe in one direction, electricity requires a return path back to the source. This is why batteries and light bulbs have two terminals. Interrupting either wire stops the flow of electricity, which is the basis for a switch. But what if you need to vary the brightness? This is where resistors come in. When inserted into a circuit, a resistor limits the flow of current. To continue the water analogy, a resistor is like a short length of narrow pipe that resists flow, as shown in Figure 23.1 There are also variable resistors—volume and tone controls—that work like a water valve or faucet.

Resistors are measured in ohms, for Georg Ohm (1789–1854), and the higher the value, the less current that will flow. There are several types of resistors. The most common type is made from a carbon compound similar to pencil lead, with several examples shown in Figure 23.2. *Wire wound* resistors are often used in high-power circuits, made from a length of thin wire that conducts less than copper, wound around an insulating form. Most modern resistors are made from metal or carbon film deposited on a ceramic base. *Surface mount* resistors (and capacitors) are becoming common in modern equipment; they look like little rectangular blocks with no wire leads at all and are soldered directly to a circuit board. We'll look at those more closely later on.

Ground

A *ground*, or *common connection*, is a way to minimize the number of wires used inside an electronic device. Even though all electrical circuits require two wires to pass electricity from one place to another, the same connection often goes to many different places. In most audio gear, the zero-volts ground wire coming from the main power supply is used as the common connection, and all other voltages are relative to that. In practice, most audio gear has two power supplies—one provides a positive voltage and the other a negative voltage—because audio circuits need to efficiently output signals above and below zero volts. The block diagram of a basic *bipolar* power supply is shown in Figure 23.3. All three connections—plus, minus, and ground—are sent to every transistor or integrated circuit (IC) that needs power.

Figure 23.1: A resistor is exactly analogous to a short length of narrow water pipe. Whereas a narrow pipe section restricts the flow of water, resistors restrict the flow of electric current.

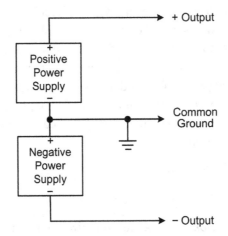

Figure 23.2: Carbon film resistors are a common building block of all electronic circuits. From top to bottom: 1/2 watt, 1/4 watt, and 1/8 watt, with a US dime as a size reference.

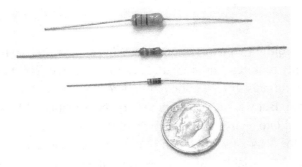

Figure 23.3: Most audio gear uses a bipolar power supply that outputs both positive and negative voltages. The common connection between them is considered *ground*, and is the zero volts reference for the entire device.

Using a common ground not only simplifies wiring, but it also makes schematic circuit diagrams easier to follow. Instead of having to draw every ground wire as a separate line back to the power supply, which adds clutter, a ground connection is shown as the three lines triangle symbol in Figure 23.3. Every wire that connects to this symbol is assumed to be connected to the device's metal chassis or to the power supply's ground. This was a fairly common construction practice in the early days of electronics when circuitry was simpler and most wiring was point-to-point rather than on circuit boards. In older tube equipment, you'll sometimes find the common terminal of jacks and volume controls connected directly to the chassis where they're mounted. With modern integrated circuit construction—particularly those having digital and analog signals in the same box—grounding that way is inadequate.

Many electronic devices are built inside some sort of metal box. This not only provides a solid mechanical structure and shields against AC hum and radio interference, but it also provides a single connecting point for any circuit that needs a ground connection from the power supply. Many circuits need that connection, so by first connecting the power supply's ground wire to the metal chassis, a short wire from each circuit to a nearby place on the chassis can be used. This method is called *chassis ground*, though it's a simplification because connecting common wires to different places along a large chassis can develop *ground loops* due to the small but finite resistance of the chassis.

Another grounding scheme uses a *bus bar*, which is a very thick wire having very low resistance. This makes sense for a large format mixing console where channel modules might be eight feet apart from one another. Another common arrangement is called *star grounding*, where separate ground wires from each subsection or circuit board are run to a common central point—usually the minus connection of the power supply, or the ground connection at the input jack. Ground problems are also avoided by using balanced wiring, which ignores ground voltage differences as shown in Figure 4.3 from Chapter 4.

Volts, Amps, Watts, and Ohms

Using the water analogy again, water pressure is equivalent to voltage, and water flow in gallons per minute is the same as current. When you turn off a bathroom faucet, the pressure is still there in the pipes. Likewise, when you unplug a lamp from the outlet on the wall, the voltage is still present at the socket, even if it isn't being consumed. Volts are named for Alessandro Volta (1745–1827), noted as V, so 10 V means 10 volts.

Current is measured in *amperes* and is named for André-Marie Ampère (1775–1836). Amperes is often shortened to *amps*, abbreviated A, though many audio circuits draw such small amounts of current they're instead rated in milliamps, or thousandths of an amp. So 10 amperes is usually written as 10 A, and 500 milliamps is 500 mA. The "m" is lower case because it *divides* the unit of amps, just as decibels are written dB because the "d" means one-tenth of a Bel. Likewise, 250 mV means 1/4 of a volt, and 1 KV means 1 kilovolt, or 1,000 volts. Here the K is capitalized because it *multiplies* the unit of volts to be larger.

Power is the amount of work that's actually done, expressed in watts for James Watt (1736–1819), and it's the product (multiplication) of voltage times current. A circuit that draws 6 amps when fed 10 volts uses 60 watts, because 6 times 10 equals 60. Likewise, if applying 6 pounds of pressure into a water pipe results in 10 gallons flowing after one minute, doubling the pressure will instead send 20 gallons down the same pipe in one minute.

It's also possible to calculate the resistance of a circuit, based on how much current it draws when a given voltage is applied. Indeed, volts, amps, ohms, and watts are directly related, in the same way as the equivalent parameters for water pressure, flow rate, and pipe size. Table 23.1 shows the relationship among these four basic electrical properties, known as *Ohm's Law*, and the Ohms Law.xls spreadsheet included with this book can calculate any one parameter from any two of the others. The superior "2" in I^2 means the value is squared, or multiplied by itself. So I^2 is the same as I times I, and $\sqrt{P*R}$ means you multiply P times R, then take the square root of the result.

If your eyes are already glazing over, don't despair. The preceding explanations and Ohm's Law formulas are included only for completeness. You don't need to remember or even understand any of them to grasp the basic concepts of electronic circuits that follow. Feel free to skip ahead at any point.

In formulas like this, the electrical letters used are not the same A for amps and W for watts as when writing their quantities. Table 23.2 shows the relationship between the unit values and their equivalent formula letters. Here, P is power, and the current symbol I is derived from the French word for *intensity*, because Ampere was French, and he got to claim the symbol. To minimize the number of zeros, common prefixes are used to indicate that a value is multiplied or divided. A small "m" means milli or one-thousandth; a small "μ" (the Greek letter mu) means micro or one-millionth; K means kilo or one thousand times larger; and M

Table 23.1: Ohm's Law Formulas

Calculate for Current (I)
$I = P/V$
$I = \sqrt{P/R}$
$I = V / R$
Calculate for Power (P)
$P = I^2 * R$
$P = V * I$
$P = V^2 / R$
Calculate for Resistance (R)
$R = P / I^2$
$R = V / I$
$R = V^2 / P$
Calculate for Voltage (V)
$V = I * R$
$V = P / I$
$V = \sqrt{P * R}$

Table 23.2: Unit Values and Symbols

Quantity	Formulas Symbol	Unit of Measurement	Abbreviations
Voltage	V	Volts	V, mV, μV, KV
Current	I	Amperes	A, mA, μA
Resistance	R	Ohms	Ω, K, M
Power	P	Watts	W, mW, μW, KW

means mega or one million times larger. So 3 μV is 3 millionths of a volt, and 2 KW is two thousand watts. The electrical symbol for ohms is the Greek letter omega, shown here as Ω.

I'm not going to belabor the math behind volts and amps, but there is one point that relates to audio that you should understand. When a given voltage is applied to a given resistance, some amount of current is drawn and converted to heat that's dissipated inside the resistor. This is a linear relation, so doubling the voltage applied to the resistor also doubles the current drawn. Since both the voltage and current are now twice as large, the power dissipated by the resistor is quadrupled:

$$\text{Power} = \text{Volts} * \text{Amps}$$

As explained in Chapter 1, this is why doubling an audio signal voltage results in a 6 dB increase, or four times more power, rather than only 3 dB, which represents a doubling of power.

Electronic Components

Capacitors, often called just *caps*, are more difficult to visualize than resistors because they pass alternating current (AC) such as audio, even though there's no physical connection through them. Direct current (DC) is similar to water comparisons because the current flows in only one direction, but AC constantly changes direction. When all else is equal, the larger a capacitor's value, the lower a frequency it will pass. However, capacitors do not pass DC, which is a frequency of zero.

Capacitance is measured in *farads*, for Michael Faraday (1791–1867), though you're unlikely to come across a 1 farad capacitor. You won't find 1 ohm resistors very often either. When Messrs. Ohm and Faraday discovered these properties long ago, they had little concept of the way their components would be used in practice. Most capacitors are measured in microfarads, nanofarads, or picofarads (millionths, billionths, and trillionths of a farad, respectively). Resistors of 10 or 100 ohms are not uncommon, but most of the ones in audio gear are measured in kilohms (one thousand ohms) and sometimes megohms (one million ohms).

A good mechanical analogy for a capacitor is a leaky bicycle pump, the type with a plunger inside a long hollow tube. Imagine the plunger is not well sealed because its gasket is worn.

If you push the plunger slowly, most of the air will leak back out. But if you increase the speed of the pumping, less air escapes compared to the amount traveling the intended path. And if you give the pump a single fast push, pressure develops, then quickly dissipates before much air gets to the tire. In a similar manner, capacitors allow current to flow, but only while the voltage is changing. Even though there's no physical connection through it, a light bulb attached to a battery through a capacitor will flash briefly when first connected. Therefore, a capacitor passes AC because it keeps changing level and polarity, while blocking DC. Further, as the frequency rises, more current can pass through until the capacitor's effective resistance is so low that higher frequencies pass fully. This frequency-selective property makes capacitors ideal for use in equalizer circuits.

A capacitor is constructed from two metal plates in close proximity but not quite touching, as shown in Figure 23.4. Obtaining a larger amount of capacitance requires more metal surface area, so either multiple plate sections are wired in parallel or one long pair of metal plates is used. Picture plastic food wrap sandwiched between two long strips of aluminum foil. A wire is attached to each strip of foil, and the whole thing is rolled up to save space, then dunked into goo that hardens. Besides passing only AC, capacitors can also store DC. In fact, a capacitor is much like a rechargeable battery. The larger the metal surfaces, or the closer they are together, the higher the capacitance. Figure 23.5 shows three common capacitor types, though there are many types including electrolytic, tantalum electrolytic, ceramic, polystyrene, Mylar, and mica, and each is best suited for specific applications.

There are also variable capacitors used to tune radio circuits. Electrolytic capacitors, as shown at the top of Figure 23.5, can pack a large amount of capacitance into a relatively small package. One trade-off is they are *polarized* and so must not have AC applied across them, though AC can pass *through* them. Electrolytic capacitors contain a conductive paste

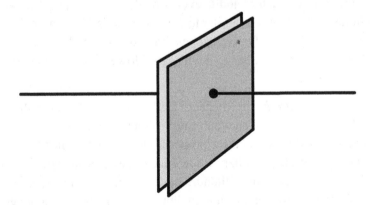

Figure 23.4: Capacitors are made from two parallel conducting surfaces in close proximity, with insulation (or an air gap) between them to prevent direct contact. Larger values are made from multiple parallel sections connected together, or long strips of foil and insulation that are rolled up to save space.

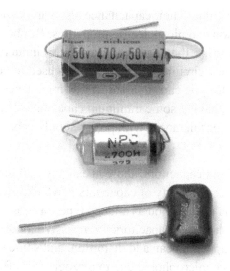

Figure 23.5: These capacitors are commonly used in audio and radio equipment. From top to bottom: electrolytic, polystyrene, mica.

similar to a gelled electrolyte lead-acid battery found in a car or UPS. They have some amount of series inductance due to their long metal strips being wound into a coil, which is sometimes a limitation. (Inductance is the opposite of capacitance and will be described shortly.) But electrolytic caps are perfect for use in power supplies, where large values are needed to filter the low 60 Hz AC mains frequency and the polarity of the applied voltage never reverses.

The middle capacitor in Figure 23.5 is polystyrene. It's also made from a rolled-up sandwich of long strips, so it has a higher inductance than types made with small parallel plates. Polystyrene and Mylar capacitors cost more than some other types, but they're very stable with temperature. They also have a low tolerance, which means their actual capacitance will be very close to the stated value. So they're a good choice in equalizers, where you need to know that a boost at 1 KHz is always 1 KHz regardless of the ambient temperature. Mylar and polystyrene caps also add very little distortion, making them ideal for audio circuits.

The bottom capacitor uses mica for insulation, which is highly stable with temperature. Mica caps are made from many parallel plate sections rather than rolled-up strips, so they have very little inductance. Mica caps are more common with radio circuits than audio because it's difficult to build them in the larger values needed for audio frequencies. They're also more expensive than other high quality types.

Disk ceramic capacitors are also made from many parallel sections, with ceramic between the plates for insulation. Ceramic caps are inexpensive, but they can differ as much as 10 or even

20 percent from their stated value. Their capacitance also varies with temperature, and they have relatively high distortion and piezoelectric properties. But being made from parallel plates rather than rolled-up strips, they have very low series inductance. Ceramic capacitors are common in high-frequency circuits, and they're often used in power supplies and audio circuits in parallel with electrolytic caps to assure a low impedance at high frequencies. If you need ceramic capacitors having some minimum capacitance, simply buy them 20 percent larger to account for their poor tolerance. With power supply capacitors, having more capacitance never hurts.

There's a lot of mystique about capacitors, and you can pay $50 for "audiophile" types to use in DIY projects. In truth, every property of capacitors is well known. Assuming you use the correct type, a 10-cent cap is just as good as a fancy type for most applications. However, accurate, low-distortion, high-power capacitors for a passive loudspeaker crossover might cost more like $5 or $10. Capacitors can add distortion, and some types add much more than others. Capacitors can also be microphonic due to piezoelectric effects, generating a voltage when jostled or exposed to loud sound waves. That's why it's important to choose the correct capacitor type for a given application.

One important property that affects a capacitor's distortion is its *voltage coefficient*, a measure of how much the capacitance changes as the applied voltage varies. Ideally, a capacitor will have the same capacitance no matter what voltage is present across its terminals. Another contributor to distortion is *dielectric absorption*, which is a "memory" effect. After discharging the capacitor fully, some of the voltage reappears soon after because some of the charge was stored within the capacitor's *dielectric*—another word for the insulation between the metal plates.

Capacitor Upgrades

There are companies that offer to upgrade your audio gear by replacing capacitors and various other components. Some types of upgrades and modifications are legitimate, but many are not. Unless your capacitors are defective and allow DC to pass through, or their value has changed significantly over time due to environmental or other factors, you're not likely to improve fidelity by replacing them. Electrolytic caps in a power supply can dry out and lose capacitance after 20 or 30 years, which increases hum. It's also true that different capacitor types are more or less suitable for various circuits, and some better types are available today than when the equipment was originally designed. So all component upgrades are not a scam. But if you think the designers of your amplifier or mixer are too stupid to have used appropriate components in the first place, why would the rest of the design be good enough to warrant the cost of "better" parts?

In fairness, some *extremely* old audio gear contains carbon composition resistors, and replacing them can yield lower noise in some circuits. But anything manufactured since 1970 will contain competent carbon film resistors. If a mixer or mic preamp is already audibly transparent, and its distortion is barely measurable, with a frequency response that's flat from DC to light, how can it possibly be made better? Again, some upgrades are legitimate and really do improve fidelity, often by replacing older integrated circuits with newer models having a wider bandwidth, less noise, and lower distortion. If a vendor promises to improve the performance of your gear with a component upgrade, I suggest asking for proof in the form of specific Before and After test data.

Inductors

The opposite of a capacitor is an inductor. This is a coil of wire, often wound on an iron or ferrite core to increase its inductance. DC passes through an inductor because it's made of wire, but its resistance to AC increases as the frequency rises. When DC is applied to an uncharged capacitor, it acts like a short circuit. The initial current flow is large, and it decreases as the capacitor charges. An inductor does the opposite: When voltage is first applied, it behaves like an open circuit. The initial current is low, then rises until it's limited by the resistance of its wire. Since AC voltage keeps changing, at some high enough frequency, current no longer passes through.

The amount of iron used for the core also has an effect, with more iron increasing the inductance. Inductors are measured in henries, named for Joseph Henry (1797–1878), an American scientist who discovered electromagnetic induction. Most of the inductors you're likely to encounter have values in millihenries or even microhenries, though you'll find larger values in low-frequency circuits such as power supplies and crossover networks. The schematic labels for resistors and capacitors are R and C, respectively, which makes sense, but the symbol for inductors is L. Go figure.

Figure 23.6 shows a variable inductor used for tuning radio circuits. The threaded rod connects to a ferrite core, which you can see just poking out the other end at the top left. As the screw is turned, more or less of the core is within the coil area, which varies the inductance. When I was growing up in the 1950s and 1960s, the pushbutton presets on car radios tuned different stations by mechanically moving the ferrite inside an inductor much like the one shown here. Ferrite is a common material for inductors, transformers, and even loudspeaker magnets. It's made from fine powdered iron, with the particles bonded together by glue to create solid pieces in various shapes and sizes. Inductors are not always used where they could be for many reasons, including the high cost of copper and iron, their susceptibility to hum pickup, and physical size. Inductors also distort at high signal

Figure 23.6: Inductors are made from a coil of wire, often with an iron or ferrite core. This one is meant for tuning radio circuits rather than for use at audio frequencies.

levels when their iron core saturates, in much the same way analog tape distorts when it can no longer accept further magnetization. Air-core inductors have no distortion, but iron is needed to get usably large values for low audio frequencies. Many equalizers use a circuit called a *gyrator* made from an operational amplifier and a capacitor to simulate an inductor, though real inductors are required in crossovers for passive loudspeakers. The simplest type of crossover adds an inductor in series with the woofer and a capacitor in series with the tweeter, as shown in Figure 18.10 from Chapter 18. The main point is that inductors and capacitors are both frequency sensitive, but in opposite ways. This frequency-dependent behavior is also called *reactance*.

Power Ratings

As you've seen, resistors are measured in ohms, capacitors in farads, and inductors in henries. But these components have another rating, too, which states how much voltage or current they can handle before blowing up or melting. If you've ever watched a Frankenstein movie, you know that when a voltage exceeds a certain level, it will jump through the air to complete the circuit. This is how lightning travels from the sky to the earth. The same thing happens with a capacitor when its voltage rating is exceeded. This occurs on a much smaller scale, but the principle is the same. Even though the metal plates of a capacitor are separated by an insulator, applying too much voltage punctures the insulation as the voltage jumps through, which destroys the capacitor. Therefore, besides a capacitance value, capacitors also have a maximum voltage rating.

Resistors and inductors can also be damaged, though in this case by applying too much current. Just as trying to force too much water through a pipe will cause it to rupture, the same happens to resistors and inductors. Even plain connecting wire must be chosen carefully when used for anything other than low-level signals. For a given wire gauge, there's a maximum amount of current it can pass without overheating. This is the basis for a fuse; many are just a short length of very thin wire. Wire insulation also has a maximum voltage rating above which it becomes punctured as with capacitors. At very high radio frequencies, wire also needs to be oversized due to *skin effect*. This is a propensity for current to flow on only the outer surface of the wire rather than through its entire cross section. When a high-powered FM transmitter is connected to its antenna, a hollow copper pipe is used instead of thick wire because a solid conductor just wastes expensive copper. There are formulas that tell how much current and voltage are allowable for a given circuit at a given frequency, though we won't bother with that. The concepts are what matter here.

Solenoids

A solenoid is an inductor with an iron or steel center plunger that's attached to a lever or other mechanical device to be moved. A common audio application is in tape transports, where a solenoid pulls the pinch roller tightly against the capstan while the tape is playing, and a spring pulls it away when at rest or while rewinding. Applying electricity to the coil creates magnetism, which pulls the plunger into the center of the coil. Figure 23.7 shows a flipper solenoid from one of my pinball machines, and my finger is pressing the lever to push the plunger upward most of the way into the coil. You can just see the thin wire spring wrapped around the plunger. This solenoid is underneath the wooden play field, and a round

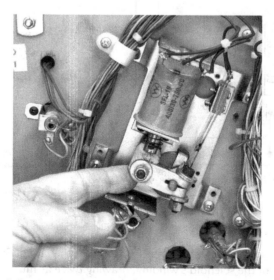

Figure 23.7: This pinball flipper solenoid is activated when a button on the side of the cabinet is pressed. When the button is released, a spring pulls the plunger back out.

pin about an inch to the right of my finger goes through the play field, which rotates to swing the flipper when the solenoid engages.

Note the *End of Stroke* switch to the right of the solenoid that opens once the plunger is fully inside the solenoid's coil. This switch is normally closed and is wired to short out half of the solenoid's windings. This causes the solenoid to draw twice as much current when the flipper button is first pressed, giving the ball an extra kick. The result is the same as applying twice as much voltage to the entire coil. But once the plunger is fully inside the solenoid, a lever capped with insulating plastic opens the switch, reducing the amount of current drawn. This clever arrangement prevents burning out the coil as the player continues to hold in the flipper button. I realize this has nothing to do with audio, but it's an interesting concept.

A switch attached to a solenoid is not unlike the switches connected to audio jacks, as shown in the earphone jack of Figure 4.9 from Chapter 4. The pinball solenoid's switch wiring is shown in Figure 23.8, and notice the diode wired across the solenoid's coil.

When voltage is applied to an inductor and then removed, a high voltage called *inductive kickback* develops briefly at the inductor's terminals. This diode shorts out that voltage to

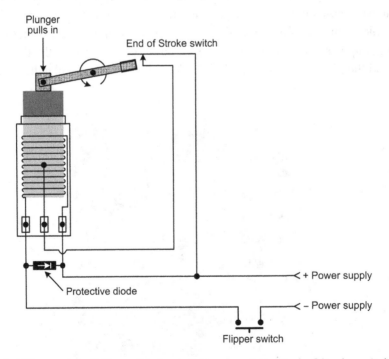

Figure 23.8: When a pinball flipper button is first pressed, the End of Stroke switch is closed. This shorts out half of the solenoid's winding, drawing twice as much current through the remaining half. Once the plunger is fully inside the solenoid, the lever opens the switch contact, drawing less current while the button stays held.

prevent damaging the transistor that drives the solenoid. For a fun experiment, touch a 9-volt battery to a solenoid and then remove it. If your fingers stay on the solenoid's terminals, you'll get a painful shock!

Transformers

A transformer is an extension of an inductor, using two insulated coils of wire—called a *primary* and *secondary*—wound around a common iron core. When an AC voltage is applied to one of the coils, a corresponding voltage is generated in the other coil through magnetic induction. If both coils have the same number of turns, the output voltage is the same as the input voltage. Although this might not seem useful by itself, having two coils that are not electrically connected can avoid hum due to a ground loop between two audio devices. Even when a circuit could operate on 115 volts directly, a transformer is often used just for safety. If a device's power supply connects directly to the AC mains voltage and a wire falls off inside and touches the metal chassis, that could expose users to a lethal shock.

Both audio and power transformers can also vary the ratio between voltage and available current, keeping the total power the same. For example, a 100-watt incandescent bulb in your home has a resistance of 144 ohms and draws 0.83 amp at 120 volts. But the same 100 watts in a car's headlight has a resistance of 14.4 ohms, which draws 8.3 amps at 12 volts. The total power is 100 watts but the ratio between volts and amperes changes. This is just like the transmission in a motor vehicle. In a transmission, different gear ratios vary the relation between speed and torque. For a given amount of horsepower you can either go slowly but with the ability to climb a steep hill, or go much faster as long as you stick to level terrain. The total available horsepower doesn't change between first and fourth gear, only the way that power is used. Nothing comes for free, and a transmission merely changes how the horsepower is distributed. A seesaw or lever is also a transformer, exchanging distance traveled for weight lifting ability. Likewise for a block and tackle that lets one person lift a heavy engine out of an automobile by pulling many more linear feet of rope than the height the engine is lifted.

Another analogy is levers and springs. The *input impedance* of a circuit determines how much of a load it presents to the device driving it. If we use the analogy of levers and springs, a high impedance load is like a loose spring that's easily compressed, where a low impedance load is like a stiff spring that requires more effort to move. The graphic in Figure 23.9 could be shown as a simple downward force on the spring, but I added a lever to show how a transformer affects the ratio between distance and torque, which are the same as voltage and current respectively. Moving the pivot point along the length of the lever varies the ratio of input to output impedance.

With the pivot as shown, nearer to the spring, the driving "amplifier" needs to move a farther distance (more voltage) in order to compress the spring (loudspeaker) fully. But very little

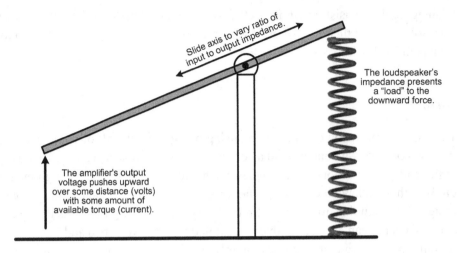

Slide axis to vary ratio of input to output impedance.

The loudspeaker's impedance presents a "load" to the downward force.

The amplifier's output voltage pushes upward over some distance (volts) with some amount of available torque (current).

Figure 23.9: Electrical volts, amps, and impedance have exact mechanical analogies in springs and levers.

force (current) is required to raise the lever, so the lever presents a high impedance load to the amplifier. Sliding the pivot to the left, farther from the spring, requires the amplifier's output to travel a shorter distance, but then more force is required to compress the spring. So the lever's "input" becomes a lower impedance load.

In the same manner, a transformer converts a voltage applied to one of its coils to produce a higher or lower voltage in the other coil. The key is the ratio between the number of windings in each coil. If one of the coils consists of 200 turns of wire, and the other has only 100, the ratio between them is two to one. So if you put 10 volts into the larger coil, only 5 volts will appear in the smaller one, though the available current will be twice as large. Further, if a load draws 2 amps from the secondary, only 1 amp passes through the primary. The total amount of power stays the same; only the relation between volts and amps changes.

Figure 23.10 shows a small audio transformer built into a round case to convert a low-impedance balanced microphone to high-impedance unbalanced—for example, to plug the mic into a guitar amplifier. The input and output voltage ratio is the same as the ratio between the number of turns in the primary and secondary. But the output impedance varies by the square of the ratio. This particular transformer has a *turns ratio* of 14 to 1, so its output voltage is 14 times higher than the input, but its output impedance rises by a factor of 14 * 14 = 196. A 150-ohm microphone connected to the primary appears as a 29.4 K source to the amplifier, and when the mic outputs 1 mV, the amplifier receives 14 mV.

Note that a transformer doesn't have an inherent impedance, even though you may find some labeled with specific values. What matters most is the turns ratio, though the amount

Figure 23.10: This transformer has a turns ratio of 14 to 1, which converts a 150-ohm microphone to an output impedance of about 30 K.

of current a given transformer can handle is determined by the coil's wire gauge, and the maximum voltage depends on the type and thickness of the insulation.

The low-frequency response of a transformer depends on the amount of iron in its core. The more massive the core, the lower in frequency the transformer can operate to. Having more iron also lets a transformer handle larger signals, because the core can accept more magnetization before it saturates, causing distortion. However, a solid slab of iron is less efficient electrically than the same mass made from smaller sections. This is mitigated by building the core from a series of thin iron plates, each individually insulated from each other with a coating of varnish. When a transformer must operate at even higher frequencies, such as radio, the core is made from ferrite. As explained previously, ferrite is made from fine iron particles that are held together with glue. So the whole of the iron mass is made up of even smaller pieces. Other materials such as nickel are also used for transformer cores. Each design and material has its good and bad points, and like any other component that comes in different varieties, there are better choices for certain applications.

A transformer can also have a single winding, in which case it's called an *autotransformer* or simply *autoformer*. While such a transformer will not isolate the primary and secondary, it can raise or lower an AC voltage efficiently. One popular type of autoformer is the Variac, used to dim lights in recording studios and other places where the electrical buzz created by solid-state light dimmers is unacceptable. In truth, Variac is a brand name, but like Kleenex and Band-Aid it has become a generic term over time. It's not practical to use variable resistors to dim light bulbs because they'd need to be enormous, would waste a lot of energy, and would run very hot.

Solid-state dimmers work by changing the amount of time the full power line voltage is applied to the light being controlled, as opposed to actually raising and lowering the voltage. AC power is supplied by the power company as a pure sine wave having no harmonics, but the dimmer's sudden switching of the full voltage generates harmonics that extend to very high frequencies. (The concept of a fast *rise time* creating high-frequency harmonics was introduced in Chapter 1, Figure 1.27.) These harmonics can get into audio gear both by radiation through the air and by traveling through the power wires directly into the audio circuitry.

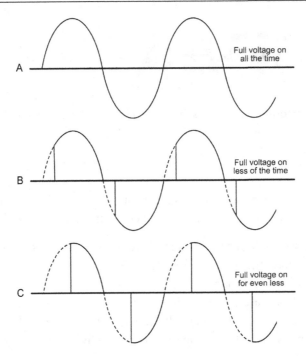

Figure 23.11: A solid-state dimmer varies a light bulb's brightness by passing the full voltage for only part of the time. The brightness knob on a solid-state dimmer determines at what point in each cycle the output voltage switches on. The steep turn-on slopes in B and C generate high-frequency harmonics that get into audio gear through the air or through the power wiring.

Figure 23.11 shows how a solid-state dimmer operates by varying the power source's duty cycle.

If you live in an apartment house, you may not be able to eliminate dimmer buzz from other apartments. But if the buzz goes away when you turn off your own dimmed lights—turned off completely, not just dimmed all the way down—then replacing those solid-state dimmers with variable transformers will solve the problem. By the way, solid-state dimmers are also called SCR dimmers. SCR stands for *silicon controlled rectifier*, and it's the solid-state device that does the actual switching. I installed seven Variac dimmers in my house, using a variety of transformer sizes based on the wattage each must control. Each transformer is mounted in a plastic electrical box inside the wall, wired as shown in Figure 23.12. I also added a standard light switch to turn the lights on and off, without having to vary the dimmer level. This was done not only for convenience but also to avoid unnecessary wear on the transformer's sliding brush contact.

Besides letting you dim your lights without getting buzz in your audio system, Variacs are useful when troubleshooting electronic gear, such as a guitar amp that keeps blowing its fuse.

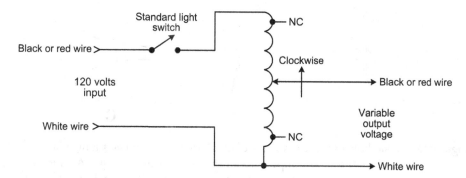

Figure 23.12: This shows how to wire a variable transformer to replace a light switch or existing SCR dimmer. Some Variacs include taps for overall voltage boost or cut, which you'll leave disconnected. These are labeled NC for No Connection.

If you raise the voltage slowly, rather than slam it with 120 volts all at once, you can see which component smokes or crackles or smells burnt. Variac dimmers can also extend the life of your light bulbs. When I was a teenager, the engineers at the local 50 KW FM transmitter let me hang around and pester them with questions. As you can imagine, the output tubes for a high-powered radio transmitter are very expensive, so transmitters use a Variac to drive the tube filaments. Each morning when the station went on the air, the engineers would raise the voltage manually over several minutes, helping the tubes last longer. Likewise, the track lights in my home studio are on Variac dimmers, which are usually kept around 70 to 80 percent brightness. In 25 years only one standard 75 watt bulb has needed to be replaced.

Acoustic Impedance

In electrical circuits "driving a low impedance" means the circuit has to deliver more current, which requires thicker wires and a more substantial power supply. For example, driving a 2 ohm loudspeaker load with 10 volts requires 5 amperes of current, versus a 10K line input that draws only 1,1000th of an amp. A low impedance is similar to a short circuit, where a high impedance is like an open circuit. If you connect a solid wire to the plus and minus terminals of a battery it will quickly overheat, and possibly melt or even explode! But with acoustics, driving a high or low impedance considers the impedance *in series* with the acoustic waves being emitted, rather than as a shunt that absorbs energy as with electrical circuits. So in this case a low acoustic impedance means sound waves can pass easily through an elastic medium such as air, versus trying to force the waves through a dense material like tightly woven fabric. The fabric presents a high impedance because it literally impedes the flow of sound waves passing through it, and that in turn increases the pressure.

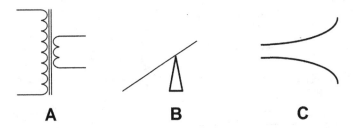

A   **B**   **C**

Figure 23.13: All three of these devices are transformers that convert a high impedance at their input on the left to a low impedance at their output on the right. A: This electrical transformer reduces the voltage but can provide more current. B: A mechanical transformer such as this lever exchanges greater vertical displacement (voltage) for more available downward force (current). C: An acoustic transformer varies the ratio between input pressure (voltage) and output wave velocity (current) such that a large amount of pressure over a small area becomes less pressure that covers a larger area.

As explained earlier, an electrical transformer is similar to mechanical equivalents such as a lever, a block and tackle, or the gears in a transmission. Figure 23.13 extends the analogy to include a flared loudspeaker horn. Here, voltage is like pressure, and current is like wave velocity (or cubic feet of air per minute if considering pure air flow). So more pressure is present at the point where the speaker driver forces sound waves into the narrow end of the horn, but there's less pressure and more wave velocity at the larger output. This is why a loudspeaker horn is a transformer. Since the output covers a larger area, the sound waves couple better to the air. This results in a better *impedance match* between the small diameter speaker driver and the larger air surface, which in turn improves the transfer of energy. Note that there's less wave *displacement* at the output, but the total energy is the same because it comprises a larger area. Not coincidentally, the larger surface area also extends the driver's low-frequency response, just as a large speaker cone can output low frequencies more efficiently than a small cone.

Finally, like all transformers, an acoustic transformer can be operated in reverse to convert a low impedance to high. Years ago, before electronic hearing aids were available, people who were hearing impaired used a device called an *ear trumpet*. This is basically a horn operated in reverse that funnels sound from a large diameter opening down to a small tube that's inserted into the ear canal.

Switches

Talking about something as mundane as a switch might seem like a letdown after more sophisticated components like capacitors and transformers. But switches are an important part of every electronic device, and there's more to some types than meets the eye. A *rotary encoder* is used on some audio gear and synthesizers. Unlike a conventional rotary switch that might select fixed EQ frequencies on a mixing console, a rotary encoder rotates continuously, outputting a series of pulses that are interpreted by digital logic circuits. This lets a single switch select from dozens of functions, without regard to its absolute position. Whatever is currently selected, turning the switch selects the next or previous choice.

Another important concept is *switch contact bounce*. When you press a typical push-button switch, it doesn't just suddenly make a stable connection. Rather, the contacts touch, then separate, then touch again, several times, over a period of a few milliseconds. For a switch that turns on a light bulb or engages a low-cut filter, switch bounce is irrelevant. But many switches, like the rotary encoder just mentioned, are interpreted by digital logic circuits. The same happens with the keys on a computer keyboard and the flat surface switches on a microwave oven. Without additional analog delay circuitry or equivalent digital logic, every time you pressed 2 on your telephone, you'd get 22222.

Diodes

The last passive component I'll address is the *diode*, also called a *rectifier*, which allows current to pass in only one direction. Diodes are solid-state devices, and they're used in power supplies to convert the AC mains to the DC needed to power audio circuits. The earliest solid-state diodes were made from a piece of wire just barely touching a semiconductive mineral such as germanium. When I was growing up in the 1950s and 1960s, building passive crystal radios was a popular hobby. A 50-foot-long antenna wire connected to a sensitive piezo earphone through a germanium diode can receive a strong local radio station loudly enough to hear clearly. Most modern diodes are made from silicon, and Figure 23.14 shows four different types.

The schematic symbol for a diode is an arrow pointing toward a line, as shown earlier in the solenoid in Figure 23.8. A diode has a *cathode* and an *anode*; the line identifies the cathode. The arrow shows in which direction current can flow, with the opposite direction blocked. However, current flow in the favored direction is not perfect; the output side of a silicon diode is about 0.6 volts lower than the input, with the voltage loss dissipated as heat inside the diode. Germanium diodes drop only 0.2 volts, making them useful for some applications, though they generally have lower current capacity.

One of the most common uses for diodes is in power supplies. All electronic circuits require a DC power source to operate, or otherwise the frequency of the AC power would modulate the audio output. For devices that connect to a wall outlet, the AC mains must be converted to a lower DC voltage the device can use. A transformer can change the input voltage to whatever is required, but these circuits need DC. This is where the diode comes in.

Figure 23.14: These diodes are meant for different purposes. From left to right: a signal diode for small voltages and currents, a 1-amp diode for small power supplies, a 3-amp diode for larger supplies, and a 35-amp bridge rectifier in a metal case that can be bolted to a heat sink.

Three basic power supply configurations are shown in Figure 23.15. One uses a single diode to block negative cycles of the incoming AC and is called *half-wave* because it passes only half of the wave. The other two are more complex but much more efficient, using four diodes arranged as a *full-wave bridge* to pass both half-cycles in the correct polarity. The large square bridge rectifier in Figure 23.14 contains four diodes connected this way internally. The third bipolar power supply requires a power transformer with a *center-tap*, which serves as the ground connection.

All three circuits pass the incoming AC mains voltage through a power transformer to create whatever output voltage is required. The top circuit uses a single diode to pass only the positive AC half-cycles and block the negative portions. A large *filter capacitor*—sometimes called a *reservoir capacitor*—is charged up by the half-cycle pulses, holding the peak DC voltage long enough to fill in the gaps between each replenishing pulse. This type of power supply is simple but inefficient. The lower full-wave circuits are more efficient and can use smaller filter capacitors because they're replenished with voltage pulses twice as often. Understand that all of these power supply circuits are very basic, though the full-wave bipolar version is useable as shown for driving an audio power amp.

A *Zener diode* is used in the reverse direction because its inherent voltage drop is much larger that way. When current flows through a silicon diode in the usual forward direction, its output is about 0.6 volts lower than its input. (The voltage increases slightly when more current passes through.) In the reverse direction current doesn't flow at all until the voltage reaches a much larger value called the *breakdown voltage*. This is the maximum voltage the diode can withstand, so the diode would normally be used in circuits where the operating voltage is

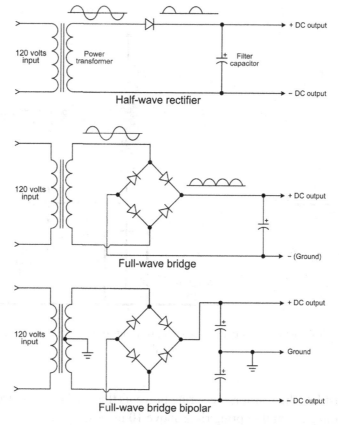

Figure 23.15: Power supplies are among the simplest of electronic circuits. The top supply is half-wave, and the lower two are full-wave. These basic power supplies lack output voltage regulation, but they're fully functional as shown.

lower. In other words, a diode allows current to pass once the signal reaches 0.6 volts in one direction, but it never conducts in the other direction until the breakdown voltage is reached. However, this reverse voltage drop can be exploited in power supplies as a simple voltage regulator.

In the circuit at the left side of Figure 23.16, as long as the input voltage is at least 10 volts, the output will always stay at 10 volts no matter how much the input voltage varies. In other words, the Zener diode "clamps" the output at a constant 10 volts. The series resistor is needed to absorb the difference between the input voltage and the fixed 10 volts output. Without this resistor, the Zener diode would draw too much current and burn out. So in a practical application you'd provide at least 12–15 volts DC input to the resistor, to ensure a steady 10 volts output even if the AC mains dips a little. A mechanical equivalent is shown at right using a spring as the resistor, with a plate whose height inside a tube casing is the same

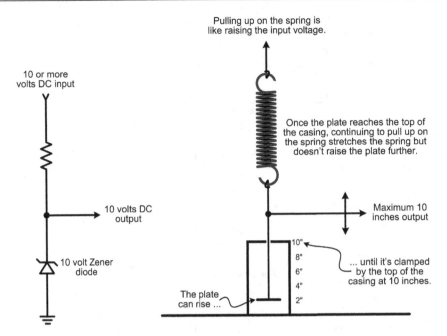

Figure 23.16: A Zener diode is used in reverse, to limit an incoming voltage to a specific output value.

as voltage. And just like the resistor, the spring prevents the mechanical pieces from damage when the input pulling up on the spring rises above 10 inches.

Parasitic Elements

All electronic components have *parasitic* properties, which means they don't possess only pure resistance, capacitance, or inductance, but also include small amounts of other elements. For example, a foil capacitor has some amount of series inductance due to the coiled arrangement of its plates. The connecting wires also add a small amount of series resistance. The wire coil of an inductor contributes series resistance, and it also has some parallel capacitance due to the proximity of the coil windings to each other. Most parasitic properties are not important at audio frequencies, but some are of concern. For example, many audio circuits have a frequency response that extends to 50 KHz or higher, so the power supply must present a low output impedance at those frequencies to prevent the amplifier from becoming unstable and oscillating.

This is where the lowly, inexpensive disk ceramic capacitor shines. These are commonly used as a *bypass* capacitor—wired in parallel with a larger, more inductive capacitor, to maintain a low impedance (short circuit) at higher frequencies. As mentioned, an electrolytic cap is equivalent to a large capacitance in series with some amount of inductance. This series inductance comes into play at higher frequencies, partially canceling the capacitance and increasing the capacitor's impedance. By putting a disk ceramic cap in parallel with an electrolytic cap, the disk ceramic takes over at high frequencies, maintaining a low impedance.

Figure 23.17 shows the true behavior of an electrolytic capacitor, or other type made from long strips of metal. The two metal foil plates are shown as long strips before being rolled into a tube. Note the small but real series resistance also present and the large but real parallel resistance, called *leakage resistance*.

Disk ceramic bypass caps are often installed on circuit boards very close to integrated circuits at their power supply connection. If a high-gain, high-frequency circuit is two feet from the power supply, even the small amount of inductance in the connecting wires can raise the power supply's effective output impedance. Placing a small disk cap right at the IC itself keeps the power supply impedance very low, even at radio frequencies, preventing oscillation.

Other parasitic effects are a resistor or capacitor changing value slightly as different voltages are applied and small amounts of resistance effectively in parallel with capacitors, which

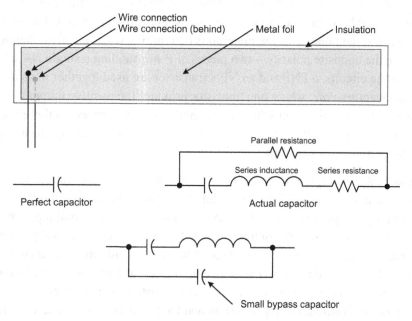

Figure 23.17: When a capacitor is made from long metal strips rolled into a tube, some amount of inductance is added in series. This reduces the capacitor's effectiveness at high frequencies. By placing a low-inductance bypass cap in parallel, the larger capacitor's low impedance is extended to higher frequencies.

discharges them slowly over time, even when nothing else is connected. Ideally, all parasitic properties will be at last two orders of magnitude (1/100) below the desired properties, though even smaller amounts are important in audio circuits where parasitic elements would add distortion.

Active Solid-State Devices

Finally we get to active solid-state components. I won't include tubes because that would require an entire chapter, and frankly I'd rather stick with technology that's more relevant in the twenty-first century. Active circuits are used wherever amplification is required. Even though a transformer might be considered to amplify, it doesn't really; as the output voltage is increased, the available current decreases. Many circuits, such as power amplifiers, amplify both the voltage and available current. Some circuits increase the available current while outputting the same voltage. That type of circuit is called a *buffer*, and it's commonly added to a line-level output stage to allow driving long lengths of wire without losing high frequencies due to wire capacitance. A buffer can also beef up a line-level output allowing it to drive headphones.

Transistors are made of the same semiconducting materials used in diodes, but they're more complex internally. Most transistors are constructed in layers, with a piece of "P" (positive) material sandwiched between two other pieces of "N" (negative) material. This is called an NPN transistor, and the polarity of the material is determined when it's manufactured. Other transistors have the opposite polarity—two pieces of P surrounding a single N—called a PNP transistor. In some circuits, a PNP and an NPN transistor are used together in an arrangement called a *complementary pair*, where one transistor provides the positive output voltage and the other handles the negative. But what's important here is the properties these materials possess. Figure 23.18 shows two typical transistor types, one for small signals and one that can handle larger amounts of current.

Continuing the water analogy, a transistor (or tube) is similar to a water faucet. In fact, the British call vacuum tubes "valves." Just as a child can control hundreds of pounds of water pressure by turning a large-diameter valve handle, a transistor uses a small input current to control the flow of a much larger output current. This is a key concept in audio circuits: An amplifier doesn't actually amplify a voltage. Rather, the small input signal controls an electrical valve that modulates a larger voltage coming from the power supply. One important transistor parameter is its *current gain* or *beta (β)*, often listed on spec sheets as h_{FE}. This is the ratio between the small amount of input current and the larger output current it can modulate. Other specs state how much voltage and current the transistor can handle without burning out.

Standard transistors have three terminals called the *base*, *collector*, and *emitter*. When using an NPN transistor, the positive power supply voltage being controlled attaches to the

Figure 23.18: The transistor on the left is a small-signal type, and the power transistor on the right handles much more current. The thick metal tab can attach to a heat sink to accommodate even higher power levels.

collector, and the input control current goes to the base. This lets a small amount of base input current control a much larger output current through the collector, which results in amplification. Figure 23.19 shows two simple but functional transistor amplifier circuits. One provides voltage gain, and the other—a buffer—has unity voltage gain, but provides additional output current for driving a low impedance load. This type of buffer is called an *emitter follower* because the output is taken from the transistor's emitter, whose voltage follows (tracks at unity gain) the input voltage at the base.

The gain of the circuit at the left is determined by the ratio of the collector and emitter resistors. The exact values used depend on many factors, including the input and output voltage and current the circuit is expected to handle. These simple circuits use only a single power supply, such as a 9-volt battery, rather than a bipolar supply that outputs both positive and negative voltages. Therefore, the input at the transistor's base must be *biased* to a DC voltage partway between ground and the power supply voltage to allow equal positive and negative output voltages. This is the purpose of the two resistors that join at the base. The input capacitor isolates the DC bias voltage, preventing it from passing back to whatever connects to the input. The additional resistor directly across the input avoids a loud pop sound when something is plugged into the circuit. Recall that connecting a light bulb to a battery through a capacitor causes the bulb to flash briefly when first connected. Likewise, the output capacitor and resistor to ground block DC and audible pops from the output when the circuit is connected to the next device in the chain.

Both of these circuits are considered *Class A* because biasing the transistor's base to halfway between ground and the power supply voltage draws DC current even when no input signal is present. *Class B* and *Class AB* amplifiers use an NPN and PNP transistor in a balanced output

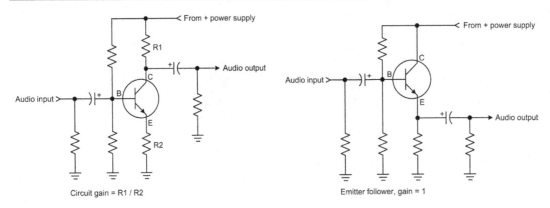

Figure 23.19: The gain of the transistor circuit at left is determined by the ratio of the collector and emitter resistors, and its output polarity is opposite the input. As the input voltage goes up, the resistance between collector and emitter is reduced, pulling the collector down to a lower voltage. The circuit at the right is an emitter follower. It has unity voltage gain with the same polarity as the input, but it can output a greater amount of current.

configuration, where each handles only the positive or negative portion of the output. When no audio is present, both are at rest, outputting zero volts and drawing little idle current. Understand that these amplifier classifications are not related to audio quality; this is simply the historical order in which the circuits were developed.

Amplifier Damping

Amplifier damping is a measure of how well a power amplifier can stop the continued motion of a loudspeaker cone after the source signal ceases. We normally think of damping as mechanical, such as gluing thick rubber mats on the floor or side panels of a car to damp body resonance. But damping can also be implemented electrically. A dynamic loudspeaker acts as both an electric motor and a generator. When voltage is applied, the coil becomes magnetized and pushes the cone forward or backward. But speakers can also act as microphones, whereby moving the coil in and out via sound waves or by pushing the cone with your hand generates a voltage called *back EMF* that's sent out the speaker driver's input terminals. Such motion also occurs after sound being played through the speaker stops; the cone and voice coil can continue to vibrate at their natural resonant frequency, which also generates an output voltage.

A well-designed power amplifier will have a very low output impedance, which helps to damp that vibration by presenting a short circuit to the speaker's output voltage. Imagine you have a small hand-cranked generator, the type used for classroom science experiments or portable radios. If nothing is connected to the generator, the handle turns easily. But when a load such as a light bulb is connected, more effort is needed to turn the crank. The greater the load on the generator, the harder it is to turn the crank. The same thing happens when a speaker driver is connected to a power amplifier having a low output impedance. The speaker's motion is damped electronically, simply because the amplifier loads the speaker while the speaker is acting as a generator.

A modern solid-state amplifier has a very low output impedance, typically a small fraction of one ohm. So if an amplifier's output impedance is 0.1 ohm and you connect it to an 8-ohm speaker, the amplifier has a *damping factor* of 8 divided by 0.1 = 80. It's common for power amp manufacturers to claim extremely high damping factors, but there's a practical limit above which higher amounts offer no further improvement. One reason is the resistance of the wire connecting the amplifier and the speaker is in series with that low output impedance. Even more important is the resistance of the speaker's own voice coil, which is in series with both. So in practice, once a power amplifier's damping factor hits about 50, higher values are swamped out by the voice coil's own resistance. Competent solid-state amplifiers easily achieve damping factors greater than 50, though tube-based amplifiers are typically much lower.

Negative Feedback

One shortcoming of transistors is that their gain is not linear. A transistor might have a gain of 200 at small signal levels, but half that when controlling large currents. Understand that an audio signal contains both large and small voltages, so if this nonlinear behavior isn't accounted for when amplifying audio, the result is distortion. The underlying reason is complex, but the solution is simple: negative feedback. Some examples of negative feedback are found inside an oven and in your toilet tank. If you set the oven thermostat to 400 degrees, the heating coil turns on fully until the desired temperature is reached, then the thermostat turns off the coil. As the oven cools, the thermostat energizes the coil again, thereby keeping the temperature constant.

Sophisticated industrial ovens use *proportional controllers* rather than simply turning the coil fully on and off. As the temperature drops, the coil voltage is increased slightly, and as it rises, the voltage decreases. It's the same for a toilet, which strives to keep the tank full. As the tank fills, the float inside rises. Since the float is attached to the valve that fills the tank, a *closed loop* is formed. The same happens with a car's cruise control, to maintain a constant highway speed. But the point here is that it's possible for a circuit to control itself by sampling a portion of its own output to ensure that it closely matches the signal at the input.

Negative feedback is the basis for a *regulated power supply* to output the same DC voltage even when the AC mains voltage varies or the connected circuits draw more or less current.

Audio amplifiers use negative feedback in the same way to ensure that the output voltage is exactly proportional to the input, thus reducing distortion. A portion of the amplifier's output is sent back to its input but with the polarity reversed. If distortion or a frequency response error inside the amplifier causes it to output more or less voltage than it should for a given input, the signal fed back to the input automatically corrects the output error. Negative feedback also increases an amplifier's damping factor. When back EMF from a speaker driver changes the voltage at the amplifier's output terminals, negative feedback immediately counters that forcing the output voltage to the correct value.

Hysteresis is a related concept that uses a controlled amount of *positive* feedback to preserve a system's state, and it's a fancy term that means "don't turn off the furnace until you get slightly *above* the temperature I set." In other words, hysteresis resists change, requiring a condition to be exceeded before the change finally occurs. Without hysteresis, your furnace or air conditioner would cycle on and off every few seconds. In most thermostats, hysteresis is implemented with a wound bi-metal spring and weak magnet. The spring tightens or unwraps as the temperature changes, and when it has opened or closed sufficiently, it touches an electrical contact to turn on the furnace. But the magnet keeps the spring from releasing the contact until the needed tension is slightly exceeded.

Hysteresis also occurs naturally in ferrous materials, and it is another source of distortion in inductors and transformers, as well as magnetic recording tape. When iron is magnetized in one direction and the magnetizing current is removed, some amount of magnetism remains until the AC current reverses and rises a little in the other direction. This causes the same type of crossover distortion as analog tape nonlinearity shown in Figure 6.1 from Chapter 6. Capacitor dialectric absorption mentioned earlier is another form of hysteresis.

Power Supplies

The basic power supplies shown earlier in Figure 23.15 convert the incoming AC power line to a lower DC voltage, but that's not sufficient when powering some types of audio circuits. If the AC mains voltage changes or the connected device draws more or less current, the power supply's output voltage will also change. The solution is a regulated power supply. A regulated supply can output more voltage than is actually required, then uses negative feedback to monitor its own output and raise or lower the voltage as needed. So if the circuit being powered draws more current, the lower output voltage is quickly raised to compensate. Likewise, if the AC mains voltage changes because your air conditioner just turned on, or off, that's compensated for as well.

The power supplies shown earlier use a power transformer to drop the incoming 120 volts down to a lower voltage needed by the subsequent circuits. But there's a problem with transformers: They're big, heavy, and expensive. The solution is a *switching power supply*.

This is a digital design that's commonly used in computers and some audio power amplifiers. Electronic components are cheaper than copper and iron, so a power supply that's more complex to design and build can still make sense if it uses a much smaller transformer. Most switching power supplies begin with an extremely crude normal power supply. No regulation, minimal filtering, and some don't even use a transformer at all; they just connect directly to the wall outlet! Then an oscillator built from high-voltage transistors converts this crude DC into a new AC voltage, but at a much higher frequency than 60 Hz. Using a high frequency reduces the size and weight of the transformer that's eventually needed, and also allows using much smaller filter capacitors because the replenishing pulses arrive more often. For the same reasons, AC power used in airplanes to operate the on-board equipment runs at 400 Hz instead of 60 Hz.

Another advantage of switching power supplies is they can provide a regulated output voltage with much less heat and wasted power than analog designs. As explained, a regulated power supply puts out slightly more voltage than needed, with the excess dissipated as heat by the regulator circuit's transistors and resistors. A switching supply does this more intelligently by varying the duty cycle of the high-frequency pulses it creates. In a switching supply, AC is derived from DC by turning the DC on and off very quickly, and the on and off times are easily varied. That is, if the output voltage is sensed as being too high, the off time is made longer, and vice versa. This change in duty cycle is similar to the way SCR dimmers work, as shown earlier in Figure 23.11. Like SCR dimmers, a switching power supply can generate substantial high-frequency noise, both radiated into the air and sent back into the power line. Switching supplies are usually put in shielded metal cases to contain their noise, and ferrite beads are often placed around the connecting wires to add inductance that prevents high frequencies from being sent back into the power line.

Passive Filters

Chapter 10 showed simple passive low-pass and high-pass filters made from a single resistor and capacitor. These basic building blocks can be cascaded (wired in series) or combined with inductors to create more complex filters. However, you can't just connect one RC filter to another because the components of each stage will interact with each other. This is another use for the unity-gain buffer described earlier to isolate the output of one stage from the input of the next. However, simple one- and two-pole filters can be built using only passive components, as shown in Figure 23.20.

The band-stop filter at the left passes most frequencies through from input to output. At high frequencies the inductor is an open circuit, and likewise at low frequencies for the capacitor. But at frequencies where the inductor and capacitor both have a low impedance, some of the signal is shunted to ground. The band-pass filter at the right is the opposite. At high and low

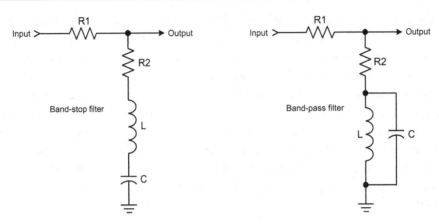

Figure 23.20: These passive filters selectively pass or block a range of frequencies, though their behavior is affected by the source and load impedance.

frequencies either the inductor or capacitor diverts the signal to ground. But at frequencies where both components have a high impedance, less signal passes through them. The values of R1 and R2 determine how much the out-of-band signals are attenuated.

Amplifiers

Figure 23.19 shows two very simple one-transistor amplifier circuits. However, most modern audio gear uses *operational amplifiers* (op-amps) because they have very low distortion and excellent frequency response, and they greatly simplify the design of complex circuits. An op-amp is basically a small power amplifier with a *differential input*—one input each for positive and negative signals. The most common type of op-amp is an integrated circuit (IC), though they can be built using discrete transistors and other components.

Figure 23.21 shows op-amp equivalents for the transistor amplifiers shown earlier, along with two common variations. Since op-amps have both a plus and minus input, they can be designed to preserve or invert the incoming signal polarity, or accept a balanced input directly to reject hum and other *common-mode* signals that are the same on both input wires. However, in order to reject hum maximally, both R1 resistors and both R2 resistors in the Differential version must have exactly the same value. In critical circuits, variable resistors are used, trimmed manually for maximum common-mode rejection. A technician applies the same signal to both inputs, then adjusts the variable resistors for minimum output using a voltmeter or oscilloscope.

Besides op-amps suitable for audio and other purposes, there are many other types of integrated circuits, both analog and digital, as shown in Figure 23.22. These are either

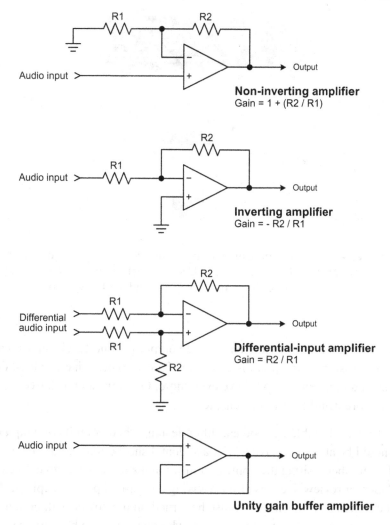

Figure 23.21: Most op-amps are integrated circuits, providing very high gain in an inexpensive self-contained package. Negative feedback sets the desired amount of gain, which also reduces distortion and improves the frequency response.

soldered directly to a circuit board or plugged into a socket, making them easier to replace if needed.

One very important characteristic of an op-amp is its extremely high *open loop* gain. This is the inherent amount of amplification, before negative feedback is applied, and is typically 120 dB at low frequencies. Negative feedback around the op-amp reduces the gain to the desired amount, which also reduces distortion and flattens the frequency response in the process. If an op-amp's open loop gain is down to 50 dB by 20 KHz, that still leaves plenty

Figure 23.22: Integrated circuits come in many different sizes and types. From top to bottom: a 68B21 peripheral adapter used to interface a CPU to external circuitry, a 7406 digital inverter to convert Ones to Zeros and vice versa, and a 741 op-amp.

of negative feedback reserve to ensure an accurate output for line-level signals that require only a little gain. Most modern op-amps—even the 25 cent types—have at least 60 dB of gain at 20 KHz, and high-speed types have even more. Op-amps and negative feedback will be explained in more detail later in this chapter.

Op-amps are also very reliable. A good amplifier design, whether an IC op-amp or a 500-watt power amp, should be able to survive a brief accidental short circuit across its output without blowing up. I remember visiting the home of a hi-fi magazine reviewer to deliver some acoustic products for review. He was also reviewing a high-end power amplifier that sells for $17,000, but it had just blown up because he turned on the power with nothing connected to the RCA line input jack. I know someone else who once blew up his very expensive tube power amp when one speaker wire became disconnected while the amp was playing. Not just the tubes, which he could have replaced himself, but the amplifier itself burned out, requiring him to return it to the manufacturer. In my opinion, any amplifier design that can't withstand such basic and common misconnection is incompetent. But I digress.

An amplifier has three primary output limitations: the largest voltage it can output, the maximum amount of current it can provide, and how quickly its output voltage can change. The maximum output voltage is determined by the power supply voltage, and a good amplifier design can output almost as much voltage as the power supply makes available. The maximum available output current is also related to the power supply—in this case, the capabilities of the power transformer, diodes, and filter capacitors.

Slew Rate

The maximum speed, or rate of change, at an amplifier's output is called its *slew rate*. This is different from frequency response because an amp goes into slew rate limiting only when trying to output large voltages. At low levels it might be perfectly flat to 50 KHz, but at high levels other factors such as stray wiring capacitance limit how quickly the output voltage can change. Slew rate limiting changes the curving slope of a sine wave into a straight slope similar to a triangle wave, and that change in wave shape adds distortion.

The formal definition of slew rate is how quickly a circuit's output voltage can change, and it's usually expressed in *volts per microsecond*. This is different from frequency response, which is independent of output voltage. If a power amplifier is 3 dB down at 50 KHz, it will be reduced by that much whether the input signal is 0.1 volts or 10 volts. However, a circuit's maximum output capability at high frequencies is a function of its high frequency response *and also* its slew rate. A preamp might be flat to 50 KHz when passing small signals, but unable to output even 10 KHz at high levels without distortion.

In order to learn how slew rate differs from regular frequency response, we first need to understand how capacitors charge and discharge over time. A typical low-pass filter circuit uses resistors and capacitors to limit how quickly the output can track the input. In the filter shown in Figure 23.23, a resistor limits the amount of current available to charge the capacitor. It takes time for a capacitor to charge or discharge, not unlike a rechargeable battery. So rapid voltage changes at the input—either rising or falling—do not make it through to the output. Eventually when the frequency is high enough, nothing at all appears at the output.

All audio circuits contain some amount of "stray" capacitance due to the close proximity of connecting wires, PC board traces, and electronic components. Indeed, a capacitor is made from two or more parallel plates of metal that are very close together, but not quite touching. So even when a circuit is meant only to amplify, and not alter the frequency response like a filter, stray capacitance limits the response at very high frequencies. Many amplifiers also include "compensation" capacitors that intentionally roll off very high frequencies above the audible range to prevent the circuit from oscillating.

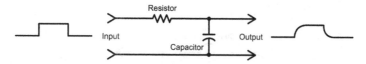

Figure 23.23: Sending audio through a series resistor and shunt capacitor as shown creates a low-pass filter. Rapid changes at the input do not pass through to the output because it takes time for the capacitor to charge and discharge.

You may wonder why charging a capacitor through a resistor creates an output voltage that rises as a curve rather than as a straight line. This is because less and less current is drawn by the capacitor as it charges. When the input voltage in Figure 23.23 first jumps up from zero volts, the capacitor acts as a short circuit across the top and bottom signal wires. So the full input voltage is sent through the resistor, which in turn draws some amount of current. As the capacitor charges, less voltage goes through the resistor, which in turn causes less current to be drawn. For example, when the capacitor is halfway charged, half of the input voltage is across the resistor and the other half is across the capacitor. So only half as much current is drawn through the resistor. When the capacitor is almost fully charged, only a very small amount of current passes through the resistor, further slowing the charging rate.

The charging rate of a capacitor is the same for small or large signals. In other words, the output reaches 80 percent of the input voltage after the same amount of time whether the input is 0.1 volt or 10 volts. However, when the input voltage becomes large enough, an amplifier's transistors or op-amps may be unable to supply enough current to charge the stray capacitance fast enough to pass high frequencies. At that point the circuit goes into *current-limiting*, which limits how quickly the stray capacitance can charge. Figure 22.24 shows the same stepped input signal as Figure 23.23, but with the resistor replaced by a *constant current source* that outputs the same amount of current regardless of the capacitor's present charge level. (It does this by raising or lowering its output voltage to whatever value is needed in order for the load to draw the desired amount of current.)

This difference in charging rates using a resistor versus a constant current source is significant because a sine wave's *rate of change* varies at different points along its cycle. As you can see in Figure 23.25, the initial portion of a sine wave rises rapidly compared to the flatter portion along the top.

When the amount of current available to charge a circuit's stray capacitance is limited, the fast rising slope of the sine wave becomes more like a triangle wave as shown in Figure 23.26. This change in wave shape creates distortion.

Just to be complete, the following example shows how to calculate slew rate. If a preamplifier has a slew rate of 1 volt per microsecond, then it can output about 10 volts peak-to-peak at

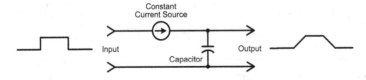

Figure 23.24: Charging (or discharging) a capacitor from a constant current source causes the output voltage to rise (or fall) at a constant rate.

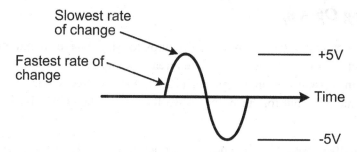

Figure 23.25: A sine wave's voltage rises and falls faster at some parts of its cycle than at others.

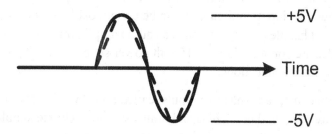

Figure 23.26: When insufficient current is available to accommodate the steep rising and falling slopes of a high frequency sine wave, it turns into a triangle wave adding new distortion frequencies.

20 KHz without adding slew rate-induced distortion. The formula to determine the minimum slew rate needed for a given frequency and output voltage is:

Slew Rate (in volts per *second*) = 2 * π * Frequency * Required Peak Output Volts

So to output 10 volts peak-to-peak at 20 KHz you need a slew rate of:

2 * 3.14 * 20,000 * 10 = 1,256,000 volts per second

Then divide the result by 1,000,000 to get microseconds:

= 1.256 volts per microsecond

One last point about power amplifiers: Earlier I explained that switching power supplies regulate their output by varying the duty cycle of high-frequency pulses for higher efficiency with less heat. Some power amplifiers work the same way, using a method called *pulse width modulation* (PWM) to obtain very high efficiency in a small package. This type of power amplifier is called *Class D*, again named in sequence after earlier letters were taken. Many amplifiers built into active studio monitors are designed this way, which is exactly what's needed when an amplifier must fit inside the small speaker cabinet it's powering while remaining cool.

Understanding Op-Amps

As explained earlier, operational amplifiers (op-amps for short) have an extremely high amplification factor, or *gain*. In theory we often treat op-amp gain as infinite, though in practice their gain is usually around 1 million (120 dB) at low frequencies. Op-amp gain falls off at higher frequencies, but for our purposes we'll consider the gain to be infinite regardless of frequency.

Op-amps are usually powered by dual power supplies, called *bi-polar* supplies, as shown in Figure 23.27. One power supply provides plus 15 volts and the other provides minus 15 volts, for a total of 30 volts. As with all circuits, the power supply determines the maximum voltage that can be accommodated. When a circuit is powered by both a positive and a negative voltage as shown here, its inputs and outputs can be either positive or negative. This is very useful for circuits that handle AC signals, such as audio amplifiers. All of the op-amps in subsequent examples are connected to +/− 15 volt power supplies, though by convention we'll omit showing those connections.

You'll also notice that an op-amp has two inputs, called non-inverting (or plus) and inverting (or minus). These are identified using plus and minus symbols. Unlike simple transistor or tube circuits that use ground (0 volts) as their main voltage reference, an op-amp amplifies the voltage *difference* between its two inputs regardless of their relation to ground. However, many op-amp circuits use the common ground as a reference for one of the inputs, as shown in Figure 23.28.

Because an op-amp's gain is so high, even a voltage difference as small as one hundredth of one thousandth of a volt causes its output to swing fully positive or negative. If the plus input is more positive than the minus input, the output goes fully positive. If the plus input is more negative, the output instead goes fully negative. Note that the maximum possible output voltage is determined by the power supply. So an op-amp running from dual +/− 15 volt supplies can output any voltage between +15 and −15 volts (or nearly so).

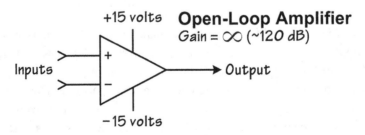

Figure 23.27: Most op-amp circuits use bi-polar power supplies to accommodate both positive and negative input and output voltages.

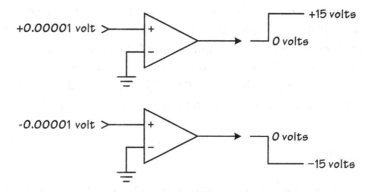

Figure 23.28: The gain of an op-amp is so high that even a tiny voltage at the input is enough to send its output fully positive or negative.

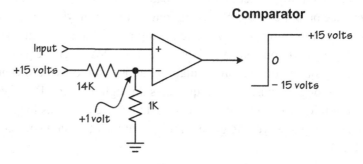

Figure 23.29: A comparator uses the full open-loop gain of an op-amp.

In Figure 23.28, applying even a tiny positive or negative voltage to the plus input causes the output to swing fully positive or negative. The minus input is at ground, or 0 volts, so the voltage at the plus input determines which polarity, or direction, the output voltage goes to. Again, the plus and minus inputs can be at any potential with regard to ground. What matters is the voltage difference between the two inputs, because that's the voltage that gets amplified. Ground is used as a reference here just for simplicity and convention.

Op-amp circuits are not usually run at their maximum open-loop gain like these examples. However, one exception is a circuit called a *comparator* because it compares two voltages. If you need to monitor a voltage and light an LED or sound an alarm when that voltage is reached or exceeded, a comparator like the one in Figure 23.29 does the job.

The minus input in Figure 23.29 is set to +1 volt by the 14K and 1K resistors. This arrangement of resistors is called a *voltage divider* because it divides the 15 volt positive power supply such that the junction of the two resistors has a voltage equal to 1/15th of the supply. As long as the plus input is less than +1 volt, the output of the op-amp is fully negative at −15 volts. If the plus input ever exceeds +1 volt, even by a few microvolts, it is then greater than the minus input so the output swings fully positive. I used a comparator with the minus input set to +0.01 volts (10 millivolts) in a preamp circuit that lights an LED when an audio signal is present. In this case 0.01 volts peak is equal to about −40 dBV, though any reference voltage could have been used.

Negative Feedback

Most op-amp applications use negative feedback to obtain a more usable amount of gain. Figure 23.30 shows an example with 100 percent negative feedback which sets the gain to 1, or unity. If the plus input is +1 volt then the output is also +1 volt. An amplifier with a gain of 1 is called a *buffer*, or a *follower*, because the output follows the input exactly. This is more useful than it might seem because it can convert a high impedance device to have a low impedance output capable of driving long wires, or maybe headphones or even a small loudspeaker. For example, the magnetic pickup in an electric guitar has a relatively high output impedance, so if you connect a long wire it will lose high frequencies due to cable capacitance. The output of a passive guitar pickup is just too weak to charge the wire's capacitance quickly enough to push high frequencies through the wire. But if you add a battery-operated amplifier with a gain of 1, you could then use a wire as long as 50 feet or even longer with no treble loss, yet the guitar amplifier at the other end still receives the expected signal level.

So how does negative feedback work, and how does it control the gain of an op-amp circuit? Looking now at Figure 23.31, imagine that when the circuit is first powered up, the output is at some unknown random voltage. In fact, that's exactly what happens when a circuit is first powered up! Depending on various factors, when the AC power switch is first turned on, the positive power supply might reach 15 volts before the negative supply reaches −15 volts, or vice versa. For now let's say the positive supply came up first, so the op-amp's output went fully positive before the negative supply kicked in and the circuit stabilized. So the output

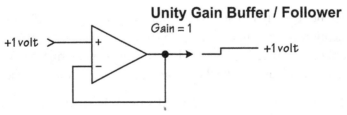

Figure 23.30: Connecting the output back to the minus input yields a voltage gain of 1, but with a lower output impedance that can drive more demanding loads.

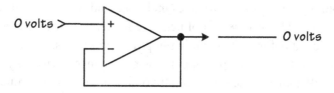

Figure 23.31: A voltage follower outputs the same voltage present at its plus input.

is at the full +15 volts from the power supply. But the output is connected back to the minus input and the plus input is at 0 volts. Since the minus input is now above the plus input, that immediately sends the output toward negative, which drags the minus input downward along with it.

Nothing in the real world ever happens instantaneously. It might take only a few microseconds for an op-amp's output to swing from positive to negative, but that's still some amount of time. Now here's the key: As the output and minus input together pass through zero volts toward a negative voltage, an equilibrium is reached at the moment the minus input reaches the same voltage as the plus input. If the minus input continues going negative and gets below the plus input, the output will start to rise again. But that raises the minus input because it's connected to the output! The output can't go higher because that would put the minus input above the plus input and force the output to reverse polarity again. And the output can't go lower either because that would create a larger difference between the plus and minus inputs, which in turn would be amplified forcing the output more positive again. So the word "equilibrium" is correct because this balance is achieved only when both inputs are at the same voltage.

This is not unlike the thermostat in a kitchen oven which also employs negative feedback as mentioned earlier. When you first turn on the oven, it begins to warm until the temperature you set is reached. Then the thermostat turns the heat off. A little while later when the temperature has fallen, the thermostat turns on the heat causing the temperature to rise again. This is pretty much what happens inside an op-amp, except the voltage difference between plus and minus inputs is sensed instead of the temperature, and the changes happen much more quickly.

As you have seen, the output and minus input of an op-amp follower are always at the same voltage as the plus input. In truth, the minus input is never *exactly* the same voltage as the plus input. If the op-amp in a follower circuit has a gain of one million, then the minus input will be one millionth of the output voltage below the plus input—there isn't quite enough gain for the output to push the minus input all the way to the same voltage as the plus input. But for all practical purposes we consider the output as doing whatever is needed to keep the minus input at the same voltage as the plus input. And since the output constantly tracks the plus input, it does so for all voltages including audio signals that can change very quickly. One important additional benefit of negative feedback is it creates a very low output impedance, as mentioned before with damping in power amplifiers. Since an op-amp's output does whatever is needed to track the plus input, you can connect nearly any reasonable load to its output and it will maintain the same output voltage no matter how much current the load draws. This is sometimes called a "stiff" output because the voltage doesn't change even as the load varies over a wide range. Indeed, most power amplifiers operate as op-amps.

Another important feature of negative feedback is it minimizes distortion. Op-amps are made from transistors, but transistors aren't particularly linear. The amount of amplification within a transistor changes with signal level, so for small voltages the amplification is different than for larger signals. Indeed, the amount of amplification changes even within the cycles of a sine wave, which changes the wave shape creating distortion. Since negative feedback forces the output to exactly match the input, any distortion occurring within the op-amp itself is reduced greatly. The amount of reduction is related to the *excess gain* of the op-amp. If an op-amp with an open-loop gain of 100 dB is used in a circuit with only 20 dB of gain, the 80 dB of excess gain is available to reduce distortion.

Op-Amp Circuits with Gain

Most op-amp circuits have a gain greater than one, again using negative feedback for control. To obtain amplification, only a portion of the output is sent back to the minus input rather than the entire output. This is shown in Figure 23.32 where a voltage divider is placed between the output and the minus input. So now in order to make the minus input equal to the plus input, the output voltage must be ten times greater than the plus input. This circuit is called a *non-inverting amplifier* because the output polarity is the same as the input. If the

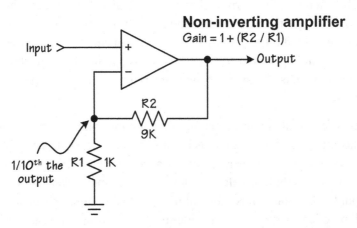

Figure 23.32: The output of a non-inverting amplifier has the same polarity as its input.

input is positive the output is positive too, and vice versa when the input is negative. This is much like the follower circuits in Figures 23.30 and 23.31, except the output voltage is larger than the input because of the voltage divider in the feedback loop.

Figure 23.33 shows an *inverting* op-amp circuit. Here the input signal goes to the op-amp's minus input (through a resistor) instead of its plus input, so the output polarity is reversed. Since the plus input is grounded, applying a positive voltage to the minus input sends the output negative. But the negative feedback pushes the minus input back toward negative, again creating an equilibrium when both of the inputs are at the same voltage (in this case zero volts).

Note that a ratio of 9-to-1 in Figure 23.32 yields 1/10th the voltage at the resistor junction, where in Figure 23.33 the same gain of 10 requires a ratio of 10-to-1. You can think of a voltage divider as a sort of ruler, where the height of an object is some fraction of the ruler's total length. In Figure 23.34, one centimeter is 1/10th of the way up the ruler. So if 10 cm is

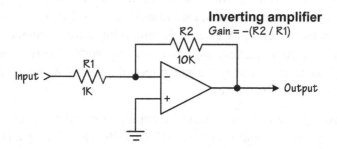

Figure 23.33: The output of an inverting amplifier has the opposite polarity as its input.

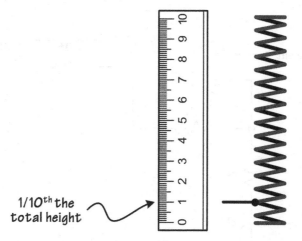

Figure 23.34: Just as a ruler divides a distance into smaller units, a voltage divider reduces a voltage to a smaller quantity. If you replace the ruler with a spring and attach a wire partway up, that wire remains at 1/10th the height as the spring is stretched and relaxed.

10 volts, then 1 cm is 1 volt. There are 9 cm above and 1 cm below, which is like the 9K and 1K resistors in Figure 23.32. But in Figure 23.33 the 10K and 1K resistors *oppose* each other rather than divide some quantity down to a smaller fraction. The input pulls up through the 1K resistor and the output pulls down through 10K. So it takes 10 times more voltage from the output to counter the 1 volt coming in through the 1K resistor. Thus, the resistor ratio for a gain of 10 in an inverting amplifier is 10-to-1 instead of 9-to-1 as with a non-inverting amplifier.

You might wonder why anyone would want an amplifier circuit that inverts polarity, other than when inversion is the intent. There are several situations where this circuit has an advantage over non-inverting types. A non-inverting op-amp circuit always has a gain of one or greater, but sometimes loss is required such as when setting an EQ circuit to cut. In Figure 23.33 you can have loss simply by making R1 greater than R2. The inverted polarity can be put back to normal with a second inverting circuit if needed.

Another common use for inverting circuits is when summing (mixing) multiple signals. Since the plus input is almost always grounded, the minus input is considered a *virtual ground*. So multiple resistors that all come together at the minus input won't interact. As explained in Figure 5.18 and its surrounding text, usually when a network of resistors are connected together, changing one resistor to vary the gain affects the gain of all the other signals. But an inverting op-amp avoids this interaction.

Of course, op-amps can do much more than just amplify or buffer voltages. One of the most important uses for op-amps in audio is filters, which are the basis for equalizers and active crossovers. We'll get to active filters shortly, but for now Figure 23.35 shows that op-amp filters employ capacitors to make their amplification frequency-dependent. This circuit is

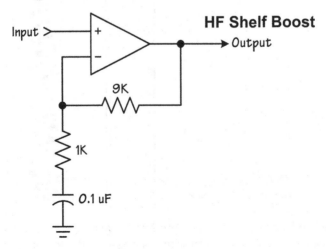

Figure 23.35: An active filter uses capacitors as frequency-dependent resistors to vary the gain.

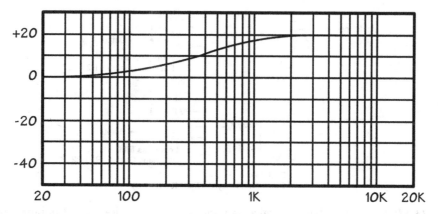

Figure 23.36: This graph shows the frequency response of the circuit in Figure 23.35.

derived from Figure 23.32, and the 9K and 1K resistors set the gain to 10 (20 dB) for high frequencies. But below a few KHz the capacitor becomes more of an open circuit, and by around 20 Hz its equivalent resistance (reactance) is around 80K. So at very low frequencies the gain is close to unity, as shown in Figure 23.36.

Active Filters

The simplest type of active filter uses amplifier circuits to buffer the input and output of a passive filter, to avoid an interaction between the resistor and capacitor values and the input and output impedance of preceding and subsequent stages. As mentioned in Chapter 5 (Figure 5.18), passive summing amps also require active buffering to avoid the volume and pan settings for different input channels from interacting with each other. If panning one channel to the right shifts all the other channels to the left, that's obviously unacceptable. But active filters go beyond adding simple buffers to isolate a series of individual passive filter stages. By replacing some of the resistors in an op-amp amplifier circuit with capacitors, many different types of active filters can be created. Further, active filters can use *simulated inductors* to avoid the distortion, high cost, and susceptibility to hum pickup of real inductors. You won't be surprised to learn that a simulated inductor is yet another op-amp circuit.

Explaining the details of active filter design is beyond the scope of this book, but several articles are available on my personal website ethanwiner.com that show schematics for active filters, equalizers, and some other popular types of audio gear. But just to give the basic idea, Figure 23.37 expands on the inverting amplifier in Figure 23.33 by adding a capacitor to create simple 6 dB per octave high-pass and low-pass filters. In these circuits,

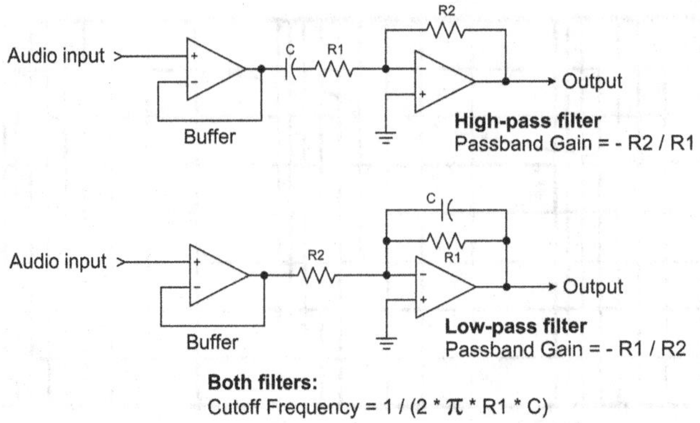

Figure 23.37: Active filters have several advantages over passive filters, most notably that the frequencies are not affected by the load impedance.

R1 and C together determine the cutoff frequency, and the ratio of R1 to R2 controls the overall volume gain (or loss). The input buffer is used only to prevent the filter from being affected by the preceding stage's output impedance. An op-amp's output impedance is very low, so an additional buffer is not needed at the output of the second filter stage.

Digital Logic

While the analog circuits shown so far are important for audio, it's useful to have a basic grasp of digital logic. Early digital logic circuits used mechanical relays to perform all of the required functions. A relay is constructed from three basic components: an electromagnet, a switch, and a spring, as shown in Figure 23.38. Whereas most switches are activated by pushing a button or throwing a lever, the switch (or switches) in a relay changes state when voltage is applied to the electromagnet's coil. This is an important concept because it lets one circuit control another, without human intervention. Voltage to the coil magnetizes the iron core, which then pulls the metal switch contacts into position. And when the voltage is removed, the spring pulls the contacts back to their rest position. Many relays contain two switches, though some have as many as four or more, offering many possible circuit combinations.

This relay is called *double-pole double-throw*, abbreviated DPDT, because it contains two independent switch poles, and each switch has two active contacts. One contact is connected, or *thrown*, when the relay is at rest, and the other closes when the relay coil is energized. These contacts are called *normally closed* (NC) and *normally open* (NO), as identified in Figure 23.39. When the relay is at rest, the center *common* conductor touches the NC contact; when activated it instead touches the NO contact. As you saw in

Figure 23.38: This DPDT relay has two switches, each with two active contacts.

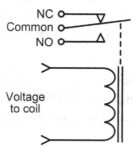

Figure 23.39: When this relay is at rest, the common switch contact connects to the normally closed contact. When the coil is energized, the common contact moves to instead touch the normally open contact. Here, "normal" means when the relay is at rest, with no voltage applied to its coil.

previous circuit diagrams, drawing a realistic picture of an electronic component is not the most efficient way to convey a circuit, so engineers instead use *schematic diagrams*. But for relays, the schematic representation is pretty close to what an equivalent picture would look like. The two solid lines next to the coil represent the iron core, and the dotted line shows that when the relay is activated the switch arm is pulled down toward the magnetized core.

One basic logic function a relay can perform is called a *latch*, shown in Figure 23.40. A latch allows a "push to start" type of operation—for example, the Play button on a tape recorder or the Start button on a microwave oven. Without it, you'd have to stand there and hold the button for the entire time the tape is playing or your coffee is warming up. Sure, you could use a regular toggle switch, but then the tape deck would keep running after the tape ran out, and the oven would not be able to turn itself off automatically after two minutes. A latch remains enabled even after its input is removed, so it can also serve as memory to store one bit of data.

Here, a double-pole relay is needed, since one of the switch poles is required just to perform the latching function, and the other can then do whatever is needed. When the push-button switch is pressed, the coil becomes energized by the power supply, causing both switches to change state. The lower switch is connected in parallel with the push-button switch, thereby maintaining power to the coil after the button is released. The relay remains activated until the voltage is removed. Incidentally, you could consider this latching action to be a form of *positive feedback*, since the output voltage reinforces the push-button input.

Negative feedback can also be applied to a relay, as shown in Figure 23.41. This oscillator circuit—commonly known as a buzzer—is created by wiring the switch contacts in series with the coil. Applying power to the coil causes the switch to change states, but the moment that happens, power to the coil is interrupted and the spring pulls the switch contact back to its normal state. This activates the coil all over again, and the process repeats indefinitely until the voltage is removed. I realize this is pretty basic stuff, but then so are most logic circuits.

Another simple digital logic circuit is the *gate*, and there are several types. Figure 23.42 shows relay circuits for AND, OR, and EXCLUSIVE OR (XOR) gates, the basic building blocks of all digital logic. Of course, computer logic circuits don't use switch contacts but

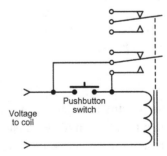

Figure 23.40: This latch circuit lets you press a button once, and then the relay remains energized until voltage to the coil is removed. The lower switch performs the latch function, and the independent upper switch is available to control another device.

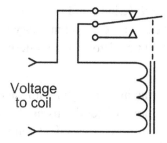

Figure 23.41: This relay is configured as a buzzer that opens and closes repeatedly for as long as power is applied.

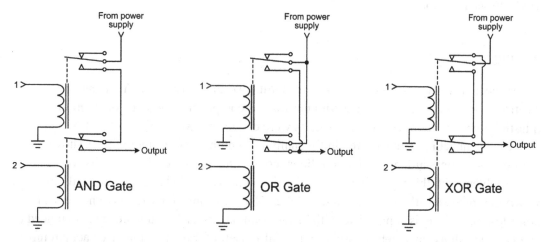

Figure 23.42: All of the basic building blocks used by computer logic circuits can be realized using mechanical relays.

instead accept and output either of two possible voltages: the full power supply or zero volts. That's why the switches in these equivalent relay circuits connect to a power supply.

Beginning with the AND gate, if Inputs 1 *and* 2 are powered, then the output will be powered, or One. Otherwise it will be off or Zero. For the OR gate, if Input 1 *or* 2 is a One, then the output will be One. All an EXCLUSIVE OR gate cares about is if the inputs are *different*, providing a One output when they are. As you can see, gates are used to make simple decisions, based on the information at their inputs. Besides these three gate types, *inverted* versions are also available that work the same but have an opposite output. Where an AND gate outputs One when both inputs are One, a NAND (NOT AND) gate outputs a Zero for the same condition. It works the same for NOR (NOT OR) and EXCLUSIVE NOR gates.

Understand that *all* digital circuits are created using these humble logical building blocks, from a simple digital stopwatch through the most sophisticated computer. The heart of every computer is its *central processing unit* (CPU), and modern CPUs contain literally millions of transistors to implement the gates, counters, and memory required. The Intel 8088 used in the original IBM PC contains 29,000 transistors, and modern CPUs employ more than one billion transistors.

Before transistors, computer logic was implemented with vacuum tubes. Even earlier computers were built using relays—lots of them!—using the same types of connections shown here. Before electromagnetic relays, mechanical computers were built with gears, levers, and pawls dropping into notches to latch (store) data. The first working relay computer I'm aware of was built at Bell Labs, completed in 1939. Incidentally, the first computer "bug" was a moth that got trapped inside a relay, causing it to malfunction. Google *"relay computer moth"* to see the photo!

Wiring

The subject of wiring might seem mundane, but it's worth a closer look. Years ago all electronic devices were hand-made with the various components mounted on phenolic (plastic) boards or terminal strips. The parts were connected using point-to-point wiring, with each connecting wire soldered individually to a component or other wire as shown in Figure 23.43. Some hi-fi and guitar amplifiers containing tubes are still built this way. The main problem with this type of construction is it can be assembled and soldered only by hand, which in turn means it's costly to build and there's more chance for wrong connections. But point-to-point wiring is appropriate with heavy components such as transformers that need to be mounted solidly on a metal chassis, and it's also useful for devices that use vacuum tubes which get very hot.

Figure 23.43: Point to point wiring is labor intensive and potentially prone to assembly errors.
Photo courtesy of Amplifier Experts.

Most modern electronic devices are constructed using *printed circuit boards*, or PC boards (or just PCBs) for short. PC boards are manufactured by bonding a thin copper sheet to fiberglass, then a photo etching process uses acid to dissolve the copper that's not needed leaving only the wiring traces. This is shown in Figure 23.44. PC boards are typically made of fiberglass and resin, with copper traces on the bottom surface that connect the various components. Some boards have traces on top too, especially when the circuits are complex and require many connections. Special holes, called *plated-through holes* can connect traces on the top to traces on the bottom, to allow a wiring route that isn't possible on only one side of the board. For small production runs the various electronic components are inserted into the board manually and soldered by hand, but machines can do both operations when larger quantities are built.

In truth, there are two types of circuit boards and electronic components. The original type in Figure 23.44 uses electronic parts with wire leads that go through holes in the board

Figure 23.44: Printed circuit boards allow rapid assembly with minimal errors, and they can be "stuffed" with parts and soldered automatically by machine. Note the two circuit boards in this photo. One shows the bottom with the wiring traces and soldered connections. The other, mostly hidden, shows the top side of another board where the components are placed. *Photo courtesy of Amplifier Experts.*

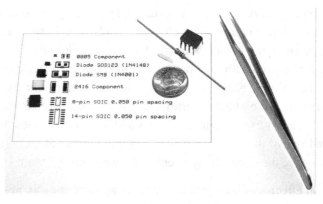

Figure 23.45: Surface mounted devices are much smaller than traditional electronic components, and many are even smaller than those shown here.

and are soldered to copper traces on the bottom of the board. A newer type uses *surface mounted devices* (SMDs) that are much smaller and connect to traces on the top side of the board. Figure 23.45 shows an assortment of SMD parts, next to a normal size 8-pin IC and 1/4 watt resistor. The dime and grain of rice give another sense of scale. The 0805 and 2416 size numbers specify the length and width, in this case 0.08 by 0.05 inches for the smallest part shown. I bought these parts for a project I soldered by hand, which requires great care and a bright light with a magnifier! Resistors and capacitors as small as 0.40 by 0.20 inches are available, but those can be placed and soldered only by precision automated machinery. Likewise for ICs that can be *much* smaller than the SOIC type I use. That type of construction is needed for things like cell phones and smart watches, or other tiny but highly complex devices.

When deciding which construction method to use for building an audio device, it's important that the conductors be thick enough to pass the required amount of current. Thin traces on a PC board are not usually a problem, even with tube amplifiers because tube circuits are high impedance—the voltage is high but the amount of current is small. So the traces shouldn't be too close together to avoid arcing, but they don't need to be unduly wide. High power solid-state power amplifiers will use wider traces where needed.

Another issue is strength and longevity. As mentioned earlier, some components are very heavy, so a PC board inside a tube guitar amplifier that's lifted and moved often can flex and eventually crack under the weight of a large power or output transformer, or a large filter capacitor. Even if the board itself doesn't crack the copper traces can break, and such "opens" can be difficult to find and repair. So additional supports should be added to handle the extra weight if needed.

Practical Electronics

If you learn how to build your own simple adapters and gadgets, you won't have to buy them, and you'll be able to do simple repairs yourself. Learning basic electronic DIY skills also lets you assemble kits to save money with less hassle and risk of failure than building devices from scratch based on web plans and without a proper circuit board. Even with complex devices, what fails most often are simple things such as switches and connectors. Two of my business telephones failed when the 9-volt battery terminals worked loose from the circuit board. Most people would have just thrown out the phones and bought new ones. Indeed, without a schematic and knowledge of the circuit, it's difficult to fix failures at the component level. But both of these phones were fixed simply by opening the case and tightening two screws.

I have a Zoom digital effects box I bought many years ago, and shortly after the warranty expired (of course), the digital display went blank. There's not much you can fix inside a modern digital device, but in this case the failure was a tiny fuse soldered directly onto the circuit board. I managed to fit a fuse holder into the very small space to use a replaceable plug-in fuse. In all the years since, that fuse has never blown again. In my opinion, selling consumer gear with soldered-in fuses should be illegal, but that's another story.

Wiring your own simple circuits is easy, but it helps to have the right tools. Figures 23.46 through 23.48 show my electronics tools divided into three categories: cutters, pliers, and soldering equipment. I've had these tools for more than 40 years, and they still work perfectly. Good tools cost more than cheap ones, but they're not overly expensive. Poor-quality wire cutters and strippers are frustrating to use and can nick delicate wires. A lousy soldering iron is also a bad investment. You can outfit yourself with a set of good quality tools for about $100. Watch the soldering video to learn how to replace a 1/4-inch phone plug

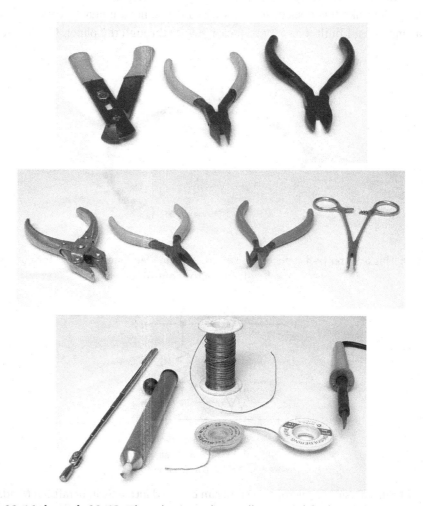

Figures 23.46 through 23.48: These basic tools are all you need for basic electronic assembly and repair jobs. From top to bottom: stripper and cutters, pliers, soldering tools.

on a guitar cord from start to finish using these tools. If you're really serious about learning to do your own repairs, it's worth investing in a decent digital multimeter that can measure AC and DC voltage and current, and resistance. Better models also test diodes and transistors and can even measure a transistor's gain.

Splitters and Pads

One of the most common studio gadgets is an attenuator, or *pad*, that requires only resistors. Figure 23.49 shows an adapter pad I made to connect an unbalanced stereo line-level output to a microphone-level input by combining the two channels to mono and reducing the level 20 dB. In this case, the resistors are built into the XLR connector. It's easy to forget what a custom-built gadget like this does or how it's wired if you need it again years later, so I made a quick drawing when I built it and then stuck the wire through the paper. Crude but effective.

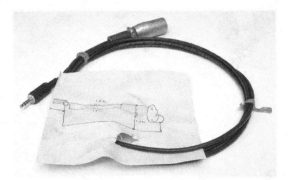

Figure 23.49: This adapter pad reduces the volume of a line-level stereo signal for connecting to a mono low-level microphone input.

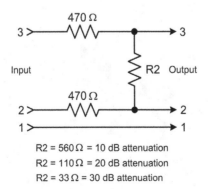

R2 = 560 Ω = 10 dB attenuation
R2 = 110 Ω = 20 dB attenuation
R2 = 33 Ω = 30 dB attenuation

Figure 23.50: This simple microphone pad can be wired into a Switchcraft S3FM adapter. The resistor values assume a mic preamp input impedance of 2 K. The numbers 1, 2, and 3 refer to the XLR connector pin numbers.

I've built a number of small passive mic- and line-level gadgets into Switchcraft S3FM adapters. This is a four-inch-long metal tube with a male XLR connector at one end and a female XLR at the other end. Whatever you can fit inside is fair game, and I've used them to build mic pads, passive mic-level filters, and even an active DI. Figure 23.50 shows the wiring for a mic-level pad, with resistor values for 10, 20, and 30 dB of attenuation. These values assume the mic preamp they'll plug into has a 2K input impedance, but a higher impedance changes the attenuation amount only slightly. If the impedance is substantially lower than 2K, you could use the resistor calculation formulas in the Loudspeaker Impedance section of Chapter 18 to calculate the value for R2 in parallel with any input impedance. Note that when wiring mic pads, it's important that both 470-ohm resistors be exactly the same value. Otherwise, the circuit balance is compromised, reducing common-mode rejection at the preamp's input. Using 1 percent tolerance resistors is typical, which ensures 40 dB of hum and noise rejection. But using 0.1 percent resistors is even better because that guarantees at least 60 dB of rejection.

When soldering wires and components, a solid mechanical connection is very important. Solder is a very soft metal, and microphone and guitar cords need to withstand rough handling. It's also important to ensure that the pieces don't shift while the solder is cooling. Then again, it's okay in a pinch to tack two parts together if you do it well, at least for temporary fixes, or for wire and gear that won't be vibrated or stressed mechanically.

Phone Patch

Another useful studio device that can be surprisingly expensive to buy is the *phone patch* shown in Figure 23.51. This lets you send and receive audio through a standard telephone line, albeit at the low fidelity telephone systems impose. The capacitor and transformer are needed to block the large DC and AC voltages present on phone lines when idle and when the phone rings, and the 180-ohm resistor provides a proper load for the line. If a transformer is the only device connected, it won't load the line enough to present an "off-hook" condition back to the phone company letting it know the line is in use.

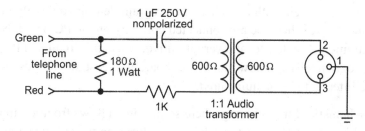

Figure 23.51: This phone patch lets you send and receive audio through a telephone line.

When I owned a professional studio in the 1980s, we recorded the then-famous "K-Tel Records Presents!" TV commercials. The ad agency that produced the spots was in Chicago, but their regular announcer happened to live in the same town as my studio. So they contacted us soon after we opened, since it was so much easier for the talent to drive ten minutes to us rather than an hour and a half into New York City. We'd record the announcer live to tape and also send it through the phone line while the ad agency's producer listened and commented. The same connection that sent audio into the phone line also received the producer's voice so the announcer and everyone else could hear his comments and instructions. Today this can be done over the Internet with much higher fidelity.

Schematics and PC Boards

As you have seen, I encourage anyone interested in audio, and technology generally, to at least dabble with electronics. Even if it's just to build a simple circuit from a magazine or DIY website, or a preamp or other device sold as a kit, this is a great way to learn a useful skill. So we'll conclude this chapter with an overview of schematic drawing and printed circuit board layout, which is the logical culmination for this part of the book.

In the old days before personal computers, both hobbyists and professionals drew schematics by hand on paper, and designed circuit boards by applying special black masking tape onto clear plastic film. Today, affordable or even free software makes both of these tasks *much* easier. If you need to adjust a small area to squeeze in an extra component, or duplicate an entire block from the left channel to the right, it's simple to move, or copy and paste, one component or an entire section with just a few mouse clicks.

Figure 23.52 shows a small portion of a very large circuit I designed to test audio equipment. This is part of the signal generator section containing four 555 timer ICs to create sawtooth waves in musical harmony. This page also contains three filters to reduce harshness, boost high frequencies, and optionally remove content above 24 KHz. A simple driver circuit lights an LED for half a second if the positive signal level approaches clipping.

I drew this schematic using the fabulous ExpressSCH program available for free from PC board manufacturer ExpressPCB. The same download also includes a PC board layout program that works in tandem with the schematic program, letting you easily verify that all of the components on the schematic are connected correctly on the PC board. Both of these programs are extremely easy to use, and very intuitive. Once the circuit and PC board are completed, the program tells you how much your board will cost, then sends the file to the company via the Internet to be manufactured.

When drawing schematics I try to arrange them so the signal flows from left to right, or top to bottom, as in this circuit. The four timer ICs in a column on the left play a musical chord,

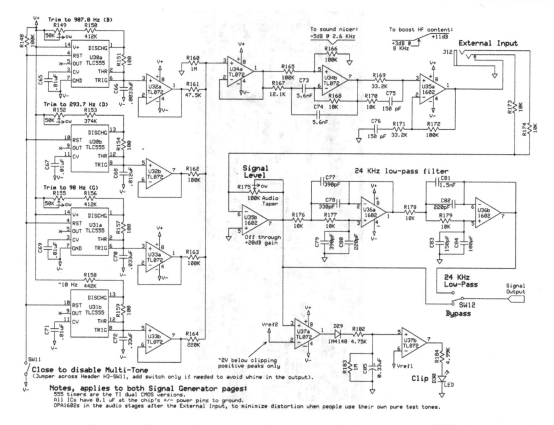

Figure 23.52: This schematic was created using the ExpressSCH program available at ExpressPCB.com.

which is more pleasing than a single buzzy sounding note or swept sine wave. Each triangle wave is buffered by an op-amp follower, then all four sawtooth tones are mixed together and sent to a filter that reduces harshness. Hey, if you have to listen to test tones for a few hours, they might as well be pleasant sounding! Another filter boosts high frequencies to ensure enough of that content for testing, and a third filter optionally blocks or passes ultrasonic frequencies depending on what's being tested. Understand that my point isn't to explain how this circuit works as much as to show what I believe is a useful *style* when drawing schematics. This includes clearly labeling all of the components, plus adding explanatory notes in case someone else ever has to make a repair.

Figure 23.53 shows part of the same Signal Generator section of the PC board. In this case, the four timer chips U30 and U31, and their op-amp buffers U32 and U33. This shows only the board itself; the text and rectangle outlines show where each part will go when the board is actually populated, and are silk-screened onto the board during manufacture. This circuit board uses surface mounted components, so the main traces are on the top and any

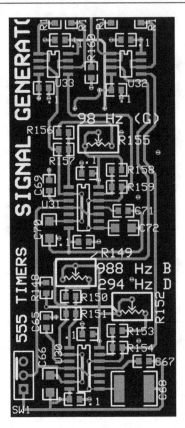

Figure 23.53: This section of the PC board corresponds to the left side of the schematic in Figure 23.52.

additional traces needed are on the bottom. This is the opposite of older style boards shown earlier in the Wiring section, where the parts are on top and most of the connecting traces are on the bottom. Although it's not visible in this view, this circuit board actually has four copper layers: the top and bottom layers shown as light and dark traces, plus two more layers sandwiched inside the board that are divided into sections to carry power and ground to different areas of the board.

Note that each timer chip holds two complete timer circuits, so only two ICs are used for all four timers. Likewise for the U32 and U33 which each contain two op-amps. Also notice the plated-through holes mentioned earlier, inside the outline of the timer ICs, and at each end of the trace running between R150 and R151. The lighter colored trace is on the bottom of the board, and metal plating inside the holes connects traces on one side to traces on the other side to allow routes that would otherwise conflict.

Figure 23.54: The electronic breadboard is used to construct trial circuits faster and more easily than soldering all the components. Each group of five holes is connected internally, and the holes are 1/10th inch apart to align with the lead spacing of standard ICs and other common components. *Photo courtesy of startingelectronics.org.*

Another program I've found incredibly helpful when designing circuits is also freeware: LTspice from Linear Technology (linear.com). This amazing program lets you draw a circuit and run it as a simulation to see its frequency response and other attributes. Just as a DAW program virtualizes an entire recording studio with effects plug-ins and even synthesizers, LTspice virtualizes a parts bin fully stocked with components, plus a workbench full of test gear. Before the availability of "spice" type programs (Simulation Program with Integrated Circuit Emphasis), engineers had to build prototypes of every circuit by hand using "breadboards" like the one in Figure 23.54. Then they'd connect signal generators, frequency counters, oscilloscopes, and distortion analyzers in order to see if and how well the circuits worked. If you accidentally hooked up the power supply backwards you could destroy dozens of expensive components and a week's worth of effort! By the way, this plastic device is called a "breadboard" because long ago trial circuits were cobbled together on real wooden breadboards using copper nails as soldering terminals!

The LTspice program in Figure 23.55 shows one of the timer circuits from Figure 23.52, along with the resulting waveform as it would appear on an oscilloscope. Many other views are also available including frequency response, phase shift, and FFT analysis. You can even save the signal at any point in the circuit as an audio Wave file! In this case the oscilloscope's "input" was taken from the junction of R2 and C1, which provides the sawtooth output. You can see that this isn't a text-book sawtooth wave with a linear (straight line) rising voltage, but instead follows the curve of a charging capacitor as was explained earlier in the Slew Rate section. However, the goal here is merely to generate both odd and even harmonics, and this wave shape is close enough for that.

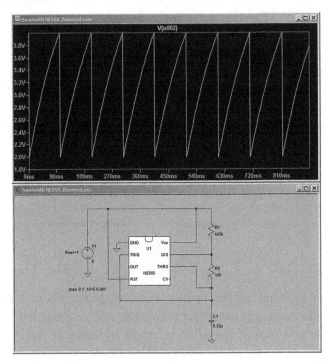

Figure 23.55: The freeware program LTspice from Linear Technology is incredibly useful because it lets designers try out circuits in a virtual electronics lab without actually building them.

Summary

This chapter explains electronic components, as well as the basics of analog and digital circuits. Just as water flows through a pipe, electricity flows through wires, and many of the same principles apply. However, electric circuits require two wires, and many of them connect to the same common point at the power supply. Common is often called "ground," but it isn't necessarily connected to the earth, though it can be. Simple power supplies use transformers to reduce the AC mains voltage, then pass the smaller AC through one or more diodes to create a pulsing DC. A filter capacitor smoothes the pulses into a steady DC by holding a charge that sustains the output voltage between the pulses. When tighter control of the supply voltage is required a regulator can be added.

You also learned about the relationship among volts, amps, ohms, and watts, and the included Ohm's Law spreadsheet lets you easily calculate any one parameter from any two of the others. But the point of this chapter is not so much about math as explaining the concepts in practical, mechanical terms. A resistor is similar to a narrow section of water pipe, and a capacitor is not unlike a leaky pump. Further, transformers are the electrical equivalent of gears in an automobile transmission, varying the relation between volts and amps rather than speed and torque. But real-world components are not ideal, and capacitors also have some amount of inductance, and vice versa. Choosing appropriate components for a given purpose is part of the art of circuit design.

I also show several very simple circuits, including transistor and op-amp amplifiers, and passive and active filters. Negative feedback is a key concept in all audio circuits, because it reduces distortion and flattens the frequency response. By feeding part of a circuit's output back to its input with a reversed polarity, the output is forced to more closely match the input. Negative feedback is also used in regulated power supplies to ensure a constant output voltage even if the mains voltage or load current vary. Where analog designs are inefficient and waste energy as heat, switching power supplies and power amplifiers are much more efficient and can be made much smaller and for lower cost.

The basic operation of digital logic circuits is also shown, using mechanical relays to explain the concepts. All logic circuits are based on AND, OR, and XOR gates, and all of these can be implemented using relays. Indeed, a complete working computer could be built today using only relays, though it would be huge and slow, draw an enormous amount of power, and would break down often due to mechanical failure.

Finally, a few small but useful audio gadgets are shown, along with the tools needed to build and repair them. Even if you have no intention of becoming a full-fledged electronics techie, knowing how to build and repair your own cables, pads, and splitters can save a lot of money over time. And sometimes, a needed adapter might not be available at all for any price.

CHAPTER 24

Test Procedures

Chapter 2 explained how audio devices are measured by professionals using modern test gear. Sadly, many people can't afford $10,000 or more for a precision analyzer, or even $400 for a basic oscilloscope. But there are many tests you can do yourself without measuring equipment. For example, the Wave files included with Chapter 3 let you assess the audibility of low-level artifacts just by listening through your own speakers or headphones. That chapter also explained how the FFT display available in many audio editor programs shows harmonics and noise at levels too soft to hear. As explained in Chapter 8, to measure the distortion of 32-bit floating point "summing" math used by most DAW software, I applied 30 sequential gain changes to a pure sine wave, then analyzed the added distortion using Sound Forge's FFT feature.

With a very good multimeter you can measure the frequency response of outboard audio gear, though oscilloscope software that uses your sound card for its input is also available. Even better is getting a used "real" oscilloscope, which will cost much less than a new one. Chapter 22 showed how an inexpensive small diaphragm omni condenser mic coupled with room measuring software can measure the frequency response of loudspeakers and microphones. So it's possible to do your own equipment testing using the tools you already own or can obtain inexpensively.

There are two very different types of audio tests: measuring tests and listening tests. For assessing raw fidelity, measuring is the better choice because it's 100 percent repeatable, and the results are highly accurate, assuming it's done properly. But there's nothing wrong with listening tests when done correctly and you account for the limitations of hearing.

The gold standard for all subjective comparisons—not just audio gear—is the double-blind test. In a sighted test, the person listening knows what source is playing and can be influenced by expectation. A single-blind test is much better, where someone else switches the A and B playback without winking or otherwise letting on. But with a double-blind test, even the person running the test doesn't know what's playing at the moment, which avoids any chance of accidentally giving a clue to the test subject. I'm satisfied using single-blind tests for my own education, and I've done that with friends many times. In my home studio, the person listening is behind me, unable to see my hands or facial expression. You can even test yourself blind, either using ABX software or by closing your eyes while clicking linked solo buttons, as described in Chapter 3.

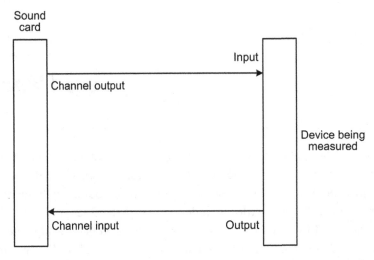

Figure 24.1: A loop-back uses DAW software to play test signals from a sound card's output through the device to be measured. The result is then recorded back to the DAW for later analysis.

Whether your intent is to measure with a meter or just listen with your ears, the first thing needed is a signal source. Obviously, music is a fine source if the intent is to identify quality differences between two devices, though you must match both volume levels to within 0.1 dB. For accurate and repeatable measuring, using test signals makes more sense. I create sine wave test tones in Sound Forge, and most other audio editor programs can do this. The "pink_noise.wav" file from Chapter 22 can be used when a noise source is more appropriate. Sound Forge can also create a sine wave that sweeps the full audio range, or just a portion. The basic setup is called a *loop-back*; the output of a sound card or external converter is sent to the hardware device being tested, then looped back to its own input or the input of another sound card, as shown in Figure 24.1.

> *A person with one voltmeter knows what the voltage is. A person with two voltmeters never really knows for sure.*
>
> **—Mike Rivers, audio journalist**

Although you might think that testing audio gear requires a meter or oscilloscope, using a decent-quality sound card and audio software is easier and more accurate than inexpensive meters. Most budget meters don't even state a frequency response, and when they do, it's often limited to 10 KHz or even lower. A competent sound card—even an inexpensive consumer model—will be reasonably clean and flat from 20 Hz to 20 KHz. Then you can record the test signal(s) once and analyze the recording in different ways later. However, when using a sound card for testing, I suggest you first assess its quality. Connect the output of the sound card directly back to its own input, then record your test signals to learn the sound card's own frequency response and how much noise and distortion it adds.

Frequency Response

The best signal source for measuring frequency response is a series of sine waves at low, mid, and high frequencies, or a slowly swept sine wave to see even more detail. Depending on what you're testing, you may not need to measure at many frequencies. Often, 20 Hz, 1 KHz, and 20 KHz are sufficient for electronic devices that are expected to be flat. If a device rolls off at the frequency extremes and you want to see the trend, you could also measure at 50 Hz and 10 or 15 KHz. When measuring gear that contains a transformer, or an analog tape recorder, using a sweep ensures you won't miss any irregularities. Either way, I generally create test files with a peak level of −6 dB. Sine waves should sustain each frequency for 5 to 10 seconds, giving you time to read the level meter later. Then load the file to a track in your DAW software and record the output of the device being tested to a second track.

Another approach uses a stand-alone signal generator program that plays tones or noise through your sound card, which you route to the device being tested. Then you can record the signal from the gear being tested using any basic audio editor. I use Vincent Burel's excellent and free LF Generator, which can output various wave shapes at any audio frequency, including sweeps, as well as pink noise.

You can also test your hearing with sine waves, but you should first verify how high a frequency your loudspeakers can produce using a decent small diaphragm omni condenser mic and room measuring software as described in Chapter 22. When playing very high-frequency tones through a loudspeaker, move your head slightly while listening to be sure you're not in a null spot. High frequencies are surprisingly directional due to loudspeaker beaming and also comb filtering that occurs naturally in rooms. Of course, using good headphones avoids these problems as long as you're certain they can reproduce frequencies higher than you can hear. Another method plays a 96 KHz recording of a tambourine or jangling keys, or other source known to contain frequencies beyond 20 KHz, as verified with an FFT. The "tambourine.wav" file from Chapter 3 is ideal for this. Sweep the frequency of a low-pass filter plug-in downward starting from above 20 KHz until you can just barely hear a difference, then note the frequency on the plug-in.

Ringing

A 20 Hz square wave will tell you more about an audio device in two seconds than all the fancy test gear in the world. Square waves are the most challenging signal for any device to pass accurately, and they reveal not only response errors but also ringing that can indicate

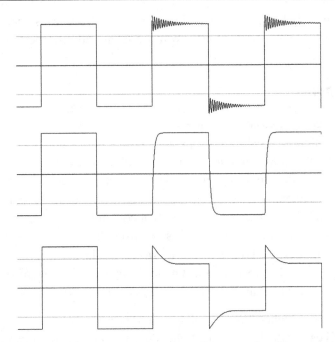

Figure 24.2: A low-frequency square wave is an excellent test signal because it readily reveals ringing as well as frequency response errors. From top to bottom: ringing, high-frequency roll-off, low-frequency roll-off.

circuit instability. Figure 24.2 at the top shows what ringing looks like, using a 20 Hz square wave before and after applying a high-Q boost at 1 KHz. The ringing is evident as each rapid wave transition creates a small sine wave that decays over time. The middle figure shows the original square wave after adding a 6 dB per octave low-pass filter set to filter content above 1 KHz. Here you can see the rounding of the edges at each transition, exactly as happens with a passive resistor-capacitor filter as shown earlier in Figure 8.2. Applying a high-pass filter, at the bottom, causes the flat top and bottom portions of each cycle to droop as the sustained "DC" component of each cycle falls off.

Harmonic Distortion

Distortion is not difficult to measure, requiring only a high-quality sine wave source and an FFT analyzer, which is included in many audio editor programs. Note that an FFT display doesn't tell you a single distortion amount in percent or dB. Rather, you see all the individual components that, if added together, would give the total amount of distortion. But an FFT shows the nature of the distortion, which often is more useful for assessing audibility than a single number.

Depending on what type of device you're testing, you might measure at 1 KHz only or over the full audio range. Transformers tend to have higher distortion at low frequencies, so you could test at 50 Hz or even 20 Hz. Absolute signal level is also a factor; some gear has more distortion at higher levels, but some devices are worse at soft levels. It's important to start with a pure sine wave having known low distortion, though most software creates reasonably pure tones. In Sound Forge, if you create a sine wave as a 24-bit project its distortion is lower than when set for 16 bits. Again, you should verify the source purity, as I did for the summing math test in Chapter 8. I examined the original sine wave with an FFT and confirmed that all the residual artifacts were safely below −100 dB. This equates to 0.001 percent distortion plus noise, which is lower than most outboard gear you're likely to test. If all of the individual artifacts are below 100 dB, it's unlikely you'll ever hear them. And that's the real point of such testing.

IM Distortion

IM distortion is more complicated to measure than simple harmonic distortion, but it's still possible using the same methods. IM distortion occurs when two or more frequencies play together and create new sum and difference frequencies. For example, if you play music or test tones containing an A note at 440 Hz and also the C# note above at 554 Hz, IM distortion adds new tones at 554 + 440 = 994 Hz and 554 − 440 = 114 Hz. Note that neither 994 Hz nor 114 Hz is a standard musical note pitch. This is one reason IM distortion generally sounds more obnoxious than harmonic distortion whose components are musically related to the source frequency.

To test IM distortion in audio gear, you'll create sine waves at two different frequencies, then mix them together and save the result in a file to play through the device under test (DUT). Standard IMD tests play 19 KHz and 20 KHz at equal volumes, then measure the level of the resulting 1 KHz difference frequency. (The 39 KHz sum is considered irrelevant, and won't be present anyway, assuming the sample rate is 44.1 or 48 KHz.) But IMD tests can use other frequencies, which is needed when measuring loudspeakers and microphones. Since two frequencies result from IM distortion—the sum and the difference—it's not difficult to determine the approximate distortion percentage. However, large amounts of IM distortion also create additional components that are harmonics of the sum and difference frequencies. Again, this is less important than seeing the big picture to verify that whatever distortion you measure is too soft to be objectionable.

When measuring the IM distortion of loudspeakers, you should choose frequencies that are both played by the woofer or by the tweeter at the same time. If a speaker crosses over from woofer to tweeter at 2 KHz, using 500 Hz and 3 KHz will not tell you the amount of IM distortion from either driver. To test the woofer, you might use 100 Hz and 180 Hz, which results in 280 Hz and 80 Hz for the sum and difference frequencies, letting you ignore the inevitable harmonic distortion at multiples of 100 Hz and 180 Hz. If you played 100 Hz and 300 Hz, the resulting 200 Hz and 500 Hz IM components will include regular harmonic distortion of the 100 Hz source at those same frequencies. Likewise, 5 KHz and 8 KHz are suitable for testing tweeters, but not 5 KHz and 10 KHz. You should also choose frequencies that are reasonable for the drivers to play at high volumes, where their distortion is greatest. A sustained loud 19 KHz tone is very demanding of a tweeter, and isn't representative of normal music. Playing such a high frequency for more than a few seconds at high volume can damage the driver, and maybe your hearing, too.

When measuring microphone IM distortion, you'll again choose frequencies that are reasonable for the speaker and mic to deal with at high volume levels. The best way to measure microphone IMD is with two speakers, with one tone played through each speaker. This avoids contaminating the resulting sum and difference frequencies with the loudspeaker's own IM distortion, which is typically higher than most microphones at low frequencies. You'll create separate files for each test frequency, panning them hard left and right for playback. The microphone should be halfway between each speaker and also on-axis to capture the flattest response from both speakers. Assuming your monitoring is set up correctly, with each speaker angled toward your ears, just put the microphone where you listen at the same height as the speakers, pointed straight ahead.

Figure 24.3 shows the IM distortion I measured for my Audio-Technica 4033 and DPA 4090 microphones using this method, by playing 300 Hz and 500 Hz through the JBL 4430 speakers in my home studio. As you can see, the AT 4033 has half the IM distortion of the DPA 4090. Adding the sum and difference signal levels, the AT is about 0.2 percent IMD versus about 0.4 percent for the DPA. You can also see harmonic distortion at various multiples of the 300 Hz and 500 Hz tones. The blip to the left of the −76 dB label is the 600 Hz second harmonic of the 300 Hz tone, and the 1 KHz second harmonic of the 500 Hz source tone is just to the right of the same label at an even higher level. It's not visible here, but when you hover the mouse cursor over the graph, a pop-up balloon shows the frequency and level at that location. This is how I identified the exact frequency and level of the various peaks.

It's difficult to measure harmonic distortion of speakers or microphones unless the distortion for one of them is known at a given frequency and SPL, or is at least known to be lower than the other device being tested. However, THD and IMD are not unrelated, so excessive IMD implies that THD will also be high, and likewise for low amounts. The 600 Hz and 1 KHz second harmonics in Figure 24.3 could be due to either the microphone or the speakers, but the 200 Hz and 800 Hz components can only be due to the microphone. Since the second harmonic from the speakers is louder than the sum and difference frequencies from the mic, it's reasonable to conclude that the speaker's distortion is greater. However, frequencies recorded by a microphone can align with a peak or null in the room's response and affect

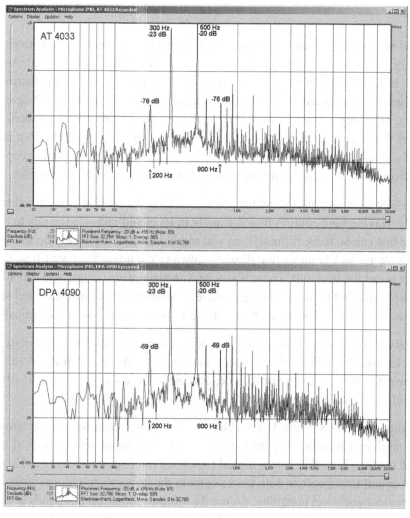

Figure 24.3: Microphone IM distortion was measured by playing 300 Hz through one speaker and 500 Hz through another, while recording the output of each microphone. An FFT display shows the levels of the resulting sum and difference frequencies.

the results, possibly by a large amount. I suggest measuring the raw response first to get a baseline for all four frequencies, then do the IMD tests without moving the microphone.

Finally, I'll point out that home loudspeaker and microphone tests as described here will never be as accurate as measurements made with real test equipment in an anechoic chamber. But you can definitely get useful data. So trust, but verify. If your results don't agree with what's expected, question the results until you're confident they're correct. Indeed, question everything, including measurements that seem too good to be true.

Null Tests

As explained in Chapter 1, a null test works by *subtracting* two audio signals to see what remains. If the result is total silence, then the signals are by definition identical. And if there is a residual, its audibility can be assessed based on the level and frequency distribution, or simply by listening. The beauty of a null test is that it reveals *all* differences between two audio signals, including differences you might not even be looking for. If someone claims playing Wave files from one hard drive sounds different from playing them from another hard drive, a null test will tell you for certain whether or not that's true. You simply copy a file to both hard drives and see if they null when aligned and played in a DAW program.

To subtract audio signals, you reverse the polarity of one, then mix it with the other at the same volume. Having both sources match exactly in level and phase is the key to a successful null test, and this can be more difficult than you might imagine. By watching the residual output on a wide-range VU meter that displays down to the noise floor, you'll tweak the level of one signal to get the best null. Then, if you're comparing recordings, you can slide one in time relative to the other to avoid time and phase differences. If the result is total silence, or at least below −80 dB, you can be confident that both sources are audibly identical.

Most null tests are done in a DAW program, placing the Before and After versions on separate tracks, with both files aligned to start at the same time and with equal volumes. If the tracks are off in time by even one sample, or their levels differ by even 0.01 dB, identical files that would have nulled to silence will yield some amount of residual.

I used null tests in my *AES Audio Myths* video[1] on YouTube to disprove two common myths: One myth is that audio plug-ins have a "sweet spot" signal level, which if exceeded harms sound quality. The other is that digital EQ cannot be countered exactly. This is a great application for null tests because both tracks contain the same source file, avoiding the need for time alignment and level matching. Figure 24.4 shows the setup that sends a mix through an EQ plug-in after raising the volume by 18 dB. The mix was normalized to peak at −1 dB, so the audio through the plug-in reaches 17 dB above digital zero. Track 1 contains the mix file, and Track 2 plays the same file with boost applied by the Track Trim to be before the EQ. The Sonalksis FreeG freeware volume plug-in restores the level after the EQ. This plug-in is great for null tests because it lets you adjust the volume in 0.01 dB steps, though only a whole number dB amount was needed here. Track 2 also has its polarity switch engaged, needed to create a null. Otherwise, when both tracks play, the output would be 6 dB louder than one track, rather than silence.

Figure 24.5 shows the setup for the second null test to confirm that digital EQ can be countered with equal but opposite settings. Here again, the same music file is placed on both tracks, aligned exactly to single sample timing. In this test, Track 1 contains two instances

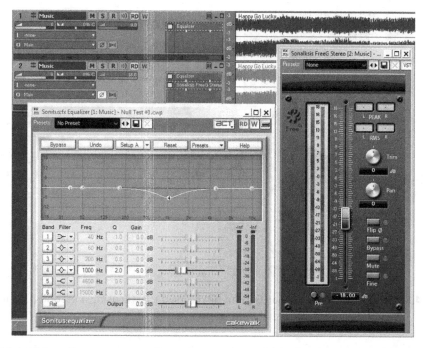

Figure 24.4: This null test setup confirms that modern 32-bit plug-ins are immune to overload, even when sent signals that greatly exceed the digital zero clipping point.

of the same EQ plug-ins, with three bands set to opposite amounts of boost and cut, with the same bandwidth (Q). Track 2 uses no plug-ins, but its polarity is reversed to cancel Track 1.

The first test proves that 32-bit floating point plug-ins are not overloaded by signals that exceed digital zero. As long as the volume is reduced somewhere later in the chain, before the mix is rendered to a file or sent out the sound card for monitoring, clipping distortion will not occur. The second test proves that EQ can be reversed exactly as long as both plug-ins are set precisely opposite. When playing either of these DAW projects, the result is total silence down to the −90 dB floor of the output bus level meter. Rendering the mix to a Wave file and examining that in an audio editor further confirms that the file contains only silence.

Null tests can also be done in real time using power amplifiers and other audio gear by subtracting the device's input and output signals. However, this is more complicated to set up, and differing amounts of phase shift within the devices can preclude obtaining a deep null. When testing something that amplifies the signal, such as a power amplifier, you need a way to make the input and output levels exactly equal. This can be done with variable resistors that are passive and thus won't add distortion, but you also need a way to reverse the polarity of one signal. That requires an additional circuit whose own distortion or non-flat

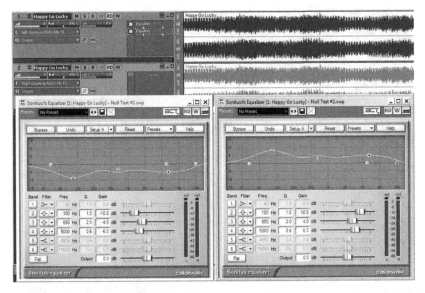

Figure 24.5: This null test proves that EQ can be reversed exactly using a second plug-in having equal but opposite settings.

frequency response can skew the test. A transformer can reverse polarity passively, but most transformers have more distortion than the solid-state audio circuits you're likely to test.

There are also situations where a null test isn't feasible at all. When someone in a hi-fi forum insisted he heard a change after demagnetizing an LP, I asked him to record an LP before and after demagnetizing. Amazingly, he did that, and he mailed me a CD with both Wave files. Alas, playback on even the best turntables varies constantly. A null test requires two signals whose samples are in exact lockstep. However, I was able to get the files to null anyway, albeit briefly. By placing the faster file a few samples ahead of the slower version in my DAW, when played the sound got softer, passed through near-silence as the samples aligned briefly, then became louder again as the tracks drifted apart.

The same happens when recording from analog tape recorders or even sound cards. The clock that sets the sample rate for a sound card is highly accurate and stable, but it still drifts a little. So you can often obtain a total null, if only for a second or two. Null tests also fail if one of the sources has been phase shifted relative to the other. Even if the amount of phase shift is small, and inaudible, it still changes the waveform, preventing complete cancellation.

Besides the fidelity tests already described, I'll also mention the fabulous RightMark Audio Analyzer software available in both free and affordable "pro" versions. This program performs all of the standard audio tests including frequency response, noise, distortion, and much more, using your computer and sound card.

Disproving Common Beliefs

Audio software can also be used to disprove other common audio myths, even without a null test. Some people wrongly believe that mixing two sine waves in a console or DAW creates sum and difference frequencies. Chapter 13 explained that amplitude and frequency modulation are linear processes that create new frequencies, but that's not the same as simple summing. To demonstrate this for an audio forum, I created a Wave file in Sound Forge having equal amounts of 50 Hz and 200 Hz, then used the FFT analyzer to display the content. The FFT showed only two blips: one at 50 Hz and another at 200 Hz, with nothing at 150 or 250 Hz.

Another simple home test assesses degradation from a seemingly transparent sound card or other device by listening only. Most modern gear is very clean, so auditioning music recorded through a device using a single loop-back may not show an audible difference. The solution is to record repeatedly through the device to accumulate the degradation. When I wanted to assess the degradation from an inexpensive SoundBlaster sound card, I recorded the same piece of music through it 20 times in succession. Each recording became the new playback source for the next, so in the end I had 20 separate files, each progressively worse than the previous one. This test is also shown in my *AES Audio Myths* video, and the result files after 1, 5, 10, and 20 passes can be downloaded from my website ethanwiner.com.

Oscilloscopes

As you've seen, many simple but useful tests can be performed using only basic audio software and a sound card. But for more serious testing, it's useful to have an oscilloscope. An oscilloscope lets you see what's really happening in an audio device at frequencies much higher than a sound card can handle, and decent models can be bought new for less than $400 or used for as little as $25. You don't need a sophisticated oscilloscope for audio testing, and even an inexpensive used model is better than a computer simulation using a sound card. Even though nobody can hear 80 KHz, power amplifiers and other circuits can oscillate or ring at those high frequencies. An oscilloscope lets you see the actual waveforms and gives an insight not attainable any other way. I used a VST oscilloscope plug-in for the *Analog Synthesizers* video from Chapter 14 to show what happens when sweeping a resonant low-pass filter. It's not possible to explain oscilloscopes in depth here, but I'll cover the basics.

The main purpose of an oscilloscope is to display signals as they change over time. A voltmeter is fine for measuring steady levels like test tones or for checking flashlight batteries, but it's impossible to observe a signal's instantaneous value—or even tell if you have a square or sine wave for that matter—because the meter's pointer can't move fast enough. Old-school oscilloscopes use an electron beam that creates a dot of light when the beam strikes the phosphor coating inside a cathode ray tube (CRT). The beam then sweeps repeatedly across the screen quickly enough to track the input waveform. Modern oscilloscopes use digital LCD displays, but you can still find old CRT models for sale used.

Every oscilloscope has both a vertical and horizontal amplifier, as well as a built-in variable frequency sawtooth oscillator to generate the recurring sweep. A CRT uses a single dot of light, so the dot must be swept constantly from left to right to create the illusion of a continuous solid line. This is how CRT televisions work, too. The signal you're watching goes to the vertical amplifier's input, which shifts the dot up or down, as shown in Figure 24.6.

Without the horizontal sweep, there's only a single dot in the center of the screen, and sending a signal to the vertical input simply moves the dot up and down. By using both the vertical and horizontal inputs, the signal's voltage can be determined by the amount of vertical deflection and its frequency determined by the horizontal position. The sweep rate is variable, letting you view any number of cycles of the input wave over a wide range of input frequencies. Both the vertical amplitude and horizontal time can be read directly from the calibrated lines, called a *graticule*, drawn on the face of the CRT as shown in Figure 24.7.

Switches set the gain of the vertical amplifier, which is calibrated in volts per division, where each division is one horizontal line on the screen. The frequency of the horizontal sweep oscillator is controlled in a similar fashion to vary how long it takes the dot to move one division to the right. When the dot reaches the right edge of the screen, the sawtooth ramp resets, quickly sending the light beam to the left edge to start a new trace. In all but the least expensive models, the dot is turned off during the retrace. Otherwise, a confusing double trace results.

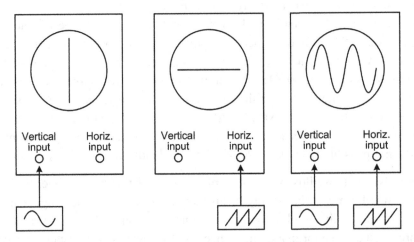

Figure 24.6: Oscilloscopes have both vertical and horizontal inputs to move a single dot quickly enough to display a waveform.

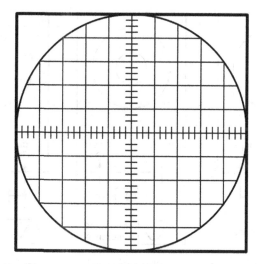

Figure 24.7: A graticule is a group of lines that identify voltage level vertically and time span horizontally.

One of the most important features of an oscilloscope is *triggered sweep*, which synchronizes the horizontal sweep to the input waveform. Without this feature the displayed waveform would flicker and wander around, because the input waveform is likely at a different part of its cycle when each new sweep begins. Instead of immediately beginning a new sweep as soon as the beam is reset to the left edge of the screen, the trigger circuit delays the sweep until the input signal returns to the same voltage as when the previous sweep began. Triggering usually occurs on the rising edge of the input waveform, though better oscilloscopes have a switch that lets you trigger on either the rising or falling edge.

Another standard feature is a continuously variable level control on the vertical input for making relative, rather than absolute, voltage measurements. If you're measuring the response of a filter, you'd set the unfiltered wave to exactly fill the screen. Then, it's easy to see when the signal is attenuated by one-half, one-quarter, or whatever after passing through the filter. In fact, many oscilloscopes have an additional dB scale printed on the graticule to allow reading dB changes directly. Likewise, a calibrated fine-tuner called a *vernier* is provided for the horizontal sweep speed as well to simplify relative frequency measurements. Every oscilloscope also includes an AC/DC switch, which inserts a capacitor into the signal path to observe only the AC component of a signal, ignoring any DC offset present. This is useful for inspecting small levels of high-frequency noise on a DC power supply's output, letting you raise the vertical gain without having the much larger DC voltage push the trace off the top or bottom of the screen.

Most oscilloscopes let you disable the automatic sweep, and most also let you feed a signal directly into the horizontal amplifier to measure the phase difference between two sources.

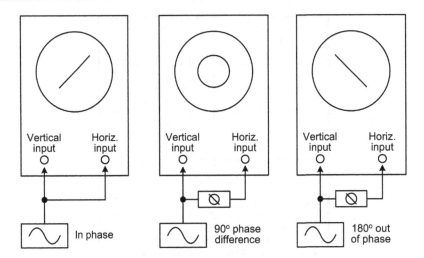

Figure 24.8: Sending audio signals separately to the vertical and horizontal inputs lets you assess the relative phase difference between them.

Figure 24.8 shows that identical in-phase signals applied to both inputs creates a diagonal line. Contrast that to the other patterns you get when different amounts of phase shift are introduced. This setup is similar to the software phase correlation meter in Figure 5.6 from Chapter 5 showing how to verify a stereo mix for mono compatibility. Indeed, this was an important use of oscilloscopes in recording studios before digital metering plug-ins were available.

Another useful feature found on medium- and high-priced oscilloscopes is dual-channel capability. This is essential if you need to view two different signals at once, such as an input and output, or both channels of a stereo device. Dual-channel oscilloscopes offer two different modes: *alternate* and *chop*. In alternate mode, the first signal sweeps across the top half of the screen, and then the second signal sweeps across the lower half. The chop mode is created by a single sweep, with the channels switching back and forth very rapidly. Depending on the frequency of the signal and the chopping rate, one mode or the other will provide a more stable pair of traces. Most dual-trace oscilloscopes also offer a differential mode. Here, different signals are sent to each vertical input, and a switch selects whether the inputs are added or subtracted as in a null test.

Once you get to the most expensive oscilloscope models, you'll find storage capability, which is a method to freeze the display even after the input has been removed. This is needed for viewing events that occur quickly only once. Early storage oscilloscopes used a charged grid inside the CRT face to retain the beam pattern, though modern units use digital memory to store the data.

Summary

This chapter explains several simple but useful tests anyone can perform using only audio software and a sound card. The two basic test types are measuring and listening, and listening tests can be either sighted or blind. Blind tests are needed when differences are subtle to be sure you really can hear a difference. For listening tests, music is a fine source, but for measuring, sine and square waves are necessary because they're precise and repeatable. The basic test setup uses a loop-back to play and record through a sound card or external converter. This is more accurate than using a budget multimeter that can't measure past 10 KHz. Further, seeing the recorded waves, or an FFT, reveals much more than a single voltage number on a meter.

Frequency response is one of the easiest tests to perform and is done by sending sine waves or a sweep through the device being measured. Low-frequency square waves are also useful because they reveal ringing that might be hidden when playing static tones. Analyzing the distortion of an audio device is equally simple, using an FFT display to see the individual distortion components. You can also measure the distortion of microphones and loudspeakers, though the distortion of one of them must be known, and preferably lower, in order to assess the other.

We also covered null tests, using two DAW projects I created to disprove the myth that 32-bit plug-ins have a sweet spot signal level above which sound quality is harmed and that EQ cannot be countered exactly. Finally, the basic principles of oscilloscopes are shown, including using the vertical and horizontal inputs to identify phase differences between two audio channels.

Note

1   www.youtube.com/watch?v=BYTlN6wjcvQ

CHAPTER 25

Computers

Try to learn something about everything and everything about something.

—*Thomas Henry Huxley*

When I entered the world of audio and recording in the 1960s, a studio owner had to know how to align tape recorders and swap circuit boards, if not repair them directly. My, how things have changed! Today, instead of aligning tape heads and repairing electronics, you have to know how to optimize your computer to handle as many tracks and plug-ins as possible, back up hard drives to protect your operating system and data files, and install drivers for a new sound card. Many musicians and studios also have their own websites, separate from social networking sites, which require yet another set of skills to maintain. Indeed, an audio expert today must also be a computer expert. I'll use Windows for the following explanations and examples, because that's what I'm most familiar with, but the concepts apply equally to Mac and Linux computers.

For most of us today, audio maintenance includes knowing how to properly set up and organize a personal computer. I've owned computers since the Apple][, and I've used every version of DOS and Windows since then. My audio computer runs all day long, yet it never crashes or misbehaves. And that's not because I'm lucky. Further, I also use my audio computer for other tasks like emailing, creating websites, image scanning and graphics design, finances, and computer programming, with no problems. Contrary to popular opinion, there's no inherent reason why a computer that records audio cannot be used for other things, too. The key is being organized: Install only programs you really need, defragment your hard drives, disable unneeded background tasks, and back up faithfully in case of disaster. By being organized and keeping your system clean, your computer will run better, and you'll be able to work faster, smarter, and more safely.

Divide and Conquer

One of the most important ways to keep your computer organized is to divide today's extremely large hard drives into separate, smaller partitions. Even entry-level computers come with hard drives that would have seemed impossibly large just a few years ago. But storing all of your files on one huge drive is like tossing all your personal correspondence

Figure 25.1: This 500 MB SCSI hard drive has five platters, and cost a fortune when new. Now it's just a really cool paperweight. *Photo by Jay Munro.*

and tax records for the past 20 years into one enormous shoebox. As each year passes, it becomes more difficult to find anything. I can't tell you how many times I've heard, "Help! I downloaded a file yesterday, but now I can't find it." Figure 25.1 shows an old hard drive I took apart just for fun.

A disk drive is much like an office filing cabinet. For example, a four-drawer filing cabinet is equivalent to a drive with four partitions, or virtual drives. The cabinet is divided rather than left as one enormous drawer, which would be clumsy and difficult to manage. Inside each drawer is a series of hanging folders, which are the same as the folders in a hard drive's root directory. And within each hanging folder are manila folders that contain both loose papers as well as other manila folders, equivalent to disk files and folders, respectively. Besides dividing hard drives into partitions, it's equally important to organize your folders and files in a logical manner. The "partitions_folders" video shows how I partitioned and organized the hard drives on my previous computer that I replaced while working on this book.

Besides helping to organize your data into logical groups, partitioning a drive has many other benefits. Perhaps most important, it's a lot easier to back up your operating system and programs separately from data files. When everything is on one drive, it's difficult to know which files have changed since the last time you backed up. There are thousands of operating system (OS) files stored in various places on a system drive, and their names are not obvious. Further, if an errant program installation or driver update damages Windows such that it won't even start, you'll be hard-pressed to recover your backup files at all unless you have an original installation DVD to first restore the OS. But most important, if you restore an entire drive from a previous backup, your audio files and other data will be lost in the process because older files will overwrite the newer current versions.

The solution is to divide your hard drives based on the types of files each partition will store. Storing files in separate partitions also reduces file fragmentation. As files are saved, deleted, and resaved, they become split up into many small pieces in various places around the drive. When you play those files later, the drive works harder as it navigates to different areas of the drive's magnetic platters to gather up all the pieces. With computers used for audio and video projects, drive fragmentation limits the number of tracks you can record and play at once. Even with modern fast computers, if a drive is badly fragmented, you might not be able to record even one or two tracks without dropouts.

Most computers include defragmenting software that consolidates all of the files into contiguous sectors on the disk. Dividing a hard drive lets you defragment partitions independently and only those that actually need it. For example, my SoundFont instrument sample files rarely change, so they don't become fragmented. But if they were on the same partition as my audio project files, it would take much longer to defragment the partition because many gigabytes of SoundFont files would be shuffled around in the process, adding hours to the process. By keeping files that don't change separate from files that change often, defragmenting is more efficient. Several small partitions can be defragmented much more quickly than one large partition, even when the total amount of disk space is the same.

Note that solid state drives (SSDs) don't need to be defragmented, nor should they be. Solid state drives have a large but limited number of write cycles, so data shouldn't be written more often than needed, as happens while a drive is being defragmented. Further, since there's no moving head traveling to different areas of a spinning platter, it doesn't even matter where the data is located. Reading and writing always occur very quickly.

Current versions of Mac OS include partitioning software, but the partitioning tools in Windows 7 and later are too limited to be useful. So for Windows computers, you'll do better using EASEUS Partition Master, a freeware program shown in Figure 25.2. This screen shows both physical hard drives in my computer, including all of the partitions on Disk 1. Disk 2 has a single partition I use to back up files from the first drive. I also have four large external USB drives for additional backup, but they weren't connected when I made this figure. An external drive that's connected only while backing up is a great precaution against ransomware viruses that destroy every file on your computer. The various operations are invoked from the menu choices at the left of the screen. I'll also mention Partition Magic, another excellent program. It's not free, but it's affordable, and it has even more features than Partition Master.

As you can see, a 108 GB Drive C: partition holds only Windows and programs. Another 40 GB D: partition holds what I call "small data"—word processor files, my income and

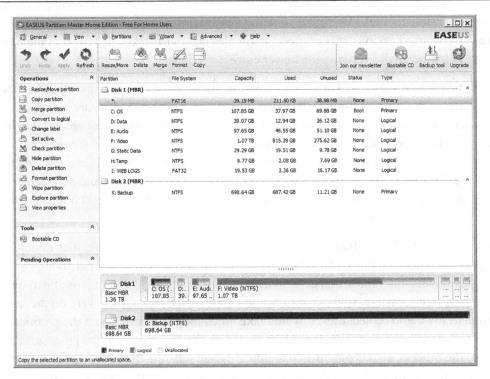

Figure 25.2: EASEUS Partition Master lets you easily divide a large hard drive into smaller partitions.

expenses database, client web pages and graphics, family photos, and so forth. The 100 GB E: partition holds all of my current audio projects and final mixes, and the large 1 TB F: partition is for video projects. Another 30 GB G: partition holds what I call "static data"—drivers and programs I've downloaded, SoundFont instrument sample sets, and other files that rarely change. The 10 GB H: partition holds temporary files I create or download, then delete, as well as temporary Internet files saved by my web browsers, and temporary storage for programs like Sound Forge and WinZip that save, then delete large files while they work. I also made an I: partition in the older FAT32 format needed for a DOS program that I wrote years ago to process my website log files. It's not necessary to divide a hard drive into this many partitions, though you should at least create a data partition that's separate from the OS and your programs. It's also a good idea to make the partitions much larger than the amount of data you intend to store in them, because defragmenting goes faster when there's plenty of free space on the drive.

I generally partition the main drive when a computer is first purchased, before installing any programs. Otherwise, once you've divided a single large drive into smaller partitions, you'll move data files that had been on Drive C: to other partitions. For Windows computers, most of these files are in subfolders under My Documents. Fortunately, most programs let you specify other locations, which you'll need when you set up folders in a separate data partition. It's also useful to move your temporary Internet files from C: to another partition, because they, too, become quickly fragmented. There's no need to actually move these files. Just change their location in your web browser's settings, and then after restarting your computer, you can delete any abandoned files left on the C: drive. Do the same for your email program to keep its files on a data drive that's backed up often.

Again, the whole point of moving data files off the OS drive is to let you back up the OS and software programs separately from your data. This way you can easily back up your data daily, which goes quickly because you copy only what's new or changed. The OS partition changes only when you install or remove programs, or change system settings, so you can back that up only occasionally by making an *image* backup of the partition to a separate internal or external hard drive. Without all of your large audio and video files and temporary Internet files, an image backup also goes quickly and takes up much less space.

Back Up Your Data

There's a saying in the computer industry that data does not truly exist unless it resides in three places. The next time you're about to power down your computer, ask yourself, "If I turn on my computer tomorrow and the hard drive is dead, what will I do?" If backing up seems like a nuisance, then devise a strategy that makes it easy so you'll do it. I use an excellent program called SyncBack from 2BrightSparks, shown in Figure 25.3, but there are others ranging from freeware to inexpensive.

As you can see, I created several different backup profiles for each group of data. This lets me alternate backing up to multiple drives. Imagine you're in some program—an audio

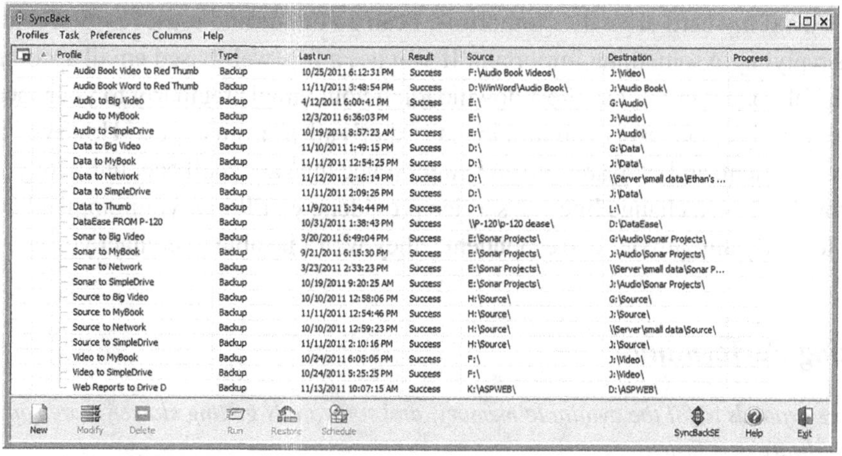
SyncBack
Profiles Task Preferences Columns Help
Profile Type Last run Result Source Destination Progress
Audio Book Video to Red Thumb Backup 10/25/2011 6:12:31 PM Success F:\Audio Book Videos\ J:\Video\
Audio Book Word to Red Thumb Backup 11/11/2011 3:48:54 PM Success D:\WinWord\Audio Book\ J:\Audio Book\
Audio to Big Video Backup 4/12/2011 6:00:41 PM Success E:\ G:\Audio\
Audio to MyBook Backup 12/3/2011 6:36:03 PM Success E:\ J:\Audio\
Audio to SimpleDrive Backup 10/19/2011 8:57:23 AM Success E:\ J:\Audio\
Data to Big Video Backup 11/10/2011 1:49:15 PM Success D:\ G:\Data\
Data to MyBook Backup 11/11/2011 12:54:25 PM Success D:\ J:\Data\
Data to Network Backup 11/12/2011 2:16:14 PM Success D:\ \\Server\small data\Ethan's ...
Data to SimpleDrive Backup 11/11/2011 2:09:26 PM Success D:\ J:\Data\
Data to Thumb Backup 11/9/2011 5:34:44 PM Success D:\ L:\
DataEase FROM P-120 Backup 10/31/2011 1:38:43 PM Success \\P-120\p-120 dease\ D:\DataEase\
Sonar to Big Video Backup 3/20/2011 6:49:04 PM Success E:\Sonar Projects\ G:\Audio\Sonar Projects\
Sonar to MyBook Backup 9/21/2011 6:15:30 PM Success E:\Sonar Projects\ J:\Audio\Sonar Projects\
Sonar to Network Backup 3/23/2011 2:33:23 PM Success E:\Sonar Projects\ \\Server\small data\Sonar P...
Sonar to SimpleDrive Backup 10/19/2011 9:20:25 AM Success E:\Sonar Projects\ J:\Audio\Sonar Projects\
Source to Big Video Backup 10/10/2011 12:58:06 PM Success H:\Source\ G:\Source\
Source to MyBook Backup 11/11/2011 12:54:46 PM Success H:\Source\ J:\Source\
Source to Network Backup 10/10/2011 12:59:23 PM Success H:\Source\ \\Server\small data\Source\
Source to SimpleDrive Backup 11/11/2011 2:10:16 PM Success H:\Source\ J:\Source\
Video to MyBook Backup 10/24/2011 6:05:06 PM Success F:\ J:\Video\
Video to SimpleDrive Backup 10/24/2011 5:25:25 PM Success F:\ J:\Video\
Web Reports to Drive D Backup 11/13/2011 10:07:15 AM Success K:\ASPWEB\ D:\ASPWEB\
New
Modify
Delete
Run
Restore
Schedule
SyncBackSE
Help
Exit

Figure 25.3: SyncBack lets you easily copy all of your data to one or more backup drives.

editor, word processor, or any program that uses files—and you save the file. But unknown to you, a software or hardware glitch caused the file to be corrupted. If you back up that corrupted file, it overwrites your only good backup copy, leaving you with two corrupted files. Admittedly such file corruption is rare, but it happens. It's happened to me a few times, and it will probably happen to you eventually, too. The solution is to reload the file after saving to verify it's good before overwriting your only backup. I did that constantly while writing this book. If I opened an earlier chapter to add or clarify something, I'd save and then reopen it again right away to verify it was okay. Most programs have a "recently used" file list, so it takes only a few seconds to reopen a file to verify its integrity.

It's also a good idea to alternate backups to separate drives. One day you back up to Drive 1, the next day to Drive 2, and then back to Drive 1 the third day. I rotate between five backup drives—a second drive inside my computer, a network server in my basement, a 128 GB thumb drive, and two external USB drives I turn on only when needed. I also have two more external drives that I leave with friends and trade every few months. This way, even if my house is robbed or burns down, I won't lose everything. While writing this book, I backed up the entire project to DVDs once a week in case I needed to refer to an older version of something. I also bought a second 128 GB thumb drive and carry it with me whenever I leave the house. There are "cloud" storage services that let you store data remotely on their web server, and many are inexpensive. I use Microsoft's SkyDrive, a free service included with Hotmail accounts. Again, you don't have to be as compulsive as I am, but not backing up is an invitation to lose every audio project you've ever done, not to mention your financial data and priceless photos.

Backing up your emails can be easy or difficult, depending on how you manage your email. As a businessperson, I keep all incoming and outgoing emails on my hard drive for at least three months, so I can refer to past correspondence. I use Thunderbird, a free email client that runs on my computer. I have it set to delete incoming emails stored on the remote server after three days. This way, if my hard drive dies tomorrow, I can go online and retrieve my recent emails to not miss replying to something important. If you use a free web-based email account such as Hotmail, Yahoo, or Gmail, I suggest downloading your emails regularly to your local data drive. A free email service owes you nothing, and that's all you can expect. However, most free services are excellent, and they are often a better choice than an email account with your local Internet provider. If you change Internet service providers, you'll lose your email address, and if you forget to notify any of your correspondents, they won't be able to contact you.

Optimizing Performance

Software expands to fill the available memory, and software is getting slower more rapidly than hardware gets faster.
—Niklaus Wirth, inventor of the Pascal programming language

The most important reason to optimize a computer is to let it use more of its computing power to process your audio (and video) data and plug-ins rather than waste resources on background tasks such as constantly verifying your Wi-Fi network's signal strength. Another is to reduce latency, which is the time it takes audio to get into and out of the computer. Modern computers are very fast and don't require all the tweaks used in years past to maximize performance. But still, there's no point in having programs you don't need running in the background slowing down your computer and wasting memory.

On Windows computers, most programs that run in the background are listed under Services, which you get to from the Control Panel. Others are listed in the Startup tab of msconfig, shown in Figure 25.4, which you can run by typing its name in the Start menu.

Don't just disable everything you don't recognize! Many needed background programs and services have names that aren't intuitive. But many are recognizable, and the descriptions in the Services list tell what they do. For example, many CD and DVD writing programs run in the background, waiting for a blank disk to be inserted. You don't need that running all the time and wasting memory. Rather than give a list of services that aren't usually needed, and that will surely change with the next version of Windows, I suggest searching the web for "disable Windows services" to find websites that list the current details. Another useful resource is Mike River's Latency article.[1]

I also suggest disabling file indexing, which runs in the background and creates a database of the contents of every file on a hard drive to make searching go faster. This is a useful feature for an office computer, but it risks slowing down your drive for audio and video projects. Right-click a drive letter in Windows Explorer, then select Properties, and uncheck that box on the General tab.

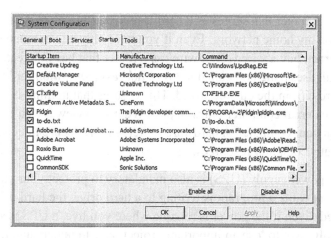

Figure 25.4: The Startup tab of msconfig lists programs that run automatically every time your computer starts. Entries at left that aren't checked are disabled.

Write caching is another "feature" that should be disabled for all removable drives because it delays writing your data. If you copy a large file and then unplug the drive, the data may not have been copied yet. Again, right-click the drive letter in Windows Explorer, then select the Hardware tab. Highlight a removable drive, select Properties, then check Quick Removal to disable write caching on the Policies tab. If there is no Policies tab, then that drive is not at risk.

Windows offers a System Restore feature that takes snapshots of the operating system and programs every few days, or when you tell it to, letting you undo an improper or unintended change. But System Restore doesn't always work, and it's yet another service that runs in the background, monitoring your files and potentially taxing your hard drive. (However, System Restore can be set to take snapshots only when you tell it to.) Making an image backup to an external drive after major changes is much safer. Current versions of Windows include an image backup utility, but I've found it to be unreliable so I use Acronis True Image which is excellent and affordable.

Practice Safe Computing

As explained earlier, I use one computer for everything. I understand that some people prefer to keep their audio computer off the Internet, but it's not really needed if you practice "safe computing," which includes keeping a current image backup of the operating system and programs, as well as backups of all your audio and other data. Further, if your audio computer is not connected to the Internet, it's difficult to install or update software that "phones home" to verify legitimate ownership. But I also have antivirus software installed to ensure that a rogue website won't infect my computer. I use the AVG antivirus program, but others are also effective. It's a nuisance to have to pay year after year for updates, but it's worth it because the software authors continually improve their programs to keep up with the latest threats.

I also have a network router that serves as a hardware firewall to block outside access to all the computers on my home network. Even if you have only one computer that connects directly to the Internet, buy a wired—not wireless—router that has this feature (many do). This type of firewall uses a method called *network address translation*, or NAT, to hide your computer's IP address from the outside world. This ensures that nothing is allowed into your computer from an external computer unless a prior request to view a web page or download a file originated from your computer. A hardware router is more secure than a software firewall, and it doesn't use any computer resources.

An uninterruptible power supply (UPS) is also mandatory to protect your computer and data. If the power goes out while a program is writing to a hard drive, the hard drive is almost sure to be corrupted. Even if the power is reliable where you live, a UPS is still a good investment. You only have to lose an important project once to understand the value of a UPS. An expensive large-capacity UPS isn't necessary; it only needs to provide power long enough to shut down properly.

Finally, a computer is not a sacred device that must be left powered on all the time. It's just an appliance! I suggest turning it off (not standby) when you're not going to use it for an hour or more. It's also a good idea to reboot before important sessions, especially if you've just done a lot of audio or video editing or web surfing.

If It Ain't Broke, Don't Fix It

I never update software unless something doesn't work right or I truly need a new feature. I know people who update to every new version when it's first released and update their OS every time a new security patch arrives, which is almost daily. I can't count how many times I've seen web forum posts where someone updated something and it made their computer worse. Many computing professionals avoid any software version number ending with ".0," instead waiting for the inevitable subsequent version that actually works as advertised. Many programs check for updates every time you power on your computer, and Windows does the same. The first thing I do after installing Windows or new programs is disable automatic updates if present. The second thing I do, when applicable, is tell the program to save its temporary files to my Temp partition to avoid cluttering and fragmenting the C: drive. Sometimes updates are desirable or necessary, and setting a Windows Restore point manually or making an image backup before installing a new program or upgrade is a good idea in case the installation doesn't go as planned.

Avoid Email Viruses

In the past I always checked my email online in a web browser to delete spam and other unwanted emails before downloading to my computer. As mentioned, I save all of my emails in and out for three months, so I prefer to avoid cluttering my data drive with unwanted emails. I currently use a fabulous program called MailWasher Pro that flags spam from both my personal and business email accounts, and lets me delete it all with just a few mouse clicks. If you spend more than five or ten minutes per day dealing with spam, you need this program. Also, my email program is set to display emails in plain text rather than as HTML, which lets spammers know your email address is valid when an embedded image links back to their website. If that's too much effort, the following guidelines will help you to avoid email viruses.

Rule #1: Never open a file attachment you receive by email, even if it comes from your best friend. Rather, save it to disk first, then open it from the appropriate program. What looks like an innocent link or photo file may install malware on your computer. Many viruses propagate by sending themselves to everyone in the infected person's address book, so the virus arrives in an email from a friend rather than a stranger. I receive virus files by email very often. The last time was from a good friend. Of course, I didn't open the attached file, but he had no idea his computer sent the email!

JPG and GIF images are always safe to open, but a virus can be disguised as an image file. By default, Windows hides extensions for common file types much as. exe,. jpg,. pdf, and so forth. So the sneaky bastards that create viruses often name them hotbabe.jpg.exe or joke. doc.exe, because most people will see only hotbabe.jpg or joke.doc, without the real .exe extension at the end. Fortunately, this is easy to fix: From Windows Explorer, go to the Tools menu, then select Folder Options and click the View tab. Find "Hide extensions for known file types" and make sure it's not checked. But even with extensions revealed, you still must be careful. If an attached file has a very long name, the extension may be hidden if there's not enough room to display the full name.

Most viruses are executable programs having an. exe file extension, but they can also use. com,. vbs,. pif,. bat, and probably others. Although .doc (Microsoft Word) and .xls (Microsoft Excel) files are usually safe, viruses can be hidden inside them in the form of self-running macros. In practice, very few people need to use Word's macros feature, so you should set Word and Excel preferences to warn you whenever such a macro is about to run.

In the long run, the best way to avoid receiving viruses by email is minimizing the number of people who know your email address. More and more web forums and online product registration forms require an email address. But unless you really want to let them contact you (perhaps to be notified of product updates), use a phony email address like nospam@ nospam.com. Or set up a free email account just for registrations, since most web forums require you to follow up to their email before you're accepted.

Finally, never click a link in a spam or other unwanted email that offers "to be removed from our mailing list." That's just a trick; the real purpose is getting you to confirm that they sent to a valid address. Once they know they found a "live" one, your email address is worth more when they sell it to other spammers.

Bits 'n' Bytes

Today's production equipment is IT based and cannot be operated without a passing knowledge of computing, although it seems that it can be operated without a passing knowledge of audio.

—*John Watkinson*

A complete explanation of computer internals is beyond the scope of this book, but some of the basics are worth understanding to better appreciate how audio software works. There are two types of computer memory: random access memory (RAM) and read-only memory (ROM). RAM is the memory used by software to store data you're working on, such as text documents and emails as you write them, and MIDI and audio clips as you work in a digital audio workstation (DAW). When you power down your computer, whatever data is in RAM is lost unless you saved it to a disk drive. ROM is permanent, and it's used for a computer's BIOS— the basic input/output system—to store the low-level code that accesses hard drives and other hardware needed to start your computer before the OS is loaded from disk into RAM. Video cards and sound cards also contain their own ROM chips for the same purpose. Some types of ROM, called flash memory, can be written to for software updates. Flash memory is also used in modern electronic devices instead of batteries to save user settings when the power is off. But most ROM is permanent and cannot be changed.

As explained in Chapter 8, the smallest unit of digital memory is one bit, which holds a single One or Zero value. One byte (also called an octet) contains eight bits, one word contains two bytes or 16 bits, and a double-word contains four bytes or 32 bits. When audio is recorded at 24 bits, three bytes are used for each sample. There's also the nybble, which is four bits or half a byte, though it's not used much today because most memory chips are organized into groups of bytes or words. These data sizes are shown in Table 25.1. The binary numbers in the second column are just for example, to show the size of each data type.

Both RAM and ROM memory chips are organized in powers of 2 for efficiency. So 1 kilobyte of memory actually contains $2^{10}=1,024$ memory locations rather than only 1,000. Therefore:

> 1 KB = 1,024 bytes
> 1 MB = 1,024 KB
> 1 GB = 1,024 MB
> 1 TB = 1,024 GB

Technically, 1 MB is one megabyte or 1,048,576 bytes; 1 GB is one gigabyte or 1,073,741,824 bytes; and 1 TB is one terabyte or 1,099,511,627,776 bytes. But not everyone uses this method—especially companies that sell hard drives—so sometimes 1 GB really means only 1,000,000,000 bytes. A mix of formats is also used, where 1 MB = 1,048,576 bytes, but 1 GB = 1,000 MB rather than 1,024 MB. The great thing about standards is there are so many of them.

Table 25.1: Digital Memory Units

Unit	Size
1 Bit	1
1 Nybble	1011
1 Byte	1011 0110
1 Word	1011 0110 1010 0011
1 Double Word	1011 0110 1010 0011 1010 0110 0001 0011

Computer Programming

Many people think that computer programming is a complex science that requires advanced math skills. Nothing could be further from the truth. Now, some types of programming require high-level math, such as coding an equalizer plug-in or a Fast Fourier Transform, or advanced financial software, but most programming is based on simple IF/THEN and AND/OR logic:

> *IF the mouse is clicked, AND it's currently positioned over the Solo button, THEN mute all of the other tracks that are playing.*

In this case, Mute is activated by storing either a One or a Zero in a memory location set aside to hold the current state of each track. So when you open a project with 20 tracks, your DAW program sets aside at least 20 bits of memory just to hold the current Mute state of each track. Then when you press Play, the program checks each location to find out whether it is to include that track in the playback. If a track is not muted, then the program reads the Wave file that's active at that time in the project and sends it through each plug-in and output bus. This is very logical, and computer programming certainly requires being organized, but it's not necessarily as complicated as many believe. Most types of computer programming require little math beyond addition, subtraction, multiplication, and simple IF/THEN logic tests.

Computer programming is also an art as much as a science. Yes, you need to know how memory is organized and other science facts, but there are many ways to accomplish the same functionality, and some algorithms are far more elegant and efficient than others. The code that sorts your address book alphabetically might occupy 500 bytes or 5,000 bytes. It might take 10 milliseconds to sort 100 names or be able to sort thousands of names in less than 1 millisecond. This is not unlike electronic circuit design, where the goal is low distortion, a flat response, and low noise, while consuming as little power as possible to avoid wasted energy and excess heat. There are a dozen ways to design a mic preamp, and some are decidedly better than others. The same is true for programming.

A *high-level language* lets you use English words like IF and THEN to instruct a computer without having to deal with the extreme detail and minutiae needed to talk to a computer in its native *machine language*. The classic first example taught in programming classes is called Hello World, and such programs go back to the 1960s when the BASIC language was invented:

```
10 CLS                  REM clear the screen
20 LOCATE 10, 15        REM put the cursor at the 10th row, in
                        the 15th column
30 PRINT "Hello World!" REM print the quoted text at the current
                        cursor position
```

In early versions of BASIC, each command was numbered so the computer would know in what order to do each operation. Lines were often numbered by tens as shown to allow inserting other commands later if needed. The REM to the right of each command stands for *remark*, letting programmers add comments to their code to remind themselves or other programmers what each command does. A program this simple is self-evident, but complex programs can be a nightmare to understand and modify a year later if they're not well documented—especially for someone other than the original programmer. This style of programming was great back when computers were text oriented. Today, modern operating systems require more setup than simple LOCATE and PRINT, but the underlying concepts are the same. For those who are curious, the Liberty Basic (Windows) source code for the ModeCalc and Frequency-Distance programs included with this book are in the same Zip files that contain the executable programs.

Coding an Equalizer

Chapter 1 gave a simplified explanation of digital filters, but in truth a digital equalizer is slightly more complex. Understand that all equalizers are based on various filter types. Computer code that implements a filter is called *Digital Signal Processing*, or DSP for short. Most digital filters emulate equivalent analog filters, and the common language for all filters is mathematics. Therefore, several trigonometry formulas are shown in the sample code that follows, and there's no escaping this! But the basic operation of the computer code that implements an equalizer is not too difficult to follow, even if you don't understand the meaning of the formulas. To keep this example as brief as possible, the code below implements a simple high-pass filter having one pole (6 dB per octave). Formulas to implement other filter types including those used in parametric equalizers are shown on the Cookbook Formulae web page by Robert Bristow-Johnson.[2]

The computer code below is written in the C programming language, which is typical for audio programs and plug-ins. In C, a semicolon is used to mark the end of each command. Note that text bounded by the markers /* and */ are *comments*, meaning it's not part of the code but rather explains what the code does. For a simple program like this filter, any competent C programmer can understand how the code works. But comments are necessary for more complex programs, especially if the code might be supported or modified later by other programmers.

The initial part of the program defines several named values that are used by the filter portion that follows. If you look at the other filter type examples on the Formulae web page mentioned above, you'll see they all use a group of values labeled a0 through b2. The specific formulas used depend on the filter type being implemented.

```
/* These floating point values are used by the filter code below */
float Fs = 44100; /* sample rate in samples per second */
float Pi = 3.141592; /* the value of Pi */
```

```
/* These floating point values implement the specific filter type */
float f0 = 100; /* cut-off (or center) frequency in Hz */
float Q = 1.5; /* filter Q */
float w0 = 2 * Pi * f0 / Fs;
float alpha = sin(w0)/(2 * Q);
float a0 = 1 + alpha;
float a1 = -2 * cos(w0);
float a2 = 1—alpha;
float b0 = (1 + cos(w0)) / 2;
float b1 = -(1 + cos(w0));
float b2 = (1 + cos(w0)) / 2;
```

This next section defines two floating point *arrays* that are used as memory buffers:

```
/* The Buffer[] array holds the incoming samples, */
/* PrevSample[] holds the intermediate results */
float Buffer[1024]; /* this array holds 1024 elements numbered 0
    through 1023 */
float PrevSample[3]; /* this array holds 3 elements numbered 0
    through 2 */
```

An array is a collection of values that share a common name. Early in the program, Pi is defined as a floating point number with the value 3.141592. Thereafter, the code can use the shorter and more meaningful name Pi, rather than having to type the same long number repeatedly. Using named values also reduces the chance of introducing errors (called "bugs") in the program due to typing mistakes. But unlike named single values such as Pi, an array defines a collection of related values that are stored in adjacent memory locations. Arrays are also named by the programmer, which is clearer and more reliable than referring to memory locations by their numeric addresses. So declaring the array Buffer[1024] sets aside enough memory to hold 1,024 audio samples. Each floating point sample is 32 bits, or four bytes, so the array occupies a total of 4,096 bytes of memory. As was explained, 1 Kilobyte comprises 1,024 bytes, rather than 1,000 bytes, because computer memory is organized in powers of 2. In this case 2^{10} equals 1,024. This is why the available buffer size options in a DAW program are always a power of 2, such as 128 bytes, 256 bytes, 512 bytes, and so forth.

The first array named Buffer holds the 32-bit floating point PCM audio samples as they're being processed. The host audio program deposits 1,024 sequential samples into this area of memory, then invokes the filter code. At a sample rate of 44.1 KHz this represents about 23 milliseconds of audio. After the filter has processed all of the samples in the buffer, the host program retrieves them and deposits the next group of 1,024 samples into the same buffer.

The second array, PrevSample, holds the previous three samples that were already processed. DSP filters process the audio samples using a "rolling" method, where math applied to the current sample depends in part on the value of previous samples before the EQ math was applied to them. Since the code overwrites each sample in the buffer one at a time as it works, the PrevSample array is needed to retain the original contents of the previous three samples. The previous samples are shuffled through the PrevSample array as each sample in the main buffer is processed, as shown in Figure 25.5.

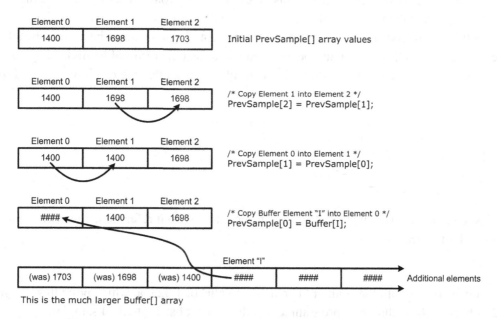

Figure 25.5: As each sample in the main Buffer array is processed, the previous samples are shuffled through the PrevSample array so the last three original samples are always available. Then the lowest array element number 0 is replenished from the current sample from the main Buffer array. The data values shown are for example only and have no other meaning.

Finally we get to the actual EQ code, which runs in what is called a *loop*. There are several types of loops, and the "while" loop shown here is very common. In this case, "While the current value of I is less than the constant value 1,024 defined elsewhere as N, execute the block of code between the open and closed curly brackets {} repeatedly:"

```
while (I < N) {/* execute the following code 1,024 times */
 . . .

 . . .
I = I + 1; /* increment the counter I by adding 1 */
} /* this is the end of the While loop */
```

The first block of code within the loop shuffles the contents of the PrevSample array as explained earlier to retain the last three values:

```
PrevSample[2] = PrevSample[1]; /* Slide the samples over one
    position */
PrevSample[1] = PrevSample[0];
PrevSample[0] = Buffer[I];
```

The next block performs a series of math calculations and assigns the result to the current sample number "I" in the main buffer. This block is a single command, but there are many intermediate calculations so it's split across separate lines to be easier to read. In plain English this line of code says, "assign to Buffer element "I" the result of b0 divided by a0 times Previous Sample element number 0. Then add to that the result of b1 divided by a0 times Previous Sample element number 1. Then subtract from that the result of . . . " and so forth:

```
Buffer[I] = (b0 / a0 * PrevSample[0]) +
(b1 / a0 * PrevSample[1]) +
(b2 / a0 * PrevSample[2]) -
(a1 / a0 * Buffer[I-1]) -
(a2 / a0 * Buffer[I-2]);
```

The last line of code increments (adds 1 to) the counter "I" so it refers to the next element in the main buffer array:

```
I = I + 1;
```

The loop continues until the counter "I" reaches a value of 1,024, at which point this program returns back to the calling host program to supply the next batch of 1,024 samples.

Note that this program has been simplified to be easier to follow. All computer programs include "header" statements that let the program access services in the operating system such as allocating memory or reading a file, among other details. Also, code like this high-pass filter would normally be set up as a subroutine, which is a self-contained block of code called by name by the host program. Another simplification is omitting code to check if the main buffer contains less than the full 1,024 samples, which is likely the last time it runs. Additional code is also needed to avoid an error when the code first runs. When the while counter "I" has a value less than 1 or 2, the statements Buffer[I-1] and Buffer[I-2] reference elements in the array that don't exist. Finally, here's the entire program in context:

```
/* High Pass Filter based on RBJ Cookbook */
/* C code by Grekim Jennings */
/* Analog Transfer Function for this filter: */
/* H(s) = s^2 / (s^2 + s/Q + 1) */
/* These floating point values are used by the filter code below */
float Fs = 44100; /* sample rate in samples per second */
float Pi = 3.141592; /* the value of Pi */
/* These floating point values implement the specific filter type */
```

```
float f0 = 100; /* cut-off (or center) frequency in Hz */
float Q = 1.5; /* filter Q */
float w0 = 2 * Pi * f0 / Fs;
float alpha = sin(w0)/(2 * Q);
float a0 = 1 + alpha;
float a1 = -2 * cos(w0);
float a2 = 1—alpha;
float b0 = (1 + cos(w0)) / 2;
float b1 = -(1 + cos(w0));
float b2 = (1 + cos(w0)) / 2;
/* The Buffer[] array holds the incoming samples, */
/* PrevSample[] holds the intermediate results */
float Buffer[1024]; /* this array holds 1024 elements 0-1023 */
float PrevSample[3]; /* this array holds 3 elements 0-2 */
/* These integer (whole number) variables are used below */
/* to process 1,024 iterations at a time */
int I = 0;
int N = 1024;
/* The code below executes repeatedly as long as the value */
/* of I is less than N. Since I was initialized to 0 above, */
/* and N was set to 1024, this code executes 1,024 times */
while (I < N) {/* beginning of the code that loops 1,024 times */
PrevSample[2] = PrevSample[1]; /* Slide the samples over one
    position */
PrevSample[1] = PrevSample[0];
PrevSample[0] = Buffer[I];
Buffer[I] = (b0 / a0 * PrevSample[0]) +
(b1 / a0 * PrevSample[1]) +
(b2 / a0 * PrevSample[2]) -
(a1 / a0 * Buffer[I—1]) -
(a2 / a0 * Buffer[I—2]);
I = I + 1; /* increment the counter I by adding 1 */
} /* this is the end of the code loop */
```

Website Programming

Social networking sites are a great way for musicians to tell the world about themselves and let fans download their tunes, and YouTube will gladly host all of your videos for free. But any musicians or studio operators who are serious about their business will also have a real website. Owning your own site is more prestigious than a free Facebook page, and it avoids your visitors having to endure unwanted ads and pop-ups, including ads for your competitors. Again, this chapter can't include everything about web design, but I can cover the basics of what studio owners and musicians need to design and maintain their own sites. Many web hosting companies include templates that simplify making a site look the way you want, but you should understand how to customize the content.

The basic language of website programming is HTML, which stands for *HyperText Markup Language*. HyperText refers to highlighted text that, when clicked with a mouse, calls up other related content on the same page or a different page. This is not limited to websites, and the first hypertext programs were like PDF files where, for example, clicking a word in a book's index automatically takes you to the entry on that page. With websites, links to additional content are typically underlined, though they don't have to be. Here is the complete HTML code for a very simple web page.

```
<html>
<head>
<title>Doug's Home Page</title>
</head>
<body>
<p>Welcome to my website!</p>
</body>
</html>
```

Most of this code is needed just to conform to HTML standards to identify the header and body portions, and page title. One line then displays a simple welcome message.

All of the visible page content is between the <body> and </body> markers, called *tags*. In this case, the welcome message. All HTML tags are enclosed in <brackets> that end with the same word preceded by a slash inside other </brackets> to mark the end of the section. For example, <p> marks the start of a paragraph, and </p> marks the end. HTML allows extensive formatting, to create tables with rows and columns, thumbnail pictures that enlarge when clicked on, and much more. A page containing many text and graphic elements can quickly become very complex.

The good news is you rarely need to deal with HTML code. Web design software lets you write text, control its font color and size, embed images and links to MP3 files, and so forth, as well as edit the underlying HTML code directly. As you type, the software adds all the HTML code and tags automatically. But it's useful to understand the basics of HTML, if only to check for unwanted bloat on your pages. For example, if you highlight a section of text as italics, the web design software places <i> and </i> tags around that text. Making text bold instead surrounds it with <b> and </b>. But sometimes after editing a page and making many changes, you can end up with empty tags from previous edits that are no longer needed:

```
<p>From this web page you can download <i></i> all of my tunes,
    and follow the links to my videos on YouTube.</p>
```

A few bytes of superfluous data is not usually a big deal, but it can make your pages load slowly if there are many such empty tags, especially if someone is viewing your site on a smart phone. My own approach is to keep websites as simple as possible. This way everyone will see the pages as you intend, even if they're using an older browser. HTML can do the same things in different ways, and some are more efficient than others. For example, you can make text larger or smaller by specifying a font size by number, or by enclosing it within <big></big> and <small></small> tags. I prefer using tags rather than specific size numbers. Not only because it creates less code, but it also lets people more easily size the text to suit their own eyes and screen resolution. On Windows PCs, most browsers let you use the mouse wheel with the Ctrl key to scale the font size larger or smaller. But some browsers won't let you scale the text when font sizes are specified by number.

There are many books and online resources that can teach you web design and HTML, so I won't belabor that here. However, just to give a taste of how this works, later I'll show code to embed audio and video files on your site, as well as create a custom error page that looks more professional than the generic "Page not found" most sites display. It's easier to embed YouTube videos into pages on your site than to embed player software manually, but again, that risks displaying unwanted ads including ads for your competitors. Also, most browsers can display the source code for web pages you're viewing. So if you see something interesting and want to learn how it's done, use View Source to see the underlying HTML.

Image Files

Web pages can display three types of image files: JPG, GIF, and PNG. The most popular format is JPG, which is great for photos, but not so great for screenshots or logos that include text and line art. JPG files use lossy compression to reduce their size to 1/10th normal or even smaller, much like MP3 files are much smaller than Wave files. But JPG files create artifacts that look like a cluster of small colored dots near sharp edges. These are less visible in a photograph, but they show clearly on images containing text or lines. GIF files are also compressed to reduce their size, but they use a lossless method that restores the original image exactly. However, GIF files are limited to 256 distinct colors, which makes them less suitable for photos. PNG files are a newer format supported by modern browsers. They are both lossless and high quality, but they tend to be larger than the other types. I suggest JPG for photographs, and try both GIF or PNG for line art and text to see which comes out smaller for a given image. Most photo editor programs let you set the amount of compression when saving JPG files, so you can balance picture quality against file size. Again, the point is to keep file sizes reasonable so your web pages load faster.

Custom Error Page with Redirect

If you click a link for a page that no longer exists, or mistype a web address by mistake, most websites display a generic "Error 404" page. Your site will look more professional if you make your own custom error page that has the same look as the rest of the site. You can even send users to your home page automatically after giving them time to read your message. After you create a custom page, change your website's control panel to specify the new Error 404 page instead of the default. Then add this line just above the </head> tag that identifies the end of the page's header, using your own site address:

```
<meta http-equiv="refresh"content="6;URL=www.your-website.com/">
```

The "6" tells the browser to stay on the current page for six seconds, before going to your site's home page. Of course, you can change the duration, or send users to a specific page. There are hundreds of useful and free programming examples on the web, showing simple JavaScript code to do all sorts of cool and useful stuff. You can even download complete web programs, such as software that adds a search capability to your site. However, integrating a complex add-on requires a deeper knowledge of web programming. It's easier to add code to let Google search your site but, again, that will display ads which is unprofessional. Further, if you make changes or add new pages, you can't control when Google will update it's search index. So it could be days or weeks before Google will find recent content in a site search. When I wanted to add searching to the websites I manage, I bought Zoom Search, a highly capable program from Wrensoft. With one click it indexes the entire site, and with another it uploads the index files.

Embedded Audio and Video

Embedding an MP3 file into a web page is pretty simple, requiring a single line with the link:

```
<p>Click <a href="mytune.mp3">HERE</a> to download the tune.</p>
```

This line assumes the MP3 file is in the same web directory as the page containing the link. If the file is somewhere else, or on another site, you'll add the complete address:

```
Click <a href="http://somesite.com/mytune.mp3">HERE</a> to
   download the tune.
```

Most people have a media player that will launch automatically when they click a link on an MP3 file. Some players start playing immediately, then stream the audio in the background while continuing to play. Other players wait until the entire file is downloaded, then start playback. MP3 files are relatively small, so your site visitors probably won't have to wait very long. But video files can be very large, so it's better to embed your own player software that you know will start playing immediately. I use a freeware Flash player, which plays FLV format videos. This player is highly compatible with a wide range of browsers, and it requires uploading only two support files to your site. Both files are included in the "web_video.zip" file that goes with this book, along with a sample web page and Flash video.

I suggest you start with the included HTM file, which is stripped down to the minimum, and add to that to make it look the way you want. However, you'll need to edit the page title, the name of your video file, and the height and width of the embedded video. This line specifies the display size for the video, and you'll change only the numbers 480 (width) and 228 (height):

```
var s1=newSWFObject('player.swf','ply','480','228','9','#ffffff');
```

This video is small—one of the SONAR editing demos from Chapter 8—and its actual display size is 480 pixels wide by only 208 pixels high. But you need to add 20 to the height to accommodate the player's Play, Volume, and other controls at the bottom. The other line you'll edit contains the name of the video file itself:

```
s1.addParam('flashvars','file=sonar_cross-fade.flv');
```

As with MP3 files, if the video is not in the same directory as the web page, you'll need to include the full web address. Once you verify that the page displays as intended on your local hard drive, copy it to your website, along with the FLV video and the player's two support files.

Finally, unless your video editing software can render to FLV files directly, you need a way to convert your video to the Flash format. I use the AVS Video Converter mentioned in Chapter 16, Video Production. This affordable program converts between many different video file types and can also extract video files from a DVD and save them in popular formats.

Summary

This chapter covers hard drive partitioning and backup and explains the basics of optimizing performance by disabling unneeded programs that run in the background. A computer whose hard drive is well organized and defragmented and runs only the programs you really need will be faster and more reliable than when first purchased. Further, by keeping operating system and program files separate from your personal data, you can back up easily and quickly. Even though most people view backing up as a burden, it doesn't have to be. A good backup program makes it easy to protect your audio projects and other files, and when it's easy, you're more likely to do it regularly. Likewise, making an occasional image backup of your system drive protects against errant programs and outright hard drive failure.

Computer viruses and other malware are increasingly common, but protecting yourself is not difficult, and it requires mostly common sense plus a few basic guidelines: buy a hardware firewall, and don't open email attachments directly. Likewise, a UPS is a wise investment that avoids losing changes made to the project you're currently working on, or possibly losing all of the data on the hard drive if the power goes out while the drive is being written to.

This chapter also explains a bit about computer and website programming, just to show that they're not as mysterious and complicated as many believe. Although modern web design software makes it easy to create web pages without learning to write HTML code, understanding the basics is useful—and even fun!

Notes

1 "Better latent than never" by Mike Rivers.
   http://mikeriversaudio.files.wordpress.com/2010/10/latency_revised.pdf
2 "Cookbook formulae for audio EQ biquad filter coefficients" by Robert Bristow-Johnson.
   www.musicdsp.org/files/Audio-EQ-Cookbook.txt

PART 7

Musical Instruments

Question: What do you get when you play New Age music backwards?
Answer: New Age music.

CHAPTER 26

Musical Instruments

This section explains the mechanics of musical instruments and how they create sound. It's impossible to cover every instrument here, but I'll describe the more popular types. Besides the construction and acoustic theory of how musical instruments work, their sound also depends greatly on how they're played; there's an intimate relationship between the instrument and the performer. For bowed instruments such as violins and cellos, their tone quality, or *timbre*, is affected by where on the string the bow is placed, how hard it's pressed against the string, and how quickly it's drawn across the string. These three bow parameters vary the timbre through its entire range of possibilities. Here, timbre describes the relative volumes of the fundamental pitch and its many harmonic multiples.

With wind instruments, the shape of your lips and mouth and how hard you blow determine the timbre. Blowing harder creates harmonics that are louder than when blowing with less force. The same applies to plucked instruments such as the guitar or banjo; striking the string harder, or nearer to one end, creates stronger harmonic content. Drum overtones also vary in level depending on where you strike them and how hard.

Bowed string instruments also respond to how they're played—not emotionally, of course, though I imagine some people might believe that. But they offer resistance to the bow as a type of feedback. At first a student musician learns to play by rote; if the violin sounds scratchy, you draw the bow more slowly or place the bow farther from the bridge. Eventually this becomes second nature, and players control the tone subconsciously using mechanical feedback from the instrument to guide them. I could tell by feel alone if my cello was making a pure tone or a scratchy sound, even if I was wearing sound isolating headphones. Indeed, "how instruments work" is as much about playing technique as how they're constructed.

Many people don't realize that learning to control the bow is the most difficult part of playing a violin or cello. The bow is the instrument's voice, and all else is subservient. To be sure, it takes years to develop the fine motor control needed to place your fingers at exactly the right place on the fingerboard—being off by just a few millimeters can make a note seriously out of tune. But controlling the bow is even more difficult. It's similar for blown instruments such as the saxophone and oboe. After a year or two, a dedicated student can learn to map notes on the printed page to the equivalent key fingers automatically without thinking. But developing the necessary breath and mouth control is much more difficult, taking thousands of practice hours over many years to truly master. Year ago my friend Phil Cramer and I were watching Jeff Beck play an amazing guitar solo on TV. I said to Phil, "Man, that's really difficult!" Without hesitating Phil replied, "Not for him." Accompanying this chapter are four videos featuring skilled musicians as they explain their instruments and demonstrate how they're played. Included are demos of the cello, saxophone, piano, and violin. The piano video shows a $100,000 Steinway as it's disassembled to reveal the keyboard mechanism! These videos, on the website for this book, give a deeper insight into instruments and playing techniques in a way not possible with the printed word alone.

I had intended to include a chapter about music theory, but it would have been huge. So instead I created a video series and put it on YouTube[1] for all to enjoy for free. (It's also on Vimeo[2] because YouTube blocks some segments in some countries.) Teaching music theory makes more sense as a video anyway, because a video can include music as audio to accompany the written examples. This video series packs a college-level course into just under three hours using detailed explanations and familiar musical examples. Segments include Notes and Melody, Intervals, Scales and Arpeggios, Musical Keys, Harmony, Chord Progressions, Musical Timing and Time Signatures, Conducting, as well as a section about musicians and musical instruments. Because of its length, the video is divided into five segments, organized as a YouTube playlist.

Instrument Types

The two basic types of musical instruments are percussive instruments, where the sound is started once, then dies away on its own, as with a piano or cymbal, and sustained instruments, where the player has continuous control of the duration, volume, timbre, and vibrato, as with a violin or trombone. For example, a violinist may start a note softly using a slow bow speed, then increase the volume by drawing the bow more quickly, then increase the brightness of the note by moving the bow closer to the bridge, and finally adding vibrato that gradually becomes faster. These variations over time make music more expressive, and thus more interesting to hear.

Percussive instruments can also be played in an expressive sustaining fashion, such as a roll on a snare or timpani drum, where the drum is struck repeatedly. A drum roll can also change tone quality over time by varying how hard you strike the drum and where, as well as the speed of the roll's repetition. Likewise, mandolin players create a sustained tremolo by repeatedly picking the same note quickly to continue the note indefinitely. But to my way of thinking, instruments fall into these two basic categories: percussive and sustaining.

Besides controlling the tone with playing technique, another factor that affects an instrument's timbre as heard by an audience or microphone is its directivity. As with loudspeakers, the vibrating wooden plates of a violin radiate more directionally at high frequencies.

Below about 400 Hz, violins radiate almost omnidirectionally, but by 4 or 5 KHz, almost all of the sound goes upward in a narrow beam, with very little energy sent forward. Reed instruments such as the clarinet and saxophone also radiate in different directions at different frequencies, as well as from different parts of the instrument, depending on which keys are pressed. In a concert hall, the full spectrum of sound reaches the audience in large part via reflections from the stage wall behind the performers and the flared ceiling and angled side walls.

> *So three bass players walked past a bar.*
> *Hey, it could happen!*

Earlier chapters explained that an FFT analyzer can display the amount of energy at different frequencies in a recording, and this tool can also analyze the output of musical instruments. Understanding the spectrum of instruments helps one to be a better synthesizer programmer, mixing engineer, orchestra arranger, or just a more knowledgeable and appreciative listener. For example, an electric bass often has less energy at the fundamental frequency than the first few harmonics, depending on where along its length the string is plucked. Much of the fullness of bass instruments is determined by the level of the second harmonic, so bringing that out with EQ can make a mix sound solid without relying on very low frequencies that are difficult for many loudspeakers to reproduce at loud volumes.

Figure 26.1 shows the spectrum of my Fender Precision electric bass playing a low A note when plucked with a finger directly over the pickup. You can see that the second and third harmonic components are louder than the fundamental. However, plucking the string farther from the bridge, closer to the center of the string, creates a mellower tone that has more fundamental with softer harmonics. Plucking a string with less force also reduces harmonic content.

The same principle applies to acoustic instruments. The harder you blow a trumpet or oboe, or strike a drum, the brighter the sound will be. Striking a drum near the edge also makes the overtones louder relative to the fundamental, compared to striking it near the center. The "drum_tone" video shows the change in tone quality of a conga and timpani when struck in the center versus at the edge. This also applies to pretty much every other acoustic sound source that occurs in nature.

Figure 26.2 shows an FFT of the same low A note recorded from a Yamaha grand piano. Again, the fundamental is softer than the second harmonic, which in turn is softer than the third harmonic. However, a piano is brighter-sounding than an electric bass played with fingers (as opposed to a pick), and this is reflected in the higher harmonics falling off in level much less quickly than an electric bass.

I also recorded a single low note on my cello, and the "cello_scope" video uses a software oscilloscope to show how the harmonic content varies with volume. As the note is bowed

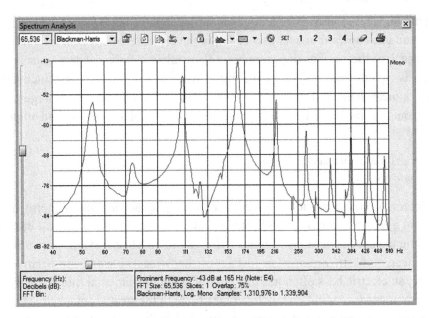

Figure 26.1: Depending on where you pluck a string, and how hard, the fundamental frequency may be softer than some of the harmonics. This FFT shows the spectrum of a low A note on a Fender Precision Bass when plucked with a finger, with no EQ applied. You can see the second harmonic at 110 Hz is about 7 dB louder than the fundamental at 55 Hz, and the third harmonic at 165 Hz is about 3 dB louder than the second harmonic.

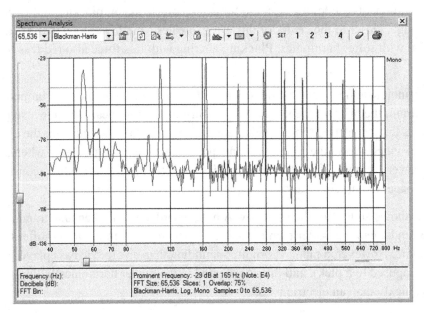

Figure 26.2: Low notes on a piano often have less energy at the fundamental pitch than the first few harmonics, though the higher harmonics fall off less rapidly than an electric bass.

more strongly, you can see more high-frequency activity; then at the end, when the bowing stops and the string continues to ring, the tone is softer and contains mostly the fundamental sine wave.

Sympathetic Resonance

The concept of *sympathetic resonance* was first introduced in Chapter 3, where a high-Q equalizer boost or strong room resonance is excited by a similar frequency from a sound source. The same thing happens with string instruments: One vibrating string can excite another that's nearby. An Indian sitar uses this principle. A group of strings (called *tarb*) are not struck, but they vibrate and sustain because they're in close proximity to other strings that are struck. This is an important part of the sitar's unique character.

Sympathetic resonance also occurs when you press the *sustain pedal* on a piano. At rest, all of the strings on a piano are silenced by felt dampers that rest against the strings. When you press a key to play a note, the damper for that note's strings is raised, freeing the strings to vibrate. As soon as the damper lifts, a felt-covered wooden hammer strikes the string, making the initial sound. While the key remains pressed, the damper is kept away from the strings, letting their vibration decay naturally over time. Releasing the key damps the strings, and the vibration dies off quickly. Pressing the sustain pedal lifts the dampers off *all* of the strings. Now, striking a single low note will cause other higher notes that align with the low note's harmonics to vibrate in sympathy.

The Harmonic Series Is Out of Tune

Vibrating strings on a violin or guitar, and vibrating air in woodwind and brass instruments, create harmonics, also called *overtones*, that follow a specific mathematical series. The lowest frequency is the fundamental, and each harmonic is a whole number multiple of that frequency. Harmonics of static waves such as sawtooth and pulse also follow the same series. Table 26.1 shows the harmonic series of a low A note whose fundamental pitch is 55 Hz, and each harmonic is 55 Hz higher than the previous one. The harmonic series of most musical instruments follows this same basic sequence, becoming out of tune at some higher frequencies. Table 26.1 shows the first 15 frequencies produced by a low A note, and you can see that the pitch of the fifth harmonic is off by nearly 1 percent. At the 11th harmonic, which can be audible on a bright-sounding instrument playing a low note, the frequency is off by 3 percent. The distance between two adjacent notes—a musical half-step—is about 6 percent, so 3 percent is seriously out of tune!

Harmonics usually become softer as they go higher in frequency, so a single note doesn't necessarily sound out of tune with itself. Even though the 11th harmonic is off by 3 percent,

Table 26.1: Harmonic Series of Low A Note

Harmonic	Frequency	Nearest Note	Nearest Note Frequency	Error
A Fundamental	55 Hz	A	55.0 Hz	0.0%
2nd Harmonic	110 Hz	A	110.0 Hz	0.0%
3rd Harmonic	165 Hz	E	164.8 Hz	0.12%
4th Harmonic	220 Hz	A	220.0 Hz	0.0%
5th Harmonic	275 Hz	C#	277.2 Hz	0.8%
6th Harmonic	330 Hz	E	329.6 Hz	0.1%
7th Harmonic	385 Hz	G	392.0 Hz	1.8%
8th Harmonic	440 Hz	A	440.0 Hz	0.0%
9th Harmonic	495 Hz	B	493.9 Hz	0.2%
10th Harmonic	550 Hz	C#	554.4 Hz	0.8%
11th Harmonic	605 Hz	D	587.3 Hz	3.0%
12th Harmonic	660 Hz	E	659.3 Hz	0.1%
13th Harmonic	715 Hz	F	698.5 Hz	2.3%
14th Harmonic	770 Hz	G	784.0 Hz	1.8%

which at half of one musical interval is a lot, on this particular piano and microphone placement it's 11 dB softer than the fundamental. It's also partially masked by all the other harmonics that are in tune. The video "harmonic_tuning" plays a low note on a synthesizer using a pulse wave, then sweeps a high-Q filter slowly upward to emphasize each harmonic in the series one by one. You can hear that many of the higher harmonics are out of tune with the fundamental. To be clear, equal temperament intentionally places musical notes out of tune compared to the natural overtone series, rather than the other way around. So technically it's the tempered musical notes that are out of tune, not the natural harmonic series.

The available fundamental notes on brass instruments also follow the harmonic series shown in Table 26.1. Consider the bugle, which is similar to a trumpet but with no valves to vary the tube length. Bugles can play only notes in the harmonic series, which the player selects by adjusting the shape of his mouth and lips, collectively known as the *embouchure*. There are many well-known "standards" in the bugle literature, such as *Taps* and *Reveille*. All bugle tunes contain only the same limited range of notes—basically the chord tones of a major key. As a bugle plays higher notes, the fundamental pitches are inherently out of tune, requiring the player to adjust his embouchure and blow harder or softer to force the correct pitch. When a player adjusts a note's pitch this way, he is said to *lip* the note.

A similar phenomenon occurs with reed instruments. In the "sax_demo" video, Collin Wade explains this as it applies to his instrument. Lipping an instrument also allows creating a *glissando* effect that transitions smoothly through a range of frequencies rather than hitting only discrete note pitches. This is used famously in the introduction to George Gershwin's *Rhapsody in Blue*, where a solo clarinet varies the pitch continuously over a range greater than two octaves.

As an aside, a single note played on a guitar can sound out of tune if the strings are old. Strings often stretch non-uniformly, creating one or more thinner portions. Depending on where the thin spots occur along the string's length, some of the prominent lower harmonics can sound out of tune with the fundamental. It's very difficult to tune a guitar with old strings!

As used here, the term "harmonic" refers to the natural overtone series of a vibrating string or the vibrating air column in a wind instrument. Another type of harmonic is created artificially with string instruments by placing a finger *lightly* on the string at specific locations while plucking or bowing. Touching a string lightly at a point halfway along its length creates a note one octave higher than the open string. Pressing the string firmly against the fingerboard creates a note having the usual overtones, but this type of harmonic is closer to a pure sine wave. Other locations along the neck do the same—common locations are 1/3 and 1/4 the length, but other integer divisions also work. In fact, doing this at 1/7 the neck's length creates the same out of tune musical 7th (plus two octaves) as the overtone that occurs naturally within a single note.

Equal Temperament

Equal temperament is a method of tuning that uses a fixed interval between adjacent notes, rather than follow the naturally occurring harmonic series. Equal temperament divides an octave into 12 equally spaced intervals, called musical half-steps. As explained in Chapter 1, the exact ratio between any note and the next higher note is 1.0595 to 1, where 1.0595 is the 12th root of 2. The frets on a guitar neck are arranged in this ratio, and all modern instrument tuning meters use this method to show if the pitch is too high or too low.

Hundreds of years ago, brass instruments had no valves. Like a bugle, early horns and trumpets could produce only a limited number of specific notes. Even though an F horn with no valves can play a high C, that note is out of tune, making the horn unsuitable for playing music in the key of C. So players either had to own several instruments, each designed to play in a different key, or they'd insert replaceable *crooks*—short sections of pipe—to vary the overall length of the tubing. Changing crooks during a performance takes time and must be done quietly, so valves were added to trumpets and French horns around 1815. This also allowed musicians to play notes in the equal-tempered scale rather than only notes in the harmonic series.

The concept of equal temperament goes back thousands of years, but J. S. Bach championed it and made it a modern reality. His 1722, *The Well-Tempered Clavier* is a series of 24 pieces—one in each major and minor key—written for the keyboard instruments of his day. This had great musical as well as technical impact because it allowed musical compositions to modulate to distant keys without some instruments sounding out of tune.

Another tuning method called *just intonation* tunes the fifths to exact harmonic frequencies. In that case, the lowest E note in Table 26.1 would be tuned to 165 Hz instead of 164.8 Hz. String players often tune their instruments this way by playing adjacent open strings while listening for *beat tones*—a slow volume undulation as two notes approach perfect unison. You play both notes at once and adjust the tuning of one until the pulsing slows to a stop.

Another tuning technique uses harmonics to achieve a perfect unison. Touching a guitar A string lightly at the seventh fret creates an E note an octave and a half higher than the open string. Lightly touching the E string below the A at its fifth fret creates the same E note. So you can play both harmonics, one after the other, letting both continue to ring, and tune either string until the beating slows to a stop. The video "beat_tones" shows an electric bass being tuned this way, with an animated cursor tracking an image of the audio to better relate what you're hearing to the recorded waveform. Tuning via harmonics is especially useful for basses because our pitch perception is poor at low frequencies.

> When performing jazz, if you play a wrong note, play it again two or three times so people will think it was on purpose.

Many musicians intentionally play notes ever-so-slightly sharp or flat for expression or effect. If not for the note's entire duration, they may slide up to a note from a lower pitch. Playing a note slightly flat can project a sense of calm, and playing slightly sharp adds tension. Again, I'm talking about extremely small amounts, on the order of 5 or 10 cents—1/100ths of a musical half-step. Playing off-pitch intentionally is easy with fretless instruments such as the violin or cello, and wind instrument players do that by varying their embouchure. Guitar players change the pitch in large or small amounts by bending the strings, which is one reason their strings can stretch non-uniformly as mentioned earlier. Musicians can also play slightly behind the beat or a little ahead—say, 10 to 30 milliseconds—to add a sense of calm or excitement.

"Wood Box" Instruments

As with listening rooms, instruments made from a hollow wooden "box" of some type also have resonant modes whose frequencies are related to the box dimensions. To prevent modal resonances from making some notes sound much louder than others, violins, guitars, and other wood box instruments are designed with an irregular curved shape. These instruments still have modes, but the continuous curves along the sides minimize a strong buildup at single frequencies and their multiples.

Also, like rooms, when a resonant mode is near, but not exactly at, a frequency being played, the result is an out-of-tune *beat frequency* caused by the frequency difference.

This produces an effect known as a *wolf tone*, and it's more prominent in cellos and acoustic basses than most smaller wood box instruments. In many cellos, a strong resonance occurs somewhere between the E at 164.8 Hz and the F# at 185 Hz. Whether or not the resonance is exactly in tune with a standard note frequency, that one note will sound louder than all others and be more difficult for the player to control with the bow. Even very good cellos are not immune to wolf tones, and various devices that attach to the cello body or a low string are available to reduce the effect.

Wood box instruments also have one or more holes in the front plate to allow lower frequencies to escape. Acoustic guitars typically have a single round hole about four inches in diameter, and instruments in the violin family have two holes shaped like a lowercase italic "*f*" called, appropriately, *F holes*. All wood box instruments create sound using the same basic principle as the cello shown in Figure 26.3. Four or more strings are stretched tightly above the *fingerboard*, with both ends anchored solidly by the *tuning pegs* and *tail piece*. The active vibrating length of the strings is between the *nut* at one end by the tuning pegs and the *bridge* that rests on the body. These also hold the strings in place side to side. The same names are used for related pieces on acoustic guitars, though on those instruments the bridge and tail piece are a single unit called a *saddle*. The cello mute shown here is a small block of rubber that's placed onto the bridge to partially damp its vibration, creating a softer and more mellow tone when that's called for in a piece by the composer or arranger.

The string tension creates a large amount of downward force on the bridge, and even more tension across the body, so cellos and guitars and other string instruments must be extremely sturdy. A typical musical string exerts a lengthwise pulling force of 15 to 35 pounds; for an instrument having four strings, that's as much as 100 pounds of tension across the body. An acoustic guitar with medium-gauge steel strings has a total linear tension of about 175 pounds! The downward pressure on a bridge is also very large—about 50 pounds for a violin and 140 pounds for a cello.

Bowed Instruments

I distinguish bowed instruments in the violin family from the broader category of "wood box" instruments, though bowed instruments are, of course, both. Most instruments in the violin family are made using spruce wood for the front plate and maple for the back, sides, and neck. The fingerboard glued to the neck is traditionally made of ebony. Many inexpensive cellos and basses are made from plywood, though better student-grade instruments made of solid wood now come from Asia. Some modern bowed instruments are made from carbon fiber, and they can be excellent with more tonal consistency from one instrument to another. Not that consistency is necessarily desirable.

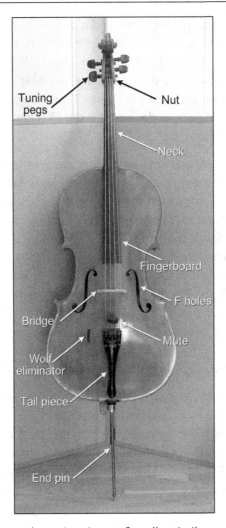

Figure 26.3: This photo shows the major pieces of a cello; similar names apply to violins, violas, and double basses.

Figure 26.4 shows a contrabass as the back plate (on the bottom) is being glued to the side and ribs. The inner plywood with the rectangular holes isn't part of the instrument. Rather, it's a form to keep the sides perpendicular to the back during assembly. After the glue sets, the plywood form is removed, and the front (top) plate is glued into place. Most luthiers use hot *hide glue* because it can be more easily unsealed for repairs when needed, compared to the more common polyvinyl acetate white glues, which are more permanent.

Figure 26.5 shows the internal construction of a violin, and the same basic design is used for violas, cellos, and double basses. The sound post is a round dowel that mechanically couples the front and back plates. When a note is played, the string vibrates the bridge, rocking it

Figure 26.4: The sides of a string instrument are made from thin strips of maple that are clamped to the interior form, with structural blocks to hold them in place. The form holds the shape of this bass during assembly until the back plate is glued to the ribs. Then the wood form is removed, and the top plate is attached, closing the instrument. *Photo courtesy Bob Spear.*

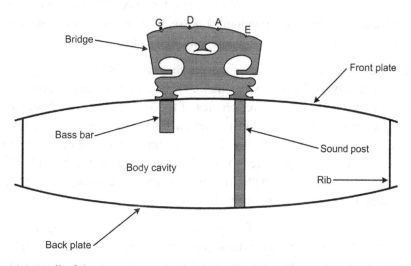

Figure 26.5: All of the instruments in the violin family are built using this basic model.

side to side, which transfers energy to the front plate. The sound post couples that to the back plate, so both plates vibrate and contribute to the overall output.

The bass bar is glued to the front plate to strengthen the instrument, and Figure 26.6 shows more clearly its shape and relative size as it's clamped in place during gluing. The bass bar is positioned under the lowest-pitched string of the instrument, which is how it got its name. Bass bars are used not only on basses but on all instruments in the violin family.

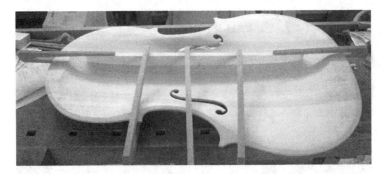

Figure 26.6: This bass bar is clamped in place under the front plate of a contrabass while the glue dries. *Photo courtesy Bob Spear.*

All violins and cellos have their own unique sound, determined by the varying thickness of their front and back plates. Handmade instruments are carved by the luthier using chisels and other woodworking tools, and each is unique. Factory-made instruments usually have the top and back shaped by a machine that follows a pattern, so they're all the same size and shape, differing only by the grain of the particular wood being carved. The size and shape of the box cavity also contributes to the tone, though most bowed instrument shapes are very similar, based on measurements of proven instruments built in the early 1700s.

The main contributor to the tone quality of a given instrument is the frequency, strength, and Q of its resonant modes. The many resonances inside an instrument's body accentuate individual fundamental and overtone frequencies of the notes being played, and the top and bottom plates also have modes. A good instrument will have many high-Q modes whose frequencies are spaced more or less evenly. This is not unlike the goal of listening rooms to have many modes that are evenly spaced to avoid overly emphasizing only a few frequencies. But unlike a room where a low Q is better because the goal is neutrality, a high Q is desired for bowed instruments to create a character of sound.

A good instrument continues to "sing" for some time after you stop bowing, and this is often used for effect by skilled players who apply strong vibrato after lifting the bow. A poor plywood cello or bass has few resonances and with a low Q because the wood grains go in different directions. So the vibrations tend to cancel each other out, causing the sound to die out quickly when you stop bowing. A poor-quality viola often sounds more like a violin because its plates are too thick, resonating at frequencies too high to support the lowest notes. When I hear a really good cello or viola, it reminds me of a wah-wah pedal set to various static positions. As a melody or scale is played, it sounds a little like "ooh eeh ahh" as each note excites a different high-Q resonance.

We distinguish human voices partly by unique strong resonances in the mouth and vocal tract. Every person's mouth is a different size and shape, and that creates different sets of acoustic filter frequencies. The principle is very similar for resonating instruments like violins, cellos, and acoustic guitars; the resonances give each instrument its own individual character, which makes them sound more interesting. Indeed, intentional resonance is an important concept in making musical instruments sound pleasing. To make this clear, I created a cheesy MIDI version of Leroy Anderson's classic *The Syncopated Clock*, which features temple blocks for percussion. I had sampled a set of real orchestral temple blocks for another project years ago, so I created the "syncopated_clock.mp3" file containing two versions of the last few bars of this piece. One version uses the simple sound of drumsticks clicking together, and the other uses the real temple block samples. Which version sounds more interesting?

A few years ago, my friend Andy Woodruff came by to get my opinion on five different cellos. He was considering buying one of them, and he wanted to hear them in my large studio room and also get my advice, since I play the cello. As explained earlier, musical instruments radiate different frequencies in different directions, so what the player hears can be very different from what someone out in the audience hears. A good-quality string instrument is a large investment, so it's sensible to get a second opinion and have someone else play while you listen critically. You can see a similar comparison of four cellos in Andy's "cello_demo" video.

The Bow

High-quality bows are made of Pernambuco wood, which comes from Brazil, with horse tail hair attached at either end, stretched fairly tightly. The hairs are made to grip the string by applying *rosin*, a sticky substance made from tree sap. As the bow is drawn across a string, it alternately grips the string, then releases it, in a motion known as *stick-slip*. This "grab and release" action happens very quickly, creating a wave shape similar to a sawtooth. The bow has a surprisingly large effect on the sound of an instrument, and players audition bows as critically as their instruments. My own bow cost $2,500 when I bought it used, and a friend of mine paid $50,000 for her bow. I can't say that the cost of a very expensive bow is proportional to its quality, but a good bow sounds vastly better than a poor bow, and it also handles better. Of course, bowed instruments can also be plucked, and that playing style is called *pizzicato*.

The Stradivarius

The "mystery" of Stradivarius violins was solved in the 1980s by luthier Carleen Hutchins, using modern modal analysis. For hundreds of years, violin makers would rap unattached front and back plates with their knuckles to determine the modes by hearing which tones rang out. If the frequencies were deemed unsatisfactory, they'd continue to carve the plates to be thinner—a process called *graduating*—until the tones were to their liking. Carleen took this analysis to a new level, using a method first published by Ernst Chladni in 1787. Her groundbreaking 1981 article[3] describes a series of experiments where she disassembled a number of Stradivarius violins and analyzed them scientifically. Imagine being allowed to borrow a bunch of million-dollar Strad violins and take them all apart!

To analyze the modes, Carleen suspended the top and bottom plates on small foam blocks, above a hole in a table over a loudspeaker playing sine waves. She sprinkled fine sand or party glitter on the violin plates to show the vibrations as different frequencies played through the speaker. For other tests she used laser interferograms, which reveal even more detail. Figure 26.7 shows the basic setup, as well as two samples of the many mode types that are revealed. At non-modal frequencies the party glitter scatters randomly over the plate's entire surface, but at modal frequencies specific patterns form to show which areas of the plate vibrate at each frequency. Many modern makers use this technique, though some continue to tune their plates the old-fashioned way: by ear. Everything that alters the tone of an instrument—the wood's thickness and density, age, and even the type of finish applied—affects these resonant frequencies, their Q, and relation to one another. There is no magic.

My own cello was re-graduated by one of Carleen's students under her guidance, using the techniques described here. My cello is not as good as a Strad for a variety of reasons, but it's a *lot* better than it was before re-graduating, with much better modal frequencies and distribution. Understand that every piece of wood is different, so it's impossible to make one violin or cello sound exactly like another. But no two Strads are identical either, and not all of them are great. Indeed, Strads sell for millions of dollars not only for their excellent sound but also their value as antiques.

Plucked Instruments

> *Question: How do you get a guitar player to play softer?*
> *Answer: Give him some sheet music.*

Most familiar Western-style instruments that are plucked instead of bowed also have *frets*—thin metal bars placed along the neck at precise intervals to facilitate playing in tune. These include the acoustic guitar, banjo, and balalaika, among many others. (Fretless plucked instruments include the original banjo and its predecessors, the Chinese sanxian and its cousin the Japanese shamisen, the Middle Eastern oud, and the Hawaiian or steel guitar.) Most plucked instruments use one string per note, though mandolins and 12-string guitars have two strings placed very close together, tuned in unison or octaves to create a unique sound. Some electric basses use two strings per note tuned in octaves, though those are rare. Modern plucked instruments have strings made of either steel or nylon, and the lower-pitched strings of either type are wrapped with a layer of bronze, nickel, or other thin metal wire to

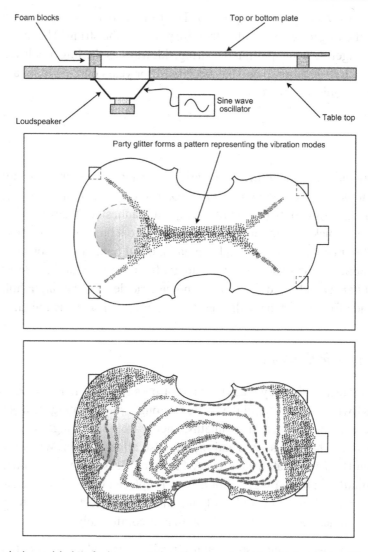

Foam blocks

Top or bottom plate

Loudspeaker

Sine wave oscillator

Table top

Party glitter forms a pattern representing the vibration modes

Figure 26.7: A shaker table lets luthiers see the vibrational modes of the top and bottom plates of a violin or cello to help guide them as they plane the wood to achieve the desired tone.

add mass. Strings can be plucked either with bare fingers, fingernails, a flat or shaped pick, or finger picks that clamp (gently!) onto the player's fingers.

Unlike fretless instruments such as the violin, vibrato on a fretted instrument can go higher in pitch only. However, guitar players will often bend a string to raise the pitch a half step or more, then add vibrato, which can then make the pitch go lower or higher. Some electric guitars have a *whammy bar* tail piece—sometimes called a *vibrato bar*, or incorrectly a *tremolo bar*—that can raise or lower the pitch of all the strings at once. Two other guitar techniques are called *hammer-on* and *pull-off*. The former sounds a note by pressing quickly and firmly with the fingering hand rather than by plucking the string. The latter is similar, again using the fingering hand, pulling the string sideways as the finger is lifted off the string to sound the note. Left-hand pizzicato has been used for years on the violin, and the "violin_ demo" video shows this technique.

Amplification

Acoustic instruments can be amplified in a variety of ways other than with a microphone. One common method uses a piezo *contact pickup*, also called a *contact mic*, attached to the front plate or bridge using putty or a similar temporary adhesive. Many modern acoustic guitars have a piezo pickup built into the bridge, wired to a built-in phone jack for connecting to a guitar amplifier or PA system through a preamp or DI box. Some acoustic guitars have a traditional magnetic pickup under the strings to get both acoustic- and electric-type sounds. In fact, some guitars are a hybrid of acoustic and electric designs, having F holes and a hollow body made from wood that's thicker than a traditional acoustic guitar.

Solid Body Electric Guitars

The model for an ideal solid body electric guitar or bass is a stiff, massive body that adds no resonance of its own. The more rigid the body, the longer the vibrating strings will sustain. If a body is flimsy enough to vibrate in sympathy, energy from the strings is removed to sustain the body's vibration. The neck's mass and rigidity also affect sustain. When I bought my Fender Precision Bass in the late 1960s, every Precision Bass came with a thick, wide neck. My hands are small, so after struggling for a year, I replaced the neck with a Jazz Bass neck, which is narrower and thinner. I could easily tell that the bass lost some of its power and sustain with the thinner neck, but at least I can play it comfortably.

Speaking of sustain, electric guitarists often use feedback to make notes continue indefinitely, rather than die out over time as normally occurs. When the amplifier's volume is loud enough, the acoustic sound in the air reinforces the string's vibration, causing it to keep vibrating in sympathy. A clever guitar accessory called the E-Bow can sustain electric guitar notes without requiring excess volume. This device has both a pickup coil and a driver coil; when it's held above a vibrating string, the vibration is picked up, amplified, and fed to the driver coil. This excites the string at its vibrating frequency, sustaining the note until it's moved away from the string.

Some guitar pickups feed back by making a nasty high-pitched squealing sound at loud volumes, rather than sustain notes in a useful fashion. This happens when the pickup coil vibrates because its windings are not secure. When I used to modify and repair electric guitars for friends as a hobby, I'd fix this by removing the pickups and soaking them in a tub of hot wax from melted candles. After the wax dried and the pickups were put back into the guitar, they'd no longer squeal when playing at loud volumes, and you could then crank the volume to get the good type of sustaining feedback.

Resonance is neither wanted nor needed with solid body electric guitars and basses, so I'm skeptical about claims that the type of paint finish affects their tone. Even with ten coats of paint, which is common, the amount of mass added by the paint to a solid body guitar is less than 1 percent. So while you might possibly be able to measure a tiny difference in tone or sustain after refinishing a solid body guitar, it seems unlikely the difference could be audible. It's difficult to prove scientifically that paint can change the sound of a solid body instrument, given the time needed to take it apart, refinish it, and put on the strings. Even if you record before and after performances, it's impossible to play exactly the same way both times, so you're really comparing the performances more than the different finishes. Of course, a fine finish on any musical instrument is aesthetically pleasing, and there's nothing wrong with that.

I've never seen a formal mute product for guitars, but some electric basses include a rubber strip that sits under the bridge and damps the strings lightly. This creates a more punchy sound, because the note has the full volume when first plucked, then dies away more quickly than usual. So it's not like a violin mute that softens the volume and tone or a trumpet mute that uses acoustic filtering to change the basic timbre. Guitarists sometimes create a similar effect manually using the palm of their hand, called *palm muting*. I've also seen guitar players stuff a handkerchief under the strings to get this effect.

Blown Instruments

As with wood box instruments, the blown instruments category consists of many very different types that create sound by vibrating air inside a tube of some sort. Although their bodies don't vibrate the way wood box instruments do, the tube's cavity size and shape affect the pitch and tone. Blown instruments use three basic sound-producing mechanisms: air blown across an opening, as with a flute or recorder (or soda bottle); single or double reeds that vibrate in free air or against each other, as with clarinets and oboes, respectively; and a cup-shaped mouthpiece that contains the player's lips as they vibrate against each other to create a buzzing sound, as with trumpets and tubas. Blown instruments can also be *overblown*—blown harder than normal—for special effect. This creates a chuffing-type sound with flutes, a growl on a saxophone, or a blatt-type sound with trumpets and trombones.

Like bowed and plucked instruments, players use their hands and fingers to play different notes, but they also must carefully control their breath. Where bowing a violin too quickly creates an unappealing scratchy sound, blowing too hard creates squeaks and other unwanted noises. Wind instrument players can also use their tongues to stop and start the air flow to better enunciate notes, which is easier and more precise than trying to start and stop the air flow using breath control alone. The tongue is also used this way to create a sequence of rapid staccato notes, using a technique called *tonguing* (or its variants *double-tonguing* and *triple-tonguing*).

Most blown instruments are built from a tube—either straight or curved—and the tube's length is varied to create the different notes. Reed instruments such as clarinets, oboes, and saxophones use a combination of normally open and normally closed holes to vary the effective tube length. A number of finger keys are connected to *pads* that expose or cover a hole, and holes without pads are normally open until a finger is placed over the hole. The overall tube length remains the same as different notes are played, but an opening in the tube creates a sudden change in acoustic impedance at that point along its length. This effectively shortens the tube, as if it ended at the open hole. Flutes work on a similar principle, though their sound is produced from air turbulence at the mouthpiece rather than a vibrating reed or buzzing lips.

Brass instruments change the physical length of the tube either with a sliding extension (trombone) or with valves that insert additional sections of tubing in musically useful lengths. The pitch of a blown instrument is created mainly by the frequency of the vibrating source—either lips or reeds—which is then reinforced by the pipe's resonant frequency, which varies as the keys or valves are engaged.

The air in a blown instrument vibrates as a column back and forth down the length of the pipe, eventually exiting at the end of the pipe if all the openings are closed, or out the first open hole closest to the mouthpiece. Figure 26.8 shows the air vibration in a flute, though the same principle applies to all blown instruments. As with rooms, a standing wave forms in the tube, causing a null halfway down its effective length. The shape of the tube's *bore*—the

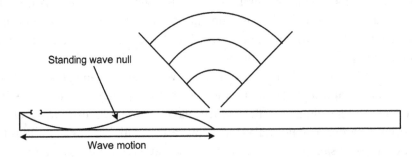

Standing wave null

Wave motion

Figure 26.8: Air in a blown instrument vibrates as a column, and the effective length of the tube ends when an opening creates a sudden change in acoustic impedance. A standing wave forms inside the tube at a frequency matching the effective length, with a null point halfway down the length, just as with rooms.

variation in diameter along its length—also affects an instrument's tone quality. The bore of flutes and clarinets is linear, as opposed to the conical bore of oboes and English horns that starts very narrow at the mouthpiece becoming wider at the far end. With the exception of the flare at the bell, a trumpet has a straight bore, while the cornet, which plays the same notes as a trumpet, has a conical bore and a more mellow tone. The varying bore of brass instruments is obvious by their flared shape, but the bores of oboes and clarinets are very different even though externally their bodies appear similar.

Many blown instruments are *transposing instruments*, meaning the pitch they produce is not the same as the written notes the player reads. Blown instruments come in many shapes and sizes, with many variations, each optimized to produce a specific range and timbre. For consistency and ease of playing, the same fingerings are used for each instrument in a family. So the fingering to play middle C on an oboe also works on the larger English horn, but the English horn instead sounds the F a musical fifth lower. The same applies to all four instruments in the saxophone family. From the player's perspective, the printed notes are all fingered the same, but the pitches produced by a given fingering are different. The composer or arranger is responsible for transposing the written notes to create the desired pitches.

The name of the instrument tells what pitch will sound when playing its lowest note, with no keys pressed, which is usually written as a C note. For example, a Bb trumpet with no valves pressed sounds a Bb when playing a written C, and an Eb alto sax sounds an Eb when playing a written C. However, not all blown instruments are transposing. The flute, oboe, bassoon, trombone, and tuba all sound at *concert pitch* to play the same notes that are written, rather than sound a whole step lower or some other fixed offset.

Flutes

The flute is perhaps the simplest orchestral blown instrument mechanically, though playing it well is not simple at all and takes years of practice! Alto flutes and bass flutes are constructed similarly but are physically larger to play a lower range of notes. The regular flute sounds at concert pitch, and the bass flute plays an octave lower, so transposition isn't needed. But the alto flute sounds a musical fourth lower than what's written. All of the flute types are characterized by a dominant second harmonic, as shown in Figure 26.9. The FFT spectrum captions that follow list the notes shown relative to middle C at 261.6 Hz.

Single Reeds

Clarinets and saxophones have a mouthpiece that holds a single reed, usually made of cane. The player blows into the mouthpiece, sending air down the length of the reed, causing it to vibrate. The principle is not unlike blowing across a blade of grass stretched between your

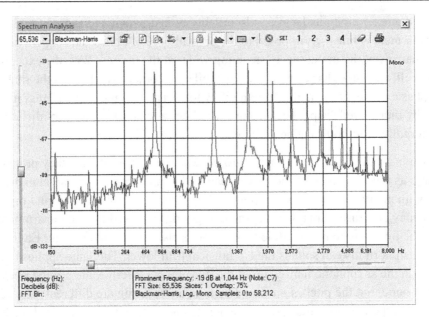

Figure 26.9: The characteristic sound of a flute is due to its strong second and third harmonics. This note is C above middle C (523.3 Hz), and its second harmonic is about 6 dB louder than the fundamental.

fingers. The tube sizes and shapes of the different instruments in the family create different sound qualities. For example, the bass clarinet has a tone quality similar to a regular clarinet but plays a lower range of pitches. Figure 26.10 shows the spectrum of a clarinet, and you can see that the odd-numbered harmonics are much more prominent than the even-numbered harmonics. This creates a tone quality not unlike a triangle wave.

Question: What's the definition of an optimist?
Answer: A bassoon player with a mortgage.

Double Reeds

Oboes and English horns use two reeds that vibrate against each other, rather than in free air as with single-reed instruments. The conical bore of these instruments creates very strong overtones much louder than the fundamental frequency. The bassoon and contrabassoon are similar, using a double reed and conical bore. Note that the English horn, also known as a *Cor Anglais*, is neither English nor a horn. Its name probably derives from *angelic*, because it looks like the type of horn played by angels in ancient religious images. Figure 26.11 shows the spectrum of an oboe note, and you can see that the first three harmonics are all much louder than the fundamental.

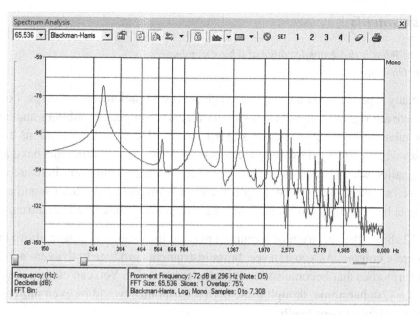

Figure 26.10: The characteristic sound of a clarinet is due to its strong odd-numbered harmonics. This note is D above middle C (293.7 Hz).

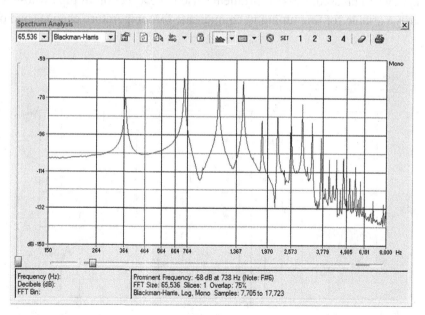

Figure 26.11: The characteristic sound of oboes, bassoons, and English horns comes from their conical bore and prominent overtones. This oboe note is F# above middle C (370 Hz), and the second, third, and fourth harmonics are all about 10 dB louder than the fundamental. An FFT of a bassoon (not shown) reveals a second harmonic that's 15 dB louder than the fundamental.

Brass Instruments

Question: What is the dynamic range of a bass trombone?
Answer: On or off.

There are many types of brass instruments, with several variations of each type. For example, there are regular trumpets and smaller piccolo trumpets, trombones and larger bass trombones, tubas and Sousaphones, and so forth. I won't list all of the variations because for our purposes what matters is how they create sound. Most brass instruments have a similar basic tone quality, which becomes brighter at louder volumes. However, there are variations in tone as well as pitch range. For example, the flugelhorn and cornet both sound mellower than a trumpet because they have a conical bore with a faster flare rate than a trumpet.

Figure 26.12 shows the spectrum of a trumpet playing the C note an octave above middle C at a moderate volume. As you can see, brass instruments produce both even- and odd-numbered harmonics that become softer at higher frequencies. You can clearly see overtones as high as the 12th harmonic, though these higher harmonics would be even more prominent if the performer played very loudly.

Most trumpets have three valves, and each inserts a short section of additional pipe increasing the total length when pressed. Brass instruments also have a section of pipe whose length is continuously variable for fine tuning to compensate for temperature changes and other

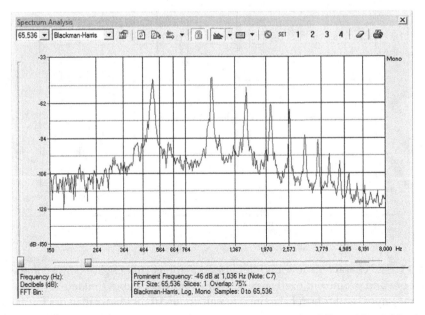

Figure 26.12: This trumpet is playing the C note an octave above middle C at a moderate volume.

factors. Another "feature" common to all brass instruments is the *spit valve*, called the *water key* in more polite circles. Over time, a player's saliva accumulates inside the instrument, producing a gurgling sound if not drained. It's common to see small puddles of saliva under the chairs of brass players after a concert.

The trombone uses a slide to adjust the pitch instead of fixed valves, though the slide varies the pitch over a range of only a musical *tri-tone*, or three whole steps. Players therefore vary their embouchure to access different groups of tri-tone note ranges.

The French horn is considered a woodwind instrument, but that's mainly historical because its soft mellow timbre blends well with clarinets and flutes. From the perspective of how it works and the harmonic spectrum generated, a French horn is a brass instrument that has valves and a brass-style mouthpiece.

Most brass instruments can be fitted with a mute to vary the volume and tone quality. Unlike a violin mute that softens both the volume and level of harmonics, a trumpet mute creates a brighter sound with more of a buzzing quality. There are also cup mutes that soften the volume and harmonic level and wah mutes that sound much like a guitar wah-wah effect. Some jazz players use a rubber toilet plunger to get a wah-type sound. In all cases, the filtering is entirely acoustic, creating low-pass, high-pass, or band-pass type filters.

Percussion Instruments

Question: What did the drummer get on his IQ test?
Answer: Drool.

Like vibrating strings and columns of air, drum heads also produce overtones. However, the overtones are not integer-related to the lowest fundamental frequency of the head. Rather, various vibrational modes create different unrelated frequencies, with each having a different acoustic radiation pattern. Bells and tubular bells also produce non-integer overtones, as do the metal and wood bars used in vibraphones (vibes) and xylophones, respectively.

Figures 26.13 through 26.15 show FFT spectrums of a rack tom in a drum kit, an orchestral timpani, and a xylophone. Figure 1.28 from Chapter 1 shows the spectrum for a tubular bell, so I won't repeat that here. In all of these graphs, you can see a fundamental pitch, with overtones at seemingly random higher frequencies. The legend at the bottom of each FFT screen states the musical note of the most prominent frequency. However, the fundamental pitch is always the lowest frequency, not necessarily the pitch that's measured or perceived as loudest.

When playing drums and most other percussion instruments, you must hit them, then pull the striker away quickly to avoid damping the vibration. Otherwise, you get a short, muffled

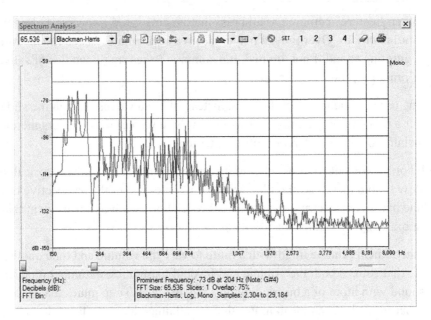

Figure 26.13: This FFT shows the spectrum for a rack tom that's tuned to produce a more or less pure tone but played loudly enough to create plenty of higher overtones.

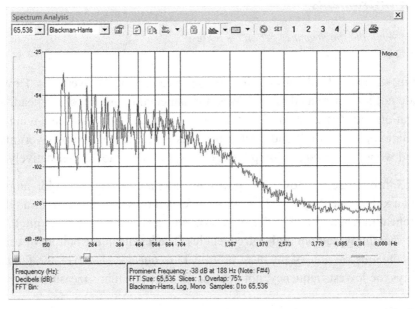

Figure 26.14: This FFT shows a timpani struck near the edge as is typical, played fairly loudly.

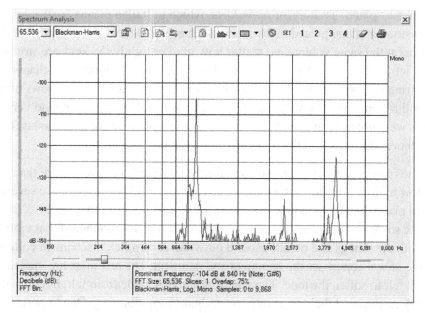

Figure 26.15: A xylophone also creates overtones that are not simple integer multiples of the fundamental frequency.

sound rather than a sustained singing tone. The "drum_tone" video mentioned earlier also shows the importance of removing a drum striker quickly.

Some drums have one head, some have two heads, and some, such as Remo Rototoms, have one head with no body at all. A timpani has one head, but the cavity below the head is sealed, which prevents the rear radiation from canceling the front at low frequencies. Other than pitched drums such as the timpani, there are a great variety of methods and preferences for tuning drums, so I won't even try to describe them. I will mention that when a tom or kick drum in a rock kit is out of tune with the music it's accompanying, it can conflict and make the music sound out of tune. Most drummers tune their drums by ear, and there's nothing wrong with that. For studio owners who are not drum experts, yet need to provide a drum kit that sounds excellent for clients, the Circular Science *RESOTUNE* from circularscience.com can make the task much easier. This electronic tuning meter is functionally similar to guitar tuners but tailored to drums and drum tuning techniques.

The Piano

I saved the piano for last because it defies categorization. Technically, it's considered a percussion instrument because sound is created by striking a metal string with a hammer. But it's also a pitched musical instrument having harmonics that follow the usual overtone series. The name "piano" is short for *pianoforte*, which in Italian means "soft-loud." Keyboard instruments that preceded the piano, such as the harpsichord, use a plectrum that plays all notes at the same volume, no matter how hard or lightly the keys are struck. The piano—first introduced around 1700—changed all that, using a complex mechanism that strikes the string in direct proportion to how hard the key is pressed. This allows music to be played at different volumes, with a wide range of tone colors from muted to very bright. Early pianos were not as loud or as bright as the modern versions we enjoy today, but they were much more powerful than other keyboard instruments of that time.

Modern pianos can play 88 different notes spanning a range from low A more than three octaves below middle C to the high C four octaves above middle C. There are several sizes and styles of piano, but all contain a cast metal frame called a *harp* that supports the extreme tension of the strings—10 to 15 tons or more—plus a wooden soundboard that vibrates in sympathy with the strings and radiates sound outward. A complex mechanism collectively known as the *action* comprises the hammers and dampers. The hammers are made from wood covered with felt to soften the tone. The dampers are also made from felt, but until they're released, they rest against the strings to prevent them from vibrating. One important feature of a piano's action is the *escapement mechanism*, which allows playing repeated notes very quickly. Each of the piano's 88 notes is sounded by one, two, or three strings, with higher notes using multiple strings tuned to the same pitch to better match the volume of lower notes.

Grand pianos also have three foot pedals named, from left to right, *una corda*, *sostenuto*, and *sustain*. The sustain pedal raises all the dampers so the strings continue to vibrate even after a key is released. Una corda is Italian for "one string," and that pedal shifts all the hammers sideways to strike fewer strings. For notes with three strings, only two are struck, and notes having two strings play only one. The una corda pedal is also called the *soft pedal*, but its real purpose is to change the tone quality more than reduce the volume. Even though the lowest notes on a piano are played by only one string, the una corda pedal still softens their tone color. When the felt-covered wooden hammers are shifted to the side, a less-used softer area of felt strikes the strings. The sostenuto pedal is similar to the sustain pedal, but it lifts the dampers for only those keys that are currently depressed. So a pianist can play a chord, press the sostenuto pedal, then release those keys and play other non-sustaining notes while the original chord continues to sound.

The "piano_demo" video describes some basic techniques used by piano players, then shows the construction and keyboard mechanism details in a Steinway grand piano to better see how it works. So I won't belabor those details here. However, I'll mention that rock 'n' roll players sometimes put thumbtacks into the hammers of a piano to create a bright honky-tonk sound. This can be taken further by putting some of the strings slightly out of tune to create an old-time sound such as might have been heard from a poorly maintained piano in

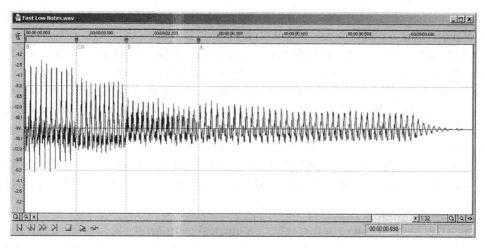

Figure 26.16: This is about as fast as I could play four low notes cleanly on my electric guitar, and each has a duration of about 80 milliseconds.

a western saloon. If you'd rather not detune your piano you can simulate that effect using a phaser effect device or plug-in with a "chorus" type setting.

I was always curious how many wave cycles of a musical note are needed in order to identify its pitch. I thought a note might be identifiable if it contains only a few cycles, but after testing this it seems we need at least 8 or 9 cycles for low notes and even more for high notes. So it's probably more about the note length, with 70 to 80 milliseconds needed to detect the pitch. I tested this by recording myself playing four guitar notes very quickly, then I noted the ruler times in Sound Forge. The online audio file is named "fast_notes.wav" and Figure 26.16 shows every note and its duration clearly. Each tick mark on the ruler line at top is 10 milliseconds, so two marks before the label 00:00:00:100 marks 80 milliseconds as the length of the first note.

As a test I tried editing out a few cycles from each note, but once there were fewer than eight cycles the notes were more of a blur than identifiable notes. I tried the same thing with sawtooth waves, assembling a similar sequence of notes each 50 milliseconds long. That too was more of a blur than music. You could tell the notes formed an A Major arpeggio, but just barely.

Mozart, Beethoven, and Archie Bell

No amount of classical music training will ever teach you what's so cool about "Tighten Up" by Archie Bell and the Drells.

It's impossible to define what makes a musical composition or performance "good" because personal preference is purely subjective. With art there can be no definitions. I believe that culture and early influences are a big factor in determining what people enjoy hearing. My mother often played classical music on our phonograph when I was growing up, and I've always loved classical music. My best friend Phil grew up hearing big band music and jazz, which he loves today, though he doesn't enjoy symphonies or string quartets. Everyone can't appreciate everything, no matter how good the quality. But just because someone dislikes opera music doesn't mean that opera music is bad. I'm not much of a Mozart fan—to my ears, many of his compositions sound like bubble gum music—but some of my friends whom I respect as musicians say he's very good, so I believe them even though I don't understand it.

Likewise, I'm not a fan of "out" jazz and other modern music containing excessive dissonance that never resolves or that relies on musical instrument sound effects. But some of my accomplished musician friends tell me that specific pieces are very good, and again I believe them. One person can appreciate only so many types of music, and just because I don't like something doesn't mean it's bad. Now I don't believe that *everything* is good. Further, there are established metrics for composition and musicianship that, when learned, help one be a more sophisticated listener. But no matter how many "rules" a composer or performer breaks, that doesn't mean the music is bad either. I think some people should show more respect for music they don't appreciate and not necessarily judge as poor everything they don't enjoy hearing.

Summary

Question: What's the difference between an orchestra conductor and a sack of fertilizer?
Answer: The sack.

Entire books have been written about violins, and trumpets, and every other musical instrument in existence. But I felt it was important to include this chapter to at least explain the basics of how musical instruments create sound and a bit about how they're played to give recording engineers and audiophiles who are not musicians some insight into this very important aspect of audio.

I divide musical instruments into two basic categories—percussive and sustaining—though percussive instruments can also be made to sustain by striking them repeatedly. Regardless of how their sound is created, the volume of the overtones in all acoustic instruments increases when they are bowed, blown, or struck with more force. With string instruments, bowing or plucking closer to one end of the string also creates a brighter sound with more prominent harmonics. Likewise, drums sound brighter when played harder and farther from their center. Indeed, it's common for the fundamental pitch to be softer than the lower few harmonics.

As shown throughout this book, an FFT display reveals the amount of energy at different frequencies, which helps us understand why instruments sound the way they do.

Most "wood box" instruments are built using the same basic construction, with strings that are stretched tightly over a wood cavity having sound holes. Like loudspeakers, the vibrating wooden plates become directional at high frequencies, affecting which frequencies arrive at the listener's ears or a microphone. The thickness and mass of the wooden plates is an important reason instruments all sound different, even when their sizes and shapes are the same. Modern modal analysis lets us measure the resonance of these plates to understand what makes one violin sound subjectively better than another. Solid body electric instruments don't rely on resonance for their sound character, so they're less affected by things like wood thickness and paint finish. However, the mass and rigidity of their construction affect how long a plucked string will sustain for.

There are a huge number of blown instrument types, and all of them require players to carefully control their mouth shape, or embouchure, as well as how hard they blow. Simple blown instruments such as the flute and recorder produce sound by sending air across an opening into a tube-shaped cavity, but others use one or two reeds, or a cup-shaped mouthpiece. Brass instruments use valves that selectively insert short sections into the overall tube length, though woodwind instruments instead vary the effective length by allowing air to escape through one or more holes along the tube's length. In all cases, the air vibrates as a column inside the tube.

Notes

1 http://tinyurl.com/ncao768
2 https://vimeo.com/album/3850080
3 "The Acoustics of Violin Plates" by Carleen Hutchins. *Scientific American*, October 1981.

Index

Note: Page numbers followed by "t" indicates table and "f" indicates figure.